RELATIVITY MATH
UPDATED AND REVISED
FOR THE REST OF US

ALL YOU WANT TO KNOW ABOUT MODERN RELATIVITY
BUT THOUGHT YOU COULDN'T POSSIBLY UNDERSTAND

LOUIS S. JAGERMAN, M.D.
Copyright 2014
Revised Edition October 2017

To My Family

TABLE OF CONTENTS

If it is your ambition to understand Albert Einstein's theory of relativity, you may be pleasantly surprised. It is an achievable goal and the rewards are rich. If you seek to grasp the basic laws of nature, relativity is indispensable. If you pride yourself on mastering profound concepts which are supposedly beyond ordinary intelligence, relativity is exhilarating. And if you relish the exercise of logic and reasoning, relativity is sensational. But then why focus on the *mathematics* of relativity? Because *relativity is fundamentally mathematical.* If you examine or apply relativity in any precise way, you will find yourself dealing with mathematics, and by avoiding the mathematical approach, much of the content, the meaning, and even the intellectual beauty of relativity is lost.

Of course your shelves may already be full of books explaining relativity, but I venture that these are either nonmathematical synopses or highly technical texts requiring a formal background in calculus, physics, and analytic geometry. For example Calder, Epstein, Gardner, and Will, among many others, write skillfully but omit the mathematical essence of relativity (see Bibliography starting on page 486). On the other hand Bergmann, Lawden, Schutz, and others cover the deepest topics but do not address the amateur reader. Even Einstein and his colleagues published "popular" expositions, but these are written in a difficult style (mostly translated from German) and they still presume extraordinary expertise. *My strategy is to rely on only fundamental geometry, algebra, and logic as the basis for a sophisticated understanding and a mathematical appreciation of relativity.* A high school curriculum strong in physics and math provides sufficient background, and I also cover the essential basics in this book and in my previous book, "The Mathematics of Relativity for the Rest of Us" (2001), though I now include recent developments and discoveries.

A major issue I must consider is the reader's level of expertise in theoretical physics–obviously including relativity–and in the kind of mathematics used in this area. First, I must emphasize that this book is intended for the non-professional "amateur" scientist or for a beginning student of relativity who seeks to fill a gap between "the basics" and advanced formal studies.

Second, I find that an effective way to appreciate and grasp the math behind relativity is to trace–at least approximately–the history of this topic. In particular, if you see the evolution of Einstein's struggles as he (with help from others) slowly reached his goals step by step, your understanding will also grow step by step. However this approach has two drawbacks: History is not always "neat" and orderly, and neither are the steps. Furthermore, modern advances (such as the use of computers and the results of recent research) are neglected.

This brings me to a third consideration. Some professional physicists/reviewers/critics provided by my publishers were impatient with my slow pace and my reliance on the various analogies that I devised. However I elect to fall back on my experiences while studying math and relativity (before I turned to medicine), and to consider what I observe when I tutor students. In essence when trying to clarify a difficult or novel concept, I would ask myself "what do I wish someone had explained step-by-step to help me understand." Or, "what did the teacher omit that made this difficult?" Or, "is there another way this can be stated or illustrated or interpreted so that it makes more sense?"

Incidentally, readers who also delve into quantum mechanics may find that there are fewer interpretations and controversies in relativity. One reason is that relativity had essentially one "father," Einstein, and most of his coworkers concurred with his thinking. Quantum mechanics had several "fathers," including Einstein plus Planck, Bohr, Heisenberg, Schrödinger, Bohm at other famous scientists, some of whom disagreed with each other vehemently and issued conflicting "interpretations of quantum mechanics." For example Heisenberg and Schrödinger each devised their own unique system of mathematics for quantum mechanics. In my opinion both systems are as difficult as the math for relativity, and more recently Feynman introduced yet another formulation, more modern but very different from the other two. No comparable chain of events occurred to mystify students of relativity.

◆

Ever since its introduction about a century ago, relativity has had a near-magical appeal. First of all, relativity has revolutionized our understanding of the universe, so that it is imperative in the study of cosmology, quantum mechanics, modern particle physics, and astronomy. In the history of physical science, relativity is a major turning point after which we could no longer treat the teachings of Galileo, Kepler, and Newton as unshakable truths. In fact, relativity represents a milestone in the development of civilization for two reasons. One, relativity exemplifies how our long-established and seemingly indisputable convictions can be challenged and replaced. Two, it underscores our diminutive and fragile status in a universe of immense size and awesome energy.

Relativity is a grand example of the power and usefulness of mathematics and geometry. It also has a philosophical side, and its concepts have been adopted in ethics, epistemology, political science, and even theology. Finally, relativity is a masterpiece of clear and forceful reasoning, in which a few basic ideas and observations are carried to astounding conclusions and predictions. According to admirers of relativity and particularly of Einstein, relativity may be the single most complex and influential creation that ever emerged from one human mind.

Parts of relativity are indeed difficult. Nevertheless practically all of its facets, even its mathematical foundations, are within the intellectual reach of an attentive and motivated reader. Relativity's reputation for being incomprehensible or utterly esoteric is not deserved. For instance, contrary to general opinion, Einstein was not the only human being capable of fathoming relativity, and he never alleged that "everything is relative." Relativity did not sanction "time travel," and it was not the blueprint for "atomic" weapons. In fact the most celebrated equation in relativity, $E = mc^2$, is easy to derive. However at the time relativity was introduced (1905-1915) it was so novel and revolutionary that few scientists could articulate it, and the news media of the day dwelt on its confusing and arcane implications, often at the expense of accuracy. A generation of writers and teachers was needed to develop the explanations, examples, and analogies which clarify relativity. In this book (revised in 2014) I address the mathematical side of this clarification, and I have updated my presentation to include newly appreciated concepts, particularly the close relationship between relativity and quantum mechanics.

CHAPTER 1: THE POSTULATES OF SPECIAL RELATIVITY

As we begin, and particularly before we consider why we can call parts of relativity "special," we should become familiar with several key terms: *Speed, velocity, uniform motion, frame of reference,* and *inertial frame of reference.* Of course speed is how fast a moving object covers a distance. We treat speed and velocity as synonymous for now (but see page 139 for the distinction). Uniform motion is straight motion at a constant speed in one direction. We assume that such motion is free of disturbances like vibrations, noise, or air currents. Synonyms for uniform motion are non-accelerated motion, rectilinear motion, and inertial motion. The concept of uniform motion is essential to relativity, even though in actuality most objects in the universe do not move uniformly. We will deal with non-uniform motion when we turn to the details of "general" relativity.

For our purposes a frame of reference is any self-contained space in which physical events can be observed, such as a room, a laboratory, or an enclosed vehicle. We will use the interiors of trains, airplanes, spaceships, boats, and elevators as examples of frames of reference. Technically, a frame of reference is "inertial" if a free body inside it (one with no forces acting on it) shows no acceleration. Practically, an inertial frame of reference is one which moves uniformly. To describe events in a frame of reference* we need some "system of coordinates," such as the set of familiar X and Y "axes" used to construct graphs; these axes are also called "coordinates" or "coordinate axes." However we will deal with events in three dimensions, so we need a Z coordinate axis. For example a room may have a height we call X, a width we call Y, and a front-to-back depth we call Z. We can then give *each location in this frame of reference*—equivalent to *each point in a system of coordinates*—a numerical value for X, Y, and Z, which can be in units such as meters or inches.

The X-Y-Z system of coordinates thus provides a way of describing where objects are located and where events occur. We can say that "event 'A' occurred at X by Y by Z" or in brief that "A occurred at X, Y, Z." For example two objects may collide at location $X = 5$, $Y = 6$, and $Z = 4$. We can even use this method to describe the events. For example in an X-Y-Z system an event may consist of an object moving from location $X = 5$, $Y = 6$, and $Z = 4$ to a new location $X = 7$, $Y = 8$, and $Z = 2$. We note that X, Y and Z can refer to coordinate axes or to numbers along these axes; the context tells us which. (At times X, Y, and Z are written as x, y, and z, particularly in certain equations. Later we will include a fourth dimension, labeled T or t.) However, to be exact, when two or more objects move at different speeds or in different directions—ergo at different velocities; please see page 139 about velocity—they each are in different frames of reference.

We also need a cast of characters whom we call *observers*, and for expedience we give them the masculine gender. We will describe objects and events as seen by ordinary observers who use instruments such as clocks, rulers, and scales. However, we will also describe objects and events as seen by extraordinary observers, ones who have the ability to travel at the speed of light, to accompany subatomic particles, and to sense four dimensions. Obviously observers may be imaginary, but they are of great help in the elucidation of relativity. In any case when one of our

---

* Though we cover the mathematical and geometric essentials as we advance through the text, the Appendix (page 481) provides a review of the underlying procedures and equations.

observers makes his observation, he reports how objects and events appear in his particular frame of reference, which means that this observer reports what has happened from his "point of view" or from his "vantage." When two observers in uniform relative motion (e.g., one is stationary and the other moves uniformly) compare their observations, in effect they are comparing two frames of reference. Whenever an observer is at rest or in uniform motion, we call him an "inertial observer."

◆

Relativity consists of two parts, "special" and "general." The full distinction will interest us later, but for now we should note that *special relativity is restricted to situations wherein motion is only uniform.* At first glance the limitation to uniform motion makes special relativity seem trivial or insignificant. However, many important objects, notably subatomic particles—including photons—move uniformly, both in nature and in experiments, and many processes—such as the passage of time—are explained only by special relativity.

Special relativity is built upon two postulates which Einstein (1879-1955) published in 1905 but which appeared in a journal under a title which does not even mention relativity. Translated from German, that title is "On the Electrodynamics of Moving Bodies." The now-famous volume in which this paper surfaced, number 17 of the Annalen der Physik ("Annals of Physics"), contains several other diverse articles by Einstein, with no hint that one of these was destined for extraordinary fame. At that time Einstein had not yet formulated the thesis that later became known as relativity, nor was this topic his only interest.

The first postulate of special relativity can be stated in many ways: The laws of nature are the same for all observers in uniformly moving frames of reference. The laws of physics are the same in every inertial frame of reference. If an observer at rest finds a law of physics valid, any other non-accelerated observer will find this law to be valid. Or, there is no privileged inertial frame of reference in which the laws of physics work better than in any other inertial frame. In short, all inertial frames and all inertial observers are equivalent.

This postulate is not new. Galileo (1564-1642) suggested it and Newton (1642-1727) applied it. Einstein expanded the idea, and he credited Newton and Galileo for it. Appropriately, a uniformly moving (inertial) frame of reference is said to be "Galilean." In fact the principle that the laws of physics are the same for everyone moving uniformly is called Galilean relativity or Newtonian relativity, but this "relativity" deals only with "mechanical" events, i.e., with the motion of ordinary objects; Einstein's relativity considers more than just ordinary objects. In any case, the next paragraph covers a very important insight.

The concept of Galilean or Newtonian relativity means that if an observer is in uniform motion, *no mechanical experiment can prove whether his frame of reference is moving or is at rest.* For example, by bouncing a ball inside a uniformly moving train, an observer cannot detect motion of the train. (Again, we must imagine that there is no vibration or swaying of the car.) The ball will behave the same—for example, it will bounce straight up and down—whether the train is at rest or in uniform motion, since the laws of physics governing this event are the same in either case. As far as these laws are concerned, the state of rest is the same as uniform motion.

Even when looking out of the window of an "inertial" train, an observer cannot prove whether the train is moving across the Earth or the Earth is moving under the train. (Both are legitimate possibilities in relativity.) For example, inside the train the ball will bounce in the same way whether the observer believes that train is moving or that the Earth is moving, simply because there is no internal indication of the train's motion. This brings us to an important corollary of the first postulate of special relativity: All that an observer can espy is *relative* uniform motion between the train and the Earth. That is to say, detection of uniform motion requires correlation with the motion of something else, and absolute uniform motion—motion without relation to any other object—cannot be detected.

Historically speaking, Aristotle would have argued that absolute rest and absolute motion do exist, specifically that the Earth is at absolute rest; if it weren't, how could a ball bounce straight up and down? Newton reasoned that although absolute *uniform* motion cannot be detected, other forms of motion are absolute. His oft-cited example is a water-filled bucket hanging on a rope. If the bucket is made to spin, the surface of the water becomes concave. But if instead the universe were rotating around this bucket, Newton reasoned that the surface of the water would stay flat. He concluded that when the water is concave, the universe must be at rest and the spinning bucket must be in absolute motion. This would mean that the laws of physics can vary between frames of reference, i.e., the laws of motion can be different for a spinning bucket than for a stationary one. As we shall see, Einstein had to show that there is no absolute motion of any kind, that all motion is relative, and most importantly that the laws of physics are always the same. For the case of the spinning bucket, Einstein had to explain how the motion of the universe alone disturbs the water.

The idea that uniform motion does not alter physical laws applies to mathematical formulations of those laws. Equations such as $RT = D$ ("rate times time equals distance"; rate is the same as speed) will work at rest or when moving at uniform velocity. For example, we can list in mathematical form all the laws of physics for the workings of an ordinary bathroom scale. We will include formulas for springs, levers, elasticity, and friction. Then we ask an observer to step on the scale under two conditions, one while at home and the other while aboard a uniformly moving train. The result? The scale works the same and shows the same weight in both cases. The inference? None of the formulae had to be altered to fit both situations. The significance? *A valid law of physics takes the same mathematical form for all observers in uniform motion.*

✦

Einstein's second postulate of special relativity is easier to put into words but harder to prove. This postulate also stimulated much scientific controversy and skepticism when it was proposed because unlike the first postulate, the second one is totally unexpected on the basis of Newtonian physics:

Regardless of the uniform motion of its source, light always moves with the same constant speed. That is, the speed of light in any medium has the same value for all observers, no matter if the source of the light is at rest or in uniform motion. In yet other words, in any inertial frame of reference the speed of light appears the same in all directions. (The speed of light, designated as "c," is very high compared to speeds we normally deal with, but it is not infinite. It is about 300,000 kilometers or 186,000 miles per second in a vacuum. Visible light is one instance of electromagnetic energy. The difference between light, radio waves, x-rays, gamma rays, and other kinds of electromagnetic energy is their wave length, but they all move at c.)

This postulate agrees with equations on electromagnetism derived by Maxwell (1831-1879), and Einstein cited these extensively, including in the 1905 paper which later became the seed of special relativity. When Maxwell's equations are combined and solved for c, it turns out that c depends on the electric and magnetic properties of the medium but *not* on the velocity of the source of the light. Maxwell's equations on electromagnetism enjoy other connections with relativity; we will cover germane details in due course.

We restate the key principle: The velocity of light is independent of uniform motion of its source. As a result, when c is measured or used in an experiment, uniform motion of the source of light makes no difference. Thus light and all electromagnetic energy are "exclusive." ("Exclusive" is in quotation marks to give the word in this context this particular denotation.) *The speed of light is like a law of physics; it is unaffected by uniform motion and it appears the same in all inertial frames of reference.* We note that now we restrict ourselves to uniform motion. Accelerated motion will interest us later. (Regarding the speed of any electromagnetic radiation, we assume passage through only one medium. If a beam passes into another medium, say from air into glass, its speed changes, which is not germane to relativity. Within one medium, the speed of light is independent of the motion of its source.)

We will also later explain that light consists of particles called photons. We may liken a photon to a lifeboat which is driven by oars; the boat transports light. Because of factors such as human strength and the resistance offered by water, the oarsmen propel the boat only at speed "c." Even if this boat is launched from a (uniformly) moving ship, it will move at speed c. The behavior of the boat is independent of the behavior of the ship that launched it. Similarly, once photons of light are launched from a moving source, their speed depends on their own energy and on the medium through which they are moving, but not on how fast the source is moving. Furthermore, even though electromagnetic energy is "exclusive" regarding its speed, this property is not unique to photons. Any particle capable of speed c shows the same phenomenon. The physical reason behind this observation will be easier to explain in later chapters. We note for future use that c is constant even if the source is in accelerated motion, because the source still has one certain speed at the instant the photons are emitted.

Relativity uses a certain logic and we ought to be able to "think like relativists." Here is an example of relativistic reasoning applied to the constancy of the speed of light: Light and electricity are forms of electromagnetic energy, and electricity can be generated by moving a magnet near a wire or by moving a wire near a magnet. The resultant electric pulse is identical whether the wire moves or whether the magnet moves. Since both cases give the same speed of the generated pulse, we conclude that the speed of electromagnetic energy cannot depend on the motion of its source.

This exclusive characteristic of the speed of light has been verified experimentally. Beams of light emanating from sources moving at different speeds appear to travel at the same speed. For instance the sun spins on an axis so that one edge is always moving toward the Earth while the other edge is moving away from the Earth. Nevertheless, light from each edge reaches us with the same speed. Revolving double stars give the same findings, confirming this postulate on a grander scale. Newton would have expected light reaching the Earth from an approaching source to be faster, and common sense dictates the same prediction. No wonder that most scientists hesitated to accept the second postulate of special relativity; it implied that Newton had erred!

✦

We now turn to an important consequence of the fact that the speed of light has a definite value. Though we do not ordinarily think about it, *the ultimate way in which we learn of any distant event is when electromagnetic radiation, usually light, informs us of such an event.* This means that since the speed of all electromagnetic radiation has a finite value—even if that value is constant and very high—there is always some delay between an event and the detection of that event by a distant observer. Hence distant events cannot be detected instantaneously. For example if a star explodes somewhere in the universe, we have no way of knowing this until some "signal" (a flash or a burst of radiation) reaches us, by which time the event has already occurred. Therefore, *the perceived time of an event—the earliest we can know of it—depends in part on the value of c.*

Before the finite speed of light had been discovered, this was highly controversial. Descartes (1596-1650) theorized that light consists of a stream particles capable of infinite speed, so that all observers everywhere, no matter how far away, should agree on when something visible happened. The corollary to this was that the passage of time acts as an absolute, invariant, and independent reference according to which all events can be defined by any observer. For instance if an observer sees a flash of light "now," this was interpreted to mean that the causative explosion, no matter how far away, also occurs "now."

Galileo, disagreeing with Descartes, tried to measure the speed of light, but he lacked sufficiently sensitive instruments. Nevertheless Galileo correctly surmised that it was not infinite. Indeed according to modern physics, including relativity, there is no absolute reference for all observers as to when an event occurred. One reason—but as we shall see not the only reason—is the finite c: Since "signals" from distant events do not cross space at infinite speed, they are inevitably delayed and by themselves do not divulge how long ago they originated. A flash seen "now" is caused by a distant explosion some time ago; that time may be very short, but it is not zero.

This is ahead of our story, but pre-relativity Newtonian science held a parallel assumption about gravity, namely that the effects of gravity between two distant objects are instantaneous. Modern physics disagrees; gravity, like light, is not infinitely fast. For example if the moon suddenly vanished, it would take a moment for our tides to start to disappear. The gravitational "signal" or "message" announcing the change, like a flash of light or a radio message, would arrive quickly but not immediately.

Let us apply the idea that light acts as a "messenger" by considering two astronauts, A and B, each in his own spaceship, who check on the synchronization of their clocks by sending hourly messages to each other. I adapted this scenario closely from pages 22-25 of Lillian Lieber's book (Einstein admired this work; please note that the issue here is synchronization, not time dilatation.) The two spaceships are so far apart that any electromagnetic signal, including a radio message, needs one minute to get from one to the other, and the astronauts know this. At this moment they share one large frame of reference.* At one p.m. on his clock, astronaut A sends a radio signal to B. Astronaut B notes the time of the arrival of this message on his own clock: 1:01 p.m. He expects a delay of one minute, so the clocks appear synchronized as expected.

However, unbeknownst to the astronauts, on the next day both spaceships are in *uniform motion* along a line from A to B, and B is leading the way, while the distance between them stays the same. Without external hints the astronauts cannot know they are moving; they cannot detect uniform motion of their now-moving frame of reference, and their motion does not disturb their clocks. Again, A sends a signal to B at 1 p.m., but while this signal is in route at speed c, a speed which happens to be independent of the speed of the source of the signal, B's spaceship has moved away from A's message. This radio signal therefore has *farther to travel at speed c*—it takes a bit more than one minute to catch up to B—and it arrives at B a few seconds after 1:01 p.m. *Astronaut B, unaware of the uniform motion in their frame of reference, concludes that A was late.*

At 2 p.m. on his clock, astronaut B tries to fix the apparent disharmony by sending a radio message to A. But in the time B's signal travels at speed c, A advances so that the message has *less distance to cover*, and this message reaches A before the expected 2:01 p.m. *Now astronaut A concludes that B was early.* The astronauts fail to agree on the timing of events—A thinks that B is procrastinating or that *B's clock is slow*, and B thinks that A is hasty or that *A's clock is fast*.

We note two lessons in this situation. First, if the astronauts had the capacity to detect uniform motion they could have drawn the correct conclusion that motion and not defective timekeeping is the cause of the discrepancy. However, the messages or signals which the astronauts traded gave no clue.

The second lesson, also explained by Lieber on page 25, is more complex: Though A and B did not move relative to each other, their agreement on time was disturbed by the common motion of their frame of reference. For argument's sake, if A's message traveled faster than c after being launched from the moving spaceship—if electromagnetic radiation gained speed when it originated from a speeding source—no disagreement would result. The same would occur if electromagnetic radiation traveled at an infinitely high speed. In reality the combination of the distance between observers, their common motion, and the speed of the signal for their messages makes it impossible to specify an absolute time, a time upon which everyone can always agree. (Again, I am not invoking time dilatation here. The crucial operative equation is rate times time equals distance.) Time on a clock does not provide an absolute indication as to when an event occurs, even for observers in one inertial frame of reference. We can say that *there exists no universal standard time.* The statement that "we never know when something will happen" can be supplemented by the statement that "*we may never agree when something has happened.*" This concept allows us to use the word "relativity" in this sense: *Our inability to agree on when events occur is governed by relativity, in that our perception of the time of an event is biased by relative motion.*

◆

_____

*As indicated by Professor Jean Fex at the Science and Loisirs academy in France.

Another important feature of the speed of light is that it measures the same regardless of the uniform motion of the *observer*. (Not just of the source; of the observer too.) This makes sense if the constancy of the speed of light is the same for all uniformly moving observers. Thus all observers in any inertial frames of reference will agree on the speed of light, no matter how it is measured or who measures it.

This phenomenon has also been verified; it is why the famous "Michelson-Morley experiments" (by Michelson [1852-1931] and Morley [1838-1923]) gave their negative results. We will look at these experiments in detail later, but they can be envisioned as having two observers compare their measurements of the speed of light, one observer examining a beam of light shone against the direction of the Earth's rotation and the other examining a beam going across the direction of the Earth's rotation. The expected finding was a slowing of the light beam going "upstream" with respect to the Earth's rotation, just as a boat slows moving upstream. The observers instead obtained the same speed of light no matter which way the beams were aimed, which indicates that light has a constant speed regardless of the motion of its source or of its observer.

This finding is doubly important in that observations of objects like bouncing balls are mechanical experiments, but Michelson and Morley were performing *optical* experiments. ("Optical" means with any electromagnetic radiation, including light.) Just as we cannot detect absolute motion by bouncing a ball, which is a mechanical experiment, so Michelson and Morley couldn't detect absolute motion by aiming light beams, which is an optical experiment. The beams behaved the same—their speed was identical—because the laws of physics, including those of optics, are not altered by uniform motion.

We conclude that absolute uniform motion cannot be detected by any physical experiment, mechanical or optical. These findings confirm the postulates that (1) *the laws of physics and* (2) *the measurement of speed of light are not changed by uniform motion. Both are the same in all inertial frames of reference and for all inertial observers.* The first, if not intuitively obvious, is easily demonstrable in daily experience, while the second, though technically more elusive and more surprising, has also been documented.

CHAPTER 2: THE RELATIVITY OF TIME

Let us consider two observers who are in uniform motion with respect to each other. We can consider the basic equation $RT = D$ (rate times time equals distance) to be a "law" of physics, and we expect it to hold for all inertial observers. We also expect the speed (rate) of light, c, to appear the same for all such observers. Therefore $cT = D$ for both observers. We know that if we bounce a ball straight up and down inside a train, the ball's behavior in the train will appear the same whether the train is at rest or in uniform motion; the ball bounces identically in the two cases. In preparation for our first math derivation, if necessary please see the Appendix.

In such a train (train A) instead of a bouncing ball, a light beam is shined up and down between two mirrors which face each other. One mirror is on the floor and the other on the ceiling of the train. An observer (observer A) aboard train A, knowing the distance (D) between the mirrors and knowing the speed of light c, can use $cT = D$ to calculate how much time (T) the beam needs to reflect or "bounce" between the mirrors. Observer A will find no difference in his results whether his train is at rest or in uniform motion; T is the same in either case. See diagram A. (Please ignore the X and Y axes in diagrams A and B for now. These diagrams will serve other purposes.) In diagram A, $T_A = D_A / c$. Solved for distance,

$$D_A = cT_A.$$

The subscripts "A" mean that this equation holds for observer A aboard train A. Thus T_A represents the time the light beam needs to "bounce" between the mirrors as measured by observer A aboard his own train A, where the distance between the mirrors measures D_A.

But another train, B, with the same dimensions as A when the two were compared standing next to each other, passes by train A on parallel tracks at uniform speed. Train B also has a beam of light, mirrors, and an observer. Observer A in train A looks out his window into the window of train B and measures the "bounce time" of the beam of light in train B. (For instance observer A may take a photograph of the interior of train B.) *To observer A, the distance the light travels in train B appears longer*—it follows a sloped course. We know that $cT = D$ for both observers, and c is the same for both observers, which means that the "bounce time" T in train B measured by observer A is *longer*! See diagram B in which the light follows an ∧-shaped path, but please keep in mind that the trains would have to be thousands of miles tall and their relative speed would have to be enormous for this effect to be so pronounced. In any case T_B is the "bounce time" as measured *by observer A in train B*. Finally we use diagram C, which isolates the segment of the light beam bouncing down from the mirror on the ceiling to the mirror of the floor.

We note the use of the *pythagorean theorem* in diagram C. In a "right" triangle (one that has a 90-degree angle), the longest side is the hypotenuse, and we label this side as "H" (which is cT_B in diagram C). If we label the other two sides of the triangle "S_1" and "S_2" (which are D and vT_B in diagram C; the order does not matter, and the subscripts for D are superfluous), the pythagorean theorem states that

$$(S_1)^2 + (S_2)^2 = H^2.$$

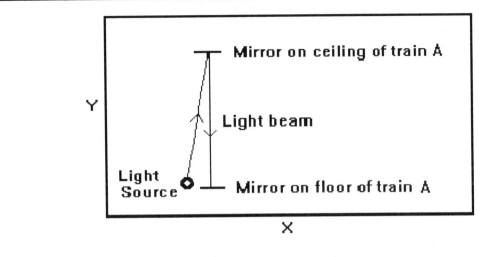

Let D_A = the distance between the mirrors.

Let c = the speed of light.

Let T_A = the time light needs to bounce between the mirrors.

Since cT = D, observer A can calculate T_A.　$T_A = D_A/c$.

Diagram A. How an event (a light beam bouncing between two mirrors) inside train A appears to observer A.

In fact the pythagorean theorem plays a major role in many of the equations we encounter in relativity, and we should keep its pattern in mind. In this case we can write

$$D^2 + (vT_B)^2 = (cT_B)^2 \quad or \quad D^2 = (cT_B)^2 - (vT_B)^2.$$

According to basic algebra, $(cT_B) - (vT_B)$ is the same as $T_B(c - v)$, which lets us solve for D^2:

$$D^2 = (T_B)^2(c^2 - v^2)$$

Therefore, solved for D,

$$D = T_B\sqrt{c^2 - v^2},$$

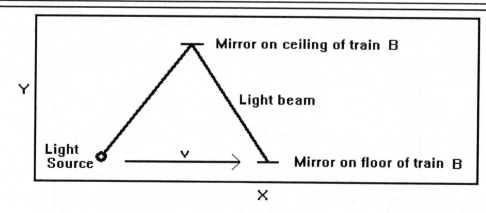

Let D_B = the distance between the mirrors as it appears in train B.

Let c = the speed of light [same for each observer].

Let T_B = the time light needs to bounce between the mirrors.

Let v = the speed of train B relative to train A.

Since cT = D, observer A can calculate T_B. $T_B = D_B/c$.

Diagram B. How events in train B appear to observer A. Observer A is looking into train B as the trains pass each other.

and with more algebra (including $c^2 - v^2 = \dfrac{c^2}{c^2} - \dfrac{v^2}{c^2} = 1 - \dfrac{v^2}{c^2}$) we can also solve for T_B:

$$T_B = \frac{D}{\sqrt{c^2 - v^2}} = \frac{D}{c\sqrt{1 - v^2/c^2}}$$

However D is the same for both trains (which is why we now can omit subscripts for D), so that

$$T_A = \frac{D}{c} \quad or \quad D = c\,T_A.$$

This means that upon substituting D by cT_A in the previous equation,

$$T_B = \frac{c\,T_A}{c\sqrt{1 - v^2/c^2}}.$$

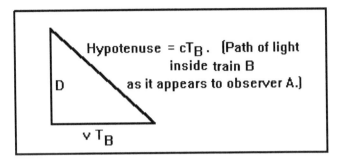

Let T_B = the time which the light needs to traverse the hypotenuse.

Let v = the speed of train B relative to train A.

During time T_B, train B moved distance vT_B.

By the pythagorean thoerem $D^2 + (vT_B)^2 = (cT_B)^2$.

Diagram C. Observer A's view of events on train B re-drawn in terms of a right triangle. To observer A the light beam inside train B appears to move along the hypotenuse of the triangle.

Two of the c's cancel out, and by solving for $\sqrt{1 - v^2/c^2}$, we can set up the ratio $\dfrac{T_A}{T_B}$:

$$\frac{T_A}{T_B} = \frac{\sqrt{1 - v^2/c^2}}{1}.$$

Here T_A is the length of time the light beam needs to "bounce" between the mirrors as measured by observer A in train A. T_B is the length of time the beam needs to "bounce" between the mirrors as measured by observer A in train B. v is the relative speed of the two trains and c is the speed of light.

The equation for $\dfrac{T_A}{T_B}$ is very important to relativity. First, it tells us that the results of measurements of time are a function of speed[*]; time is not invariant. This means that *our determination of how long an event takes depends upon how fast we are moving or on how fast the event moves past us.* Second, the equation implies that speed influences the ratio between time measured at rest and time measured during relative motion; it does not implicate absolute time. Third, if v is zero (no relative motion) the ratio is one, so there is no difference between time measured for the two moving events. But as v increases, that ratio changes, and when v approaches the speed of light, the discrepancy between the time measurements becomes very large. Finally, we note that v is only uniform motion; non-uniform (accelerated) motion is not covered by this equation, and we will revisit this important stipulation at the end of this chapter.

We must keep the following expression in mind. It is crucial in any mathematical discussion of relativity, and we will use it repeatedly. We call it the *special relativity factor*:

$$\boxed{\sqrt{1 - v^2/c^2}}$$

Please note that the numerical value of this expression is at most 1, never less, and it can only be zero if $v = c$.

The special relativity factor appears in the literature in various formats. Some authors (e.g. Pauli, p. 10) let the Greek letter β stand for $\dfrac{v}{c}$, so that this factor is written as $\sqrt{1 - \beta^2}$. Meanwhile Einstein and most writers let β represent the entire factor, so that in The Principle of Relativity, p. 46, we find $\beta = \sqrt{1 - v^2/c^2}$. Menzel (p. 378) surprisingly uses a different Greek letter for the entire factor, he also lets β stand for $\dfrac{v}{c}$, and he expresses the square root in exponential form (see Appendix), so that $\gamma = (1 - \beta^2)^{-1/2}$. Moreover, some authors (e.g. Weyl, p. 172) prefer $c^2 - v^2$ to $1 - v^2/c^2$, and for convenience c can be assumed to equal 1, as in Schutz, pp. 5 and 20, so that the special relativity factor becomes $\sqrt{1 - v^2}$.

---

[*] We will use the term "function" repeatedly. Our kind of function is a mathematical relationship between two quantities, such that if one quantity changes, the other changes in a predictable manner, and this relationship is described by an equation. For example "distance traveled is a function of speed." At greater speed, more distance is traveled, and an equation can predict the distance from the speed. In more formal terminology, a change in the independent variable (e.g. speed) causes a change in the dependent variable (distance). The equation here is ST=D; speed times time equals distance, though traditionally we say that rate times time equals distance. Several independent variables (speed, time) can determine a dependent variable. Usually the words "a function of" can be replaced by "varies with and depends on," and an equation exists which defines the dependence.

Obviously such variations in mathematical style may perplex the unwary reader and/or student, but *the following fairly standard convention is important to know.* Under many circumstances one observer is considered to be at rest (not moving) or to be the observer with respect to whom another observer is moving. For instance in the above diagram in effect A is at rest, and B is moving with respect to A. Furthermore, in most physics journals and books, time is symbolized by "t," and the time noted by the stationary gets the subscript "o" or "r." The time noted by another relatively moving observer usually gets no subscript. Ergo, the above boxed equation is often written as

$$\frac{t_o}{t} = \frac{\sqrt{1 - v^2/c^2}}{1}.$$

Another presentation is also useful,

$$\frac{t}{t_o} = \frac{1}{\sqrt{1 - v^2/c^2}}.$$

Again, some prefer the more explicit "r" for "rest," as do I often, in place of the "o." I will also continue to use the more visible "T" for time.

Here we glimpse a mathematical detail, peculiar to special relativity, that has a huge physical significance: Given that the maximum possible speed v is c, we see that t_o can only be the same as or less than t.

Another symbol is popular and useful for just this expression, where we see the Greek lower case letter gamma (which is unfortunate, since this letter is used in other contexts):

$$\gamma = \frac{1}{\sqrt{1 - v^2/c^2}}.$$

What follows the equal sign is the Lorentz factor, named for its appearance in Lorentz' transformation equations, though Einstein "put it on the map" of relativity. Please note that $\gamma = 1/\beta$.

Readers, and especially students of special relativity, will therefore also encounter

$$t = \gamma \, t_o,$$

which deceptively says a lot: To calculate t, which can be how long a moving event takes, we take how long this event takes according to the non-moving observer (t_o) and multiply by the Lorentz factor γ. What's more, we have just answered a virtually ubiquitous question on special relativity exams; students, keep this in mind.

✦

Please note carefully: In line with the first postulate of special relativity, our two relatively moving observers agree that $cT = D$; and, in line with the second postulate, they agree on the constancy of *c. As a result, they do not agree on T for the same event!* Observer A concludes that *time passes slower* in train B as train B moves by. The same is true reciprocally: As long as the two trains are moving with respect to each other, observer B will likewise conclude that the "bounce time" is longer and therefore that the rate of time is slower in train A. This means that *any observer finds that time flows slower for a relatively moving event.* Moreover this "relativistic effect" on measured time is permanent. That is to say, the discrepancy between clocks does not vanish after the readings are compared. For instance if, as a result of uniform relative motion, B's clock is 1 second behind A's and the clocks are brought together, B's clock will continue to be 1 second slower.

Of course we suspect a contradiction: Suppose observer A determines that an event consumes 10 seconds on his clock but 11 seconds on B's clock, and similarly observer B determines that the event consumes 10 seconds on his clock but 11 seconds on A's clock. Then what do the clocks read if they are brought together and compared? Which one has lost the one second? The answer depends on the nature of motion. To compare the two clocks side by side, the motion of at least one of them must change (e.g. one train must turn around and come back), *which breaks the restriction to uniform motion*, and which means that only one of them lost time. This issue arises in the "twins paradox" which we will consider later.

Furthermore, as either observer looks out into the other train, he cannot prove which train is moving. The results of his experiment are the same whether he believes train A is moving and train B is at rest, or vice versa. This means that absolute uniform motion cannot be detected by any experiment, be it mechanical (e.g. bouncing a ball) or optical (e.g. "bouncing" a light bean). Similarly, an absolute rate of time cannot be detected; how much time an event (e.g. one "bounce") requires *depends on relative motion*, and it does not matter whether an observer considers himself to be in motion or considers the event to be in motion. How an event appears is "relative" to its observer.

Even if observer A looks outside, notices motion with respect to the ground, but considers his train to be at rest and the Earth to be moving under it, the beams of light aboard the trains behave the same to this observer. Again, only relative uniform motion between the trains can be perceived, but such motion alters the observer's determination of $\dfrac{T_A}{T_B}$. We can say that the beam of light (and for that matter the bouncing ball) "does not know" and "does not care" that someone believes that the Earth moves under the train.

But are we really claiming that a train can be at rest while the Earth is moving under it? Yes, we can treat an object as stationary and the environment as moving. Relativity does not say that the engine of the train must cause the Earth to move. It merely says that an observer cannot prove which is the case: the train is moving or the Earth is moving. Either possibility complies with the same laws of physics.

Also, neither observer can claim to be more correct. Each notes nothing unusual or special about measurements taken in his own train, but each finds a "bounce" to take longer on the other's train. They are both correct, as each observer can consider his vantage to be a frame of reference, and no physical experiment can prove his choice to be wrong; the laws of physics appear equally valid in any frame. For example, if there are 10 such trains all moving past each other at 10 different (but in this setting uniform) speeds, there exist 10 different times for the same event. Each measurement is correct in its own frame of reference, and none is absolute or privileged or more legitimate. This conclusion has found its way into philosophy and ethics to support the notion that there may be no absolute truths or absolute standards, but relativity itself is only concerned with the much narrower question of measured time for physical events.

In principle each observer is unaware that someone else who is timing the same event has a different result, and the discrepancy is manifest only when the observers compare their findings. This means that each observer has his own "private" reckoning of time which is unaffected by someone else's motion or by someone else's observations. Such a personal determination—for instance T_A with respect to observer A—is called "proper time," also often expressed by our aforementioned t_o or t_r.[*] The idea of proper time ought not to surprise us: Since we cannot sense our own uniform motion, we cannot detect the relativity of time purely on the basis of our own clocks. Thus if T_A is 1 second, it remains 1 second even if T_B turns out to be 1 hour.

In as much as "proper time" is the time measured on a clock in its own frame of reference, an alternative term is "local" time or relatively-stationary time. We will use the term "local" frequently. It means confined or limited to the immediate proximity. Therefore "local" space or "local" time is a measurement with no relative motion involved. Our own age and the time we read on our pocket-watch are proper or local measurements, since we experience such times only within our own small frame of reference. This is why I stated on page 1 that—"relativistically" speaking—objects moving at different velocities have different frames of reference.

In contradistinction, the time measured by a non-local or distant observer may be called "coordinate time" or relatively-moving time (or just "t"). For us a coordinate time is what is measured on a clock carried (moved) by someone else and then revealed to us. For observer A, T_A is the proper time and T_B is the coordinate time. This distinction clarifies the objection to the notion of "absolute time." The term "absolute" implies the existence of a measurement that all observers agree upon, despite changing physical conditions. Thus if measurements of duration in any experiment never varied, time would be absolute and invariant. Obviously, *since relative motion causes coordinate time and proper time to diverge, time is not absolute or invariant.* In fact *time is relative.*

It is important to note that the conversion factor between proper and coordinate time is the special relativity factor, or more completely, the Lorentz factor. ($\gamma = 1/\beta$.) Since v is never greater than c, coordinate time is never less than proper time. Imagine that every person in the universe carries a clock designed to tick every second. Your own clock ticks at your "proper" "local" rate, once per second. As far as you can tell upon comparison (and only upon comparison), another clock—a "coordinate" clock—ticks slower, and the more mobile you or the other person are, the greater the

---

[*] This "proper" derives from the French adjective "propre" which can mean "one's own."

disparity. Upon cross-comparison, *you will find that own the fastest clock in the universe.* Furthermore, as perplexing as it is, every clock-owner in the universe can make the same claim! *Every* observer in uniform relative motion who has a "proper" clock sees slowing of every "coordinate" clock, and within each (inertial) frame of reference ideal clocks run at the same rate. In short, just as all motion is relative, all observations are relative. No observer can claim to be absolute, but all cross-comparisons reveal the same effects. However, this context—namely special relativity—deals with just uniform (un-accelerated) motion; the reason for this caveat will emerge on page 73, where we consider accelerated motion and general relativity.

The notion that every inhabitant of the universe thinks that his clock is the fastest brings up an interesting argument. Today even a grade-school student might chuckle at the "egocentric" doctrine that the Earth is at the center of the universe, although at one time this was an accepted tenet of cosmology. As science became more sophisticated, it became clear that neither the Earth, nor the sun, nor even our galaxy can claim that exalted position. However as far as the relativity of time is concerned, perhaps such egocentricity is not so naïve after all. If we go by the rate of time on our clocks compared with the rate of time on anyone else's clock, our rate of time appears to be unique and all others rates of time appear to be different. In this sense, *special relativity entitles each of us to believe that we are the center of the universe.*

This concept raises a practical issue: It is clear that all observers are entitled to the same laws of physics. What is not as obvious—but as Einstein realized—is that as long as there is relative motion and relative time, the duration of any event covered by these laws cannot be the same for all observers. Therefore, any law of physics involving time (even "$RT = D$") must be adapted to relativity. That is to say, *the laws of physics must be written so as to work the same for any observer even though time passes differently for different observers.* Indeed Einstein forced us to espouse new laws, and as we shall elaborate, many exciting, intricate and even philosophic ramifications of relativity stem from its intent to ensure that even though time varies, the laws of nature do not vary.

✦

In the case of the two tethered spaceships, we saw that there is no such thing as universal time. In particular, because of the fixed speed of light it is impossible to synchronize all clocks everywhere to one reference. Earlier, using events aboard our moving trains as examples, we have reached the more significant conclusion, which is the essence of special relativity: *The measurement of duration is altered by uniform relative motion.* Not only is there no universal time, but there is no such thing as a universal *rate* of time. We reiterate what it means that *time is relative*: Because of relative motion, we not only disagree on *when* events occur, but we also disagree on *how much time events consume.*

We should not assume that the relativity of time is an intangible figment of theory with no practical meaning. It is an observable and measurable fact of nature, and indeed its authenticity empowers the rest of relativity. Still, it is derived mathematically using only four ingredients. The first two draw upon basic physics and plane geometry: $RT = D$ and the pythagorean theorem. The third ingredient is the postulate that the laws of physics are the same for all observers in uniform motion. The fourth ingredient is Einstein's postulate that for all observers the speed of light is independent of the uniform speed. Einstein enlarged this concept to say that the speed of light is the same for all observers independent of their own uniform motion, and, as we shall detail later, Michelson/Morley unwittingly demonstrated this. All Einstein had to do was grasp the significance of combining these insights. He did just that around 1905 and thereby changed the universe!

Historically, this idea evolved quite logically: First Galileo and later Newton argued that there is no physical way to distinguish rest from uniform motion, so that the laws of physics ought to be the same for an observer at rest as for an observer in uniform motion. Maxwell showed that the speed of light is constant, so Einstein, along with Poincaré (1854-1912), realized that the speed of light would also appear the same whether an observer is at rest or in uniform motion. Einstein surmised that when two observers in relative uniform motion with respect to each other compare an event described by $cT = D$, they are forced to conclude that T is not the same for the two observers. Rather, the speed of light and the laws of physics are constant for all inertial observers, and hence the *rate of time must be relative to uniform motion*. (Poincaré, and not Einstein, introduced the term relativity, referring to the idea that the laws of physics should appear unchanged during uniform relative motion.)

◆

This is not the first revision of our understanding of time and motion. The classical Greek philosopher-scientists accepted absolute rest; the Earth, for example, was considered to be set in the center of a universe in which all moving objects tended to come to rest. In the sixteenth century Copernicus and Kepler forced the abandonment of the idea that the Earth is stationary and that it is at the center of everything. Shortly thereafter Galileo and Newton forced the abandonment of the idea that there is such a thing as absolute rest. In fact, Newton's laws are (approximately) valid precisely because objects tend to stay in motion, but Newton implied that absolute time exists. Finally Einstein bade us to forsake even absolute time.

In short, *we have discovered that we can alter how fast events proceed into the future*. Fortunately this stunning notion can be easily stated with ordinary language: Time does not flow equally for all observers. Relative motion changes measured time; motion stretches time; motion bends time; motion dilates time; time is not independent but depends on speed; time flows at different rates in vehicles which move at different speeds; an event consumes less time at rest than while moving; a moving observer will measure a longer time for an event than if he is not moving; how long a process takes depends on how fast it is moving, or how long an activity takes depends on how fast an observer moves by it; if either moves faster, it lasts longer.

There are even non-scientific analogies, notably that a boring event seems longer, or enjoyable time passes faster. However such notions are not relevant because relativity deals with actual mathematical measurements of time rather than with feelings. The distinction is quite real: A psychological event may seem to take hours even though a clock marks off only a few minutes. A physical event may take one hour on one clock and more or less time on another.

What makes the relativity of time difficult to accept is that it sounds irrational. The notion of relative time represents a revolution in attitude and mind-set as much as a scientific advance: Obviously motion changes location, but now we are told that motion changes duration. In other words, if we have just moved we quite naturally ask *where* we are, but relativity demands that we should also ask *when* we are! Speed changes when we are. Even our language stands in the way of expressing relativity; does anyone normally ask "when am I?"

It's not that we are too unobservant to ask "when are we." It's that we do not move fast enough and we cannot measure accurately enough. If we lived in a world where speeds near c are usual and/or where timepieces routinely read off nanoseconds, Newton and Galileo or even Aristotle, Archimedes, and Euclid would have already figured out relativity. As Einstein himself was aware, outstanding scientists of the past certainly were intellectually capable of grasping relativity. However they lacked the means to detect its subtle effects, and when they acquired these means in modern times, many still hesitated to abandon traditional notions and to acknowledge the unsettling implications. Relativity requires that we set aside our instincts and that we accept wider perceptual horizons, for once we admit the possibility that our measurement of time is changed by motion, the door is open to other even more dramatic and profound tenets of relativity.

✦　✦　✦

Let us explore time relativity further. We imagine that two airplanes, A and B, leave at noon from Chicago to fly (about) 200 miles to Detroit. The trip takes two hours, and another hour is lost crossing into the next time zone. Each plane carries an accurate clock. At take-off the two clocks are synchronous. Plane A flies directly, but plane B (capable of very high speeds) instead circles the Earth 20,000 times before landing in Detroit. Both planes land at 3:00 P.M. Detroit time, plane B having moved much faster than did plane A during the past two hours. On arrival, A's log says "flight time two hours," but B's log says "flight time 1 hour and 50 minutes." Indeed after an adjustment for the change in time zones, the clock on plane A reads 3:00 P.M., *but the clock on plane B reads 2:50 P.M.* We already know why the two clocks are no longer synchronous: The clock on the faster plane slowed compared to the clock on the slower one. In terms of relativity, planes A and B underwent relative motion; their flights were relatively moving events.

If we assume that each clock ticks and that each tick by itself is a separate event, then the time gap between each tick is longer in plane B, so there are fewer ticks of clock B during A's two hours. But when the term "clock" is used here, it encompasses any process which takes time: not only a tick of a clock, but also a vibration of a tuning fork, the undulation of an electromagnetic wave, the swing of a pendulum, the decay of an unstable subatomic particle, a heart beat, or a human lifetime. In other words relative motion slows all time-consuming events when measured by a relatively stationary observer. For example, during the flight the pilot in plane B ages less if (and only if) compared with his colleague in plane A; B's trip cost him less age. Or if we define "aging" of a clock as its recording of the passage of time, then clock B has "aged" 10 minutes less than clock A.

We can even have a pilot pace up and down the aisle during the flight in one of the planes. For example this pilot might own two clocks, one which he carries along while he walks in the aisle and another which stays put in the cockpit. In theory the clock carried while walking loses some time.

We emphasize that relativity does *not* mean that pilot B feels time slowing during his trip. Even while flying at great speed, he will feel and measure a seemingly normal hour and 50 minutes. If his heart rate is 60 beats per minute, then each pulse still occupies one second on his clock. Or if pilot B needs one minute to read a page in a book, his reading speed seems the same to him before, during and after the flight. After all, in his frame of reference (the airplane) there is no relative motion between the pilot's heart or brain and his clock. In other words B's clock (his local clock and his every time-consuming process) records B's proper time; he notes no slowing of his clock.

However when pilot B compares how many minutes passed, how many beats his heart made, or how many pages he could read during the flight, he finds smaller numbers than pilot A. Conversely, pilot A does not detect any anomaly during his flight. Furthermore the 10-minute difference is not the same as one trip taking 10 minutes less because it was faster. Rather *the difference appears between the rate of a stationary clock—in the given inertial frame of reference—and the rate of a clock moving relative to that frame.* Moreover as we pointed out already, this difference is permanent.

Let us add another feature: At 2:00 pm Detroit time there is a fire in Detroit observed by two nearby pedestrians who happened to possess accurate clocks. At that moment plane A is 100 miles east of Detroit, and plane B, having made half of its 20,000 orbits, is 100 miles above Detroit. Both pilots notice the flames and note the time on their respective clocks. Plane A's log reads "Fire in Detroit a small fraction of a second" (100 miles/c) after 2:00 pm, but plane B's log reads "Fire in Detroit a small fraction of a second after 1:55 pm."

The situation is this: The two "local" pedestrians in Detroit agree that the fire occurred at 2:00 pm, but the two pilots disagree on when the same event occurred. *They disagree on the simultaneity of the event.* When the two pilots compare their measurements of time, they conclude that there were two fires about five minutes apart. This discrepancy does not stem from a difference in distance across which light must travel from the fire to each pilot; the planes were each about 100 miles from the fire. The discrepancy is from their relative motion. In short, simultaneity is relative.

Let us change the scenario and imagine not one but two fires, one on each plane. Each pilot notes the clock reading, say 2:00 pm, when their respective fires start. As before, one plane flies much faster than the other. When their flight logs are compared, the pilots conclude that their two mishaps were simultaneous (at 2 pm). However if each pilot had sent out an immediate distress call, the air traffic controllers in Detroit would have received the calls about five minutes apart! What was found to be simultaneous by the pilots was experienced as consecutive by the controllers.

So, if time is relative and its rate is altered by relative motion, then simultaneity is also relative. This phenomenon appears in various settings: The same event which according to local observers is single may have occurred at two different times according to observers in relative motion to each other. Or, relatively moving observers disagree on which events are simultaneous. Or, two events seen synchronously by one pair of observers may be successive to other observers. We note that the only requirement is uniform motion; it need not be accelerated (though it could be).

Incidentally, in this section we imply that the lack of absolute simultaneity is an appendage to the idea of relative time, but historically Einstein developed these concepts in reverse. He first proposed relative simultaneity and from there proceeded to relative rates of time. Indeed when the issue is approached non-mathematically, it is easier to deduce relativistic dilation of time from the relativity of simultaneity: Observers who disagree on "when" readily disagree on "how long."

We should reiterate that discrepancies in the measurement of time or in the simultaneity of events appear only when the observers correlate their findings. Again we note that each observer, no matter how fast he moves, will detect no change in his own proper time, in his own clock, heart beat, life span, simultaneity of local events, etc. Any "clock" obeys the laws of physics, and these laws appear the same to all observers at rest or in uniform motion, each in his own frame of reference. As we said, such clocks do not "know" or "care" that other clocks aboard moving events run slower or that other observers may disagree with the sequence of local events.

✦ ✦ ✦

A flaw has been proposed in this scenario, representing a celebrated criticism leveled against special relativity, one which Einstein himself addressed: If there is no absolute motion and either frame of reference can be considered to be moving, why don't two moving clocks each measure a slower time? In our airplane example, each pilot could claim to have aged less, or both pilots' logs might show 5 minutes less flight time. Yet relativity claims that after relative motion, one set of time measurements differs from another in a specific direction and by a specific amount; only one is the slower (or the other the faster), and the discrepancy in the passage of time can only go one way.

This issue is called the "twins paradox" from an analogy with a pair of twins, one of whom goes on a speedy trip while the other stays home. After this trip, which twin is younger? Skeptics claim that since motion can only be relative, relativity permits either twin to end up younger than the other once they are reunited. The reply has been stated in various ways. One explanation is that the stationary twin did not move or moved less with respect to the universe, and if the universe is considered the frame of reference, only the other twin traveled faster.

Another explanation is more satisfactory but more complex: To be separated and then rejoined, at least one of the twins had to undergo accelerated motion to turn around, so that this twin did not move exclusively in uniform manner. (If both moved only uniformly and linearly, they could never meet again anyway.) A term used here is that the "histories" of the two twins differed—specifically one accelerated, and one did not—so that their actions were not really symmetrical. Thus one of them exhibits slower measurements of time than the other. The math would be quite complicated, but we could calculate their unequal "proper" ages. We will say more about this when we get to accelerated motion and general relativity, where this explanation is more convincing.

In any case, although special relativity is based on the notion that time does not pass uniformly, it does not allow a return to the past. The fast-moving twin, like our fast-flying pilot, may appear to age more slowly than his counterpart, but he will *not* grow younger. Let us say that the twins' age was 40 when twin A left upon a long journey at about 90% of the speed of light. If the pair is reunited (e.g. twin A turned back) when twin B is 50 years old, then twin A will be only 45; the calculation of course uses the equation $t = \gamma \, t_o$. However, no matter how long the trip lasts nor how fast twin A travels (the maximum speed is c), he will not return younger than 40; his aging is delayed but not reversed. Incidentally, this calculation is another favorite for exams on relativity.

Neither does relativity permit the order of casually related events to be reversed. No observer will ever report that a twin returned before he left, even though observers may disagree on their ages. Likewise no observer will ever report that a fire went out before it started, though observers may disagree on when it occurred and how long it lasted. We will elaborate on this issue in another chapter, where the concept of orderly causality—a one-way link between causes and their effects—comes into focus.

✦

No matter how bizarre the concept of relative time seems to us, modern physics has ample evidence for the slowing of a moving clock or other timing mechanism when measured by a stationary observer. In a particle accelerator, for example, the duration of unstable subatomic particles such as muons (a.k.a. mu-mesons) can be altered: Moving muons are found to decay more slowly than stationary ones when observed from the vantage of these stationary muons; the half-life of muons is increased (their life span is extended) if their velocity is increased. In other words the decay of a muon is a physical event which appears measurably slower if the muon moves faster.

In fact a particle accelerator is not even needed to demonstrate this fact. We know that muons decay very quickly. According to Newtonian calculations, muons moving at close to the speed of light should decay after only several hundred yards. Yet natural muons created in the atmosphere over a mile from the surface of the Earth reach the ground traveling close to speed c. Relativity explains why: This situation allows ordinary local observers to measure time by the muons' "clocks," which is their rate of decay. Muons last longer when their natural "clocks" slow down at high speed.

Because extremely accurate "atomic" clocks and fast airplanes are now available, the slowing of measured time during motion can be confirmed even more directly. Such a clock placed aboard an airplane indeed will lose time during a long flight (mostly at a uniform speed) when compared against a similar stationary clock on the ground. We will cite other practical instances of this observation, but clearly, how long events take is influenced by how fast they move.

We shall next examine a crucial consequence of the relativity of time. Let us revisit our trains, but we now aim the beams of light horizontally along to the lengths of the trains. Observers agree that when the two trains stand still, they are of equal lengths; their D's are the same. The observers know that $cT = D$. Observer A aboard train A uses his clock to measure T_A, the time it takes for a beam to traverse the length of train A. From this he calculates the length of this train using $cT_A = D$. We call this length L_A.

Let us again have the two trains move uniformly past each other on parallel tracks, but *observer A looks into train B*. He uses the reading of clock B—not that of his own clock, or else he is not observing coordinate time—to calculate the length of train B. To observer A time appears slower in train B, so that his reading on clock B (T_B) will be *less* than T_A. Solving $cT_B = D$ for the value of D (the length of train B), *observer A concludes that train B has become shorter* than his own train. In other words D, here called L_B, is no longer the same; L_B is less than L_A.

The reciprocal occurs for observer B measuring train A. As long as there is relative motion between them, the two observers disagree on the lengths of the trains; upon comparison, the *other* relatively moving train appears shortened. We note that the discrepancies only involve the dimensions of the trains *in the direction of relative motion*, which in this situation are the lengths, not the vertical dimensions or the widths. Moreover such discrepancies are often more difficult to detect experimentally than disparities in time-measurements, because of technical obstacles in determining the exact size of objects moving at very high speeds.

What if observer A tries to measure the length of train B with a ruler rather than by timing a light beam? Here relativity necessitates that he be precise as to *which* ruler to use. If observer A calls upon observer B's cooperation and uses a ruler on board train B, observer A will *not* report any relative shrinkage of train B. This is because B's ruler, if laid parallel to the length of train B, also shrinks as a result of relative motion (another favorite tricky point on physics quizzes). To elicit the relativistic change in D, an observer must compare gauges in different frames of reference; as he did with the clocks, observer A must use a ruler aboard train A and compare it with a ruler aboard train B which is in relative motion. The point is that observer A finds that everything aboard B's train that takes up space—even observer B himself—appears smaller when compared along the direction of relative motion, just as he found all events aboard B's train to be slowed.

Recalling the previous chapter, we see that *moving lengths appear to shorten and moving durations appear to stretch*, and we say that because of uniform motion relative to us, in those moving frame of references *time dilates and space contracts*! In fact the "official" wording of these phenomena uses these very terms: Time dilation and space contraction.

Moreover the ratio between length at rest and length in motion is similar to the case of time, but it is inverted:

$$\frac{L_A}{L_B} = \frac{1}{\sqrt{1 - v^2/c^2}}$$

For instance L_A is the length of train A at rest. L_B is the length of train B as measured by observer A. v is the relative uniform speed between A and B, and c is the speed of light. Note that if v is more than zero, L_B is less than L_A. But again, L's must point in the direction of motion.

As noted in the previous chapter with regard to time, physics literature usually applies the symbol l_o (less commonly l_r) for length measurements with respect to the stationary observer, and no subscripts for the moving counterpart(s). The key equation then looks like this:

$$\frac{l_o}{l} = \frac{1}{\sqrt{1-v^2/c^2}}$$

Using the same abbreviations, we also often find $l = l_o \beta$ and $l_o = l \gamma$.

Analogous to proper time, we can think of "proper length" or "proper space." If an observer is at rest with respect to an object, what he measures that object to be is its proper length, and this measurement does not match that made by an observer who is *not* at rest relative to the object. If we call the length as it appears to that other observer "coordinate length," then again the conversion factor between proper and coordinate lengths is the special relativity factor or the Lorentz factor.

Imagine that every person in the universe carries a yard-long ruler, and you are sure that your "proper" ruler measures one meter. As far as you can tell upon comparison, every other "coordinate" ruler is shorter, and the more mobile the other person, the greater the difference; in theory, even walking by a ruler shortens it, but only in the eyes of the walker. (Again, when two people walk by each other, each is in a separate frame of reference. Although you walk, anyone can claim that you are the "proper" observer, and that the ruler you walk by is doing the moving.) Just as from your vantage yours is the fastest clock in the universe, *yours is the longest ruler in the universe, yet every ruler-owner and every clock-owner proclaims the same from his vantage.*

Einstein conceived an ingenious way of predicting this effect. He pointed out that to measure length accurately, observers must know the location of the ends of an object at the same time. But there is no such thing as absolutely "same" time for different observers. As two observers in relative motion disagree on the "sameness" of time, they will also disagree on the "sameness" of length; if they differ on "when," they may differ on "how much time" and on "how much space" (which is why it is easier to deduce relativistic dilation of time and contraction of space from the relativity of simultaneity). *Length is relative because time is relative*, and neither is absolute. Since length is a measure of space, *space is relative.*

Thus special relativity shows us that the measurements of *both time and space are altered by relative uniform motion.* The size of physical objects as well as the duration of events is related to how observers are moving, and motion makes space and time elastic. We also stress that the extent of this elasticity, expressed in the special relativity factor or the Lorentz factor, flows only from $RT = D$, from the pythagorean theorem, and from the postulates that the laws of physics and measurement of the speed of light do not change during relative uniform motion.

Another way of looking at this concept will be more meaningful when we get to general relativity: Two relatively moving observers agree that a certain law of physics ($RT = D$ in this example) appears valid, but when they apply this law to one physical event, they obtain differing results. Similarly, two identical but relatively moving events give dissimilar results when investigated by using the same law. How can apparently universal laws yield conflicting outcomes? The answer: Because *the measurements of time and space are changed by relative motion.* The implication: *In order to be universal, the laws of physics must be written so as to heed relative motion.*

✦ ✦ ✦

Relativity forces us to accept the notions that duration can vary and that distance can vary, both as a result of relative motion. The latter may seem even more bizarre because it means that the size of presumably rigid objects can be different to different observers. Of course common sense resists both conclusions, and Newton would have argued that time and space should be the same for all observers regardless of their motion. Time is "absolute" and it is "without relation to anything external" according to Newton. And if time is absolute, so is space, and therefore in Newtonian science relative motion cannot alter the appearance of physical objects or events.

How could Newton have missed this fact? The key is that he was not aware that light has a finite speed which is independent of uniform motion. We shall reconsider this property of light in the chapter covering quantum mechanics, but now we note a difference between light and an ordinary slowly-moving object. Imagine that an observer, running at 10 mph (v_A), can throw a dart so it flies forward at 30 mph (v_B). If we could ask Newton, or Galileo for that matter, how fast the dart flies relative to the Earth (V), he would not hesitate: 40 mph, since he has no reason to deny that

$$V = v_A + v_B.$$

But instead let the running observer turn on a flashlight and have the beam shoot forward. How fast does the *light* move relative to the Earth? If Newton knew only that the speed of light is c, he would say "c + 10 mph," but Einstein says "still only c," implying the seemingly illogical statement that

$$c = c + v_A.$$

The same holds for the runner who aims the flashlight behind him; Newton predicts "*c* - *10* mph" but Einstein maintains "*c*." This is how sunlight reaching Earth from one edge of the sun appears to have the same speed as light from the other edge, even though the sun is spinning: Motion of the source does not add to or subtract from the speed of the emitted light. We say that the speed of light is *not additive; it is "exclusive."* Of course this notion raises a question about speeds that are between slow (e.g. 40 mph) and very fast (e.g. c). Are these speeds additive or not? We will consider this point shortly.

We also stated earlier that c appears independent of the uniform motion of an observer, so that if the runner meets an oncoming beam of light and measures the speed with which that beam passes by him, he will find it to be c. Again, Newton would have expected the result to be *c + 10* mph, but now we are in a better position to explain why this is not so. Let us assume that our observer-runner has a way of determining the speed of light, and that he can measure how much time an oncoming beam needs to get past him while he is standing still and while he is running at 10 mph. In effect he is comparing an event—a stream of photons passing by—as it appears at rest against how it appears in uniform relative motion. Since his motion causes his measurement of time to be increased while he runs into the light beam, light takes more time to pass by him, so that in net-effect c remains constant and not additive.

That is to say, the runner's own speed *adds* to his measurement of the speed of the oncoming beam to give him a result of *c + 10* mph, and his relative motion *decreases* his result by 10 mph. In short, $c + 10 - 10 = c$. We can say that Newton adds 10 mph and Einstein takes it back. Hence the observer finds light to a have speed which is independent of his own motion. (Actually, 10 mph would generate an extremely small and practically imperceptible relativistic effect, but the speed of the Earth orbiting the sun is substantial, which is why the negative results of the Michelson-Morley experiments are so important. What confounded them is that c is not additive!)

Since speed is distance (space) per time, the same result ensues through the relativity of space: While running into the light beam at 10 mph, the *distance* over which the runner measures speed appears less, so that his measurement of the speed of the beam is decreased by 10 mph. Thus his measurement of c remains unaffected whether he uses the fact that time is lengthened or that distance is shortened by relative motion.

We note a surprising inference: A 40-mph dart can overtake a 10-mph runner by 30 mph. But if we assume for the moment that c is 1000 mph, then it is *not* true that a beam of light overtaking the 10-mph runner appears to him to pass at 990 mph; to his reckoning the beam passes him at the full 1000 mph. And if the runner could reach 1000 mph, the photons in the beam do *not* appear to be at rest with respect to this remarkable athlete. They still appear to zip by him at the full value of c, 1000 mph in this analogy. In fact, as illogical as it seems, no amount of effort will suffice to close the gap, and no matter how slow or fast we go, to us light moves at c and light goes faster than us by c. As Jones says it (p. 12-13), "...you can never 'catch up' with a light beam....[and] there is no *reference frame* in which light is at rest." Incidentally, the latter statement looms prominently in quantum mechanics, specifically in modern particle physics: Photon, having a rest mass which is zero, have a maximum possible speed, which is c.

◆

We can now consider whether speeds that are high but less than c are additive. Intuition suggests that any speed of any object is possible. Therefore it seems that two velocities, $v_A + v_B$, imparted to an object should be additive, and that there should be no limit to the net speed V. Again, we expect that

$$V = v_A + v_B.$$

However, in the case of light, relativity holds that not only is the speed of a beam unaffected by the speed of the source, but that V, as measured by an observer in uniform relative motion, cannot exceed c. Specifically, where A and B are uniform speeds,

$$V = \frac{v_A + v_B}{1 + \frac{(v_A)(v_B)}{c^2}},$$

derivable from the special relativity factor and from the "Lorentz transformation." Often we see the speeds A and B expressed as "*u*" and "*v*," and this equation is the key to typical exam problems.

For the moment, let us just accept this equation and for example imagine that two spaceships approach each other, each at 3/4 the speed of light ($v_A = 0.75c$ and $v_B = 0.75c$). Viewed from the Earth the spaceships appear to pass each other at 1.5 times the speed of light, which does not violate relativity because neither is traveling beyond c. However, measured from either one of the spaceships, the closing speed (V) always appears to be less than $v_A + v_B$ and is always less than c. *This means that the higher the speeds, the less additive they are and the more "exclusive" is their behavior.* In other words as speeds rise to approach c, their sum approaches $\frac{v_A + v_B}{1 + \frac{(v_A)(v_B)}{c^2}}$. In fact

if we solve this equation, their relative speed is only about 0.96c and cannot reach 1c. Of course since c^2 is a very large number, this effect is imperceptible when speeds are low (such as 40 mph), and their sum appears to be $v_A + v_B$, which means they appear to be additive, just as they did to Newton and Galileo. (Students of special relativity should note the use of the above equation.)

This is not a trivial point, for it reveals a profound harmony in physical nature. *Slow objects do not obey different laws than fast ones*, and the effects of relativity apply to both. Photons obey the same laws as darts and spaceships, but their speed happens to be the fastest that anything can move. The exclusive behavior of light is not an anomaly but a consequence of its great speed. The same idea is accessible in terms of the relativity of time, specifically the slowing of time during relative motion. The closer either of the spaceships gets to c, the more time is slowed, and theoretically the last bit of acceleration needed to attain c requires an infinite amount of time. In the truest sense, ordinary objects can *never* achieve c, and things that can reach this highest of speeds, such as photons (and forms of electromagnetic radiation), are indeed the most exclusive.

The notion that added velocities of objects can never add up to c lends itself to dramatic experimental confirmation in modern "atom smashers." Nonetheless such an experiment was successfully performed by Fizeau 28 years before Einstein was even born.[*] See Gamow in <u>The Great Physicists from Galileo to Einstein</u>, pp. 182-183. As Einstein nurtured his ideas he found

---

[*] Fizeau had also determined c itself quite accurately.

inspiration in Fizeau. For example he devoted a short chapter to him in <u>Relativity</u>, pp. 38-41 and 49, and he considered Fizeau's result to constitute "fundamental" support for special relativity.

We note that events may have velocities beyond c, like the crossing of the two spaceships. However no object actually exceeds that maximum. Similarly the speed of many non-material objects can exceed c, and hence we stipulate that no *material* object can be accelerated to a velocity of c. For example see Epstein, p. 75: A beam of light can cross a surface (even a wall) at a speed beyond c, but the light itself does not strike the surface with greater-than-c speed.

✦ ✦ ✦

As we already mentioned, Newton thought of time, or more precisely the rate of time, as absolute. He held similar views regarding space. To him, space was a rigid three-dimensional framework in which physical events occur, and any spatial measurement should appear the same to everyone. However, as we also have said, Maxwell calculated, and experiments confirm, that the speed of light is constant. For the sake of argument let us revise our calculations abroad our moving trains, and let us imagine that the speed of light is additive by assuming that this speed jumps to c + v aboard the train that is moving a speed v. We will then calculate *no* discrepancy between A's and B's measurements of the time for one event. (T_A and T_B will turn out to be equal.) In other words we will conclude, as did Newton, that the rate of time is absolute, which implies the same for space.

But why couldn't Newton and Maxwell both be right? Why couldn't time be absolute and c constant? Because then the observer aboard one of the trains finds that RT does not equal D, signifying that *the laws of physics are different if we are moving*, and that one of the frames of reference is endowed with better laws. Neither Galileo nor Newton nor Einstein would accept that possibility. Please examine the following table, reading down the three vertical columns.

Laws of physics are fixed?	Yes	No	Yes
Speed of light is constant?	No	Yes	Yes
Rate of time is absolute?	Yes	Yes	No
Result?	Not true	Anarchy	Relativity

(We ignore meaningless possibilities, such as No/No/No). We are saying that if the laws of physics are fixed for all observers in uniform motion and the rate of time is absolute, then the speed of light cannot be constant, which contradicts experimental results. If the speed of light is constant and time is absolute, then the laws of physics cannot be fixed. In that case not only is nature chaotic, but uniform and absolute motions should be detectable, which they are not (and we will deal with possible exceptions later.) Finally, in the crisp and clear logic of special relativity, if the laws of physics are fixed and the speed of light is constant, then time must be relative.

Moreover we see that relative time implies relative space, which raises an admittedly academic but important argument: We discern the relativity of time by measuring time intervals on a clock or on any other device that quantifies duration, and (even if technically difficult) we discern the relativity of space by measuring objects with a ruler or with any other device which quantifies space. These facts temp us to infer that relative motion induces an alteration in such devices. In other words we might conclude that the effects of special relativity constitute a change in the nature of clocks and rulers. As illogical as it seems, we should think of relative motion as causing *a change it the nature of time and space itself*, while clocks, rulers, and similar devices—and indeed all objects and events—merely occupy that time and that space. We will come back to this point, but let us be aware of just what is "relative" in special relativity: Fundamentally, it is time and space.

We can also touch on a particular consequence of special relativity. Ordinary physics tells us that RT = D, or *rate equals space divided by time R = D/T*), while special relativity tells us that the rate of light, rather than space or time, is absolute. If we let the rate R equal *c*, then

$$c = \frac{space}{time}!$$

We note this ratio: *c*, which is the speed of light and which is a constant, defines the relationship between space and time, which are relative. This again suggests that space and time are closely interrelated, and indeed we will make much use of this concept. In contrast Newton not only assumed that time and space are absolute, but he also followed common sense by treating time and space as separate and independent components of physics.

✦ ✦ ✦

We also note that for the basic equations of special relativity to work properly, the velocity *v* in the special relativity factor, in γ, and in β is constant. However, non-constant velocity but *constant acceleration* is useful in applied physics (and in science fiction). For instance when space vehicles are sent to other planets in our solar system (and potentially to distant stars), they can be designed to travel in outer space (far from strong gravitational fields) *while accelerating at a constant rate*. Is there a counterpart of the special relativity factor that is valid under these conditions?

The answer is yes, but the main mathematical difficulty is that the trajectory of a constantly accelerating object is a hyperbola, entailing specialized trig functions, namely sinh, cosh and tanh. (These are defined and/or derived in any detailed writing or web site on interplanetary motion; e.g. http://math.ucr.edu/home/baez/physics/Relativity/SR/Rocket/rocket.html.) At least so that readers can recognize them, here are some basic useful equations; gravitation and the expansion of the universe are neglected. Constant acceleration is *a*, and the speed of light is *c*:

To calculate coordinate time *t* from proper time *T*: *t = (c/a)(sinh[aT/c])*. Clearly this is the constant-acceleration counterpart of $t = \gamma t_o$ where *v* is constant, but with different symbols.

To calculate the distance *d* covered after a (proper) time period *T*: *d = (c² /a)(cosh[aT/c]-1)*.

To calculate the speed *v* reached after a (proper) time period *T*:, *v = c(tanh[aT/c])*.

CHAPTER 4: THE RELATIVITY OF MASS

The key to this chapter is the set of "laws of conservation" or, in short, the "conservation laws." These include the law of conservation of mass, of energy, and of momentum. In closed systems (nothing can leak in or out), mass cannot be created or destroyed; nor can energy, nor momentum. In each case a similar equation applies: The change (Δ) in the total amount (Σ) of mass (m), of energy (E), or of momentum (mv) is zero:

$$\text{Conservation of mass: } \Delta\Sigma m = 0.$$

$$\text{Conservation of energy: } \Delta\Sigma E = 0.$$

$$\text{Conservation of momentum: } \Delta\Sigma mv = 0.$$

Conservation of mass, for our purposes, is the same as conservation of matter. The law of conservation of energy has a corollary, the conservation of kinetic energy. Kinetic energy, expressed as $\frac{1}{2}mv^2$, is the energy an object has by virtue of its motion, and the change in the amount of kinetic energy remains zero:

$$\text{Conservation of kinetic energy: } \Delta\Sigma\frac{1}{2}mv^2 = 0.$$

Students of differential calculus will recognize kinetic energy as the derivative of (rate of change of) momentum. Other laws of conservation appear in the sciences.

The law about momentum is of interest to us now. It states that if one object strikes another, its momentum is transferred completely; the net change is nil. This is readily visible in a game of billiards, although it also applies to all moving objects, even subatomic particles. Later we will refer to "angular" momentum, but till then we are concerned only with "linear" momentum. Under many circumstances momentum is a measure of how difficult it is to stop a moving object. It is an indication of the inertia of an object quantified by "inertial mass" times velocity—*mv*. Momentum is abbreviated as *p*; hence

$$p = mv.$$

(Momentum abbreviated as *p* is not to be confused with density, abbreviated by the Greek letter ρ; both appear in our equations.) Inertial mass is what a moving object shows when we try to alter its course, or it is what a motionless object shows when we try to move it. Thus inertia is resistance to acceleration. Inertial mass can be distinguished from gravitational mass; the latter simply appears as weight. (Inertia is discussed later.)

Applying these principles about momentum and inertia to subatomic particles raises a practical problem: How do we weigh a moving subatomic particle? The answer is that we assume that inertial and gravitational mass are equivalent (and later we will justify this assumption), which means we can determine a mass of a particle by how difficult it is to accelerate it in an experiment.

But is $\Delta\Sigma mv = 0$ correct? This equation contains v, which is distance/time, yet we know that *time is relative*. In other words given the relativistic effects of (uniform) relative motion on measured time, we must distinguish mass in motion, labeled m_{motion}, from mass at rest, labeled m_{rest} and called "rest mass." Then the law of conservation of momentum should state that

$$\Delta\Sigma m_{motion}v = 0,$$

but in fact when there is relative motion between the object and observer, $p = mv$ does not hold, because according to special relativity m_{motion} need not be the same as m_{rest}. To reconcile the conservation of momentum with special relativity—which in effect means to heed the postulates that the speed of light and the laws of physics are constant—we need *a relativistic form of momentum*, one which contains the special relativity factor. Instead of $p = mv$,

$$p = m_{motion}v = \frac{m_{rest}v}{\sqrt{1-v^2/c^2}}.$$

Eliminating one v from each side of the equation, the ratio between the mass at relative rest to mass in relative motion is

$$\frac{m_{motion}}{m_{rest}} = \frac{1}{\sqrt{1-v^2/c^2}}.$$

This kind of ratio should look familiar. It means that the mass of a moving object is greater than it was at rest; relative motion increases mass! In other words in adapting "$\Delta\Sigma mv = 0$" to special relativity, we find that *mass is relative*. We may appreciate the parallel here: While adapting "$RT=D$" to special relativity we found that time is relative. (As compellingly, space is relative.)

As we can expect, the above equation is commonly, though less explicitly, written

$$\frac{m}{m_o} = \frac{1}{\sqrt{1-v^2/c^2}}.$$

The similarity to its relativity-of-time counterpart is obvious:

$$\frac{t}{t_o} = \frac{1}{\sqrt{1-v^2/c^2}}.$$

Please also notice that the ratio is reversed for the relativity-of-space case. Thus we have the abbreviated equations, $m = \gamma\, m_o$ and $t = \gamma\, t_o$, but $l_o = \gamma\, l$. In the first two cases, for mass and time, the local observer will find that *m* and *t* (in a moving frame of reference) can only be larger that his own measurement, whereas *l* is contracted in a moving frame of reference. In other words, as a result of relative uniform motion we see dilation of time *and mass*, but contraction of space.

In line with the notions of proper time and proper space, we can also call m_{rest} or m_o the "proper mass." For example when you stand on a scale, you are determining your proper mass. We then call your mass as it appears to another observer "coordinate mass," and the conversion factor between the two is the special relativity factor. Relative motion between us does not change my "proper" or "rest" mass, but for me it changes your coordinate mass. Now we can imagine that an observer at rest will find that he has the shortest 1-second time interval in the universe, the lightest 1-gram weight, and the longest meter stick. On comparison with everyone else in relative motion, all others have slower ticks, heavier grams and shortest meters!

Incidentally it is easy to see that density, which is mass per volume, is also relative. Indeed we can say that density is "doubly" relative: In the presence of relative motion, space (e.g. length) shrinks so that volume shrinks, while mass increases, and the effects are multiplicative.

Getting back to the relativity of mass, for our next step an approximation of the special relativity factor is available, based in the fact that normally *v* is much less than *c*:

$$\frac{1}{\sqrt{1-v^2/c^2}} \approx 1+v^2/2c^2$$

(This approximation is based on the algebraic power series expansion:

$$\frac{1}{\sqrt{1-x}} = 1 + \frac{1}{2}x + \frac{3}{8}x^2...,$$

but when $x = v^2/c^2$, the third and subsequent terms on the right side are so small that they become negligible.) Relativistic momentum can therefore be expressed as the approximation

$$m_{motion}v = m_{rest}v\,(1 + v^2/2c^2).$$

If we again eliminate one *v* from each side, we have an equation for inertial mass,

$$m_{motion} = m_{rest}\,(1 + v^2/2c^2),$$

which can be rearranged into

$$m_{motion} = m_{rest} + \left(\frac{1}{2}m_{rest}v^2\right)/c^2.$$

But the term $\frac{1}{2}m_{rest}v^2$ is kinetic energy, so the implication of this equation is that imparting an object with kinetic energy raises its mass from m_{rest} *to* m_{motion}.

Again we see that *relative motion increases mass* and that the amount of increase is approximated by the special relativity factor or, as in the commonly used equation

$$m = \gamma \, m_o$$

by the Lotentz factor.

By convention again we can dispense with the subscript for m_{motion}; just m suffices. Similarly m_{rest} can be labeled as m_r or m_o to express zero relative or proper or local motion. Thus, upon multiplying the previous equation by c^2,

$$mc^2 = m_r c^2 + \frac{1}{2} m_r v^2.$$

By rearrangement we obtain an explicit equation for the effect of the *change* from *rest* to *motion*:

$$\frac{1}{2} m_r v^2 = mc^2 - m_r c^2$$

The left side is the kinetic energy for getting from rest to motion, but we can rearrange this equation as follows:

$$mc^2 = \frac{1}{2} m_r v^2 + m_r c^2$$

And in this mathematical form we see a profound statement, the importance of which can hardly be overstated: *Total energy equals kinetic energy plus rest energy*. This insight, rather obvious once we say it, leads to the equally obvious conclusion dealing with E, the total energy:

$$\boxed{E = mc^2}$$

Einstein introduced this concept in 1905 in his famous but brief paper (only three pages), "Does the Inertia of a Body Depend upon its Energy-content?" (See The Principle of Relativity, pp. 69-71.) L is the symbol used in this paper for the energy in light waves. Einstein reasoned that when a radiating (e.g. glowing) object gives off energy, it loses mass by the amount L/c^2. The key mathematical ingredient was the special relativity factor which he had already derived. $E = mc^2$ did not appear in the literature of physics in this form until about two years later.

Surely $E = mc^2$ is the most familiar equation in all science. As even the casual student of physics knows, $E = mc^2$ asserts that energy and mass are equivalent and interchangeable. The factor c^2 which relates E to m is in itself remarkable in several ways. First of all it makes the formula work in the units used to express E and m. (Usually E is joules, m is in kilograms, and c is in meters per second.) Moreover, since the speed of light is constant during uniform motion, c^2 is unaffected by such motion. Finally, c^2 is a very large number. Since mass is linked to energy by such a number, a very large amount of energy can be recovered from small amounts of matter. This is what happens in thermonuclear and similar "atomic" reactions. In fact because of $E = mc^2$, we can consider all mass (matter) to be merely very concentrated energy, and we can say that *energy has mass*. Conversely, under certain conditions energy can be turned into mass, and we can also say that *mass, even non-moving mass, has energy*. Said succinctly, "energy weighs."

The notion that E is linked to m can be reached by non-mathematical reasoning. Let us recall our two trains, A and B, passing each other while an observer aboard each train looks at the other train. Since neither observer can detect absolute motion, relativity allows both possibilities to be equally valid: Train A is moving and B is at rest, or B is moving and A is at rest. If the observer in train A accepts the first case, he ascribes kinetic energy to train A but not to the other. If this observer accepts the second case, he ascribes kinetic energy to train B but not to his own. But how can a train have kinetic energy and not have kinetic energy? The only way is if mass and energy are interchangeable. Conversely, if mass and energy were not equivalent and interconvertible, it would be possible to prove that only one of the trains is moving. In other words, if E were not linked to mc^2, absolute motion would be detectable.

We found $E = mc^2$ by adapting the law of conservation of momentum to special relativity. However $E = mc^2$ can also be readily derived using Newtonian terms without resorting to relativity: By definition, energy is the capacity to do "work," and in physics, work equals force (F) times distance (D), or FD:

$$E = FD$$

Newton's "first law" states that force equals mass times acceleration ($F = ma$). The equation for rate times time equals distance can be written $v = D/t$ which, as we will show later, means that acceleration (a) is distance divided by the square of time (for example meters per second per second). Hence

$$a = D/t^2$$

and by substitution in $F = ma$,

$$F = mD/t^2.$$

Combining $E = FD$ and $F = mD/t^2$ gives

$$E = m(D/t)^2.$$

If the pertinent speed is that of light so that $D/t = c$, we reach

$$E = mc^2.$$

Clearly, Newton had all the mathematics needed to derive $E = mc^2$. What he lacked was any sign that light has a constant speed and that nature might evince this equation. Instead Newton surmised that gravitational force is proportional to mass—if the Earth doubled its mass, objects should fall twice as forcefully (see page 58)—but he could not reach the next logical step: *Energy is proportional to mass. If m doubles, E doubles.* This is why Einstein's pronouncement that *imparting an object with energy for motion raises its mass* was so astonishing. However it is another conclusion of relativity which can be said in a few simple words: Moving objects or particles appear heavier; or measured mass is altered by speed; or mass depends on relative speed. In theory, merely walking by an object makes it weigh more—but only from the vantage of the walker. And as we have already seen, walking by a clock makes it slow and makes it shrink, and now we add that walking by a clock makes it heavier! Such is the world of special relativity, but we re-emphasize that only basic algebra, geometry, and physics suffice to predict and quantify these incredible but demonstrable effects.

We also note that mass is altered by any kind of kinetic energy. We need not move an object to increase its mass. It suffices to warm it (so its atoms vibrate more), to wind it up (in the case of a coiled spring), to charge it (in the case of a battery), or to stretch it (in the case of a rubber band). For instance a battery is slightly harder to move when it is charged. Conversely a cooling object, an unwinding spring, a discharging battery, and a relaxing rubber band all become slightly lighter and easier to move.

As with other relativistic effects, under conventional conditions these changes in mass are undetectable, but they carry great significance. For example we know the amount of energy in the electric charge of an electron. That amount, divided by c^2, turns out to equal the mass of an electron. This is quite important, since electrons are basic to the structure and chemistry of all matter. It means that the energy of any electron accounts for its mass, or conversely that its mass explains its energy—in this case its electric charge.

Two comments are in order here. First, popular opinion often contends that $E = mc^2$ was intended for the conversion of matter into "atomic" or nuclear energy. This is not accurate. In the original context $E = mc^2$ denoted only that a change in energy is associated with a change in mass, and initially Einstein discounted the feasibility of releasing energy from matter. The latter prospect was appreciated only years later, and even though nuclear weapons would not work if E did not equal mc^2, knowing this equation was not an essential ingredient in their development. Second, although Einstein did not participate in the building of nuclear weapons, he raised this possibility before World War II, fearing that Nazi Germany was working toward this goal. He communicated his concerns to President Roosevelt in 1939, though later he beseeched Roosevelt not to use the atomic bomb.

✦

We will now dwell on a point that is often overlooked, to the detriment of students in poorly taught courses on relativity. The equation $E = mc^2$ actually refers to total energy, and it is *not* the same as $E = m_o c^2$. (clearer as $E_o = m_o c^2$.) The latter quantifies rest energy, and it has an implication that would have startled Newton: Objects at rest have some energy; in fact they have enormous energy! Of course an "atomic" (nuclear, really) explosion demonstrates this very violently, but the same kind of energy is released naturally and subtly in the annihilation of subatomic particles. Indeed we can say that rest energy is the energy of an object by virtue of it existing.

Let us look at this distinction mathematically: We found that $E = mc^2$ signifies that mass, like time and space, is relative. This was implied in the equation $\dfrac{m_{motion}}{m_{rest}} = \dfrac{1}{\sqrt{1 - v^2/c^2}}$ by which we can predict the effect of relative motion on the masses (weights) of two objects. Now let us recall that the Lorentz factor (the right half of this equation) can be conveniently abbreviated as γ. In that format, using the common subscript for a rest mass, and no subscript for the moving mass, we will often see (as earlier above)

$$m = \gamma \, m_o.$$

To grasp the full meaning of this equation in the spirit of special relativity, we note that the plain m describes a real, weigh-able mass. However, the m_o is more subtle, *as it represent mass that has intrinsic energy*. The astounding implication is that since γ must be at least 1 (and it can be infinite when $v = c$), giving a mass a "kick" or a "boost" (so that it moves) *raises its weight*. So from this point of view, m is the true total mass of an object, consisting of m_o mass by virtue of its existence, and of additional mass stemming form its motion.

We draw on simple algebra to multiply both sides of $m = \gamma \, m_o$ by c^2, which gives us

$$m \, c^2 = \gamma \, m_o c^2,$$

which turn, again surprisingly, has converted masses into energies. Indeed, total energy is such that

$$E_{total} = \gamma \, m_o c^2.$$

(The subscript for E is to emphasize "total energy.") However, in the same format, rest energy is such that

$$E_o = m_o \, c^2 \text{ (or } E_r = m_r c^2 \text{)}.$$

In other terms the energy in an object at rest is $m_r c^2$, while its total energy, including the kinetic source is mc^2. Again for emphasis,

$$kinetic \ energy \ = \ mc^2 \ - \ m_oc^2,$$

which can be rearranged to read

$$mc^2 \ = \ m_oc^2 \ + \ kinetic \ energy$$

and which is often abbreviated (with skipping the redundant subscript $_o$) as

$$K.E. = \gamma \ mc^2 - m \ c^2.$$

These are akin to the approximated equation (with terms reversed) we used to derive $E = mc^2$,

$$mc^2 \ = \ m_oc^2 \ + \ \frac{1}{2}m_ov^2.$$

We can replace each mathematical term by words:

Total energy = intrinsic energy + kinetic energy.

Clearly the total energy of a moving object has two sources. One is energy-of-motion, kinetic energy, which we quantify via momentum and velocity. The other is intrinsic energy contained in matter even while it is at rest.

We note how kinetic energy fits into the picture. By intuition, of course a heavy projectile can be destructive if we throw it so it moves fast, but by the teachings of special relativity a thrown heavy projectile becomes even more destructive because—from the vantage of thrower—it gains weight! (From the vantage of the projectile, no weight has changed.)

For instance a 10-kilogram object, if thrown hard enough, becomes an 11-kilogram object. Now a question, particularly for student-readers who take physics tests: What is this object's kinetic energy? Easy: It's 1 kilogram. Yes, mass has been "dilated" to 11 kilograms by the addition of 1 kilogram, but by just common sense that's weight and not energy. Really? What about $E = mc^2$? Just multiply by c^2, and we have kinetic energy. *In short, the increase in mass due to motion, when multiplied by c^2, is kinetic energy.*

Again for students of relativity, what is that kinetic energy if the speed of this projectile is 60% of the speed of light? Well, $m = \gamma \ m_o$, γ is the Lorentz factor, the "rest" mass was 10 kg, and now we are told that $v = 0.6c$. The algebra to get m is not hard: $m \ = \ (10) \ \dfrac{1}{\sqrt{1-0.6^2}}$ which yields 12.5 kg. The total mass has been dilated from 10 to 12.5 kg. And what is the energy of this total mass? It's $E = mc^2$! In fact we can give the answer in Jules, knowing that (approximately) $c - 3 \times 10^8$ meters/second: $(2.5)((3 \times 10^8)^2) = 2.25 \times 10^{17}$ Jules. This is the total energy of a 10 kg object with enough kinetic energy to move at 6/10 of the speed of light. We can also calculate γ if we know, for instance, that the velocity v is 0.6c; the hurdle is finding the square root of $1-0.6^2$.

Lest a student think that all this is fictitious theory (projectiles thrown close to the speed of light?), machinery does exist that emits proton beams for cancer therapy. We know that the rest mass of a proton is about 1.67×10^{-27} kg, and here a desirable velocity of a proton can be 6/10 of the speed of light. Considering the design of the machine, how much energy must be exerted on a proton? The answer is to solve K.E. $= \gamma\, mc^2 - m\,c^2$, and every variable is known: $m = 1.67 \times 10^{-27}$ kg, $v = 0.6c$, $c = 3 \times 10^8$ meters per second, and energy can be expressed in Jules, as done above.

Incidentally, again for students, exam questions may give you the rest mass and total energy, and ask for the speed. Here is a rarely taught equation solved (by tedious algebra which we can skip) for *v*: "Small" and "large" can refer to either the masses or the energies (they are proportional!); just do not invert the ratio and remember "rest squared divided by dilated squared."

$$v \;=\; c \,\sqrt{1 - small^2 / large^2}$$

To obtain all the values, you may have to calculate the masses; remember rest mass + mass from motion = dilated (relativistically increased) mass. And be prepared to astound your teacher.

✦

Despite its notoriety, the equation "$E = mc^2$" is not the most useful format for expressing the equivalence of matter and energy, especially in the context of general relativity and quantum mechanics. Let us derive better alternatives, in which we will find the concept of momentum to be essential.

We will see details in another chapter, but please be aware now that modern quantum mechanics encompasses particle physics, requiring that, for example, electrons be described in terms of their special-relativity features, particularly mass and momentum. The reason is simple: Electrons often move at "relativistic" speeds (close to c), including in modern high-energy experiments. Therefore the equations of particle physics would fail without heeding special relativity. But at the same time electrons show behavior that can only be described by quantum mechanics, such as "spin," and these equations would fail without including such "quantum" features. Indeed brilliant scientists and mathematicians (Dirac comes to mind first, working in the 1920's and 1930's) managed—for the first time in history—to derive a successful physics equation that represents a confluence of special relativity and quantum mechanics. This equation is named for Dirac, and it accomplishes even more, as we will see on the next page: He anticipated that there exists antimatter, and this too appears in his equation; please see next footnote.

Incidentally, such a confluence has not been achieved for general relativity and quantum mechanics. For one thing, if the former has subatomic particles for gravitation (like photons for light), they have not been definitively detected yet. For another, gravity is much weaker than, and very different from, the forces explained by particle physics. We will look into this issue later.

In any case, we combine the equation (let's use the subscript r to be more explicit)

$$E = mc^2 \text{ or } m = E/c^2$$

with

$$m = \frac{m_r}{\sqrt{1-v^2/c^2}}.$$

Solving for E/c^2 gives

$$\frac{E}{c^2} = \frac{m_r}{\sqrt{1-v^2/c^2}}.$$

For identification we call this our "alternative equation A."

Squaring this equation (c^2 squared is c^4; see appendix about squares) is gives

$$\frac{E^2}{c^4} = \frac{m_r^2}{1-v^2/c^2}$$

or, upon rearrangement,

$$E^2(1-v^2/c^2) = E^2 - E^2v^2/c^2 = m_r^2c^4.$$

(If $\dfrac{A}{B} = \dfrac{C}{D}$, then $\dfrac{A^2}{B^2} = \dfrac{C^2}{D^2}$. Also, $\left(\sqrt{1-v^2/c^2}\right)^2 = 1-v^2/c^2$.)

Similarly we re-use $E = mc^2$ in squared form [*],

$$E^2 = m^2c^4,$$

which we insert in the previous equation to give

$$E^2 - m^2c^4v^2/c^2 = E^2 - m^2v^2c^2 = m_r^2c^4.$$

By moving some of the terms, this can be written as

---

[*] $E^2 = m^2c^4$ allows the equation $-E = mc^2$ (since $(-E)^2 = E^2$). The possibility of negative energy and negative mass arises in quantum mechanics in the context of antimatter and in the aforementioned Dirac's equation. Antimatter is crucial in quantum particle physics, and yet its mathematical inspiration arises in special relativity. I discuss this in detail in chapter 20.

$$E^2 = c^2(mv)^2 + m_r^2c^4,$$

which we call our "alternative equation B."

We can restate alternative equation B by the definition of momentum $mv = p$ but in squared form,

$$m^2v^2 = (mv)^2 = p^2.$$

This gives us an equivalent equation,

$$E^2 = c^2p^2 + m_r^2c^4,$$

which can be written as

$$E = c\sqrt{p^2 + m_r^2c^2}.$$

Alternative equation B and its two equivalents each claim that *the relativistic energy of an object stems from its kinetic energy expressed as momentum, plus its rest mass expressed as m_r.*

We can say that one way or another, the value of E must be increased by relative motion. In $E = mc^2$ this requirement is implicit. In equation A this requirement is met explicitly through the special relativity factor; if v is greater than zero, E is increased. In equation B and its equivalents, this requirement is met explicitly through momentum, which makes these equations more useful; that is, if mv or p is greater than zero, E is increased. Either equation A or B reiterates that relative motion adds energy; indeed energy *is* relative.

Meanwhile the equation

$$E = c\sqrt{p^2 + m_r^2c^2}$$

suggests two possibilities: p can be zero, and m_r can be zero. The former is the simpler case, wherein there is no momentum (i.e. no motion and no kinetic energy). Under such conditions all the energy of an object resides in its mass; p drops out, leaving

$$E = c\sqrt{m_r^2c^2} = m_rc^2.$$

Common sense suggests that no energy resides in a stationary object, but here we see again that *a non-moving object owns a clearly defined amount of energy.* Indeed the notion that even immobile objects are vast storehouses of energy revolutionized our concept of matter, and we have ample proof—for instance an "atomic" bomb, even when immobile, can be far more destructive than the fastest bullet. We can even say that from a historical point of view, this conclusion—that motionless objects contain an enormous amount of energy—is the most important consequence and the most dramatic significance of special relativity.

We can reach the same conclusion from our conception of motion. The reasoning is somewhat pedantic, but if we grant that some object somewhere in our ceaselessly active universe is always moving, *no object is ever absolutely at rest and therefore no object we can examine is ever absolutely free of kinetic energy.* To be practical we can treat objects as if it they are at rest and we can label them as m_r, yet in a large enough frame of reference m_r is an illusion in that this term implies absolute rest, which—like absolute motion—does not exist. In short no m is ever free of E because each m is always in some relative motion.

This brings us to the second possibility, that $m_r = 0$, which is more complicated because it presupposes "objects" which have *no mass at rest.* We assume that a photon has this property, in which case $m_r^2 c^2$ drops out, leaving some energy as

$$E = c\sqrt{p^2} = cp = cm_{motion}v.$$

(We use $m = m_{motion}$ to avoid ambiguity.) This means that for a photon, all of its mass is m_{motion}. We also have Einstein's quantum-mechanical equation $E = fh$: The energy of a photon or particle is reflected in its frequency (or in its wave length, so that $E = h/\lambda$); see Al-Khalili, pp 40-47.

Let us tie this possibility to the finding that according to uniformly moving observers, a photon always appears to be moving at speed c, and let us recall the notion we cited earlier that "there is no reference frame in which light is at rest." A photon, even with zero m_r, must comply with

$$\frac{m_r}{m_{motion}} = \frac{\sqrt{1-v^2/c^2}}{1},$$

which is only possible if v = c. That is to say, as long as m_r is zero, v can only be equal to c. If m_r is not zero while $v = c$ in the equation

$$p = m_{motion}v = \frac{m_r v}{\sqrt{1-v^2/c^2}},$$

then m_r is infinite. (Any number, except zero, divided by zero gives infinity.) Indeed, as we shall discuss shortly, an object with any rest mass must have infinite weight at speed c. The exemption is when $m_r = 0$, as in the case of a photon. Our point is that the possibility that an "object" may have no mass is elegantly explained by relativity, and, even more importantly, relativity implies that *"objects" with no rest mass, including photons, must move with the speed of light.* We hasten to add that the term "no rest mass" is misleading. Under normal circumstances photons do show mass, but they never show mass at rest since, as already stressed, *they never are at rest.* This raises the question how a particle which under some circumstances appears to have mass can also show speed c, but we shall defer that issue. We also note for our discussion of quantum mechanics that even though a photon has no rest mass, it always has momentum; hence *light has momentum.* Thus the equations (often the key to exam equations) for massless particles like photons,

$E = pc = hf$ and momentum is given by $p = h/\lambda$.

Suffice it for now that very large mass and very high speed are incompatible. *Maximum speed requires minimum mass.* Once we think it through, this concept is practically self-evident, even without experimental support: Maximum mass at maximum speed implies the availability of unlimited kinetic energy, but as far as we can tell the universe contains only a finite amount of mass, which, by virtue of $E = mc^2$, means that it cannot provide an infinite supply of energy. (A universe with infinite energy would be infinitely massive and/or infinitely hot and/or infinitely bright.) The idea, based in the conservation of energy, is quite transparent: Given limited energy, only the lightest objects can be fastest, and if somehow our universe had been endowed with more available energy—perhaps as the result of a "bigger bang"—c might be even greater and photons even faster. No wonder then that photons do not show infinite speed; they do not have access to infinite energy. No wonder that c is the fastest that we can detect distant events; instantaneous notification requires infinite energy. No wonder that there is a time lag between a cause and an effect; otherwise the past, the present, and the future could occur in zero time. We will pick up these threads later, including the point that quantum mechanics may allow exceptions to the idea that there must be a time lag between a cause and an effect.

Meanwhile, we note the logical connection between two ideas: *An objects with no rest mass must move with the speed of light*, while *the speed of light (the speed of photons) does not depend upon the motion of the source.* We also note that any object, subatomic or otherwise, if brought to speed c, will possess the same "exclusive" feature, which means that any object that can move at c in one frame of reference will move at c in all inertial frames of reference. In other words the faster something moves, the smaller is the effect of the motion of its source, and if something can reach c, the motion of its source makes no difference at all. In terms of m_r and m_{motion}, the less of the former and the more of the latter, the more "exclusive" its behavior. Or, in terms of the addition of speeds, the closer a speed is to c, the less additive it is. As we pointed out earlier, the "exclusive" behavior of a photon is a consequence of its great speed, and now we see that a photon has great speed, namely c, because it has no m_r.

An obvious question is whether "objects" with zero m_r other than photons exist in nature, and in fact it appears that one kind of subatomic particle, the neutrino, may fit the description. This discovery has inspired a very exacting demonstration of special relativity: An exploding star releases a burst of neutrinos which can be detected on Earth. In such an explosion some parts of the star, i.e., some sources of neutrinos, move away from us and others move toward us, which means that we might expect the emission of a long stream of neutrinos, some faster than others. A neutrino-emitting stellar explosion located thousands of light-years from Earth has been observed, yet all the neutrinos which reached us did so *within a few seconds of each other*. This confirms that despite originating in an explosion and then traveling for thousands of years over trillions of miles, neutrinos with no rest mass all shared the same speed, namely c. (One light-year is the distance covered by light in one year of travel; 5,878,000,000,000 miles.)

We can carry the logic further: Using Einstein's terminology, the energy content of a mass is linked with its inertia. Therefore not only does mass have the property of resistance to acceleration, but so does *energy*. Energy has inertia! And if all forms of energy are interchangeable, *all energy has inertia*. This notion raises intriguing issues which we will eventually address: How can energy and mass share this property? If light is a form of electromagnetic energy, does light have inertia? Can light behave like mass, particularly in a gravitational field?

✦

Further elaboration is needed about the special relativity factor $\sqrt{1-v^2/c^2}$. We see that this quantity compares a "Newtonian" time interval, a length (space), or a mass (weight) with its "relativistic" counterpart upon uniform relative motion. When the Newtonian number, such as time or mass, is less than the relativistic number, the ratio is

$$\frac{Newtonian}{Relativistic} = \frac{\sqrt{1-v^2/c^2}}{1}.$$

In the case of length, the ratio is

$$\frac{Newtonian}{Relativistic} = \frac{1}{\sqrt{1-v^2/c^2}}$$

because the Newtonian value is greater than the relativistic. In the case of mass,

$$\frac{Newtonian}{Relativistic} = \frac{m_{rest}}{m_{motion}} = \frac{\sqrt{1-v^2/c^2}}{1}.$$

If for example relative motion is 86% of the speed of light ($v = 0.86c$), an object appears to be about half as long but twice as heavy as a similar stationary object. If $v = c$ could be reached, $\frac{\sqrt{1-v^2/c^2}}{1}$ would be zero and $\frac{1}{\sqrt{1-v^2/c^2}}$ would be infinite. Thus an object with any non-zero rest mass moving at the speed of light would appear infinitely short, infinitely heavy, and, in the case of a clock, infinitely slow. (Yes, a faster-moving clock ticks slower, and a theoretical "c-moving" clock would stop ticking when observed by a non-moving observer.)

This means that *in reality no mass can be accelerated beyond the speed of light*, i.e. nothing physical can be faster than c. As strange as it sounds, *added kinetic energy adds mass* until an unsurpassable speed is reached. Many experiments confirm this, wherein an accelerated particle becomes progressively harder to move. This fact accounts for the main problem in generating particles fast enough to "split atoms." If special relativity were invalid, and in particular if E did not equal mc^2 and $\frac{m_{rest}}{m_{motion}}$ would not equal $\frac{\sqrt{1-v^2/c^2}}{1}$, high-school physics labs could build "atom smashers!" Yet even a flashlight can generate light beams with speed c because photons have no rest mass.

Still, the idea that nothing with mass can be accelerated beyond the speed of light seems unreasonable. Imagine that a rocket moving very fast, say at speed 0.8c with respect to the Earth, is given a small "boost" (a kick) to speed 0.9c. Wouldn't just two more such boosts give it a speed over 1.0c? No! Each successive boost requires far more energy; we note that the velocity v in the special relativity factor is squared. The final boost needed to exceed c, no matter how small, would require infinite energy. Indeed viewed from the Earth's frame of reference, boosting from 0.9c to 1.0c is vastly more difficult that from 0.8c to 0.9c. (Use $E_{total} = \gamma \, m_o c^2$ as above to compare the energy-cost going from 0.8c to 0.9c vs. going from 0.9c to 1.0c. The algebra boils down to contrasting $0.9c \, \gamma - 0.8c \, \gamma$ against $1.0c \, \gamma - 0.9c \, \gamma$, as often needed in physics exams.)

We used the same reasoning to explain the non-additive and "exclusive" characteristic of light. If the sum of relative speeds were $v_A + v_B$ rather than $\dfrac{v_A + v_B}{1 + \dfrac{(v_A)(v_B)}{c^2}}$, any speed could be easily achieved, and the physical nature of our universe—not to mention "atom smashers"— would be quite different. Of course in the absence of relative motion ($v = 0$), the special relativity factor is 1, and there is no need to consider special relativity. On the other hand, since v cannot exceed c, the special relativity factor cannot be negative; i.e. it cannot be less than zero. And if v is very small compared to c, as is in events we normally encounter in our environment, this factor is very close to 1. Therefore, "Newtonian" and "relativistic" calculations usually give us practically identical results. For example, today's space travel achieves what we might consider very high speeds, yet the added "relativistic" weight of our rockets is negligible because their velocities are still minuscule compared to those of light. This is why the effects of relativity were discovered only recently, why for two centuries Newtonian physics remained unquestioned, and why the latter is still useful.

✦

We are treating "$E = mc^2$" as a law of nature, implying that it obeys the first postulate of special relativity. However if two observers in relative uniform motion examine the same object, their values for p, m, and E are different. Is $E = mc^2$ nevertheless valid for each observer? In other words does relativity's definition of mass and energy hold in different frames of reference?

We have the answer: The derivation of $E = mc^2$ incorporates the special relativity factor which provides a compensation for uniform relative motion. Thus while mass and energy are relative, the relation between them is constant, and $E = mc^2$ holds for all inertial observers. In other terms, if we consider mass and energy separately, relatively moving observers will disagree, but *if we consider the totality of mass and energy*, there is no discrepancy. In yet other terms, the ratio between energy and mass ($\dfrac{E}{m}$) has a constant of proportionality, c^2, that is immune to the effects of uniform relative motion. We can write that $c^2 = \dfrac{energy}{mass}$. Meanwhile we recall another ratio: $c = \dfrac{space}{time}$. The parallel is clearer when we use quantities as $space^2$ and $time^2$ in the ratio $c^2 = \dfrac{space^2}{time^2}$. The concept is striking if not exhilarating: Like four paths converging on a key intersection, space, time, mass, and energy are linked by one universal quantity, the speed of light.

✦

We opened this chapter with a list of conservation laws which apply in Newtonian physics. Then how do they appear in special relativity? The relativistic version of the law of conservation of momentum is easy to write; we simply divide by the special relativity factor:

$$\Delta\Sigma\frac{mv}{\sqrt{1-v^2/c^2}} = 0$$

For the laws of conservation of matter and energy, the requirement is more subtle. In Newtonian science, energy and mass are two separate and autonomous features of an object, whereas special relativity sees them as linked; e.g. adding energy increases weight. Hence special relativity unites the two laws: Matter and energy are "conserved" together, not independently of each other. The total amount of mass *and* energy in any closed system remains constant. Accordingly

$$\Delta\Sigma mc^2\sqrt{1-v^2/c^2} = 0.$$

The physical and chemical significance of a unified law of conservation of mass-and-energy is quite concrete. Let us assume that a machine burns 10 units of fossil fuel, and we expect it to convert this fuel into 4 units of water and 6 units of carbon dioxide. At the same time, let us assume that these 10 units of fuel contain 100 units of energy, and we expect to retrieve 51 units of kinetic energy and 49 units of heat-energy. According to pre-relativity thinking, two independent laws of conservation are in action here, one for mass which predicts that the 10 units we put "in" equal the $4 + 6$ which we get "out," and another for energy so that the 100 units "in" equal the $51 + 49$ "out."

Relativity shows otherwise: Given sufficiently sensitive measurements, the mass that comes "out" might be $3 + 5$ or 8 units, while the energy retrieved is $52 + 50$ or 102 units. The *combined* law correctly predicts that $10 + 100$ "in" yields $8 + 102$ "out;" i.e., mass-plus-energy "in" matches mass-plus-energy "out" because *some of the incoming mass is converted to energy.* We conclude that rather than

"$m_{in} = m_{out}$" and independently "$E_{in} = E_{out}$,"

the true picture is

"$m_{in}c^2 + E_{in} = m_{out}c^2 + E_{out}$."

Indeed we can say that because of $E = mc^2$, conservation of matter compels conservation of energy and vice versa. Of course this fusion of the basic laws of nature adds to the aesthetic appeal of relativity, as it reveals a previously unappreciated unity and clarity in how our universe works.

✦ ✦ ✦

We see that special relativity touches physics on a very profound level. All basic formulae of physics ultimately reduce nature to three parameters: time, space, and mass. A parameter is a characteristic that describes something in a quantitative manner; it is a measurable quantity. (For example parameters of a child's growth might be height. In other contexts which we will encounter later, the terms parameter and parametrization have a more specific mathematical meaning.) Thus when we apply a law of physics we basically determine how long something takes, how much distance something covers, and/or how much something weighs. Earlier we asked why apparently universal laws of physics yield conflicting results simply as a result of relative motion. We can now give a more specific reason: *Because these laws deal in time, space, and mass, while the measurement of each of these parameters is affected by relative motion.*

Let us consider the principle that pressure increases temperature (based on Gay-Lussac's and Charles' Laws, usually applied to a gas.) Pressure is generally expressed as mass per space (e.g. grams per area). Temperature is a measure of the kinetic energy of molecules, and kinetic energy is the ability to do work by way of creating motion. But work is expressed by force times distance; force is mass times acceleration; acceleration is change in speed over time; and speed is space per time (e.g. centimeters per second). Relativity thus redefines the very ingredients of the law of pressure and temperature—which applies to just about every object in nature. This feature makes relativity important and useful in areas such as subatomic and high-energy physics, where we study how very small but very rapidly moving objects behave.

This is also why special relativity is so controversial. Ordinary experience sustains the common-sense prediction that speed does not influence the rate of time, size, or mass. As we said, we do not naturally expect moving events to take longer, nor do we expect moving objects to shrink or to gain weight. However relativity forces us to see physical reality differently. What we thought was stable and obvious turns out to be elastic, irrational, and indeed incorrect.

✦ ✦ ✦

We used diagrams A, B, and C for the case where one event is investigated by two relatively moving observers, meaning that the event appears in two different frames of reference. This can be described more formally utilizing Cartesian coordinates—the set or "system" of three coordinate axes; the horizontal axis is X, the vertical axis is Y, and the front-to-back axis is Z. These coordinates naturally correspond to our familiar three dimensions of space: height, width, and depth. However on paper we usually show only two axes, X and Y, simply because of the difficulty in drawing three-dimensional objects.

Various other terms are used for "Cartesian" systems of coordinates, such as "Galilean" and "Euclidean." Their common feature is that the axes are straight, which is expressed by the term "rectilinear" coordinates. These coordinates can have axes that cross at right angles, in which case they are "orthogonal," or at other-than-right angles, which are called "oblique." For example ordinary graph paper lends itself to a system made up of two rectilinear orthogonal coordinates because its lines are straight and meet at right angles. We also have "curvilinear" systems, obviously with curved axes, and these are also called "Gaussian" systems. (Large-scale maps of the Earth may have curved coordinates to display the Earth's roundness.)

When Cartesian coordinates are used to compare how one event appears in two frames of reference, we say that a transformation of Cartesian coordinates is being performed. For example comparing diagrams A and B is such a transformation. Because of the restriction to *uniform* relative motion between observers or between events, this procedure is known as a "Galilean transformation of Cartesian coordinates" or simply a Galilean transformation. We emphasize that a Galilean transformation is usually shown in rectilinear orthogonal coordinates. However since not all coordinate systems are Galilean—as not all motion is uniform—not all transformations are Galilean. In fact we will find that transformations between curvilinear (Gaussian) systems are indispensable to general relativity, in part because general relativity deals with non-uniform motion.

Let us examine certain details about transformations. Mathematically, a transformation is a substitution of one system of coordinates for another, but we can think of a transformation as a shift from the point of view of one observer to that of another. For example, navigators use systems with latitude and longitude acting as X and Y coordinates, so let us imagine two geographic points which we label as "A" and "B." One navigator uses an English map, wherein the prime meridian (zero longitude) runs through Greenwich near London, and he finds that A and B are 100 miles apart, with A northwest of B. Another navigator has an equally accurate American map, one with its prime meridian running through Annapolis, Maryland. This scenario demonstrates a transformation of coordinates—a shift from the vantage of one navigator to that of another.

Of course common sense predicts that *even though the second navigator uses a different set of coordinates*, he should reach the same conclusion, 100 miles distance and northwest direction between points A and B; the relationship between A and B should not change. We say that an *invariant* result is expected "under" (to use the standard mathematical term) this transformation. "Invariant" in this context means "same" or "constant." The outcome is unchanged.

Let us further assume that a ship sails from A to B, and each navigator tries to reckon how much time (T) the trip requires. On the basis of the first postulate of special relativity, we expect all rules of physics, including $RT = D$, to remain valid despite the transformation of coordinates, and of course common sense predicts that both navigators forecast the same answer—i.e. that T remains invariant. However we can foresee the difficulty: What if one of the navigators is in uniform motion? Relativity predicts that under those circumstances, the two navigators *fail to agree on how long it takes to sail from A to B*. In other terms under certain conditions, some outcomes are indeed changed; specifically in the presence of relative motion, measurements of the time (T) consumed by an event time cease to be invariant "under" this transformation.

As we already explored, the underlying reason for this effect resides in the nature of the speed of light. Before the introduction of relativity, we implicitly assumed that during uniform relative motion, any speed, including that of light, changes under a transformation of coordinates. In consequence, measurements of time were expected to be invariant under a Galilean transformation. However, special relativity tells us that since the measured speed of light is independent of the speed of its source and is the same in all inertial frames of reference, *moving observers do not agree on the duration of one event.* Nor do they agree on distances and masses found in this event.

We conclude that because of special relativity—because time is relative—a Galilean transformation describing an event seen by relatively moving observers gives an erroneous result. This effect may be subtle, but it has far-reaching significance: *Traditional mathematics, here represented by a Galilean transformation, cannot be used to predict how an event will appear in the presence of relative motion.*

In terms of the history of science, this simple line of reasoning shook the foundation of physics as it was accepted before the twentieth century: Newton's laws had been expected to hold under Galilean transformations of coordinates, and if such transformations are defective, the astounding implication was that our most respected and apparently logical laws of physics are fundamentally faulty.

✦

To deal with the non-validity of the Galilean transformation, special relativity provided a correction entailing none other than the special relativity factor. The originators of this correction were not Einstein but Lorentz (1853-1928) and Fitzgerald (1851-1901). Its derivation has an interesting history involving the "Michelson-Morley experiments," to which we already alluded.

The pertinent issue is as follows. At the time Michelson and Morley performed their experiments (beginning in 1886), science believed that the universe is pervaded by a substance called "ether" through which all movement of matter and energy occurs (also spelled "aether," a word taken from Aristotle's work but, in this context, ascribed to Descartes). One reason the ether theory was subjected to experimentation is that a few years earlier Faraday and Maxwell had proposed that an ether-like medium transmits electromagnetic energy, much like air transmits sound. For its presumed capacity to transmit light this ether was labeled "luminiferous," and in particular the luminiferous ether was believed to be able to "drag" light, like a river drags a boat.

Furthermore physicists thought that this ether fills space so as to provide an absolute and rigid frame of reference for all motion. Thus if an observer walks on a train at 1 mph and the train moves at 50 mph, then relative to the Earth the observer moves at 51 mph. Meanwhile the Earth is spinning at about 1,000 mph, so the observer's speed is really 1,051 mph. But the Earth revolves around the sun, and the sun is part of our moving galaxy, etc, etc. The ultimate frame of reference which does not move could be the ether, and Michelson and Morley performed experiments to confirm the presence of the ether by measuring the motion of the Earth through it.

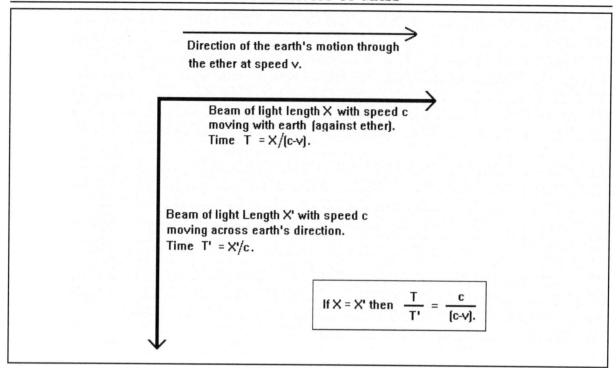

Diagram D. The Michelson-Morley experiment in diagrammatical form.

With this theory in mind, let us look at these experiments more closely. Michelson and Morley focused on the motion of the Earth because it was the fastest available moving object. They ingeniously arranged two beams of light at a right angles to each other. The two beams had exactly the same length, but one beam ran parallel to the Earth's motion, presumably in line with motion through the ether, while the other ran across (perpendicular to) the Earth's motion, presumably across the ether. The goal was to detect a difference in how much time light spent in the two beams.

This situation can be analyzed with Cartesian coordinates, each one housing one of the beams, as in diagram D. We designate the speed of light as c and the speed of the Earth as v. The length of the parallel beam is X and the length of the crossing beam is X', and X = X'. T is the time light spends in the parallel beam; the time for the other beam is T'. Using the rules of *Galilean* transformations, the key assumption is that light in beam-X running in the Earth's direction loses some speed, specifically from c to c-v, much like a boat is slowed by the drag of the water while going upstream. Presumably T and T' should differ; here is the math:

Since $RT = D$ or T = D/R, the experimenters expected that T = X/(c-v) in one of the systems of coordinates and T' = X'/c in the other. And (as in diagram D) since X = X', algebraic substitution of X for X' leads to

$$\frac{T}{T'} = \frac{c}{(c-v)}$$

By determining T/T', the experimenters hoped to find v, the absolute speed of the Earth through the ether. T was expected to be greater than T', so the ratio T/T' should have turned out to be more than 1. Much to everyone's consternation, what they obtained instead was T/T' = 1 and T = T'. This of course implied that c = (c-v), that v = 0, and that the Earth does not move through the ether!

Various explanations were proposed for this result. One was that Aristotle and the Vatican critics of Galileo had been right after all; the Earth does not move around the sun. Or perhaps here RT did not really equal D. A more serious explanation offered by Lorentz and Fitzgerald was that the ether put pressure on the parts of the experiment (length X) which were moving "upstream" and that this pressure contracted the experimental apparatus just enough to mask the anticipated results. This proposal seemed plausible because Lorentz had already theorized that electrons may be flattened while moving. If electrons can be deformed, why not entire objects? The main problem with this idea was the absence of any theory on how ether could deform matter, and Poincaré, himself a champion of relativity, criticized this explanation. On the other hand Poincaré pointed out that the results confirmed that Galileo and Newton had been right: uniformly moving observers, Michelson and Morley in this case, cannot detect their own motion. (The Earth's motion around the sun is practically uniform.) Seeing that a Galilean transformation of coordinates did not yield correct answers when applied to these experiments, Lorentz and Fitzgerald issued a mathematical modification of the transformation, dubbed by Poincaré as the *Lorentz transformation*.

It is important to examine Lorentz' accomplishment. To do so, let us go back and describe Galileo's and Newton's treatment of uniform relative motion, which is easier using our two moving trains rather than the Michelson-Morley experiment. Again please look at diagram A. We call the motion of light from its source to the mirror an "event." Let us label the vertical border of the box in diagram A as the Y axis and the horizontal border of the box as the X axis. We can omit a Z axis, but we designate the time needed for this event as T. This arrangement allows a description of the event in the frame of reference of observer A by using the Cartesian coordinate system X, Y, T. For example according to observer A, the beam strikes the upper mirror at location X_A, Y_A and at time T_A. In the frame of reference of observer B, the beam hits the upper mirror at location X_B, Y_B and at time T_B.

We can imagine that train A has a tinted glass window. Observer A uses this window to see the reflection of the event in his own train; he sees diagram A. But he can also look out of his window through the glass into train B. As the trains move past each other at uniform speed v, he sees diagram B. (Sliding diagram A to the left mimics what is shown in diagram B.) *As observer A compares what he sees in his own train with what he sees in train B*, in effect he executes *a transformation of coordinates*. First of all the height of each train appears the same, so that observer A notes that

$$Y_A = Y_B.$$

Secondly, and very importantly, observer A may see the event in his train as simultaneous with the corresponding event in train B. Therefore observer A *expects* that

$$T_A = T_B,$$

and we note that *this equation represents and implies absolute time*; the determination of time ought to be the same in A's and B's frames of reference. If something occurs at or lasts T seconds according to A, it should occur at or last T seconds according to B.

Thirdly, when the beam reached the upper mirror of train B, that mirror moved a certain distance along the X axis. Since $RT = D$ and here R is v, that distance is vT_B, and since the positive direction along the X axis is to the right, vT_B is a positive quantity. So observer A in his frame of reference *expects* that

$$X_A = X_B + vT_B.$$

If we include a Z axis, along which there is no motion, and if use the traditional order in which these equations are presented, the entire set reads as follows:

$$X_A = X_B + vT_B$$

$$Y_A = Y_B$$

$$Z_A = Z_B$$

$$T_A = T_B$$

These four equations constitute a Galilean transformation. This is how an event was expected to be "transformed" from one system of coordinates (A) used by one observer in one frame of reference to another system of coordinates (B) in another frame of reference, when these two are in uniform relative motion. If we reverse the situation and solve for observer B's coordinates, these four equations stay the same except the first reads

$$X_B = X_A - vT_A.$$

The term vT_A is negative because from B's vantage, the event has moved left, "down" the X axis.

On the surface this scheme is quite logical: For either observer the event should appear the same except for the effect of horizontal motion over a span of time along the X axis. But is this entirely accurate? As we know, in reality it is not. The event under scrutiny is the motion of a beam of light, and *its speed, c, is the same in both frames of reference.* This crucial fact, shown in theory by Maxwell and in practice by Michelson and Morley, invalidates the Galilean transformation.

Let us examine the crux of Lorentz' improvement upon the Galilean transformation: Y_A still equals Y_B and Z_A still equals Z_B because there is no motion along the Y and Z axes. But for observer A,

$$X_A \ does \ not \ = X_B + vT_B.$$

Since X_A and $X_B + vT_B$ are lengths, Lorentz applied the ratio containing the special relativity factor,

$$\frac{1}{\sqrt{1-v^2/c^2}}.$$

Hence the first equation in the Galilean set, $X_A = X_B + vT_B$, is recast in "Lorentzian" form as

$$X_A = \frac{X_B + vT_B}{\sqrt{1-v^2/c^2}}.$$

This equation tells us that when we transform an event from one system of coordinates to another during uniform relative motion along X, the distance traversed by that event is not what Galileo or Newton assumed. *It must be adjusted for relative motion.* Lorentz interpreted his finding as the effect of motion through the ether upon the *length* of matter, which means that motion modifies our measurement of *space*. (This effect is also called the Lorentz contraction or the Fitzgerald-Lorentz contraction.)

But what about *time*? The critical fact is that *Lorentz had to adjust the time consumed by that event by incorporating the special relativity factor.* Hence the fourth equation in the Galilean set, $T_A = T_B$, is replaced by an equation in which T_A and T_B no longer need to be the same. In other words *time* is not absolute; motion also modifies its measurement. This new "Lorentzian" equation is

$$T_A = \frac{T_B + vX_B/c^2}{\sqrt{1-v^2/c^2}}.$$

We also note for later consideration that to determine T_A, contributions from measurements on the other coordinates, T_B and X_B, are invoked; that is to say, time and space are linked in the transformation. For now we recognize the special relativity factor in the right side of this equation, but another term, vX_B/c^2, has been added in the numerator. This term will be easier to understand written out as $(X_B/c)\,(v/c)$.

To appreciate the above equation we must recall that our experience of the timing of a physical event is subject to two effects: One, motion in relation to the speed of light (v/c) alters our measurements, which is the key relativistic effect incorporated into the special relativity factor as v^2/c^2. Two, as explained on page 5, a time-delay ensues as the "signal" from the event crosses space at the speed of light in order to reach an observer. That time-delay is (X_B/c), based on $RT = D$ which in this case is $(c)(T_{delay}) = X_B$. Then (X_B/c) is also adjusted by (v/c), giving us vX_B/c^2. Incidentally, the Lorentz transformation can be derived in detail in various ways. For instance compare Shadowitz, pp. 67-68, with Einstein in <u>Relativity</u>, pp. 115-118. As we already mentioned, direct verification of these derivations is difficult (one needs to measure an event moving uniformly near the speed of light), so that Einstein's reliance on the Lorentz transformation constitutes a tempting criticism of special relativity. However indirect evidence does exist; see Bergmann, pp. 36-38.

Here are two instructive mathematical maneuvers: First if we let c be infinite, we end up with the Galilean transformation which tacitly presupposes an infinite speed of light, and in which $T_A = T_B$ is the mathematical equivalent of the Newtonian assumption that time is absolute. Second, solving the Lorentzian equations together yields $X_A = cT_A$ and $X_B = cT_B$, showing that c is not infinite and is same in two inertial frames of reference. This clearly is Einstein's second postulate!

We see that by modifying the Galilean transformation to fit what Michelson and Morley had found, Lorentz was forced to compose *an equation for how measured time is changed by uniform motion*. Of course this is the most interesting feature of the Lorentz transformation for relativity: In trying to explain the contraction of space, Lorentz had to include an adjustment for measured *time*. Since Lorentz' work in this area was published several years before Einstein's first paper on relativity, one might assume that Lorentz had discovered the relativity of time. However Lorentz did not grasp the significance of what he had done. The implications seemed so preposterous to him that he ascribed the differing values of time to an artificial mathematical quirk rather than to any real physical effect. Like Michelson, Morley, Fitzgerald, Poincaré, and others, Lorentz was "so close and yet so far." Then Einstein, still unknown in their circle, literally burst upon the scene with a fresh and innovative approach which soon resolved the issues. He essentially stated that both Poincaré and Lorentz were right! Absolute uniform motion really *is* undetectable, yet measured time really *is* relative and not absolute.

As Einstein worked out his mathematics he too derived the special relativity factor—which is why we call it that. However in Einstein's work, this factor quantifies a *real "deformation" of time*. Einstein and Lorentz thus devised the same formulation but for different reasons. Lorentz tried to explain Michelson-Morley's results, while Einstein tried to show how relatively moving observers disagree on their assessment of the same events. To Lorentz, matter moving through ether shrinks, which appears to retard time; to Einstein, moving time indeed slows, and consequently the measurement of time and space is altered. Even after Einstein explained and confirmed that Lorentz' mathematics is correct, the latter still did not accept this concept.[*] Moreover, it is not clear whether Einstein was aware either of Michelson and Morley's results or of Lorentz' derivation at the time he first worked out the special relativity factor, but historians agree that Einstein would have come to his conclusions on his own.

---

[*]This issue did not interfere with the great friendship between the two. Indeed biographers stress the warm personal relationships between Einstein and his contemporaries, even with those with whom he disagreed and those who resisted his ideas.

In fairness to Lorentz we should note that despite his personal reluctance, he was not naïve to the fact that his derivation had generated two gauges of time. For example he clearly distinguishes between "general time" and "local time" and gives them different labels, t and t' (p. 26 of his paper in The Principle of Relativity). We should also mention that the first hint that the measurement of mass, like that of space and time, is influenced by motion via the special relativity factor appeared in the writings of Lorentz, not Einstein. Nor should we presume that Einstein declared instant victory. Even he was puzzled at first. For example (ibid. p. 49) when he predicted that relatively moving clocks become dissynchronous, he used words which translated mean "peculiar consequence."

In any case the Lorentz transformation enables us to revise key laws of physics so that they apply even during relative uniform motion, i.e. so that they give all uniformly moving observers comparable results. In other words with the Lorentz transformation the laws of physics can survive the effects of special relativity, notably the relativity of time. In formal terminology, unlike in the case of the Galilean transformation, *the laws of physics are invariant (or "form-invariant") under the Lorentz transformation*. This means that the correct form of an equation expressing a law needs no revision for use in another inertial (uniformly moving) frame of reference. No terms have to be changed, added, or deleted.

Let us exemplify how the Lorentz transformation achieves this end. Here is a "law" for the addition of velocities, which, by common sense, appears as

$$V = v_A + v_B.$$

Again we let velocity be synonymous with rate; V and v are R in $RT = D$. We say that in this form, this law is *not "Lorentz-invariant."*

Meanwhile here are the four equations of the Lorentz transformation solved for A's vantage:

$$X_A = \frac{X_B + vT_B}{\sqrt{1 - v^2/c^2}}$$

$$Y_A = Y_B$$

$$Z_A = Z_B$$

$$T_A = \frac{T_B + vX_B/c^2}{\sqrt{1 - v^2/c^2}}$$

Solved for B's vantage, these equations are identical except the two plus signs are minus signs. Either way, this set of equations discloses how physical events appear to different observers moving uniformly relative to each other. Let us look at the first and fourth equations: The first relates X_A with X_B, which are length or space (D in $RT = D$). The fourth relates T_A with T_B, which are time (T in $RT = D$). If we re-label and rearrange $RT = D$ into $v = \dfrac{D}{T}$ for observers A and B, we can divide the first equation by the fourth, which yields

$$v_A = \frac{v_B + v}{1 + \dfrac{(v)(v_B)}{c^2}},$$

where v is the relative velocity between two observers or between two events viewed by one observer. As we previewed on page 26, we see that the relativistic refinement of $V = v_A + v_B$ is

$$V = \frac{v_A + v_B}{1 + \dfrac{(v_A)(v_B)}{c^2}}.$$

We now have a "law" of physics—the law for the addition of velocities in this case—in a revised form, a form that grants any inertial observer equal success in calculating the addition of velocities. We have made this law invariant under the Lorentz transformation; it is "Lorentz-invariant." Its mathematical form is the same at rest as in uniform motion, and hence this law is form-invariant. Only the values of the variables (the v's) change, and students of relativity should memorize this!

We note—and students should also keep in mind—that $\dfrac{(v_A)(v_B)}{c^2}$, *representing the difference relativity makes*, can be zero or more than zero, which is why V is either the same as or less than $v_A + v_B$. We also note that both v's, multiplied together, contribute to this effect. Finally we note the relationship between the term $1 + \dfrac{(v_A)(v_B)}{c^2}$ and the special relativity factor. In essence this term represents the special relativity factor multiplied by itself (the square root sign is absent), reflecting that each of the two v's brings the effects of special relativity into the equation.

What about a more familiar law such as Newton's $F = ma$? Let us consider our two uniformly moving trains, each of which contains an object with mass m. As we expect, experiments prove that $F = m_A a$ in train A, and that $F = m_B a$ in train B. However if an observer in one train studies the objects as the trains move past each other, the m's are found not to be identical. We recall that the special relativity factor quantifies the effects of relative motion on the measurement of mass by the ratio

$$\frac{1}{\sqrt{1 - v^2/c^2}}.$$

This allows us to revise *F = ma so that it is Lorentz-invariant*:

$$F = \frac{ma}{\sqrt{1-v^2/c^2}}$$

(The equation is somewhat different depending on whether motion is parallel or perpendicular [Lawden, p. 60], but that is not pertinent here. The *a* in *F = ma* refers to accelerated motion, which is also not pertinent here. *F* is force.) Only the above form of this law is valid under transformations between inertial frames of reference, and conversely this law is invalid during relative motion without an adjustment for special relativity. We re-emphasize: *The Lorentz transformation enables laws of physics to be written so that their forms are independent of—i.e. they remain valid during—uniform relative motion.* Special relativity makes Newton's second law form-invariant! We also recall that this is true because the Lorentz transformation allows uniformly moving observers to experience a relative rate of time but a constant speed of light.

Of course for an observer in his own frame of reference *v* is zero, in which case the special relativity factor reduces to 1, and *F* does equal *ma*. In other words locally in each frame of reference, *m* is the "proper" measurement of mass expressible as m_{rest}. However when the observers compare each other's measurements while *v* is not zero, they encounter the discrepancy between "proper" and "coordinate" measurements of mass.

What if *v* is high enough to be close to the speed of light, *c*? In that case the Lorentz-adjusted equation shows that *F* approaches infinity, which, as we already indicated, means that no mass can be forced to move faster than *c*. Since our normal speeds are much less than *c*, this has no every-day significance, but it has a conceptual meaning: In its original form *F = ma* means that a mass resists a change in motion; it resists acceleration; it has *inertia*. The Lorentz-adjusted form—i.e. the Lorentz-invariant form—tells us that uniform motion further increases the resistance to acceleration. In other words more kinetic energy means more inertia, which, we recall, is the rationale behind $E = mc^2$, and which elegantly insinuates that energy has inertia.

We add that Newton's law of gravitation (detailed later) cannot be successfully adjusted by the Lorentz transformation; it is not Lorentz-invariant. Gravitational motion is always accelerated, while the special relativity factor only adjusts for uniform motion. We also note that only the most basic laws of physics, those which do not rely on a medium, need to be invariant. For example sound depends on the presence of a medium, so that acoustic laws need not be invariant. In fact, to be very precise we could qualify the first postulate of special relativity for the current context: The laws of nature are the same for all observers moving in non-accelerated frames of reference when the phenomenon in question requires no fixed medium.

◆

Let us return to Michelson's and Morley's unexpected results. Working independently, Einstein solved the mystery: The ether theory is needless, and light does not require an unmoving medium such as an ether; nature provides no absolute frame of reference, nor is there absolute time, in which case absolute uniform motion cannot be detected, using even the optical methods devised by Michelson and Morley. However contrary to popular opinion, Einstein did not prove that ether cannot exist. Relativity maintains only three notions about the ether: That its presence cannot be demonstrated by any physical experiment, that no valid laws of physics require it, and that no valid laws can be derived from it. In this sense the water in our lifeboat analogy (page 4) may exist but really is superfluous. No medium is necessary to explain the behavior of a "boat of photons."

With benefit of hindsight, the fact that ether cannot be detected is rather self-evident. An ordinary object dropped to Earth falls "down" despite the Earth's (west-to-east) motion *because there is no ether to drag falling objects sideways*. We can think of a Michelson-Morley experiment as a case in which photons are allowed to "fall" as if they were ordinary objects, and it should come as no surprise that these particles likewise are not dragged by an ether. By analogy, a boat made of photons has the same speed whether it is rowed upstream or downstream; it acts as if there were no water; its physical behavior is "exclusive."

As for the mathematical conclusion of Michelson and Morley that the speed of the Earth is nil (that $v = 0$), relativity's explanation is that if v could be measured, an absolute motion of the Earth could be determined, which would mean that uniform motion could be detected and that the entire structure of relativity is faulty. Though history is not so tidy, we can say that Michelson and Morley set out to discern the ether but encountered relativity instead. In their experiments v turned out to be zero not because the Earth has no motion but because it has no absolute motion.

Let us review the effects of special relativity with an eye on the questions they suggest. Uniform motion alters our measurement of space, time, and mass but not our measurement of the speed of light, nor does uniform motion alter the local validity of the laws of physics. Lorentz' adjustment allows these laws to be successfully applied despite uniform relative motion between frames of reference. However the Lorentz transformation says nothing about accelerated motion, while accelerated motion obviously is important to physics, since falling and orbiting objects do not move uniformly. Therefore we have two interconnected issues to consider: One, if uniform motion alters our measurements of time, space, and mass, what are the effects of accelerated motion? Two, if the laws of physics need to be restated to make them invariant during uniform motion between frames of reference, how ought these laws be restated for accelerated motion? The replies are provided by the mathematics of general relativity, but our approach must be indirect, and we will next examine certain preliminary issues.

CHAPTER 5: BACKGROUND FOR GENERAL RELATIVITY

To introduce general relativity we must say more about Newton's laws of physics. The "first" of these is the law of inertia, which has two components: Objects stay at rest if they are stationary, and if they are in motion they tend to maintain a straight path at uniform velocity. The latter is the more pertinent to relativity—as relativity deals primarily with motion—and therefore we should keep in mind that *moving objects seek uniform motion.*

When motion is not uniform, we say it is accelerated. Technically, any change in speed or direction of motion over a period of time is acceleration, including starting, speeding up, slowing, stopping, veering, and turning. Even a spinning motion, such as orbiting or rotating, is a form of acceleration. In mathematical terms velocity describes motion in a certain direction, and acceleration is any change in velocity. In this context Newton's first law avers that *objects resist acceleration.*

Contrary to the Aristotelian view, Newton stressed that objects do not spontaneously or naturally seek rest. Force is needed to change motion into rest, to change rest into motion, and to change the speed and/or the direction of motion itself. When an object resists acceleration, we say that it exhibits *inertial effects,* such as the feeling of added weight when an elevator starts to ascend. Similarly, a motionless bowling ball resists a deviation from its state of rest, yet once the ball is rolling, its inertia causes it to resist deviation from straight-line uniform motion. We note that *acceleration creates inertial effects.*

This brings us to Newton's "second" law, which we already met. Force equals mass times acceleration:

$$F = ma.$$

This equation means that acceleration of a mass requires force, and that heavier objects require more force to achieve acceleration. Conversely in the absence of force, objects move in straight lines—they do not accelerate.* Since momentum is mass times velocity (mv), another interpretation of this law is that force is needed to change momentum.

The principles behind Newton's first two laws had been conceived by Galileo and refined by Descartes before Newton formalized them. Still, the pronouncement that "$F = ma$" was as dramatic an event in science as Einstein's "$E = mc^2$." From a historical perspective, "$F = ma$" may even be the more significant because it embodied the renaissance concept that nature is rational, predictable, and explainable by precise physical laws which are accessible to human intellect. We also pointed out that $E = mc^2$ can be derived from $F = ma$ with little effort.

---

* In effect Newton's first law verbalizes what happens if $F = 0$ in his second law. Therefore Newton's critics point out that the first law is actually redundant. $F = ma$ says enough.

Newton also formulated the "law of universal gravitation," or "law of gravitation" for short: All objects attract each other by a force G equal to their masses m_1 and m_2 multiplied, and divided by their distance D squared:

$$G = K\frac{m_1\, m_2}{D^2}$$

K is a constant known as Newton's universal constant of gravitation. It makes the equation applicable for any mass, and it unites the units of both sides of the equation. Historically, the idea behind the law of gravitation belongs to Kepler (1571-1630).

When we say "gravity" or "force of gravity" or "gravitation" in everyday parlance, we mean *the effects of gravity*, such as weight, falling, and orbiting, and we will find that orbiting (in the astronomical sense) is an important instance. We underscore that *when any object manifests gravity, such as when it falls or orbits, it accelerates*. Furthermore the most conspicuous form of natural acceleration in the universe is gravitation. We also know that acceleration causes inertial effects, so that Newton's laws suggest an important link between inertia and gravity. Acceleration plays an integral part in both.

The law of gravitation describes the behavior of spatially separated objects, and thus it implies gravitational "action at a distance." As we shall see, Einstein accepted that all objects exert a gravitational effect on all surrounding objects, but he replaced Newton's doctrine as to why and how gravity acts across distance.

✦

To appreciate Einstein's reasoning about gravity we focus on the fact that both Newton's second law and his law of universal gravitation describe *forces*. The former deals with the force related to inertia (F) and the latter with the force of gravity (G). A modern wording of this idea is that when an object is accelerating and experiences inertia, *it is in an inertial field*. When a mass is under the influence of gravity—i.e. it is in the "pull" of gravity—*it is in a gravitational field*.

Furthermore the equation for each law contains at least one "m." In the case of

$$F = ma,$$

the m means inertial mass, which is a measure of an object by how it resists acceleration. For example how difficult it is to get our bowling ball to roll is a measure of its inertial mass. When Einstein formulated $E = mc^2$, the m he was referring to was inertial mass.

In Newton's law of gravitation m appears twice. The order may be reversed but for clarity we can think of m_1 as a the gravitational mass of a falling object, in which case m_2 may be the mass of the Earth. For example we know a bowling ball's gravitational mass, its m_1, by weighing it on a scale located on Earth's m_2.

We can treat parts of the right side of Newton's law of gravitation, namely the m_2, the distance D, and the constant K, as a unit that quantifies the "strength" of gravity. In the modern scientific paradigm, these quantities together describe *the intensity of the gravitational field*. This is clearer if we solve for $\dfrac{G}{m_1}$, which is the force (G) per mass (m_1):

$$\frac{G}{m_1} = \frac{K\, m_2}{D^2}.$$

In this arrangement

$$\frac{K\, m_2}{D^2}$$

represents the intensity of the gravitational field at this location. Since the equation for Newton's second law also deals with force, we can set the two equations equal to each other ($F = G$), so that

$$ma = m_1 \frac{K\, m_2}{D^2},$$

which raises a critical point: *Inertial mass, the m in ma, should equal gravitational mass, m_1.* (We moved m_1 for clarity.) In that case these two m's cancel out of the above equation, leaving

$$a = \frac{K\, m_2}{D^2}.$$

This means that since acceleration alone generates force, *acceleration alone generates gravity*. We therefore say that

Acceleration = Intensity of the Gravitational Field.

Before going further let me detour to discuss "fields," though I focus on gravitation. (The concept of physical fields was first developed for electro-magnetism by Faraday in the 1840's). Newton had treated gravity as a force that, in the case of gravity, attracts two objects. This approach worked well between, say, the sun and the Earth, but it becomes cumbersome when applied to the several members of the solar system, and it is hopelessly complicated for a large collection of objects or particles. A better approach is to consider (for example) the solar system as a field, herein each point in its space has a gravitational intensity "felt" by an object at that point. In theory the outcome is the same; instead of computing what each object "feels" at its location, we compute a field and hence can predict how an object behaves at any location in space. Imagine several downhill skiers; by knowing how steep the hill is here we can easily estimate how each skier is moving here. In general, thinking of gravitation as an expression of energy, if we know about the energy available at each point in space-time, we can determine what should happen there.

Let's get back to the idea that *acceleration = intensity of the gravitational field*. We note that the right side of the equation for "*a*" contains only "m_2," which here designates the mass responsible for the gravitational field. The mass of the falling object (m_1) which is the object "being gravitated," is irrelevant. In short, *the gravitational acceleration of an object is independent of its own mass.* [*] This concept is also quite old; Galileo had argued that without air resistance, all objects fall to Earth together. (He corrected Aristotle's teaching that heavier objects fall faster.) More precisely, falling objects with different masses accelerate equally. Newton surmised that the force of gravity causes acceleration of a mass as it falls, but this did not explain why objects of different mass should accelerate equally under the influence of gravity. Newton knew that inertia resists acceleration, so he had to assume that heavier masses have more inertia. In fact Newton's explanation demands that the inertial resistance to falling be *exactly proportional* to the "pull" of gravity. For instance, the Earth pulls harder on a heavier object but a heavier object needs more pull, just enough so that when it falls, it shows the same acceleration as a lighter object, and in this way doubling the weight of an object doubles its inertia.

This proportionality between weight and inertia can be confirmed experimentally. Galileo and Newton tried, and more recently Eötvös demonstrated it very precisely using a torsion-balance. This device measures inertial mass and allows an exact comparison with gravitational mass. The same results have been obtained with more sophisticated equipment. *Inertial mass invariably matches gravitational mass*; resistance to acceleration always offsets weight. We should add that such experiments appear to give the same results in all frames of reference, which is surprising. On the basis of the relativity of mass, we might expect relative motion to disturb the results, yet the equality of inertial and gravitational mass turns out to be invariant. The implication is that there is never a discrepancy between an inertial and a gravitational field.

But why should inertial mass always balance gravitational mass? Is there some natural reason, is it coincidence, or is Newton's underlying concept faulty? Newton himself already suspected that this was rather cumbersome logic. For example he surmised that gravity on the moon is weaker than on Earth, but if his laws are to hold for both locations, the inertia of an object transported to the moon must also become weaker. How can an apparently inherent property as inertia be changed just by placing an object somewhere else? Einstein realized that the assumption that gravitational mass

---

[*]The modern extension of this idea is that fields can exist independently. E.g. an energy field does not require things to energize, and a magnetic field can cohabit a gravitational field..

is always exactly countered by inertial mass is an artificial explanation of our observations, and he sought a better answer. However we must set this issue aside for a moment and just be aware that gravity appears to be a basic property of mass, that there is a close relationship between inertia and gravity, that inertia is a consequence of acceleration, and that accelerated motion is very important.

✦

Having proposed that uniform motion is relative, Einstein was aware of Newton's arguments that some forms of motion appear to be absolute. Newton envisioned a water-filled bucket suspended on a rope. As the bucket is spun, the water becomes concave, which is an inertial effect—the resistance of the water to the bucket's acceleration. Newton reasoned that if the bucket were at rest and the universe spun around it, why should the surface of the water be concave? Accordingly, the universe must be absolutely stationary, and the spinning bucket can only be in absolute—not relative—accelerated motion.

A more modern scenario for this idea is a spaceship during lift-off. An astronaut feels "g's," which is also an inertial effect—the resistance of his mass to the spaceship's acceleration. He can thus sense accelerated motion without looking outside his spaceship. After all, if the spaceship were at rest and the universe accelerating the other way, why would he experience inertia? Yet when the spaceship is coasting, the astronaut cannot detect uniform motion without looking outside. Does this mean that unlike uniform motion, accelerated motion takes place in an absolutely stationary universe? Are we forbidden from treating the universe as mobile and the spaceship as fixed? Most, critically, is some motion absolute after all?

✦ ✦ ✦

We now have three main questions before us: First, why is gravity exactly offset by inertia? Second, even if an observer cannot detect absolute uniform motion, can he detect absolute accelerated motion? Third, how does accelerated motion, rather than just uniform motion, affect the laws of physics? Einstein worked on these quandaries for several years. He believed in the basic simplicity of nature, disagreeing with scientists who conceded that uniform motion is relative but that accelerated motion is not. He refused to accept that the laws of physics can be different for different sorts of motion and that frames of reference with accelerated motion are fundamentally different from inertial frames. He also refused simply to grant that gravity is always offset by inertia. In 1915 Einstein proposed general relativity which explains these inconsistencies, but which, as we shall see, goes much, much further.

In a narrow sense, special relativity is to uniform relative motion what general relativity is to accelerated relative motion. However this statement is oversimplified, for the key concepts of the two are quite different. The main tenet in special relativity is that time, space, and mass are relative. We will see that the main tenet in general relativity is that space and time are curved by matter. General relativity is also more complex, its mathematics and geometry are far more complicated and challenging, and its concepts are intellectually even more profound. We are ready to tackle the first postulate of general relativity.

CHAPTER 6: THE EQUIVALENCE OF GRAVITY AND INERTIA

Reminiscent of the structure of special relativity, general relativity is built upon two main postulates, espoused by Einstein around 1911. The first is *the principle of equivalence of gravity and inertia*, asserting that gravity and inertia are one and the same; not similar or related or proportional to each other, but the same.

To follow Einstein's reasoning we imagine an elevator in outer space far from any external source of gravity. An engine—never mind what kind—which is adjusted to match acceleration from gravity on Earth, can propel this elevator "upward" toward its ceiling. (Gravity at the Earth's surface causes falling objects to accelerate about 10 meters per second every second; the engine of our imaginary elevator is set to produce this much acceleration.) Observers aboard this elevator are not told whether they are on Earth or whether they are accelerating in space. They are asked to determine whether their observations in the elevator are the result of the Earth's gravity or the actions of the engine. They are free to use any experiment they wish, but they cannot look outside.

The outcome is that there is no way the observers in the elevator can tell. (Please hold your objections until page 65.) *The acceleration of the elevator gives inertial effects which are indistinguishable from the effects of gravity.* The observers in the elevator can only tell that they experience all gravitational phenomena found on Earth—weight, falling, bouncing, pouring, sinking, swinging pendula, tilting balances, rising smoke, even orbiting if it were feasible. Indeed all physical events appear the same in the accelerating elevator in space as they would in an elevator under the influence of gravity on Earth. For example as the elevator accelerates, the mass of an observer offers resistance, and if he is on a scale it will register his weight, but he cannot tell whether this is the result of the "push" of the elevator's acceleration or from the "pull" of the Earth's gravity. Likewise if he lets go of an object in the elevator and the object falls to the floor, he may conclude either that he is on Earth and the object moves "down" because of gravity, or that he is in outer space and the object stays still because of inertia while the floor accelerates "up."

If the observers guess that they are on Earth, they declare that they are in a gravitational field. If they guess that they are in space and that what they observe is resistance to acceleration, they consider themselves to be in an inertial field. However, these observers cannot be sure; *they cannot ascertain whether the behavior of objects represents inertial (uniform) motion in a gravitational field or accelerated motion of an inertial field.* They cannot distinguish between inertia and gravity. To them, inertia can duplicate gravity and vice versa; the common link is acceleration, in this case engine-produced acceleration.

Because the observers cannot prove or disprove that acceleration engendered their of gravity, they cannot determine whether they are accelerating or at rest (and "rest" is indistinguishable from uniform motion). Therefore *they cannot detect absolute accelerated motion.* Furthermore, their inability to tell inertia from gravity means that *the laws of physics in a frame of reference at rest but under the influence of gravity must be the same as they are for a frame of reference in relative accelerated motion.* For example, the scale obeyed the same laws in either case.

The same logic holds if an observer inside the accelerating outer-space elevator compares his findings with those of an external observer, one who knows for certain that an engine accelerates the elevator. The observer in the elevator can say that he has gravity. The observer outside the elevator says that this is inertia. Since both observers are entitled to deduce the same laws of physics, gravity is entitled to be considered inertia, and vice versa. This is also what the experiments by Eötvös and others showed: The intensities of gravity and inertia are equal. We also mentioned that such experiments give the same results in any frame of reference, even in the face of accelerated relative motion; the equality of inertial and gravitational mass appears to be invariant.

A more realistic scenario than our powered elevator is a spaceship designed to mimic the effects of gravity and to give astronauts a more normal environment. The spaceship is made to spin, inducing a centrifugal force. Such spinning is a form of acceleration which generates an inertial field inside the spaceship. Again, the astronauts cannot tell why they experience weight, falling, etc, aboard their spinning vehicle. No experiment they can perform indicates how the effects of gravity are created. Looking outside is also not decisive. The astronauts may see objects or horizons moving past their portholes, but they cannot be sure which is spinning, their ship or the universe, so that they still cannot detect absolute accelerated motion. They can only detect the relative motion between their frame of reference (the spaceship) and the universe (the surrounding stars, etc.). If the astronauts select the universe to be their frame of reference, they are saying that the motion of the spaceship created an inertial field. If they select the spaceship to be the frame of reference, they are calling this a gravitational field created by the universe. Either selection is equally legitimate, even if we know of no way to accelerate the whole universe. We emphasize this point because it reveals the importance of the choice of frames of reference; more on this later, but clearly the presence of gravitation depends on the frame of reference.

We note the critical implication in the case designated as a gravitational field: *Acceleration of the environment, which is a legitimate possibility in general relativity, does "create" gravity.* The reasoning is clear: Nothing bars us from ascribing accelerated motion to the universe, while gravity can be a consequence of acceleration. As we already deduced by a simple rearrangement of Newton's laws:

Acceleration = Intensity of the Gravitational Field,

even if the environment "does the accelerating." In other words by virtue of acceleration a field comes into existence, whether we call it gravitational or inertial—i.e. whichever way we identify relative accelerated motion. Since accelerated motion is so ubiquitous, the concept of a field must be critical to general relativity, and indeed we find this to be so.

Now we can re-examine the problem posed by the water-filled spinning bucket. Newton argued that the surface of the water would not be concave unless the bucket were in absolute accelerated motion. Relativity replies that if the universe revolves (accelerates) around the bucket, *a gravitational field is generated*, one which affects the bucket and its content. The water behaves the same whether we consider it or the universe around it to be stationary, and this scenario does *not* necessitate absolute motion after all.

It may not make sense that if the universe is considered in motion and the bucket considered at rest, gravity is being created, but this is analogous to the inability of observers to know whether they are under the influence of inertia or gravity. The water in the bucket likewise is unable to "tell the difference." All the water "knows" is that something is altering its behavior, and that something can be inertia—the bucket "really" spins—or it can be gravity—the universe "really" spins. Both are legitimate and equivalent choices.

This scenario really has two elements, an object—the water—and all other existing objects—the universe. A physicist and philosopher, Mach (1838-1916), deliberated the question what would happen if Newton's bucket were the only object in the universe: He argued that if all other objects vanish, the concavity of the water should vanish too (assuming the water is contracted to a solitary mass-point so it really is one object; Pauli, p. 179, or Pais, p. 285). Mach's argument suggests that all motion in the universe, including that of Newton's bucket, occurs only in relation to all other objects in the universe. Einstein applied this concept in the form of "Mach's principle": The concavity of the water is the consequence of all the other objects around it, which means that *the distribution and motion of matter in the universe is responsible for inertia.* This concept unites the principle of equivalence with the idea that all motion, even the accelerated kind, can only be relative: Both inertia and gravity stem from accelerated motion, but it can only be motion with respect to other objects. It must be *relative* accelerated motion. Otherwise, if we ascribe any motion to only one object—if we grant absolute motion to that object—gravity and inertia cannot exist Later we will see whether this idea can be validated and experimentally supported, but we already drew a similar conclusion on the basis that ever-present kinetic energy in the universe implies ever-present relative motion (page 40).

Not only can acceleration reproduce inertia or gravity, but the two can be made to cancel each other. Let us envision another elevator in a tall building on Earth. If this elevator is allowed to fall freely, *an observer in it feels weightless.* Galileo predicted this on the basis that without air resistance, all objects (here the elevator and its contents) fall together, and Newton's analysis would be that the elevator is accelerating downwardly while at the same time an observer in the elevator resists this downward acceleration. Gravity pulls down, inertia resists, and the observer feels weightless because the two effects balance each other. In Newton's view again inertia must offset gravity, in this case to explain weightlessness during free-fall.

Einstein's interpretation is that inertia replaces gravity in a freely falling frame of reference because gravity *is* inertia. Each is a consequence of acceleration, and here two accelerations cancel each other. Objects in the elevator float freely, and observers in this frame of reference cannot tell which is the case—no inertia or no gravity. Indeed it was this situation, free-fall, that impelled Einstein to recognize the equivalence principle: *During free-fall, zero-gravity is indistinguishable from zero-inertia. Therefore gravity is indistinguishable from inertia.*

The idea of free-fall is not as esoteric as it sounds. In a sense we and all objects in the universe are constantly in free-fall, but the solidity of the ground usually restricts us; take away the Earth or put us in "space," and we do "fall" freely. Free-fall also interested Einstein because in a frame of reference like the loose elevator, the familiar effects of gravity (such as the accelerated motion of falling) are absent, and yet the postulates of special relativity still apply. Therefore, for instance, absolute uniform motion remains undetectable in the falling elevator. This suggests a link between special and general relativity in the sense that the events in a falling frame of reference provide a starting point from which the laws of physics dealing with gravity can be derived. As we shall see later Einstein exploited this notion extensively.

✦ ✦ ✦

Now we can raise an objection. Couldn't an observer detect *convergence* of gravity towards the center of the Earth? We know that if two strings are hung one meter apart from a ceiling on Earth, their lower ends are very slightly closer than one meter because each points to the Earth's center; see diagram U on page 322. This convergence would not appear in an accelerating elevator in space. Similarly, the astronauts aboard the spinning spaceship could notice that hanging strings converge. Isn't this a way an observer could distinguish inertia from gravity? Isn't this a refutation of the principle of equivalence? In reply, we can argue that even if observers can tell something about the geometric structure of their gravitational or inertial field—for example, even if astronauts accelerating in outer space can tell they are in a field with no center—they still cannot tell how this field is created.

A mathematically more precise solution to this objection is to consider only a small region of space, so small that the structure of the mass responsible for gravity does not matter. For example we could limit the space to that occupied by only one string. Then the gravitational field appears center-less, and we say that the field is uniform and acceleration is uniform (though this is a different use of the word "uniform"). The principle of equivalence can then be stated with the provision that it applies only where *the uniform gravitational field always is equivalent to uniform acceleration.*

Let us elaborate on this point for upcoming discussions: The two strings in a natural non-uniform gravitational field reveal *two effects of gravity.* One, the strings are being pulled down, or, if permitted, they would fall.

But in addition their dangling ends tend to converge toward a center, and in this context we can consider the latter to be a "side-effect" of the gravitational field. This side-effect actually has striking natural manifestations on Earth, namely lunar and solar tides. For the moment let us think of the Earth as orbiting in the gravitational field of the moon (rather than the reverse, as is more intuitive). Not only does this gravitational field create the obvious attractive effect on the Earth, but it also creates a tidal effect which deforms the Earth slightly. (The Earth, particularly its watery parts, is made less spherical in the process.) The tidal effects which draw the "waist" of the Earth together are comparable to those that draw the hanging strings together. The significant point for relativity is that in addition to the obvious major effects of gravity—orbiting, falling, weight—we must deal with *tidal change-of-shape* effects. When we revisit this issue we will reapply string analogy.

✦ ✦ ✦

As we said, acceleration can account for gravity. This easily explains why different objects fall with the same acceleration. Imagine several different objects dropped together from the ceiling of our accelerating elevator in outer space. All objects experience only one and the same acceleration, so they hit the floor together (with no air resistance). It may be easier to think of this in reverse. The several objects are not "dropped" but are lined up together on the ceiling of the elevator when the elevator starts "rising." The accelerating floor reaches each object at the same time!

If Newton were inside this elevator and unable to peek out, he would say that because of the force of gravity, the objects are attracted downwardly, presumably to the Earth, while the proportional retarding effect of inertia is just enough to account for the equal acceleration and simultaneous fall of these objects to the floor of the elevator. In short, Newton would invoke a law of gravitation plus a law of offsetting inertia. But if the reality of the situation were revealed to him—the elevator is far from Earth and is accelerated by engines—he would say that inertia imitates gravity because inertia happens to equal gravity. He would invoke only a law of inertia. Even if we knew nothing of physics, this scenario would arouse skepticism. How can one observer need two laws and later, because of a new vantage, need only one law to explain the same event?

Einstein asserted that there is no need to invoke a force of gravity and of inertia, nor an offsetting balance between the two. Since gravity and inertia are identical, two laws are not needed. We are witnessing only one phenomenon of nature, and accelerated motion alone suffices to explain our observations. The following sounds like a play on words, but it is a serious scientific generalization: Newton taught that acceleration can be the result of gravity, which happens to equal inertia. Einstein taught that gravity is a result of acceleration, as is inertia. A change in frame of reference—"elevator is in outer space powered by engines which create inertia" to "elevator is on Earth under the influence of gravity"—does not require different laws, because inertia and gravity are not different from each other.

This idea points to another critical difference between Newton's and Einstein's interpretations. Newton had assumed that gravitating objects attract each other. Except for a measure of the distance, the space separating them was considered irrelevant. He proposed that equal acceleration of different falling masses is set by the properties of each object; their masses determine their gravitational force, as in $G = K\dfrac{m_1 \, m_2}{D^2}$ and $F = ma$, while their inertia matches their gravity.

Einstein proposed that the properties of each mass are not relevant; different falling masses move similarly because each is influenced by a property of its surroundings. That property is the *gravitational field*, and this view espouses that any acceleration between objects exists by virtue of a field that contains these objects. Each point in this field has a certain gravitational potential (which we will discuss later) which stems from the nature of space-time (which we will also cover later). This field-concept elegantly explains how an apparently inherent property of an object, namely its inertia, depends on its location: *Changing its location, e.g. from Earth to moon, places it in another field. The object has not changed, but the field is different.* As we shall see, general relativity goes much further with these ideas, but clearly it shifts the explanation of gravity away from a study of masses and towards a study of the intervening environment. In short, general relativity provides two interrelated simplifications. One, the equivalence of inertia and gravitation allows one set of laws to suffice for both cases. Two, the concept of gravitational fields allows one set of laws to suffice for various locations.

We will say more about fields—particularly about their mathematical characteristics—but we can already see how Newton's law of gravitation has been obviated and supplanted by relativity: There is no need to ask why masses attract, only why they accelerate. The equivalence principle thus leads to *a new perception of gravitation.* Newton's gravitational-force-and-inertia model is replaced by Einstein's acceleration-in-a-gravitational-field model. This development may imply that Einstein belittled Newton. To the contrary, he honored and praised Newton, suggesting that if Newton had had access to more modern experiments, he would have surely reached Einstein's conclusions.

The role of acceleration in Einstein's explanation of gravity of course leads to another question: If inertia and gravity both are a product of acceleration, whence this acceleration? As we intimated, the reply involves the gravitational field and the gravitational potentials at each point in that field, but the details must wait for now. The immediate concern still is the significance of the equivalence between gravity and inertia.

✦ ✦ ✦

Let us summarize: Galileo, Newton, and Einstein concurred that gravity matches inertia, although Einstein gave a different reason—gravity *is* inertia. An inertial field is created by accelerating a mass, and this duplicates a gravitational field. Only our choice of the frame of reference determines which we call it. Because inertia and gravity are indistinguishable, absolute accelerated motion cannot be detected, and hence the laws of physics must be the same with gravity as they are with accelerated motion. To paraphrase Einstein loosely from his lectures, inertia and gravitation have numerical equality and unified nature because they are equivalent, and this makes nature much simpler and makes general relativity a better way of explaining gravitation.

We note the symmetry, which prompts us to label relativity as "beautiful" and elegant: Special relativity showed that absolute uniform motion cannot be detected in a moving frame of reference because physical events appear the same with or without uniform motion. Now general relativity shows that absolute accelerated motion cannot be detected in an accelerating frame of reference because physical laws and events appear the same whether the frame is moving or the universe is moving. In short, because rest is indistinguishable from uniform motion, uniform motion can only be relative. Because inertia is indistinguishable from gravity, accelerated motion can only be relative. All motion is relative.

The implication of the first postulate of special relativity—inertial frames are equivalent—together with the principle of equivalence—accelerated frames are equivalent—is critically important: *The laws of physics should be written so as to be valid for all observers in any frame of reference, at rest or in any motion, uniform or accelerated.* Observers at rest or moving uniformly see the same laws; observers accelerating or in gravity see the same laws. The accord within relativity, and the harmony within physical nature which Einstein admired, appear to exist.

CHAPTER 7: GRAVITY BENDS LIGHT

We have cited mechanical experiments, such as using ordinary objects on a scale in an accelerating frame of reference, which demonstrate the principle of equivalence. However we have not considered optical experiments—using light or other electromagnetic radiation—which do the same. (The Michelson-Morley experiments are "optical" but do not address the principle of equivalence.) Given the importance of electromagnetism, the equivalence between gravity and inertia would be of little consequence in modern physics if it did not apply to optical events. In fact general relativity would then be invalid.

For this issue let us envision a narrow beam of light as a stream of photons which we can "throw" from one wall to the opposite wall across our outer-space elevator. We also assume that we can see photons. We imagine that while the elevator is in outer space with its engines off, we stand against one wall and "throw" photons so they strike a target on the opposite wall. For instance we aim a flashlight onto a spot on the opposite wall. We should observe a straight stream of photons. An analogous "mechanical" scenario is squirting a stream of water inside a space shuttle while in orbit. (Imagine an astronaut shooting a water gun across the cabin.) Because of weightlessness, the stream, be it optical or mechanical, appears as a straight line.

If the elevator is then made to accelerate upwardly by use of an engine (it is accelerated by the engine), the photons will strike the wall below our target, because as the elevator gains speed, the point of impact of the photons shifts lower and lower. In theory, as long as the elevator is accelerating, we see the stream of photons form a curved trajectory arching across the elevator. Again in a mechanical analogy, a steam of water would arch "down" during acceleration of the elevator. In either case, the stream is bent.

Meanwhile observers outside the elevator see the whole elevator accelerate, but if they can see through the walls, they find the beam to be straight. That is because the external observers see the photons cross the elevator in a straight line while the target wall moves up. These observers recognize an inertial field in which the motion of the photons is uniform (and hence straight) while acceleration causes internal observers to see the same path of photons as curved. If all observers pool their observations, they conclude that the observation of a curved stream of photons inside the engine-powered elevator indicates that *inertia bends light*.

However the observers inside the elevator cannot tell whether they are really accelerating by means of an engine or they are witnessing a natural gravity while the elevator is at rest on the Earth's surface. But how can gravity cause bending of the beam if the elevator is standing still? The walls of a non-moving un-powered elevator do not appear to be accelerating relative to the beam, so shouldn't the beam appear straight inside the elevator? Isn't this proof that the elevator is stationary in what can only be a gravitational field? In fact isn't this an optical experiment that identifies inertia over gravity and that refutes the principle of equivalence? And if so, doesn't this experiment reveal different laws of optical physics for inertia than for gravity?

Einstein's dramatic reply was that *gravity bends light*, just as does inertia. If this can be verified, we have an optical experiment clinching the principle of equivalence, which would mean that gravity and inertia are equivalent in both mechanical and optical experiments. Therefore *the key to validating the principle of equivalence for application in general relativity is showing that gravity bends light*. This was a critical juncture Einstein had reached when he announced his theory in 1915 and published it in 1916. The paper appeared in volume 49 of the Annalen der Physik under the (translated) title "The Foundation of the General Theory of Relativity," which is Einstein's first comprehensive exposition using the term "general relativity." (He began using this term earlier, and it appears in the title of a paper co-authored with Grossmann in 1913; we discuss Grossmann's role later.)

An opportunity to demonstrate Einstein's assertion, inspired by Britsh astronomers Dyson and Eddington, arose in 1919 during a solar eclipse, as visible light from a distant star passed very close to the sun. Because of the sun's gravity, this beam was bent—the images of stars were displaced—by approximately the amount that Einstein had predicted. This experiment was quite complex, and we will return to its details, but should remember that gravity is associated with mass, in this case the mass of the sun; hence this experiment showed that *the presence of mass bends light*.

Yet here again arise grounds for an objection. While discussing special relativity, we indicated that light can travel at velocity c because it itself has no innate or "rest" mass. Then how can gravity affect something with no mass? Moreover if light experiences inertia while crossing our accelerating elevator, it again must have some mass. How can light have mass and not have mass?

To answer this objection, two notions apply. First, light is a form of energy, and energy is equivalent to mass via $E = mc^2$. Thus the energy in light is the same as a small amount of mass, and both are susceptible to inertia as well as to gravity. The second notion is that light can behave as a weightless wave of energy *or* as a particle of mass. Huygens and Maxwell proposed that light is a wave. Newton proposed that light consists of "corpuscles." Einstein refined the latter theory, concluding that light has mass in the form of moving subatomic particles which exhibit inertia or gravity; light has momentum. (This work led to Einstein's Nobel Prize in 1921.) These particles are the "photons," and the dual nature of light is the subject of quantum mechanics, which we will discuss in a separate chapter. Light also has "exclusive" properties, such as a speed which appears not to be additive, but nevertheless it can behave like any other object. Indeed now we can answer our question (page 42) whether briefly light can behave like a mass: Yes, in the form of photons, *light has weight*. As we said, photons lack *rest* mass, but they do not lack momentum because they are never at rest. (Photons come to rest when they hit an opaque surface such as an object or certain molecules in the rods and cones of our retinae, at which moment they cease to exist and their energy is transformed to some other form, such as heat or an electron-mediated nerve impulse.)

We are asserting that a stream of water is mass *and* energy (more obvious if the stream contains some plutonium), and light is energy *and* mass (demonstrated in experiments in quantum mechanics). Once we grasp that in some respects light is a stream of mass-containing particles, the idea that gravity bends a beam of light loses much of its mystique. It is no more surprising than the observation that when we water a lawn or shoot a water gun, the squirted stream arches to the ground. From this point of view, the principle of equivalence *must* apply to mechanical as well as optical experiments, just like the same laws of physics apply to all masses—big and slow ones like droplets of water, or very small and fast ones like photons.

(Four comments here: First, the idea that gravity can bend light had been proposed by Newton as he was trying to explain how optical lenses work, but his formulæ do not give the results predicted by general relativity. Second, a one spin-off from Einstein's work with light waves is the principle of the laser. Third, as important as relativity is, its discovery did not reap a Nobel Prize. Einstein won his prize in another area of physics. Fourth, our pythagorean diagram \ showing the relativistic effects of uniform motion aboard a train is not accurate in that the light beam would be bent by gravity. If we wished to be more precise, the train should be in "outer space.")

Thus, no matter how we look at it, there is no physical reason that a light beam will not behave the same whether we call the effect of acceleration "gravity" or we call it "inertia." In our elevator, for example, the beam is bent the same in an inertial field in outer space as in a gravitational field on Earth. Inertia and gravity are indistinguishable from each other no matter which the frame of reference. We already emphasized that if this is so, absolute accelerated motion is undetectable. Hence properly composed laws of physics, mechanical or optical, are valid for relatively moving observers, uniform and accelerated.

This conclusion of course further satisfies Einstein's notion that there is simplicity and unity in nature. What holds for special relativity and uniform motion should hold for general relativity and accelerated motion. What holds for mechanical events should hold for optical events *Absolute motion of any sort can never be detected.* Incidentally, the idea that all absolute motion is undetectable means that no physical event is measurable without some frame of reference; this point will be important later. Now we can appreciate why it was so important for Einstein to show that all motion is relative; the reasons are not merely aesthetic. Only if absolute accelerated motion cannot be detected by any method are inertia and gravity truly equivalent. Hence *acceleration can create a gravitational field, gravity can bend light, and that the presence of mass can bend light.*

Although similar more accurate experiments on the bending of light have been carried out since, agreeing even more closely with Einstein's prediction, the widely published 1919 results launched him into instant fame. This event, rather than his Nobel Prize, established Einstein as one of the foremost scientists of all time. Firm evidence that gravity bends light placed general relativity into the mainstream of modern physics and silenced any serious doubt about the validity of Einstein's ideas. As we shall see, the fact that *mass can bend light* is crucial to the rest of the logic of general relativity, and this notion has captured public imagination and scientific attention with corollary concepts such as space-time and black holes.

An intellectually gratifying feature of relativity is how various basic concepts can be combined to yield vital new insights. This chapter presents several instances, starting with the following: Since uniform relative motion slows measured time, it stands to reason that accelerated motion has an analogous consequence. But since accelerated motion induces the effects of gravity, *gravity itself should cause measured time to slow*. In fact this previously unforeseen phenomenon, named "gravitational time dilation," can be demonstrated.

To see how accelerated motion affects time, let us reuse our imaginary accelerating elevator in outer space. Two observers, one sitting on the floor of the elevator and the other clinging to the ceiling, each have a clock which, rather than ticking, emits regular pulses of light. These pulses travel only at speed c because the speed of the source does not matter, even if that source is accelerating. The two clocks and both observers are accelerating upwardly with the elevator, so the pulses emitted by the clock on the floor arrive at the ceiling at ever-increasing intervals; each new pulse has farther to travel as the ceiling "rises" faster and faster. We can even imagine that the clock on the floor "throws" clumps of photons at the ceiling, but while they are in flight at speed c, the ceiling moves up, so that each successive pulse takes longer to reach the observer on the ceiling. Meanwhile, according to the observer on the floor, the pulses moving from the ceiling to the floor arrive at ever shorter intervals; acceleration of the floor shortens the distance for each successive clump of photons as the floor rises faster and faster. We can think of the observer on the floor as rising to meet the photons, so that as long as photons maintain their speed—which they do—each clump strikes him sooner than the previous. The observer on the ceiling therefore judges that the clock on the floor is slowing down, while the observer on the floor finds that the other clock is speeding up. Since each observer finds nothing different about the behavior of his own "proper" clock, they surmise that *accelerated relative motion alters the rate of time*.

An incidental question might be what if the elevator were accelerated to as close as possible to the speed of light. Then the observer on the ceiling would find that the clock on the floor has nearly stopped, or, to state it dramatically, that time nearly "stands still." Of course the elevator cannot be brought to speed c because that would require an infinite amount of energy, but a fair additional question is what if the elevator could somehow accelerate to beyond the speed of light. Then this observer should see the other clock as running backwards! Since relativity prohibits speed beyond c, in effect it prohibits reversed time.

But we shall pursue a less exotic issue: These observers again cannot tell that they are in an accelerating frame of reference in space far from any "real" gravity. Therefore they cannot ascertain whether inertia or gravity altered their measurements of time. All they can tell is that while the elevator is accelerating at about 10 meters per second every second, the physical properties of Earthly gravity are duplicated. For example the observer on the ceiling is free to assume that the elevator is in a gravitational field, and that the clock on the floor, which to him is the slower one, is closer to the Earth's center of gravity. He then infers that *gravity* alters time and in particular that *measured time dilates in a gravitational field*. In short, *gravity "bends" time*.

In the development of general relativity, a different line of reasoning led Einstein to gravitational time dilation. He knew that light can appear as an undulating wave of energy, and that the amount of energy carried by a wave of light is proportional to the frequency of its vibration. When a wave escapes from a gravitational field, it gives up some of its energy, and consequently the frequency of the vibration decreases. (This does not involve the speed of light, only the frequency of its wave-undulations.) On this basis Einstein predicted that the stronger gravity of the sun should reduce the wave-frequency of sunlight reaching the Earth. Such an effect is called a red-shift, because as the wave-frequency of light is decreased, its color shifts towards the red end of the spectrum. In this context this is also known as the "einstein red-shift." Although difficult to perform, experiments on sunlight indeed reveal this red-shift.

We will revisit the einstein red-shift once we have the appropriate mathematical tools, but four comments are needed here. First, the same effect occurs in reverse, so as to reveal a "blue-shift" if a beam of light could be sent from the Earth and observed on the sun. Likewise, when a shift is very large, the resulting beam need not remain visible. For example it is possible for light to show an "x-ray shift." Nevertheless by tradition we call this phenomenon the red-shift.

Second, the bending of light and the shift of its frequency supplement each other regarding the dual nature of light. The bending of a beam presents light as particles of mass, while the red-shifting presents light as waves of energy. (This red-shift is not the same as the astronomic Doppler effect seen in the light that emanates from stars that are moving away from us [as the universe expands]. The astronomic Doppler effect does not entail relativity.)

Third, anything which measures time—anything which has a measurable frequency—will show a shift. Thus the observers on the floor and ceiling of an accelerating elevator may detect shifts in the color of the pulses of light. In particular, just as they see a "shift" in the rate of the pulsating clocks—one observer sees less frequent pulses while the other sees them as more frequent—one observer sees redder light while the other sees it bluer. However, it still does not matter whether the acceleration of the elevator is from gravitation or from being propelled by an engine; the principle of equivalence applies, in that inertia and gravity *both* affect the time it takes a clock to tick (to flash, in our example) as well as the time it takes a wave to undulate.

Incidentally, let us note that frequency—how many crests of a wave arrive in a given time period, usually one second—is the arithmetic inverse of how much time elapses between the arrivals of crests. We can label the time interval between crests by "dt." (The symbol dt appears in differential calculus, which we will use later.) For example a wave with a frequency of 60 cycles per second has a dt of 1/60th of a second, and a lesser frequency, as seen in a red-shift, has a longer dt.

The fourth comment is the most intricate. A legitimate question—which we could have raised earlier but we had not yet said enough about energy—is why events must slow down when moving out of a strong gravitational field. In the einstein red-shift, why is the frequency of the wave of sunlight reduced and its dt lengthened? In reply we should think of all time-consuming events as *processes that require energy.* (The link between frequency and energy in waves also appears in quantum mechanics.) Just like crossing the space of a gravitational field is an energy-consuming event, *"crossing" time is an energy-consuming process,* and we can say that the passage of time in which an event occurs relies on available energy. Later we will see how we can treat the dimensions of space on equal footing with the dimension of time and why "motion through time" is an energy-requiring event. The point now is that a relatively faster-moving object, including a clock, expends more of its energy for crossing space, *leaving less energy for "crossing" time.* Consequently its time-measuring activity—namely ticking—slows down.

Thus the photons emanating from the sun give up some energy while escaping from the sun's gravitational field, and as a result they have less energy available for the passage of time. On the other hand we already pointed out (page 41) that the universe does not provide an infinite amount of energy, and we know that all physical events obey the laws of conservation. We explained that this is why photons do not move with infinite speed, but now we add that the frequency of the wave of sunlight in the einstein red-shift is reduced (its dt is lengthened) because an infinite supply of energy is not available for "crossing" time. We can say that all time-consuming processes share in a limited supply of energy, be they the undulating of waves of light, the ticking of clocks, or the aging of twins. Indeed seen in this way, the relativity of time seems much less enigmatic.

✦

As mentioned, extremely accurate clocks are now available, facilitating the study of the effect of relative uniform motion on measured time. Similarly, gravitational time dilation can be detected by comparing very sensitive clocks *placed at different altitudes off the Earth's surface.* A clock at ground level, where gravity is stronger, actually *runs slightly slower than an identical clock atop a high building.* We note that the clocks do not have to "move" in the ordinary sense to show gravitational time dilation. Stronger gravity alone slows time. The two clocks at different altitudes in effect are in relative accelerated motion with respect to each other, based solely on the difference in gravity at different altitudes—unless of course the two clocks are in "outer space" where there is no gravity. Thus general relativity asserts that the rate of time depends on relative *location.* In other words how long an event takes is influenced by *where it is relative to gravity.* Therefore in one gravitational field, clocks at separated altitudes cannot run synchronously. (This is because a gravitational field is not homogenous; more on this later.) For instance your wristwatch slows down slightly if you put it on your ankle. Even the Shangri-La story is backwards: In theory inhabitants of high plateaus age relatively faster, not slower, and please note this "gravitational dilatation" of time. An observer's own clock runs *slower* that a clock at a lower altitude, but this effect is opposite from the inference of special relativity that each proper clock seems the *fastest*; please see page 16.

But let us carry our logic through another step and again invoke the concept that *gravity is associated with mass.* After all, "altitude" in the present context means distance from the Earth's mass, and by way of

$$a = \frac{K\,m_2}{D^2}$$

we know that the intensity of a gravitational field depends on distance (*D* in this equation). Now if gravity is a consequence of acceleration, and if the presence of mass is associated with a gravitational field, then general relativity leads us to conclude that *the rate of time increases with distance from a mass,* and that *nearby mass slows measured time.* In short, as odd as it sounds, *mass "bends" time.* And sometimes concurrently, special relativity accounts for its own effects of relative uniform motion on measured time.

✦ ✦ ✦

We have reached a momentous scientific vista: Light can behave like a particle with mass, so that gravity bends a beam of light. This leads to the discovery that gravity bends time. But gravity is found near mass, so mass bends time. Still, this pivotal concept can be worded in ordinary terms: Clocks run slower closer to a mass. An event takes longer where gravity is stronger. Our measurement of time depends upon gravitation. Such statements may not make sense to us by intuition, and they certainly wouldn't have to our ancestors. How could they have detected a speeding up of timepieces atop a building? How could anyone before this century do experiments in outer space where gravity is nearly absent or on the moon where it is weaker? In other words, as is true for special relativity, the effects of general relativity are not mysterious or supernatural, but they are very difficult to recognize.

The slowing of time during relative accelerated motion as well as during relative uniform motion resolves the "twins paradox" mentioned with special relativity on page 20. This paradox is the contention that if each member of a pair of twins traveled relative to the other, each would age less and their clocks would remain synchronous, contrary to the teachings of relativity. After all, skeptics argue, which ever twin went away, their relative motion, by definition, is identical; twin A leaving B is the same as twin B leaving A. The general-relativistic reply is that in order to have these twins travel with respect to each other and then compare their aging, *one of them has to accelerate* when turning around in order to rejoin his brother. (Turning necessitates "accelerated" motion.) According to the principle of equivalence, only the turning twin experiences gravitation, which also slows time. Hence their experience is not symmetric and identical; twin A leaving and then turning around is not the same as B leaving A. After the trip their "clocks," biologic and otherwise, are asynchronous in only one direction, namely A has aged less than B. In terms of the consumption and conservation of energy, we already pointed out that our universe does not provide infinite energy, including for the passage of time, and that what is available is conserved. A's turning around consumes extra energy, which slows this twin's aging even more. We can say that acceleration has "cost" twin A some energy, which "cost" him some time.

But what if the twins accelerate equally in opposite directions from one place and both return to that place? Then we actually have two similar experiments with their take-off place as the stationary frame of reference. A triplet who stays home appears to age more than either traveling sibling, so that the relativity of time is still demonstrable. To confirm this, again we can observe real clocks on fast airplanes. When "twin" planes circle the Earth in opposite directions, one moving with the Earth's rotation and the other against it, their clocks do indeed end up non-synchronous, because with respect to the rotating Earth their trips are not symmetric. Furthermore, the discrepancy is a complicated combination of effects: High speed (uniform relative motion) slows a clock, whereas high altitude (less gravity, which in effect is accelerated relative motion) speeds it up.

This fact has practical applications. For example thanks to the location of certain stars such as Polaris, it is easy to calculate latitude, but to determine longitude a navigator must know the exact time and must compare time on his clock with another reference clock. The more synchronous the clocks, the more accurate his navigation. Likewise modern navigational aids, notably the Global Positioning System (GPS), rely on very accurate clocks placed aboard Earth-orbiting satellites. Since these clocks are in uniform as well as accelerated relative motion, they must be adjusted for special as well as general relativity. In chapter 17 we will see some of the mathematics behind this dual adjustment.

✦ ✦ ✦

General relativity predicts that measured time slows during accelerated relative motion, i.e. in the stronger gravitational field closer to a large mass. Meanwhile we recall special relativity: Since measured time slows during uniform relative motion, length also shrinks when measured by a stationary observer. Shouldn't measurements of *length* likewise change during relative *accelerated* motion, and therefore shouldn't measured length also change in a gravitational field closer to a large mass?

Our reasoning could come from another example with two trains which have the same lengths measured while the trains are side by side. Train A is then placed atop a mountain, and observer A measures the length of train A using a beam of light, a clock, and the rule that $cT = D$. He then looks down at train B below, and he measures the length of train B using B's clock. Since clock B is in a stronger gravitational field and therefore runs relatively slower, *observer A concludes that train B is shorter*. We note again that the trains do not have to "move" with respect to each other in the ordinary sense; unequal gravity is their only difference. Furthermore, gravity is unequal in this case because one train is closer to the center of mass (the Earth) than the other. Just as we concluded that the presence of mass changes measured time, we expect that *the proximity of mass distorts the measurement of length*.

We know that in geometry length is one dimension, but traditional geometry recognizes three dimensions of space. Earlier we indicated that uniform motion affects the measurement of space in only one dimension (the dimension which is the direction of motion), but now we are dealing with accelerated motion. We recall that accelerated motion is any change in speed or a change in any direction of motion. Therefore, accelerated motion should affect space in three dimensions, and *we expect gravity and nearby mass to distort the measurement of space in three dimensions*. Such reasoning can further expand our appreciation of why Newton's bucket has a concave water level even if we consider it relatively stationary: The theoretically spinning universe induces a gravitational field (a consequence of accelerated motion) which distorts not just one dimension but three dimensions, i.e. space and shape, including the shape of the surface of water in the bucket. In short, a gravitational field alters all space.

✦

Please imagine that you are seated next to an enthusiastic teacher of relativity while reading this chapter. A flash of comprehension appears in your mind. You turn to your mentor to ask, "*doesn't that mean that a nearby mass alters space as well as time?*" You should not be astonished if, upon hearing your perceptive question, the relativist doesn't leap with joy and yell out "yes, precisely!" Why the excitement? Because once you achieve this insight, you enter the heart and mind of general relativity, and your understanding of how the universe works will be changed forever.

We are ready for a tantalizing and important aspect of relativity: *four-dimensional space-time*. (The term is also spelled spacetime.) Visualizing an event in three dimensions is no problem for us: An object may move to the right, while moving up, and while moving forward. But the object is also "moving" through a fourth dimension, *time*: It not only has a location in three dimensions of space but it also has a past, a now, and a future. It has a temporal "location," one that indeed is incessantly changing. It is always "moving through time," and though we can alter how much time an event containing that object appears to consume, the object is inexorably aging.

Space-time, as we simply call it, is a mathematical construct which was invented for use in relativity to calculate where an object is in space and in time simultaneously. In saying that space-time was "invented," we do not mean that anything new or previously unknown had to be fabricated. Even in the mathematics of space-time, time is still only a matter of duration or simultaneity, and space is still only a matter of location and distance. Space-time merely pools this data, so that

knowing an EVENT = knowing its SPACE + knowing its TIME = knowing its space-time.

Said in more detail, *space-time combines the measurement of time* (commonly expressed as T) *with the measurements of space* (commonly expressed as X-Y-Z) *into one system which allows the complete measurement of physical events*. In this way events can be displayed in diagrams equipped with coordinates X, Y, Z, and T, though obviously a lucid four-dimensional diagram or graph is impossible to draw. However by omitting some of the space coordinates, we can still include of time as a dimension, and in any case events can be analyzed mathematically in equations containing coordinates, X, Y, Z, and T. (T can be algebraically negative; more on this later.)

We note why space-time has four dimensions: it is easy to fully describe location in *three* dimensions, and time flows *one* way. Therefore the description of an event calls for four items of information. If somehow time were measured forward as well as side-to-side, we might describe space-time in five dimensions, and if we lived in a totally flat environment without an up or down, fewer than four dimensions would suffice. Use of exactly four dimensions also has some theoretical-mathematical advantages, though in the chapter covering quantum mechanics we will touch upon theories that implicate more than four dimensions. Also see Barrow and Tipler, pp. 258-276, and Hawking, pp. 164-165 of A Brief History of Time.

Unfortunately visualizing space-time is difficult. We often resort to spatial analogies which omit time, such as surfaces of ordinary flat or curved objects. We should keep this in mind when, for example, we use the sphericity of the Earth to represent the curvature of space-time. Yet even if mixing time and space into one entity is intellectually difficult, there is no need to exaggerate this problem, nor to allow it to confound relativity. Except in science fiction, space-time has no mysterious, supernatural, or occult properties.

Moreover, quite intuitively and unconsciously, we use some notion of space-time in many common activities. Saying "meet me in New York" forces us to deal in the four dimensions of space-time in order to define the "when-and-where" of the planned event: Our date might be on the corner of

Fifth Avenue (one dimension) and 42nd Street (second dimension) on the second floor (third dimension) at four pm (fourth dimension). In this context the term "place-time" or "time-and-place" may be less troublesome that "space-time.".

On the other hand a stumbling-block is thinking of time as a dimension. We are accustomed to describing space with lines which easily depict dimensions, but when we describe time we ordinarily do not draw lines. Nonetheless history books routinely treat time as a dimension in the form of time-lines. The notion of a time-line can be applied in special relativity. We described scenarios where two observers start out with synchronous clocks, then they move relative to the other, and subsequently their clocks are no longer synchronous. Initially these observers agreed that one brief nearby event occurred at one moment, so on a time-line they can show this event as *one* point. After their relative motion, they disagree on when an event occurs, and as each observer marks this event on a time-line, they end up with *two* points some distance apart. The gap between the points, though it is a distance, reflects an interval of time.

In the sciences time can be used to express distances, and for this purpose it is customary to use the speed of light as a standard. Not only is this a convenient way of representing great distances, but it has the merit that the speed of light is the same for all uniformly moving observers. Therefore distance units such as light-years are used in astronomy, and a distance can be expressed as ct or cT (the speed of light multiplied by time) in relativity and in other branches of physics.

The observation that four dimensions are required to define an event does not fully disclose why *relativity* needs space-time. We already cited some reasons space-time explains certain phenomena better than separate time and space, but here is a more complete list:

1.   Newton's law of gravitation makes no provision for time, only for space, implying that the force of gravity acts through space without consuming time. By uniting time with space, relativity can accommodate the observation that *gravitational events require at least some time.*

2.   Relativity calls upon space-time because space as well as time are subject to alteration by relative motion and by the proximity of matter; i.e. *both space and time are relative.* Suppose our date at 4 pm in New York is with an extra-terrestrial visitor who is flying to Earth at close to the speed of light. Unless he adjusts for *the effect of relative motion on all four dimensions*, he will not arrive at our 4 pm in time, nor at our designated location in space.

3.   We based our conclusion that length is relative on our conclusion that time is relative, and the relativistic relationship between space and time is fixed as a reciprocal: Relative motion increases measured time and decreases measured distance, as suggested in the equation

$$c^2 = \frac{space^2}{time^2}.$$

4.   While both space and time are relative, events within the combined entity we call space-time are *not* altered by relative motion. This may make little sense now, but we will cover this very important point a little later in this chapter.

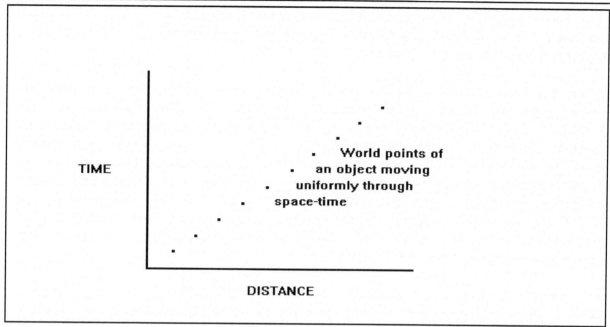

Diagram E. A simplified Minkowski diagram. "DISTANCE" represents the three dimensions of space combined into one coordinate axis. "TIME" of course represents the time axis.

✦ ✦ ✦

The mathematician Minkowski (1864-1909), who introduced the idea of combing space and time in the study of relativity, devised graphic ways to show four-dimensional space-time. These are called Minkowski space-time diagrams. A simplified example (diagram E) condenses the three traditional dimensions of space into one, labeling it simply "distance." A graph can be drawn with a horizontal coordinate axis for "DISTANCE" and a vertical axis for "TIME." In such a diagram a series of dots represents how an object or particle moves through space while it moves through time; i.e. *the string of dots represents motion through four-dimensional space-time.* Here motion occurs only relative to the axes. If the object or particle is in uniform motion, the dots are set in an oblique but straight line. If motion is accelerated, this line can be drawn curved, though as we shall learn shortly, we can keep the line straight but bend the TIME and DISTANCE axes. If the object or particle has no motion at all, it is depicted as a straight vertical line; it "moves" only through time. (We note that in this context we can use the terms "axis," "coordinate," and "coordinate axis" interchangeably.)

A dot or point in a space-time diagram represents something which happens in a certain place at a certain time, and it is called a "world-point." An entire series of points represents the existence of an object or particle and is called its "world-line," though we usually draw world-lines as solid lines rather than as a series of dots. The reason we say that we can be describing objects *or particles* is that light can be considered to consist of photons, so that we can draw a world-line for a beam of light, and this is a useful procedure in relativity. In either case, considering ordinary objects or subatomic particles, we must visualize that *a world-line signifies the progress of an object or particle in four-dimensional space-time. When such progress involves only uniform motion, the world-line is straight. When there is accelerated motion, the world-line is curved, as in the case of gravitational behavior such as falling and orbiting.*

There is nothing enigmatic about world-lines or world-points. For example a map with all of Marco Polo's voyages, dates included, shows his world-line, and if we pinpoint where he was at one specific moment we show a world-point. Similarly, if we note that we are in New York City at 4 pm we are specifying a world-point on our world-line. Thus *to exhibit a long-lasting event we use a world-line. To designate a momentary event we use a world-point.* Einstein called world-points "point-events."

In a narrow sense meaningful momentary events occur only where at least two world-lines meet or cross. For example Marco Polo's meeting with the Great Khan can be perceived as a crossing of Polo's and Kahn's world-lines in space-time. For our purposes a noteworthy event can be what happens at one world-point, or what happens along a world-line, or what happens at an intersection of world-lines of objects or particles. We shall soon find that thinking of events in these ways allows us to work with points in space-time, and this is essential in the mathematics of general relativity.

We see that the term "event" is used in two ways, and the distinction is often critical: An event can mean what happens at one point in space-time, in which case this event consumes infinitesimally little time and fills infinitesimally little space, and it corresponds to a world-point. "Event" can also refer to a longer occurrence, one which consumes a measurable amount of time and space; i.e. one which involves two distant points in space-time. We can consider a prolonged event to consist of a series of many infinitesimal adjacent events in space-time, which then form a world-line. When the distinction is significant, we will use adjectives such as momentary or brief for the former sense of "event," and we will use terms such as prolonged or time-consuming for the latter. (We can think of a prolonged event as a chain of world-points, but they do not interlock.)

Even though Einstein eventually embraced Minkowski's notion of space-time, he initially opposed the concept and saw no need for it. Nevertheless Minkowski developed sophisticated graphic methods to explain and illustrate space-time, and in one form or another his diagrams appear in most studies of relativity. Minkowski and Einstein jointly published some work in 1908 and 1909 while Einstein was turning his attention to general relativity. In fact Minkowski's contributions to relativity suggest that he might have pre-empted some of Einstein's ideas, but he died at age 45 in 1909. Incidentally, Minkowski had been one of Einstein's teachers who thought that Einstein was a bright but lazy student.[*]

One lesson we will glean from space-time diagrams is that if an observer cannot sense all four dimensions simultaneously, his observations will differ from another more perceptive observer.

---

[*]Like others who met Einstein in his youth, Minkowski failed to foresee any greatness.

Normally we heed two or three dimensions of space at the same moment, and we can also sense time, but we do not consciously sense four-dimensional space-time. An approximation we can make to visualize physical events and objects in four dimension is by "stepping back" and imagining that we watch them from afar over some period of time.

For example an observer spins a yo-yo around over his head while he is inside a uniformly moving elevator (and the elevator has transparent walls). Here we must be imaginative and think like relativists: *As the elevator rises, it is moving from the past to the future; it is moving through time.* We ask how the path of the yo-yo looks to this observer inside the elevator: Of course like a circle. But how does the same path of the same yo-yo appear to another observer who is looking into the moving elevator from afar? Like a helix or spiral. The distant observer sees the same event but in more dimensions! This means that as far as relativity is concerned, our ordinary three-dimensional perception is limited, and the way we use our natural senses does not give us a complete appreciation of physical reality.

At the same time we must recall that two observers in relative motion may disagree on their measurements of time and space (and also mass, which is not pertinent now). Is there a connection between their disagreements and their inability to sense four dimensions? Do relatively moving observers disagree *because* they cannot naturally "see" reality in space-time? *Might their disagreements be resolved if by some means the observers could "see" events as they appear in four dimensional space-time?* We further recall that the laws of physics should be written so as to accommodate the effects of uniform and accelerated relative motion. Might this goal also be served by considering events in space-time?

To answer these questions let us begin by reviewing the three graphic methods presented so far to illustrate relativity. First were diagrams A, B, and C which are unrealistic but suffice to show the relativity of time. Second were the more formal Cartesian diagrams and transformations which we detailed because they are part and parcel of the history and vocabulary of relativity. Third, Minkowski diagrams were introduced to depict space-time. We can also adapt and combine features of various kinds of diagrams to create what we will call Minkowski-like diagrams.

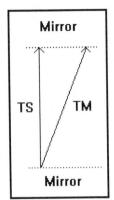

Line TS is the time needed for light to get from one mirror to the other measured in the stationary frame of reference.

Line TM is the time needed for light to get from one mirror to the other measured in a relatively moving frame of reference.

The difference between the lengths of the vertical line TS and the slanted line TM represents the discrepancy in measured time when an observer compares an event in a stationary frame of reference with the same or similar event in a moving frame of reference.

Diagram F. Measured time compared in two relatively moving frames of reference. Their motion is uniform. The event is a light beam traveling between two mirrors.

Let us reconsider our two trains outfitted with mirrors on their floors and ceilings. Diagram F depicts time alone. The length of lines TS and TM signifies the time needed for the light beam to "bounce" from one mirror to the other. Lines TS and TM thus represent time intervals—clock readings taken on two relatively moving events or taken by two relatively moving observers—even if it may be technically impossible to time a short light beam. To conform with convention, the coordinate axis for time is vertical. By virtue of relative motion, *TM is longer than TS*. With no relative motion, the two lines would be superimposed and have identical lengths.

In the case of accelerated relative motion, line TM would be curved but it would still be longer; for simplicity, this curvature is not shown in diagram F, and in fact when Minkowski introduced his ideas on space and time, he was concerned with the conditions of special relativity, namely uniform motion. Our point now is that diagram F illustrates that *an observer comparing two time-consuming events which appear synchronous at rest will detect unequal duration during relative motion between these events. Similarly, two relatively moving observers disagree on how much time the same event consumes.*

This diagram is similar to diagram F, but it represents distance (length) rather than time.

Line DS is the distance measured by a stationary observer.

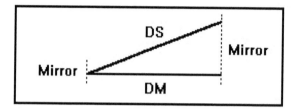

Line DM is distance measured by a relatively moving observer.

As in diagram F, the difference in measurement of distance is represented by the difference in the lengths of lines DS and DM.

Diagram G. This diagram is similar to diagram F, but it represents distance (length) rather than time.

Diagram F can show us more. As relative speed *increases*, the angle between the lines *widens*, and as TM tilts further clockwise, TS will shorten relative to TM. In our example, if the trains pass each other at higher speeds, the larger discrepancy in time-measurements is associated with a larger angle between TS and TM. We can show this effect by increasing v in the formulæ in diagram C, since the ratio between TS and TM includes the special relativity factor. Later we will explain the angle between TS and TM mathematically.

Diagram G depicts space alone. For simplification, only one spatial dimension is compared, and the "distance" coordinate axis is horizontal. Lines DS and DM represent space intervals, such as measurements on a ruler taken on two relatively moving events or taken by two relatively moving observers. Here *DM is shorter than DS*. Again accelerated relative motion would be shown by a curved line; this is not included in diagram G, but the effect is the same: *Objects of similar size at rest appear different during relative motion, or two relatively moving observers disagree on the lengths of one object*. Again the ratio and angle between DM and DS is linked to the special relativity factor.

This diagram is a compilation of diagrams F and G. Time and space are combined into space-time. Time is represented on the vertical axis and space on the horizontal axis. Line TS-DS represents the world-line of the beam of light. To make line TS-DS single, some geometric manipulation is needed, so that greater relative speed is depicted by clockwise rotation of BOTH axes. [Essentially, diagram F was turned mirror-image.] Line TS-DS also represents the time interval AND the distance (length) as these are measured by an observer moving with the beam of light as the beam travels from one mirror to the other.

Note the effect of increased relative speed: A distortion of time and space, without a change of the world-line in space-time. Time projected on the time axis lengthens, and space projected on the distance axis shortens.

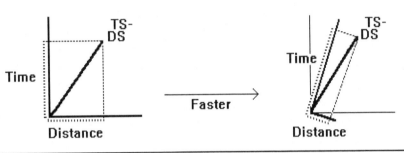

Diagram H. The effects of increased relative speed on the measurements of time *and* of space.

Diagram H is particularly important, as it displays the advantage of considering time and space together for our understanding of relativity. Again for simplicity space is reduced to one dimension. We transferred the information in diagrams F and G with some re-arrangement (the essentials in diagram F were reversed like a mirror image) so that the oblique lines TS-DS in diagram H are the world-line of this prolonged event, which is a light beam moving between two mirrors. *Greater speed results in clockwise rotation of the time coordinate axis as well as of the space coordinate axis.* (Of course these axis-rotations are imaginary. No observer can look at a physical event and actually see the coordinates or their tilting.)

In the lower-left corner of diagram H, the distance along the Time axis is the duration of the event as measured by an ordinary observer who *cannot* "see" the world-line of this event in four-dimensional space-time. This observer can only measure the *projection* of the world-line on the time axis. The way to picture this is by imagining that the world-line casts its shadow on the time axis, and *when the observer measures the amount of time consumed by the event, in effect he measures the projected length of this shadow.*

If an observer moves faster past the event depicted by TS-DS and compares his measurements with another observer, the result is shown in the lower-right corner of diagram H; the projected time along the new (rotated) time coordinate axis is elongated, so that the event appears to take more

83

time. Picturing this as a shadow of the world-line cast on the time axis, this shadow is longer. The same process holds for space, except that the ordinary observer detects a shortening of distance (space) at greater relative speed. The shadow of the world-line on the rotated coordinate axis is shorter. Clearly, *both time and length change with relative speed*; i.e. a moving prolonged event appears to take longer but cover less distance.

Now we should imagine that an extraordinary observer can somehow accompany a photon as it travels between the mirrors in our train, and this observer can measure the time as well as the length of his trip from one mirror to the other. He essentially follows the world-line of this prolonged event in space-time. (Some authors call him a "co-moving" observer.) The key feature in this scenario is that his experience of time and distance will be the *same* whether the train is at rest or in uniform motion; the tilt of the axes is of no consequence to him. This important point suggests that *measurements obtained within space-time are immune to the effects of relative motion* (even if it is difficult to imagine an observer accompanying a photon).

The same conclusion is clearer in a somewhat more reasonable scenario. We imagine an observer in a stationary train tossing a clock (e.g. a stopwatch) into the air so it just reaches the ceiling and then falls back into the observer's hand. The clock travels, say, up 2 meters, falls down 2 meters, and altogether takes 1 second to do so. In short, this event requires 4 meters of space and one second of time. We also imagine an extraordinary ant clinging to the clock, and we assume that this ant can vouch for these measurements. However *if the train is moving uniformly, the ant will confirm that the event appears the same—it still requires 4 meters and one second*, which are measurements of the ant's proper space and proper time.

Indeed no matter how fast the train moves, the ant's experience is undisturbed. In effect the ant "does not care," nor does the clock change its behavior, just because another observer outside the train detects the relativistic effects of relative motion, specifically that on his watch and ruler the event does not take 1 second and 4 meters. Still the external observer, whom we can consider to be an ordinary observer, does not know what the ant perceives—in particular he does not know that the ant's experience is impervious to the effects of relative motion. The external ordinary observer, who is not privy to accompanying the event, only knows what he can measure from his vantage. He gauges coordinate time and space.

No need to pursue this point, but with two relatively moving observers, *three* sets of measurements arise, one proper and two coordinate. E.g. in terms of time, the ant claims the event took 1 second, an observer moving with respect to the train at 80% of c claims it took 1.7 seconds, and an observer moving at 99% of c claims it took 7 seconds. (Solving the special relativity factor will confirm these numbers.)

We note that if we share the experience of observers traveling on a photon or on a clock through space-time, time and space are not altered by uniform relative motion. But if another observer who cannot accompany a photon or clock examines the same event, the measured intervals of time and space (on the Time and Distance axes in diagram H) are altered by relative motion. This concept is critical enough to reiterate: Even though measurements of time and space for an event have been changed by relative motion because the shadows of its world-line on the rotated coordinate axes are altered, *the ordinary observer does not perceive that the world-line of this event is unaffected*. Only his impression of the event is affected.

We should note that, obviously in theory, if a weightless observer (having zero rest mass) could accompany a photon moving at speed c, the observer would experience no passage of time. Indeed anything with the least possible rest mass, zero, must travel through space-time with the greatest possible speed, c. Please recall page 40 and preview on page 111.

However, these discrepancies in measurement appear when the non-accompanying observer compares his measurements with an imaginary extraordinary observer who was able to accompany the event, or as is more realistic, when two relatively moving ordinary observers compare their measurements. The same holds for one observer comparing his measurements of two similar events in relative motion. If relative speed increases, the turning of the coordinate axes increases and the discrepancies widen, but we re-emphasize that in four-dimensional space-time the world-line itself is *not* changed; it is immune; it is *invariant*.

This concept can be envisioned in another analogy, one that will be useful later: An object in a house casts a shadow on a wall so that if the object should fall to the floor, an ordinary observer looking at the wall can track this event by the motion of the shadow. Meanwhile something causes the wall of the house to tilt (without affecting the fall of the object). As a result of the wall's new posture, the shadow behaves differently, and the observer of the shadow can detect the difference. However if another observer, analogous to the extraordinary observer on a photon, watches the object itself, he finds no difference, because from his vantage the tilting of a wall has no impact on his experience. In other words the falling object itself does not reveal when the walls tilt unless someone watches the shadow.

In this analogy the fall of the object is a prolonged event in space-time, one with a world-line. A wall may portray space, and the tilting of the wall represents relative motion which causes a shrinking of the shadow as it is projected onto a coordinate axis. Although harder to imagine, another wall may portray time, and the tilting is such that relative motion appears as a lengthening of a shadow. However in both cases the motion of the object, representing the event itself, does not change. *In short, moving events appear to shrink in space and expand in time, even though in four-dimensional space-time the events themselves do neither.*

We ought not to overlook how simple this notion is: The appearance of a shadow changes when the surface on which it is projected is at a different angle. Hence if we cannot see the object directly, the appearance of its shadow can deceive us. The extrapolation of this notion is that *as long as we cannot "see" an event directly, the "shadows" we perceive can misinform us; our ordinary experience does not faithfully reveal what happens in space-time.*

Of course the extent to which we are deceived depends on how fast we move, but relativistic "deceptions" are normally very small, because compared to the speed of light we hardly move at all. Nevertheless the projection-of-a-shadow analogy supports the philosophical notion that our familiar perceptions are merely shadows of the real world, and that we are restricted to perceiving subjective and misleading images because we cannot observe objects and events directly.

The analogy is even more meaningful through the idea that each of us—each "ordinary" observer according to our terminology—possesses his or her own "walls" for the "shadows" of events, and whenever we use clocks and rules and similar devices to study relatively moving objects and events, we are merely determining *projections of duration and size on our walls*. Since there is relative motion between observers, the angular orientation of the "walls" is different for each observer, which means that the "shadows" visible to one observer are not identical to those seen by another. In geometric terms the effects of relative motion stem from seeing the same event *at different angles*. Observers disagree when these angles disagree. (In our context the "orientation" of something is the direction in which it faces.)

A key point needs emphasis here: The challenge of relativity is not just that we must deal with *four dimensions*; it is also that *our ordinary measurements are altered by relative motion*. In effect we face two problems, the amalgamation of time and space, and the relativity of time and space. In terms of space-time diagrams, the key issue therefore is not only *four* axes; it is four *tilted* axes. This means that precise knowledge of physical events begins with two basic considerations, *four dimensions and four tilts*. We will find that the mathematics of relativity addresses both. However we should add that further complexity will spring from the curvature to which we alluded. Not only can the four axes of space-time tilt; they can also curve.

Let us return to our diagrams. The two portions of diagram H can be combined into one, as in diagram I. To make it less crowded the rotation is shown counterclockwise, but this has no physical significance. Rather than labeling the prolonged event or world-line as TS-DS, we call it simply "EVENT." In this setting an *EVENT is nothing more than an interval between two points in space-time: a "space-time interval."* While we naturally do not think of physical events as space-time intervals, doing so is essential for the purposes of relativity. (Diagram I shows that the EVENT need not touch the vertex of the axes, as it does in H.)

Diagram I concentrates on the time-aspect of the EVENT and underscores three important points:

1.   Ordinary observers do not measure events, *only projections or "shadows" of those events*. The projections provide us with our everyday experience of such events. For example if we find that an event takes 1 second, all that we have found that this events is projected as 1 second on our time coordinate.

2.   As long as there is motion between observers, the appearance of an event *varies depending upon which observer* is doing the measurement. For example an observer in another frame of reference may find that this event takes 1.2 seconds. Why? Because his time axis is at a different angle.

3.   The event itself is unchanged despite the differences noted by moving observers. That is to say, our one subjective time-measurement and our three subjective space-measurements are *relative*, while *a four-dimensional space-time interval is invariant*. Soon we will show equations which quantify these ideas.

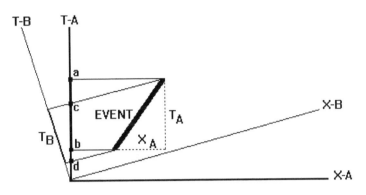

An event in space-time is measured by two observers who are in relative motion with respect to each other. The time and space axes of observer A are labelled as T-A and X-A. The corresponding axes of observer B are labelled as T-B and X-B. The time consumed by the event appears less to observer B $(T_A > T_B)$. The EVENT is a space-time interval.

A right triangle can be drawn so that $(EVENT)^2 = X_A^2 + T_A^2$.

Diagram I. Two observers in relative motion measure the time consumed by one event. The event, a space-time interval, is represented by the heavy line. We note that the EVENT need not touch the vertex of the axes. (We will use points a, b, c and, d later.)

Diagram J shows the same concept applied to simultaneity. We depict two momentary events as two dots (world-points A and B) placed next to each other so that they project as one on the time coordinate axis of an observer. Thus A and B are deemed to be simultaneous by that observer. From the vantage of another observer in relative motion, the coordinate axes are tilted—clockwise in this example—so that the two events are projected as separate points on the time axis of the second observer.

Again the locations of the two dots and the interval between them has *not* changed in space-time, but how the corresponding events are perceived—together or separate—is determined by the relative motion between the observers.

✦

We find that systems of coordinates and frames of reference are vital in relativity. First of all, since motion is never absolute, the study of motion implies a frame of reference; i.e., all meaningful physical events occur relative to some reference. Furthermore *altering a set of coordinate axes, while it portrays relative motion, is equivalent to a change in the frame of reference.* Thus in diagram I the axes labeled T-A and X-A represent the frame of reference of one observer, while axes T-B and X-B, which are tilted relative to T-A and X-A, represent the frame of reference of another observer. We can say that any moving event studied by two observers, in effect, occurs in two different frames of reference and is projected on two different systems of space-time coordinates. Furthermore, since events themselves are invariant during relative motion, *relatively moving observers differ only in that they experience events in different frames of reference and they measure these events using different coordinate systems.*

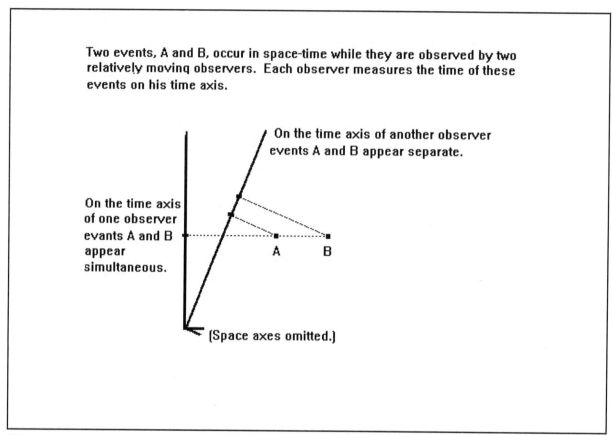

Two events, A and B, occur in space-time while they are observed by two relatively moving observers. Each observer measures the time of these events on his time axis.

On the time axis of another observer events A and B appear separate.

On the time axis of one observer events A and B appear simultaneous.

A   B

[Space axes omitted.]

Diagram J. The effect of relative motion between two observers on simultaneity.

We must also be aware that the very act of measuring events using different coordinate systems constitutes a transformation of coordinates. This means that *any case of relative motion is tantamount to a transformation of coordinates.* In this regard we recall that Lorentz showed how an event could be analyzed in Cartesian coordinates while viewed by two observers, A and B, in relative uniform motion with respect to each other. The Lorentz transformation is based on two systems of coordinates "sliding" past each other along their X axes at speed *v*, like two trains passing by each other on parallel tracks. This form of transformation is called "translation," and its two key equations, here solved for X_B and T_B respectively, are

$$X_B = \frac{X_A - vT_A}{\sqrt{1 - v^2/c^2}}$$

and

$$T_B = \frac{T_A - vX_A/c^2}{\sqrt{1 - v^2/c^2}}.$$

The first equation calculates how different the length of an object or event appears; X_A versus X_B. The second equation does the same for time; T_A versus T_B.

Meanwhile our space-time diagrams depict the effects of relative motion *by a rotation or tilt of coordinate axes* around the point where they intersect. They do not show a sliding or translation of axes seen in the Lorentz transformation. This point raises two questions: Does a geometric *rotation* of axes (as in diagrams H, I, and K) truly represent the physical difference between two frames of reference? If so, what are the corresponding equations? In reply, we will see that we can derive and verify *rotational* transformation equations *which contain R, the angle of axis rotation, which is the amount of tilting*:

$$X_B = X_A \cos R + T_A \sin R$$

$$T_B = T_A \cos R - X_A \sin R$$

Since this was worked out by Minkowski, we can call a rotation of axes the Minkowski-like rotational transformation. It suffices to detail the derivation of only the first rotational equation, as they are similar, but we need some familiarity with trigonometric ratios; see Appendix if necessary. For example we can re-examine diagram C and select the angle formed by cT_B and vT_B to be the angle in question. The sine of this angle is D/cT_B. Its cosine is vT_B/cT_B. Its tangent is D/vT_B.

Now we use diagram K, a modification of diagram I in which only one point, P, is considered; P might represent a momentary event. *Angle R is the angle of rotation.* Observer A finds P projected

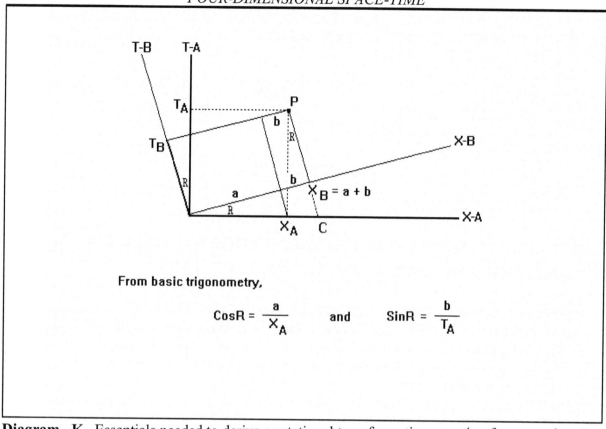

From basic trigonometry,

$$CosR = \frac{a}{X_A} \qquad and \qquad SinR = \frac{b}{T_A}$$

Diagram K. Essentials needed to derive a rotational transformation equation for one point. R is the angle of rotation. Note that X_B = a + b. Point C will be used later.

at X_A on his X-A axis and at on his T-A axis. B finds P projected at X_B and T_B on axes X-B and T-B. We let X_B consist of two parts, a and b. The equations for *cosR* and *sinR* (in the diagram) can be rearranged so that

$$a = X_A \; cosR \quad and \quad b = T_A \; sinR.$$

Since X_B = a + b, we easily reach our goal, the rotational transformation equation

$$X_B = X_A cosR + T_A sinR.$$

If angle R happens to be zero, its sine is zero (sin 0 = 0), eliminating T_A *sinR* and causing b to vanish. Simultaneously the cosine of R is 1 (cos 0 = 1), causing X_A *cosR* to become just X_A. This leaves X_B = X_A, showing that zero discrepancy between A's and B's assessment of P corresponds to zero rotation between A's and B's axes.

Our earlier diagram I applies the same concepts to a prolonged EVENT. Observer A deems that on his coordinates the EVENT needs X_A-amount of space and T_A-amount of time. In accommodating relative speed, B's set of coordinates tilt R degrees, and B finds that the same EVENT needs X_B-amount of time and T_B-amount of space. As the observers compare their measurements, they find that X_A is different from X_B and that T_A is different from T_B, unless R is zero.

Since v represents the "amount" of relative motion, the mathematical relationship between R and v—not surprisingly—hinges upon the special relativity factor:

$$cosR = \frac{1}{\sqrt{1-v^2/c^2}}$$

We must interpret this equation carefully. It spells out the connection between the rotation of axes and uniform relative speed, and of course speed has real physical meaning. The value of *cosR* rises with higher speed. However in this equation *cosR* can be greater than 1, which is impossible in "real" triangles and which means that this R is an "imaginary angle." (Cosine is defined as adjacent side/hypotenuse, but in a "real" right triangle, the adjacent side cannot be longer than the hypotenuse.) Indeed R is only an abstract geometric quantity in a space-time diagram, which in turn is an abstract geometric device to show how an event is projected on observers' coordinate axes. Still, we conclude that *R-amount of axis rotation faithfully represents the effects of v-amount of relative motion.* Thus in the absence of relative motion ($v = 0$), the special relativity factor is 1 and $cosR = 1$, in which case $R = 0°$; there is no rotation of axes. When $v = c$, which is the maximum speed, $cosR = 0$. Now $R = 90°$, which is the maximum rotation in this kind of diagram. (We can set up our diagrams and equations to avoid an "imaginary angle." For instance we can derive

$$cosR = \sqrt{1-v^2/c^2}$$

where *cosR* is always 1 or less than 1, but this denotes a different relationship between R and v.)

An alternative protocol for diagrams such as I and K has the T-B axis rotated clockwise rather than counter-clockwise. Then relative motion is shown by a converging of axes T-B and X-B, like blades of a scissor when cutting something. This system avoids imaginary quantities but introduces other complications. After Minkowski, others—notably Amar, Brehme, Loedel and Synge—revised space-time diagrams, again partly to avoid an imaginary angle. Loedel rotational diagrams, somewhat different than ours, are particularly precise illustrations of transformations. See Shadowitz, pp. 12-26; also Mook and Vargish, p 218. Lawden covers imaginary angles on pp. 10-11.

Let us recall our goal that the laws of physics appear alike to all observers, and we found that the Lorentz transformation helps secure that goal for inertial conditions. The Minkowski-like rotational transformation also has that attribute. *The same rotational equations tell an observer how this event appears in the coordinates of another inertial observer.* We note four other features of our rotational equations:

1.   These transformation equations are the same for either observer A and B; they are *form-invariant*, which is vital for using these equations in laws that are designed so as to appear alike to all observers.

2.   The mathematics for rotation is essentially the same in four dimensions as in two, but the Y and Z axes are included. Not surprisingly, to calculate, say, X_B, in the four-dimensional case, observer B needs contributions of data from *all four axes* of observer A. In fact one advantage of rotational transformations (over "sliding" transformations) is that how one observer sees an event can be easily expressed as a sum of contributions, or as we shall call it, as a sum of "partial" contributions. Therefore, when we get to the mathematics of general relativity, we will find rotational transformations with their angle R easier to use. We touched on this point on page 51 and we will apply it on page 269.

3.   The time coordinate works the same as a space coordinate. In other words in these transformations, time acts like a dimension.

4.   If we square the two equations

$$X_B = X_A cosR + T_A sinR$$

$$T_B = T_A cosR - X_A sinR$$

by means of algebra (the applicable expansion is $(C + D)^2 = C^2 + 2CD + D^2$) we obtain

$$X_B^2 = X_A^2 cos^2R + 2X_A T_A sinRcosR + T_A^2 sin^2R$$

and

$$T_B^2 = X_A^2 sin^2R - 2X_A T_A sinRcosR + T_A^2 cos^2R.$$

With trigonometry (the trig identity here is $cos^2R + sin^2R = 1$) these equations are simplified to

$$X_A^2 + T_A^2 = X_B^2 + T_B^2.$$

(The *T*'s are negative in most four-dimensional derivations, and we will come back to why.) By including Y_A, Y_B, Z_A, and Z_B, we can embrace four dimensions:

$$X_A^2 + Y_A^2 + Z_A^2 + T_A^2 = X_B^2 + Y_B^2 + Z_B^2 + T_B^2.$$

As we will discuss shortly, this result is critical. It affirms that the Minkowski-like rotational transformation respects the invariance of an event in four-dimensional space-time for two observers. It also confirms that a Minkowski-like "tilting" of axes corresponds to the "sliding" Lorentz transformation and that both transformations can describe the effects of relative motion.

The use of rotation in space-time diagrams raises a fine point: what if relative speed is so great that the tilt of the axes overtakes the line used to show the event? Here we invoke the principle that

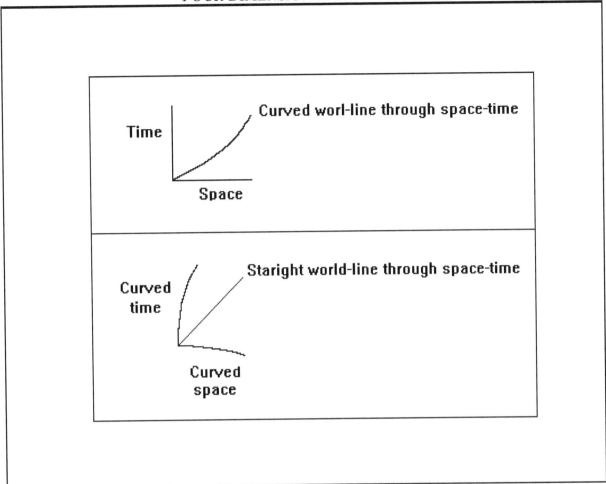

Diagram L. Two ways of representing accelerated relative motion: A curved world-line in unaltered space-time, or a straight (uniform) world-line in distorted space-time.

✦

speed beyond that of light is impossible. Customarily rotational diagrams are set up so that the world-line of a beam of light, which of course has speed c, is at 45 degrees. In this system the allowable rotation of the axes is 45 degrees or less. (We select units so that if a photon goes twice as far, it also "goes" twice as long. I.e. with a 45-degree world-line, a change in location in space is associated with an equal change in time.)

✦

We shall now consider an important detail of the rotational transformation: Our implicit assumption has been that the amount of rotation (angle R) is *constant*, and this corresponds to the stipulation that relative motion (v) is *uniform*. A corollary to this stipulation is that coordinate axes as well as world-lines are *straight*. However we must deal with the prospect that if relative motion is not uniform, angle R may be *variable*.

In other words our space-time diagrams—and later our equations—must provide for accelerated motion and for *curvature*. In that case we can draw a curved world-line and straight coordinate axes, but we can also re-draw the situation by *curving the time and distance coordinate axes*, so that the world-line is straight, as in diagram L. To show greater acceleration, we bend the axes more, but the world-line as it appears in four-dimensional space-time nevertheless can remain straight and unchanged; let us use diagram L. We emphasize "curvature" because various forms of curvature arise in the mathematics and geometry of general relativity.

Here we may challenge the idea that we can arbitrarily switch between two ways of designing space-time diagrams, but the justification goes back to the principle of equivalence between inertia and gravity. In the example of water in a spinning bucket, if we consider the universe to be stationary and the bucket to be in motion, a space-time diagram with a curved world-line and straight axes is more appropriate. If we consider the bucket to be stationary in a moving universe, a space-time diagram with a straight world-line and curved axes is more appropriate. But just as either view—a spinning bucket or a spinning universe—is legitimate in relativity, so is either graphic method. Hence we can safely consider the world-line of a prolonged event in space-time to be straight and unchanged, while *the coordinate axes for time and space are deformed*. We are getting ahead of ourselves, but there is a physical basis for considering world-lines straight and coordinate axes curved rather than in reverse, and this is related to the fact that objects tend to stay in straight motion. On the other hand we will encounter circumstances in which it is easier to consider the coordinate axes as straight and world-lines as curved. (We may ask whether all four axes must be curved. Can we allow the SPACE axes to be straight but the TIME axis to be even more curved, or vice versa? Do objects accelerate to cover the same distance in less and less time, or do they take the same time to cover more and more distance? In theory either view is legitimate, but we avoid this complication by envisioning that either all four axes are subject to curvature while the world line is not, or all four axes are straight while the world-line can curve.)

◆

At this juncture we should re-examine a situation of great interest in general relativity, free-fall, and we can link the notion of free-fall with the observation that our perception of physical reality depends on our vantage; we recall our analogy (on page 80) with the yo-yo inside a transparent moving elevator. We can use a freely falling but transparent elevator to reveal the difference coordinate systems and frames of reference can make on the aassessment of a physical event. Let us assume that this elevator contains an unattached object, say a suitcase standing on the floor. We imagine that an *outside* observer can measure the time and distance which an object (the suitcase) in the freely falling elevator traverses along time and distance coordinate axes. This observer finds the object to be in *accelerated* motion, and he may surmise that in his frame of reference a gravitational field accounts for his observations. Meanwhile an observer *inside* the falling elevator studies the behavior of the same object. He finds that on the coordinate axes of his own frame of reference the object has *no* motion and that its acceleration is *zero*. In particular, the suitcase seems weightless; if the internal observer picks it up and then lets it go, it will not fall.

94

Obviously the object in the elevator does not stop falling wherever the inside observer looks at it, and it does not re-start whenever the outside observer looks at it. In four-dimensional space-time the world-line of the object continues invariantly. Yet for each observer the difference between their frames of reference induces an striking disparity on the assessment of the same event. One observer sees the suitcase falling and heavy while another sees the same object as immobile and weightless. Only one observer witnesses gravity.

We surmise from this scenario that *gravitation is relative to the frame of reference in which it is observed*. From a local vantage (that of the internal observer) the effects of gravity are canceled, while from a wider more "global" vantage of the same event, gravitation is evident. However the key difference between these two vantages—between the two frames of reference—is relative accelerated motion; one observer accelerates along with the event, and the other does not. This conclusion raises an obvious question: How can the laws of physics be formulated to embrace the observations that physical events can appear different in different frames of reference? In particular, how should the equations for physical laws be structured so as not to yield conflicting results simply because of relative accelerated motion? We already have hints as to how to answer this question. The elasticity of our ordinary measurements of time and space during relative motion appears to be linked with our inability to directly discern events in four-dimensional space-time, yet intervals in space-time appear to be invariant. The implication is that to predict physical events consistently we must study them *as they occur in space-time*. Moreover we must admit *curvature* in the sense we discussed already.

We also recall that relative motion is tantamount to transformation between frames of reference and between systems of coordinates, and that we can treat space-time as a collection of points. Altogether this means that we can treat gravitational motion as a series of events at various points in space-time while each such point has its own system of coordinates. In this view, as we shall find later, relativistic mathematics of gravitation is largely a question of point-to-point transformations, since these tell us how events appear when we change coordinates.

These concepts give us a clear direction: *If relativity is to render the laws of physics the same in any frame of reference, even in frames with accelerated motion, then a guiding principle must be that upon four-dimensional transformation between frames of reference, intervals in space-time are invariant*. The traditional statement is that *space-time intervals are invariant under the Lorentz transformation*. Moreover our equations for such laws must accommodate curved world-lines and curved coordinates. An analogy we already used can be adapted here: We can think of space-time as the interior of a wobbly and pliable building which houses prolonged but invariant physical events, such as free-fall of objects, but we analyze motion point by point. We must find a way for us to "peek" into the building so as to bypass the effects of its tilting and curved walls. Then we can ensure that all physical events, together with their laws and equations, appear independent of any relative motion and hence appear the same for all observers.

We can also foresee how the notion of free-fall compliments the logical coherence of relativity. Special relativity predicts that when an observer studies physical events within an inertial frame of reference, he has no sensation of—nor can experiments reveal—the *uniform* motion of that frame.

Now we enter general relativity: When an observer studies events within a freely falling frame of reference, he has no sensation of—nor can experiments reveal—the *accelerated* motion of that frame. The effects of uniform motion vanish in an inertial frame of reference, and the effects of accelerated motion vanish during free-fall. Absolute uniform motion does not appear in the former case, and absolute accelerated motion does not appear in the latter.

Diagram I (page 86) reveals several mathematical points which we now need to consider. The oblique line represents a prolonged "EVENT," and for observer A the EVENT requires T_A-amount of time and X_A-amount of space. Observer A can apply the *pythagorean theorem* to say that

$$(EVENT)^2 = X_A^2 + T_A^2.$$

However, the EVENT, which after all is *an invariant space-time interval*, is unchanged by relative motion. Hence observer B can say *with equal certainty* that

$$(EVENT)^2 = X_B^2 + T_B^2.$$

We note that *the same EVENT can be treated as the hypotenuse of different right triangles for the pythagorean theorem.* Likewise the motion of the light beam aboard our trains is an EVENT, and it is not coincidence that we drew right triangles and used the pythagorean theorem in diagram C.

The above equations consider only two coordinates and two dimensions, X and T. Accordingly we add the other two, Y and Z, so that space-time is fully embraced, and there is no mathematical objection to applying *the pythagorean theorem to four dimensions*. Furthermore even in four dimensions the EVENT is equivalent to an invariant space-time interval, so that if observer A measures an event, the equation becomes

$$(EVENT)^2 = X_A^2 + Y_A^2 + Z_A^2 - T_A^2,$$

while observer B finds that

$$(EVENT)^2 = X_B^2 + Y_B^2 + Z_B^2 - T_B^2.$$

As was the case in two dimensions, observers A and B may disagree on the values of X, Y, Z, and T. In particular, in the presence of relative motion X_A may not be the same as X_B, Y_A may not equal Y_B, Z_A may not equal Z_B, and T_A will not equal T_B, but still the EVENT is the same in these equations. We note that algebraically the three spacial dimensions increase EVENT but the time dimension decreases it.

Here the space terms (X's, Y's, and Z's) are positive and time terms (T's) are negative, reflecting the reciprocal effect of relative motion; when A and B move with respect to each other, time lengthens but space shrinks, so that EVENT can be invariant. The basic reason for the negative time term is the constant speed of light, which causes time and space to be relative in opposite directions. Thus, to be precise, we need not say that "space and time make space-time." We are justified in saying that "space minus time makes space-time." At the appropriate juncture we will deal with the issue of negative time. (Since in Euclidean geometry the pythagorean theorem has all positive terms, we are now working in "pseudo-Euclidean" geometry, in which some terms are negative; see Kenyon, p. 26, and Pauli, p. 62. This point will also justify "pseudo-Riemannian" manifolds.)

There is a more pressing difficulty with the notion of EVENT if we are to employ the pythagorean theorem. This theorem only applies when a right triangle is located on a flat surface. As mentioned, an EVENT can be prolonged and can have a *curved* world-line, as in diagram L. This point raises an imperative question: *How do we use the pythagorean theorem on a triangle with a curved hypotenuse?* We have a way. First let us adopt the customary symbol for a space-time interval, the letter s. We can think of s as the amount of space-time between the beginning and the end of a prolonged EVENT, or as the separation between any two points along a four-dimensional world-line. In diagram J (page 88), the space-time separation between A and B is an s, and in diagram I (page 86), EVENT is an *s*. Thus,

prolonged EVENT = prolonged space-time interval = *s*.

Now we can write for observer A

$$s^2 = X_A^2 + Y_A^2 + Z_A^2 - T_A^2,$$

and for B

$$s^2 = X_B^2 + Y_B^2 + Z_B^2 - T_B^2.$$

Like EVENT, the s is invariant, but the problem remains that s can be curved.

Rather than being prolonged, an EVENT can also be momentary. We can use diagram K to enlarge upon this notion. Even though this diagram covers only two dimensions, P can be an EVENT which is infinitely brief in time and infinitely short in space; for example P may be only the start of the trajectory of a falling object. In that case, considering all four dimensions, the set of quantities X_A, Y_A, Z_A, T_A can represent *one point in space-time* as ordinarily detected by observer A; the where-and-when of P from A's vantage. Similarly X_B, Y_B, Z_B, T_B can represent the same one point in space-time as ordinarily perceived by observer B; these four quantities represent B's where-and-when of the same P. Here too, A and B may disagree; $X_A \neq X_B$, $Y_A \neq Y_B$, etc. (The symbol $\neq$ means "is not equal to.") Algebraically the three squared spacial dimensions increase the space-time interval *s* but the squared time dimension decreases it, as was the case for EVENT. Without the negative T^2 terms, the above equations would simply not hold. I.e., special relativity would be violated; an EVENT could precede its cause in some frames of reference. I.e., we cannot have a negative EVENT and, as we shall see later, we avoid a negative space-time interval s. (Basically the minus sign here has the same mathematical origin as the minus sign in the spacial relativity factor; Lorentz and Minkowski derived these equations with their negative terms.)

We will next distinguish in a mathematical way between prolonged and momentary events. We already considered any prolonged event as *a series or chain of infinitesimally short events*. To do this we take *the smallest possible segments* of s, so small in fact that *any curvature of s is imperceptible*, just like when we take a small step while walking, the overall curvature of the Earth is imperceptible. We call each segment of s "*ds*."

Small segment of s = *ds*.

Similarly we can preface *X, Y, Z, and T* with *d*'s. The above pythagorean formulæ are then re-written as

$$ds^2 = dX_A^2 + dY_A^2 + dZ_A^2 - dT_A^2,$$

$$ds^2 = dX_B^2 + dY_B^2 + dZ_B^2 - dT_B^2,$$

and again like EVENT and s, *ds*² *is invariant for both observers*. Again the time-dimension is negative, as noted above. We can also write a general equation,

$$ds^2 = dX^2 + dY^2 + dZ^2 - dT^2.$$

These equations will mean more to us later, but meanwhile we can think of each *ds* as an event requiring as close as possible to zero time and zero space—the shortest possible world-point or the smallest possible segment of a straight or curved world-line. Practically then, *ds* stands for a space-time interval short enough to be treated as the hypotenuse of a right triangle for the pythagorean theorem *even when s is curved*. Similarly the dX's, dY's, dZ's and dT's are the smallest conceivable space distances and time increments accessible, at least in theory, by ordinary observers. We can then break the path of a falling objects into tiny segments and express the length and duration of each by a set of dX, dY, dZ and dT. By dealing with such a prolonged event and its measurements in infinitesimal pieces, we will be able to handle the curvature of its world-line (though $ds^2 = dX^2 + dY^2 + dZ^2 - dT^2$ will need additional modifications to fully accommodate curvature).

We must emphasize that in our context *ds* describes gravitational motion in space-time. Thus an object may start at location

"X by Y by Z at time T"

and the object may fall to location "X-plus-a-short-distance-along-X by Y-plus-a-short-distance-along-Y by Z-plus-a-short-distance-along-Z, arriving there a-brief-moment-after T." *Having crossed ds-amount of space-time*, the object's location becomes

"(X + dX) by (Y + dY) by (Z + dZ) at time (T + dT),"

which is like saying the object was "on Fifth Avenue at 42nd Street on the second floor at four pm " and it moved through ds-amount of space-time to "Sixth Avenue at 43rd Street to the third floor at 4:01 pm." This means that when we ask "how does a falling object behave here," *we seek the solution of equations for ds*.

◆

The equation

$$ds^2 = dX^2 + dY^2 + dZ^2 - dT^2,$$

though based on little more that the pythagorean theorem, represents several major aspects of relativity. To glimpse why, let us make the following correlation: In three dimensions, if dX, dY and dZ are shadows or projections of a (very small) *object*, we can calculate its length in space by using the equation

$$ds = \sqrt{dX^2 + dY^2 + dZ^2}.$$

But in four dimensions, if dX, dY, dZ and dT are our ordinary measurements of a (very brief) *event*, we can calculate its *ds* in space-time by using the equation

$$ds = \sqrt{dX^2 + dY^2 + dZ^2 - dT^2}.$$

Meanwhile we found that the experience of two relatively moving observers A and B (or the measurements of two relatively moving events A and B) can be linked by transformation equations. Since our equations can be solved for the invariant ds^2, we can say that

$$ds_A^{\ 2} = ds_B^{\ 2}$$

and hence we can write a basic transformation equation showing that

$$dX_A^2 + dY_A^2 + dZ_A^2 - dT_A^2 = dX_B^2 + dY_B^2 + dZ_B^2 - dT_B^2.$$

Before proceeding we must address a detail we ignored so far: We must give each T-containing term a conversion factor, namely c, so that all terms are compatible. As a result dT_A^2 becomes $(cdT_A)^2$ and dT_B^2 becomes $(cdT_B)^2$. (More on this use of c later, but we note that dT is one quantity which can be squared, so that $(cdT)^2$ can be written as $c^2 dT^2$. [d^2T means something else.]) Upon this substitution,

$$dX_A^2 + dY_A^2 + dZ_A^2 - (c\,dT_A)^2 = dX_B^2 + dY_B^2 + dZ_B^2 - (c\,dT_B)^2.$$

We know that if we compare a (very short) time interval on clock A with clock B, then dT_A does not equal dT_B. Likewise if we compare a (very short) distance on ruler A with ruler B, then, if relative motion occurs along the X axis, dX_A does not equal dX_B. But we see that if the set of four squared measurements for A is compared with the set of four squared measurements for B, then the two sets of data *do* agree. In other words *moving observers differ on the quantity of time for events and on the quantity of space for events but not on the quantity of space-time for events.* For each momentary event there may be many different time intervals (dT's) and many different space intervals (dX's, dY's, dZ's)—in theory as many as there are observers—but *for each momentary event there is only one space-time interval (ds).* In short

$$dX^2 + dY^2 + dZ^2 - (cdT)^2$$

for any momentary event is invariant. That we are considering infinitesimally small segments of space and time is immaterial to this conclusion, nor does it matter that we consider a prolonged event to be a series of momentary events. Since both observers A and B find the same *ds* and the same c, the implication is that all inertial frames of reference are equivalent and that all observers in uniform relative motion can use the same equation—they can agree on the form of a law of physics. By inserting the terms of the Lorentz transformation (page 53), we can say that *all physical events in the universe with uniform motion appear such that*

$$ds^2 = \left(\frac{dX + vdT}{\sqrt{1 - v^2/c^2}}\right)^2 + dY^2 + dZ^2 - c^2\left(\frac{dT + vdX/c^2}{\sqrt{1 - v^2/c^2}}\right)^2.$$

If an observer measures an event with respect to which he is not moving, this equation holds because v is zero. (When $v = 0$, the first term in the right side becomes dX^2 while the fourth term becomes $(cdT)^2$. More to the point, if he measures an event moving along the X axis, this equation holds because v is relative uniform velocity. In fact this is the approach Einstein took to justify the Lorentz transformation in <u>Relativity</u>: <u>The</u> <u>Special</u> <u>and</u> <u>the</u> <u>General</u> <u>Theory</u>, pp. 118-120.

Recalling our comments on page 86, so far we have addressed three main issues, namely *relative motion, four dimensions*, and *coordinate axes*. Nonetheless a key element is still missing: We have not made sufficient provisions for accelerated motion; v is still constant and uniform, which means we still need to consider transformations of systems of curved space-time coordinates. In fact we will work hard to meet this requirement.

✦

The issue of relative motion is clearer with actual numbers and with all four dimensions made positive: Observer A says to B, "You and I are moving with respect to each other while we examine an event, namely an object falling inside a house. We can only measure the projections of the object on the walls and on the ceiling. However this house also has a 'time wall.' During this event I noted four numbers for the change in location of this object as I see the fall of this object projected on the walls and ceiling. I know that you have done the same for the same event as you see it. I find that projection dX on the south wall is 3 units of length; projection dY on the west wall is 4 units; projection dZ on the ceiling is 5 units; and projection dT on the 'time wall' is 3.87 units of time. From this I know that the object in the house moved a little more than 8 *ds* units:

$$ds = \sqrt{3^2 + 4^2 + 5^2 + 3.87^2} \quad or \quad \sqrt{65} \quad or \quad about \quad 8$$

(All terms are positive here for show, and quared negative numbers are positive.) "However, to you each wall and the ceiling are tilted through an angle (such as angle R, the magnitude of which depends on our relative motion). You find 2 for dX, 3 for dY, 4 for dZ, and 6 for dT, but using

$$ds = \sqrt{2^2 + 3^2 + 4^2 + 6^2} \quad or \quad \sqrt{65}$$

your result is the same, about 8 units. Thus while we are moving, we disagree on the projected measurements of distance and duration for this event, but *if we apply appropriate mathematics, we can agree.* You might call this event 'small-but-fast,' I might call it 'big-but-slow,' but we both call it 'ds' and we agree that it measures about 8 units.

We can lump the dX's, dY's and dZ's together as *SPACE*, and T can be *TIME*. In general

$$ds^2 = (EVENT)^2 = (SPACE)^2 \text{ with } (TIME)^2.$$

Then $SPACE_{my}$ does not match $SPACE_{your}$ and $TIME_{my}$ does not match $TIME_{your}$, but

$$(SPACE_{my})^2 \text{ with } (TIME_{my})^2 \text{ does match } (SPACE_{your})^2 \text{ with } (TIME_{your})^2.$$

Clearly a space-time interval, which we now express as

$$(SPACE)^2 \text{ with } (TIME)^2,$$

is invariant.

"We see that space is not the same for the two of us and time is not the same for both of us, but *space-time is the same for any of us.* When each of us considers all four of our private projections and uses the pythagorean theorem, we agree on the behavior of the object in the house. Our guiding principle, despite what common sense dictates, is that *time and space are relative but space-time is not.* A Lorentz transformation, with its special relativity favor representing relative motion, changes an interval of time and an interval of space but *not* an interval of space-time.

"After all, then, events in space-time are not so elusive; we *can* 'peek' into our impenetrable space-time house. It is not necessary for us to observe events directly; we need not 'see' the *ds*'s. It suffices that we have access to ordinary space measurements, like dX, dY and dZ, and to ordinary time measurements, like dT. With these data each of us can calculate what happens in space-time, represented by *ds*. In other words even though we each only measure our own projections (shadows) of events, the mathematics of relativity endows us with the power *to know and to agree upon a physical reality.* This concept parallels an appealing idea in philosophy: You and I see the world differently, but if we gather sufficient evidence and apply the proper method, we will not reach conflicting conclusions about one and the same truth."

✦

In this context we draw a parallel between the equivalence of mass and energy on the one hand and the unity of space and time on the other. Because of $E = mc^2$ we do not have conservation of mass and separate conservation of energy (page 44). We have a merged conservation of mass and energy. In particular, rather than

$$\text{``} m_{in} = m_{out} \text{''} \quad \text{and independently} \quad \text{``} E_{in} = E_{out} \text{,''}$$

we find

$$\text{``} m_{in} c^2 + E_{in} = m_{out} c^2 + E_{out} \text{.''}$$

Likewise the true picture is not

$$(TIME_{my})^2 = (TIME_{your})^2 \quad and \quad (SPACE_{my})^2 = (SPACE_{your})^2$$

but rather

$$(SPACE_{my})^2 - (TIME_{my})^2 = (SPACE_{your})^2 - (TIME_{your})^2.$$

Just as we must fuse matter and energy into one quantity, we must fuse space and time into space-time. Now the postulate that the laws of physics should appear the same to all observers follows naturally: One way or another all of physics deals in space, time, and mass (page 45), and even mass is described by the behavior of objects in space and time. This means that all valid laws of physics are either equivalent to or compatible with solutions of four-dimensional pythagorean equations for the space-time interval *ds*. We label such laws *Lorentz-invariant*. The ultimate reason these laws appear the same to all observers is that the space-time interval is the same to all observers in any frame of reference. In short, *the laws of physics can be invariant because ds is invariant.* (Please be forewarned: In Einstein's formulation of general relativity the term "covariant" arises.)

Several notes of explanation are in order here: The procedure by which we obtain *ds* from s, dX from X, dY from Y, etc. is in the realm of differential calculus, which was invented independently by Leibnitz (1646-1716) and Newton. This branch of mathematics is used on quantities which *vary* or *change*. A curve is a changing line, so differential calculus is used when studying a curved line. The basic idea is to subdivide a line into infinitely small parts, each of which can be analyzed mathematically and geometrically. Newton needed this calculus since his laws involve changing quantities. When differential calculus describes the rate of change of something, it gives its *"derivative"*; the process of *"differentiation"* finds a derivative. Often we describe a rate of change over time—for example the rate of change of velocity during a short period of time. In that case we say that we "differentiate with respect to time," and the resulting quantity contains the term dt (as we will see in the next paragraph). Another common process is differentiation with respect to space-time, in which case the quantity contains ds. Incidentally, such quantities look like fractions, but they are not.

Let us cite an example. When we move at a certain velocity (or speed), we change our location (or position) at a certain rate during an interval of time. Therefore *velocity is the derivative or the differential of location with respect to time.* Since location can be described on an x coordinate axis, we can define velocity v as $\frac{dx}{dt}$, and of course this pattern is important in relativity. Moreover differential calculus also allows derivatives of derivatives or differentials of differentials, a.k.a. *second derivatives*, and we will use these as well. For instance velocity itself can change, and the rate of change of velocity is acceleration. Hence acceleration a is defined as $\frac{dv}{dt}$, which can also be written as $\frac{d^2x}{dt^2}$. This means that the derivative of location is velocity and the derivative of velocity (also with respect to time) is acceleration—i.e. we have a derivative of a derivative, or the rate of change of a change. We can also say that velocity is the first derivative of location and acceleration is the second derivative of location. Other examples are a slope, which is the (first) derivative of a line with respect to distance (how steeply a line rises or falls), and a child's growth rate, which is the derivative of height with respect to time (how fast the child grows). Since children do not grow at a constant rate, their height is a consequence of a secondary derivative. In fact we can find examples of third derivatives, but these are rarely important in relativity. In any case we often abbreviate first and second derivatives by using commas. The first derivative of x can be written $,x$ (which could be velocity) and the second derivative of x can be written $,,x$ (which could be acceleration). However use of such abbreviations requires that we know the parameter, such as time or distance, with respect to which we are differentiating.

For readers new to differential and integral calculus (we shall use the latter shortly), let us digress into a familiar case: If we are to figure how long it takes to walk one mile at 2 miles per hour, we do not need calculus; knowing that $RT = D$ suffices. However the presumption is that our speed is constant, with no consideration of getting up to speed or slowing down. To encompass non-constant (non-uniform) motion—i.e. to handle acceleration—we need calculus, by which we splinter the mile-long trip into many small steps. If the steps are small enough, RT still equals D in each step despite non-constant speed. When we use this notion to work from the vantage of the whole trip toward finding the speed in any small step, we are using differential calculus; we are finding a derivative. When we work from the vantage of the small steps toward analyzing the whole trip, we are using integral calculus; we are integrating.

A way of picturing differential calculus in our present context is by imagining that we are looking at a curved world-line of an event in space-time under a microscope. If our microscope is strong enough, we will see only one small area—one "point"—in space-time, so that the overall curvature will not be apparent. (Similarly, by looking at a small area of the Earth, its curvature is not apparent.) How long or curved the world-line is does not matter. At each such point we can imagine a minuscule right triangle built from ds, dX, and dY. It is harder to picture this in four

103

dimensions, but mathematics allows us to include dZ and dT. The slope of the world-line at each point can be seen as the slope of a *ds* which is the hypotenuse of a right triangle (as in diagram I, page 86), and at each microscopic point the pythagorean theorem is thus adapted to calculate *ds*.

The reader may object: How can a microscope, even by analogy, reveal a small "area" of space-time, especially an area marked by a triangle? How small need an area be to be a point, and how can a microscope be used to "see" time? Here we must resist the temptation to envision something both four-dimensional and tangible. The procedure is entirely abstract. We must simply trust that triangles can be "point-sized"; that they can have four dimensions; that time can be one of those dimensions; and that any point in space-time can accept such a triangle. Unlike our minds, mathematics is not perplexed by such ideas. Thus the equation

$$ds^2 = dX^2 + dY^2 + dZ^2 - (cdT)^2$$

is valid even if it describes something unimaginable. Our infinitesimally small triangles, which we can call "differential" triangles, provide us with a geometric method by which to study space-time point by point, and if each triangle at each point is small enough, we can apply the familiar rules of geometry, such as the pythagorean theorem, *despite any curvature* of world-lines. However we must presume that our triangles populate a collection or a "mosaic" of infinitesimally small and infinitesimally close points with smooth and gap-free passage from one to the next, and if this is so, we have a "continuum" which we will define more completely when this term is pertinent.

Let us focus on the "line" in world-lines. When a line is straight, we have no special term for it; "straight line" will do. But when it resides on a curved surface, it is called a "geodesic." A straight line is the shortest distance between two points on a flat surface, and *a geodesic the shortest distance between two points on a curved surface*; both kinds of lines curve as little as possible. We already touched on this subject on page 21, and we will elaborate upon and apply this concept later, but at this juncture let us embark upon a detour into aspects of geodesics, world-lines, space-time intervals, and causality.

Geometrically speaking, *the world-lines in our Minkowski-like diagrams are geodesics*. In space-time, individual points (world-points) along geodesics represent momentary events, and *longer events, such as the falling of objects, are represented by the geodesics themselves*. Although it is not immediately obvious, relativity restricts our notion of a geodesic, so that not all geodesics in space-time are the same. Let us amplify: In Newton's and Galileo's physics all events are deemed to be either past, present, or future, and there is no reason any observer cannot detect any past or present event. Relativity is more complicated: It asserts that *not every past or present event is accessible to every observer* and, more significantly, *not every cause can have effects*. The key to this idea is the limited speed of light together with the notion of a space-time interval.

Let us imagine a pair of momentary events. Event A is the explosion of a star, and event B is an observer taking a photograph of the sky. First we assume that events A and B are close enough so that the light flash (photons traveling at speed c) can leave A and reach the site of B before B ends, in which case the photograph will show the explosion. That is to say, event B can be reached from event A on a geodesic in space-time without traveling beyond the speed of light. The interval between A and B is then called a "time-like" interval, and the path of light from A to B forms a "time-like geodesic." Actually it matters whether the gap between A and B can be crossed at only the speed of light or it can be crossed by traveling at less than the speed of light, but let us temporarily consider these two cases as one; a time-like interval can be crossed without exceeding c. In other words, for now, no matter how A affects B, the maximum speed for this to occur is c. The significance of this stipulation is that *we must define our notion of causality: Effects follow causes only when they are close enough.* By "close enough" we mean in space and in time, and we mean close enough for something from A to reach B before B ends. We call that "something" a "signal," which here is a flash of light moving geodesically in space-time. That is to say, *in the "time-like" case a causal link between A and B is possible*, and an observer at B can determine that A occurred. Rucker (p. 86) uses the apt statement that "B lies in A's future."

We can also use a simple analogy: A gun was fired and a person was shot. We cannot assert whether that gun could have been the causative weapon for the resulting wound without three facts: the distance between the gun and the victim, the time interval between the shot and the hit, and the speed of the bullet. In other words we need to know space, time, and speed (how much space can be covered per time). For instance if the two momentary events, gun fired and person hit, were 100 meters apart, we cannot be certain that "person hit" was the result of "gun fired," even if we also know how long it takes to cover the distance. The bullet might cover the 100 meters at 1,000 meters per second, but if the victim was hit in 1/100 of a second, then this was not the causative gun; 1/100 of a second was not enough time. And if the two events were separated in time by, say, 1/10 of a second we again cannot be certain that they are causally linked. I.e., this gun was not causative unless it was close enough (100 meters) to the victim to cover the distance in 1/10 of a second. We can say that the bullet—representing a signal—has a fixed and limited speed, and likewise *causality has a fixed and limited speed*. (In this analogy 1,000 meters per second represents c, but we ignore the question whether bullets must follow geodesics.) This means that if the bullet had a speed of 1,000 meters per second, it can only cause certain victims to be shot—for example only those who are at 100 meters and are hit in 1/10 of a second. In short, because the bullet has a set speed, *only certain events can share causality*.

In particular, if the speed is 1,000 meters per second, then the combination of, say, 99 meters and 1/10 of a second is a time-like interval; it is possible for the bullet to have linked the two events, the firing of the gun and the hitting of the victim. This idea is intimated in

$$(space\text{-}time\ interval)^2 = (SPACE)^2 - (TIME)^2.$$

Here the bullet acts as a "signal" in space-time, and causality between two events exists only when a "signal" can link them. When the "signal" from A can reach B, $(TIME)^2$ is numerically greater than $(SPACE)^2$ because there is ample time to cover the space-time distance—hence the term "time-like." We note that because we consider *SPACE* to be positive and *TIME* negative, the value of this "$(space\text{-}time\ interval)^2$" turns out to be negative. A negative space-time interval presents a dilemma which we will address shortly.

But what if momentary events A and B are so far apart that light or any "signal" from A cannot reach B before B occurs? Such an interval is called "space-like" (or spacelike), and no "signal" from A is fast enough to influence B, even if that signal travels at speed c. As far as event B is concerned, event A does not exist; B cannot appear in A's future; A cannot cause B, and in the case of the exploding star, the photograph taken at B shows no explosion at A. *The space-like geodesic prevents causality.* In other words if two events are separated by a space-like interval, there is not enough time for the amount of space between them (or too much space for the amount of time) unless a "signal" has velocity greater than c, which is incomprehensible in the realm of special relativity. Likewise, since no signal or material object can exceed the speed of light, no material object can have a space-like world-line in the eyes of special relativity. The eyes of quantum mechanics may see this otherwise, and we will return to this issue.

In our bullet analogy, the combination of, say, 101 meters and 1/10 of a second is a space-like interval. No 1,000-meter-per-second-bullet can be fired from the gun and hit the victim at this interval. There is too much *SPACE* or not enough *TIME*. However we note that by virtue of

$$(space-time\ interval)^2 = (SPACE)^2 - (TIME)^2,$$

a space-like world-line is associated with a positive $(space-time\ interval)^2$.

When we speak of "signals" between events, we include all forms of energy or matter, and that is what makes this concept so profound; it applies to *all* events. For example signals for causality can be electromagnetic or gravitational, they can be sound waves or heat, and they can be tangible missiles—even bullets. The ubiquitous and most prompt "signal" we can obtain is via light, and we can call this the "first signal." In the case of the exploding star, a resulting event is the arrival of a cluster of photons—the reception of a "first signal." However it may also be a the detection of a ripple in space-time because of the change in the distribution of the star's matter, it may be the hearing of a "bang" if an atmosphere exists, and/or it may be the arrival of a fragment of the star. Although these cases are very different, the space-time interval must at least allow a "first signal" to bridge the gap. If the interval fails to do so, it is space-like.

Space-like intervals are not just hypothetical. Let us consider two events in one frame of reference which occur exactly simultaneously but at some distance from each other. For example two stars are each 100 million miles from a non-moving observer, and they are 1 million miles from each other. They each explode, and the observer perceives the two flashes at the exact same instant. If nothing bent the paths of light, the observer can assert that the simultaneity of the flashes must be coincidental. They cannot be causally related to each other because no signal can cross space (the million miles between the two stars) in zero time. The physical limit on c prohibits one star from having triggered the other explosion.

Since the two events (the explosions) are not causally linked, relativity permits observers in other frames of reference to detect these two flashes in any sequence. That is, in a space-like circumstance there may be relatively moving observers who claim that a certain flash came first, others who claim that the other flash came first, and still others who claim the flashes were simultaneous. All three are possible if, and only if, the interval between the two initial events prohibits a cause-and-effect link between them. In our bullet analogy, if a non-moving observer sees gun A fired and victim B 100 meters away being hit at the same instant, then he knows A cannot be the culprit for B. The A-B interval is space-like, and the firing of gun A and the hitting of victim

B are not causally linked. However in this case nothing bars another observer from detecting B hit before A is fired, B after A, or B with A.

On the other hand if two events are indeed causally linked (e.g. one star disturbs the other, or a gun shoots a victim, or a fire starts and goes out), they can never appear exactly simultaneous to a non-moving observer, and *in no frame of reference can they appear as reversed*—i.e. in time-like circumstances no observer anywhere, moving or not, sees the flash before either star exploded, or sees the victim hit before the gun is fired, or sees a fire go out before it started. Let us use diagram J on page 88 to illustrate this point. Assume B is a momentary event caused or influenced by A. We know that space and time are coupled because as a moving event appears longer in duration, it appears shorter in size. Therefore if a signal crosses any space (e.g. if a burst of photons, or even a bullet, moves from A to B) it must consume some time—unless the signal moves infinitely fast, which nature prohibits.

What if two observers, C and D, in motion with respect to each other detect causally linked events A and B? Although in effect a set of coordinate axes is rotated as in diagram J, the causal link from A to B is invariant; rotation does not change the A-to-B interval in space-time; since that interval was time-like, it remains time-like. However relative motion changes the way observers perceive A and B. Observer C may "see" A and B as simultaneous—projected as one point on a time axis—but then that observer will not "see" A and B in the same place, as A and B must be projected as two points on a space axis. Or observer D may "see" A and B at the same place, but then they cannot appear at the same time—A can only appear before B. In fact no observer can be found who "sees" A and B at the same time *and* at the same place, and no observer can be found who "sees" cause A *after* effect B. We should add that since observers themselves are limited in their maximum velocity (after all, as we noted earlier, observers are "objects" too; they cannot exceed c), observers themselves can only move on time-like geodesics. Of course any clock or other time-measuring device carried by such observers will manifest each observer's proper time.

Meanwhile relativity maintains that our ordinary perception of events is unreliable because we only assess our subjective space and time axes. For instance if an observer sees momentary event A before B, how can he distinguish the two possibilities? Is he seeing two causally *un*related events which happen to be projected on his time axis as two points? Or has A caused B, and there exists no time axis on which B might be projected before A? In order to discriminate these two possibilities the observer must "peek" into space-time by solving

$$(space-time\ interval)^2\ =\ (SPACE)^2\ -\ (TIME)^2,$$

which in theory means by measuring A-to-B along four axes so as to give dX, dY, dZ and dT, and then solving

$$ds\ =\ \sqrt{dX^2\ +\ dY^2\ +\ dZ^2\ -\ (cdT)^2}.$$

If $(TIME)^2$ is numerically greater than $(SPACE)^2$, i.e. if

$$(cdT)^2\ >\ dX^2\ +\ dY^2\ +\ dZ^2,$$

then the space-time interval A-to-B, expressed as *ds*, is time-like, and the observer knows that it is physically possible for A to have caused B. In other words because "signals" moving from cause to effect need time to do so, the ultimate reason we have causes and effects in nature is that there is enough time to transfer information. Or we can say that the reason effects follow causes is because nature provides signals, such as photons, which are fast enough to do the job, but not too fast. If signals were infinitely fast—if special relativity allowed additive velocities—we would see causes simultaneous with or even after effects. Indeed under such conditions all causality could be instantaneous, and the entire history of the universe might have occurred in one instant. On the other hand if signals were too slow—if enough intervals were space-like—nature might grind to a halt. Moreover relativity is sometimes imputed to allow time reversal, but the opposite is the case: The sequence of causality—effects follow causes—and the promptness of causality—how soon effects follow causes—are set by space, time, and the speed of light. In this way relativity ensures the orderly perception of the sequence of causally-related physical events. In particular the existence of time-like intervals in space-time allows effects to follow causes in logical fashion, and it even explains why the universe took so long to develop.

Such logic and order were absent in pre-relativity physics: On the basis of Newton's law of gravitation, it seemed that falling objects should move faster and faster to reach infinite speeds; there was no reason to suspect an upper limit such as c for anything, even a "signal." The implication was that two distant gravitational events can be instantly linked. For example let us imagine that a large meteor suddenly lands on the moon (event A) and that this significantly increases the mass of the moon. Newton's law suggests that a tide on Earth (event B) should rise at once; the time interval between the causally-related events A and B should be zero. In other words this law expects instantaneous "action at a distance."

We reiterate this point for good reason: Newton's equation

$$G = K \frac{m_A \, m_B}{D^2}$$

has a provision for the mass of the moon (call it m_A), for the mass of the Earth (m_B), and for their separation (D^2), but it has no provision for the time it takes for a signal to get from m_A to m_B across D; it says nothing about the possibility of a fourth dimensions nor of relative motion; *T* is not in the equation. This means that at least for a moment of time, Newton's equation for gravitational force is not reliable in both locations. In other words Newton's law implies that the values of the *m*'s are known simultaneously. Likewise, the D in the equation dictates that any observer know where the two masses are simultaneously. However we know that observers may not agree on mass nor on simultaneity. Therefore this law is incontrovertible only in frames of reference in which the two masses or observers are at relative rest, and the equation for this law does not hold true in *all* frames of reference. Yet as far as we can observe, the laws of nature appear the same everywhere. Thus, so as to account for the logical causality of gravitational behavior, today's physics rejects Newton's instantaneous force and looks to relativity for the clarification.

Nevertheless certain experiments on subatomic particles in the realm of quantum mechanics suggest that some "signals" can pass between such particles faster than at the speed of light. (In the chapter covering quantum mechanics we will emphasize that relationships between subatomic events are guided by probability rather than causality, and that these relationships may appear to take no time. As a preview, if two causally related events act as if they are adjacent to each other, causality can appear to be immediate.) Still, *material objects and signals which participate in observable events follow only time-like geodesics through space-time.*

◆

We should now consider certain objections. First we may argue that perhaps light (or any "signal") can just turn back, but this violates the conservation laws: The "signal" of causality is made of matter and/or energy, so that whenever one event causes another, some transfer of mass and/or energy occurs which obeys the conservation laws. Thus when the sun sends us photons, it gives up some of its mass/energy; not only do we become warmed and the sun cools, but the sun becomes slightly lighter and we become slightly heavier. This even applies to gravitation. As the sun affects the surrounding gravitational field so as to keep us in its orbit, it surrenders part of itself, and it becomes even lighter. This implies that objects radiate particles of gravity just as they radiate photons, and we will cover this issue in a later chapter. More to the point, we know of no case in nature where causality is *not* a matter/energy-transferring process that obeys the conservation laws. In fact we can quantify the energy transferred in the einstein red-shift, and the result shows transfer *only from cause to effect in orderly fashion.*

A broader objection is this: Even if light cannot "turn back," perhaps the space and time axes can be so curved, as in the letter "U," that light is sent back. In that case cause-to-effect signals might make "U-turns." The reply is that there is a limit to how bent the axes can be and still allow effects to flow from causes; this phenomenon appears in black holes. In theory, just as a black hole prevents the escape of light, it prevents the "escape" of all signals; more on this later.

◆

Let us reconsider the issue of the "(*space-time interval*)^2" along time-like geodesics. (We still multiply *TIME* by c so that cT appears in units of distance, as do X, Y, and Z.) The equations with positive (*SPACE*)^2 and negative (*TIME*)^2 can then be written as

$$(space\text{-}time\ interval)^2 = ds^2 = X^2 + Y^2 + Z^2 - (cT)^2.$$

What if an object moves with a speed less than c? The value of c does not change, but less space-time is crossed. This of course is the circumstance associated with our time-like interval, wherein

$$(cT)^2 >> X^2 + Y^2 + Z^2,$$

and most natural objects do cover much less distance per time than a photon. However in this case the value of (*space-time interval*)^2 is negative, and whenever we take the square root of ds^2, we find ourselves in the awkward position of dealing with an imaginary number. (The square root of a negative number is an imaginary number.) This implies that the object is on an "imaginary" geodesic, which is a conclusion we wish to avoid, since a natural time-like interval, with real physical meaning, should not be mathematically imaginary.

To circumvent this problem, we may make the space-values negative and time positive, so that

$$(space\text{-}time\ interval)^2 = (cT)^2 - X^2 - Y^2 - Z^2,$$

or

$$(space\text{-}time\ interval)^2 = (TIME)^2 - (SPACE)^2.$$

We now have negative *SPACE*, which, though peculiar, is mathematically legitimate. In other words the price of having these equations yield a mathematically real space-time interval is considering space to be negative. Indeed depending on the requirements, we often set up these equations in this fashion. Later we will see another mathematical remedy to the dilemma of "negative space" or "negative time."

✦

Let us now consider the possibility that momentary events A and B are at a distance so that *only* something capable of the speed of light can link them causally. That is to say, the only "signal" that can bridge the A-B interval is one with velocity c, the paragon of which is a photon. This interval is called "light-like." The light-like interval is important in that both electromagnetic and gravitational cause-to-effect processes propagate at c, but it has additional meanings which we will now examine.[*]

The possibility of a light-like interval has a remarkably simple mathematical basis. The ordinary pythagorean theorem can be written as

$$s^2 = X^2 + Y^2 + Z^2.$$

This equation can be applied to a real physical event, a flashbulb emitting a burst of photons. We can designate the moment the flash occurs as $T = 0$. The photons form an ever-enlarging sphere which grows at speed c. Geometry tells us that in Cartesian coordinates the radius (r) of a sphere is such that

$$r^2 = X^2 + Y^2 + Z^2.$$

If we allow time T to pass, the radius of our light-sphere is cT (it is a distance like s), in which case

$$(cT)^2 = X^2 + Y^2 + Z^2.$$

This can be rearranged to yield zero is two ways,

---

[*] Time-like, space-like, and light-like intervals can be displayed graphically by using "light-cone" diagrams, but this author has elected not to do so. Versions appear in many sources, including Shadowitz, p. 74, and Penrose, p. 194.

$$X^2 + Y^2 + Z^2 - (cT)^2 = 0$$

or

$$(cT)^2 - X^2 - Y^2 - Z^2 = 0.$$

We note that the sum of the space terms $X^2 + Y^2 + Z^2$ equals the time term $(cT)^2$. In other words in the case of light, the numerical value of $(SPACE)^2$ is neither larger nor smaller than that of time $(TIME)^2$. Rather,

$$(SPACE)^2 = (TIME)^2,$$

which, incidentally, is why we set a 45-degree limit on how far the axes of a space-time diagram can be rotated and why we give light a 45-degree world-line in such diagrams. Here the space-time interval, which we call the light-like interval, is zero, as in

$$(SPACE)^2 - (TIME)^2 \quad or \quad (TIME)^2 - (SPACE)^2 = zero.$$

This *zero space-time interval* is also called the case of the "null" interval, and it is associated with a light-like or "null" geodesic. It means that *a photon, which of course travels at velocity c, always follows a null geodesic.* We note that now we make a distinction between ordinary objects, which cannot reach velocity c, and photons, which do so routinely. The former's world-lines are limited to time-like geodesics.

An implication of a light-like or null geodesic is that if an observer could travel at the speed of light, he would experience *no passage of time*; proper time would be zero. (Pages 40 and 85 please.) It does not matter how long the trip appears to another observer. For this observer the interval between every event on his world-line is a null interval. We note that the speed at which this is expected to occur is not infinite, as common sense might suggest—it is c. Indeed when we think it through, another instance of the logical coherence of relativity is revealed: Any clock or other time-measuring device which accompanies an observer a geodesic reveals proper time, and *the quickest possible trip, which needs zero proper time, is the one with greatest possible speed, which is c.*

The notion of zero proper time addresses the question how human beings can live ling enough to make intergalactic trips. One answer is to reach speed c as much as possible, which minimizes "proper aging". Of course in practice we still cannot harness enough energy to accelerate a vehicle to such speeds, but at least relativity provides the theory. This notion likewise applies to the "twins" scenario (page 20) in which a fast-moving twin ages less: If he could move at the maximum rate, which is c, he would age at the minimum rate, which is "null." His elapsed proper time would be zero; he would not age at all compared to his stay-put twin!

The light-like interval also confirms the constancy of the speed of light, and the mathematics is far simpler than the proof based on Maxwell's equations. In general for any kind of interval in differential form,

$$ds^2 = dX^2 + dY^2 + dZ^2 - (cdT)^2.$$

For the light-like interval, the terms $dX^2 + dY^2 + dZ^2$, which together represent the *SPACE* crossed by a photon, must equal $(cdT)^2$, which represents the *TIME* consumed by a photon crossing that interval. In that case

$$dX^2 + dY^2 + dZ^2 - (cdT)^2 = 0,$$

which means that in the case of light, we have a null ds:

$$ds^2 = 0.$$

Of course speed is defined as Distance/Time or $\dfrac{SPACE}{TIME}$. When $SPACE = TIME$,

$$c^2 = \frac{dX^2 + dY^2 + dZ^2}{dT^2},$$

which means that c is a constant, and which also neatly reiterates our conclusion on page 44:

$$c^2 = \frac{SPACE^2}{TIME^2}.$$

We see that the brief equation $ds^2 = 0$ is actually the general-relativistic analog of the second principle of special relativity. *It indicates that the speed of light in any inertial frame of reference is the invariant c.*

We may compare this argument with that on page 40, where we concluded that because

$$m_r = 0$$

for photons, the velocity of light is c. Here we assert that because $ds^2 = 0$, the velocity of light is constant. We see that we can reach the second postulate of special relativity by two paths, via

$$E = mc^2$$

and via

$$(space\text{-}time\ interval)^2 = (TIME)^2 - (SPACE)^2,$$

not to mention via Maxwell's equations.

We should add that only something like a photon crosses *SPACE* and *TIME* proportionately to keep the ratio $\frac{SPACE}{TIME}$ the same. However the effect of ordinary matter on space-time is such that measured *TIME* lengthens but measured *SPACE* shrinks. This argument corresponds to the notion that in theory any object—photon or otherwise—brought to speed c will show the same speed for all inertial observers. The faster an object moves, the smaller is the effect of the motion of its source, and if an object can reach c, the motion of its source makes no difference at all. In short, all photons have maximum velocity for all inertial observers. Hence photons are maximally "exclusive."

We also note that the notion of invariant space-time intervals only applies in four dimensions, which leads to a further important deduction from the null-geodesic nature of light: A photon follows a geodesic in four dimensions, not in three, and we recall that a geodesic is the shortest possible path. *Since a curved world-line—one associated with accelerated motion—forms a geodesic, the path of light must be curved by a gravitational field, and four dimensions are required to define gravitational geodesics.* Einstein used this line of reasoning (p. 92 of The Meaning of Relativity) to predict that a light beam must be curved in a gravitational field, and it led him to conclude that space-time gives the gravitational field its structure; more on this shortly. Moreover the path of *any object or particle* is curved by a gravitational field, and four dimensions are needed to fully describe any physical event. This means that the concept of geodesics in four-dimensional space-time is essential to physics in general as well as to gravitation in particular. In short, photons travel on geodesics in space-time, but photons are particles, so *everything travels on geodesics in space-time.*[*]

Finally we note that time-like, space-like, and light-like intervals arise in the context of uniform motion and straight coordinate axes. However we cannot single out inertial observers as the only ones whose universe appears orderly. As we shall see, the principles of causality hold even if motion is accelerated and axes are curved, though the math is far more complex. That is to say, if a gravitational field bends the paths of photons in a predictable manner while photons are our "signals," then the kind of space-time intervals (time-, space-, or light-like) safeguards our experience of orderly causality among all physical events.

We summarize our detour: We encounter three kinds of space-time intervals. The significance of time-like intervals is that natural cause-to-effect processes are mediated along time-like geodesics, and that ordinary moving objects and observers follow geodesics which provide enough time to cover their space. The significance of space-like intervals is that immediate causality between distant events is prevented. The significance of light-like intervals is that cause-to-effect processes can be mediated along light-like geodesics, as in the case of electromagnetic and gravitational events. However nature prohibits a reversal of cause-and-effect relationships, and we must heed all four dimensions and the maximum-and-constant velocity of light in order to understand causality.

✦ ✦ ✦

---

[*] The idea that light travels along the shortest possible path is hardly new; While studying the reflection of light, Heron of Alexandria (a.k.a. Hero) proposed this notion in the third century B.C. If he had known that light can behave like particles, he might have surmised even then that everything travels along the shortest path, one which we now call a geodesic.

Let us step back a moment and review four ideas Einstein and his colleagues had assembled: Physical events as they appear to ordinary observers using every-day measurements of space and time are altered by motion; hence relativity. Space and time should be considered together; hence space-time. Physical events can be drawn in coordinate systems in which time is treated like a fourth dimension; hence space-time diagrams and four-dimensional pythagorean equations. Lastly, physical events as they appear in four-dimensional space-time are themselves not altered by relative motion; hence the invariant interval.

We emphasize the concept that time, space, and mass are perceived differently by different observers because of relative motion, be it uniform or accelerated, but in four-dimensional space-time, all appears the same to all observers. We can therefore say that in relativity, only the laws of physics, the speed of light, and events in space-time are absolute, but the word "absolute" is out of place in a discussion of relativity, so we use "invariant." However Einstein himself, in illuminating the distinction between Newtonian and relativistic absolutes, used the Latin phrase *continuum spatii et temporis est absolutum*: The continuum of space and time is absolute. (The Meaning of Relativity, p. 55.) But if crucial parts of nature appear invariant, why call this system *relativity*? As mentioned, this term was coined in French by Poincaré in 1904 and was later applied to Einstein's thesis by other scientists. Einstein himself initially favored a term which translates from German as "invariance theory." Clearly while Einstein stressed the invariables which he appreciated in an orderly and rational universe, others were more impressed by the relative facets of nature that he had discovered.

We skimmed over the contribution of Descartes, who gave us a powerful tool for studying physics and geometry. The key feature of Descartes' "Cartesian coordinate systems" is that they allow us to assign mathematical symbols and numerical values to geometric locations—i.e. to solve geometric problems with algebra and algebraic problems with geometry—as we do with the pythagorean theorem. In fact coordinate systems have become the framework in which the laws of physics are expressed. For example parts of special relativity can be explained in Cartesian coordinates. The unstated assumption had been that the configuration of these systems does not change during motion, until relativity suggested that *motion itself* basically alters the coordinates on which we study physical events. That is to say, *the very act of moving influences any system of coordinates we use to observe and measure physical events*.

The skeptic might now point out that relativity merely tampers with the constancy of time and space so as to accommodate discrepancies in our ordinary measurement of events. The relativist will reply, *yes*, precisely. In order to reflect reality, *space-time must be elastic*, and it can be measured in *elastic coordinates*. The logic is compelling: Moving observers or observers of moving events find that motion itself distorts their measurements of space and time, but simple geometry tells us that motion does not distort intervals or events in space-time. The implication here is spectacular and is in fact a key assumption of general relativity: The relativistic effect of uniform or accelerated motion on our measurements of space and time is mediated by *an alteration of four-dimensional space-time*. Motion distorts space-time!

Indeed, the relativist will argue, the discrepancies we detect during relative motion are artificial. They arise and mislead us because we use inappropriate methods to analyze physical events. Thus when different observers measure physical events using their ordinary senses and relying three-dimensional space and separate time, erroneous results are obtained. Only the use of appropriate geometry and mathematics in four-dimensional space-time corrects the errors.

We can now return to an earlier statement which in this context carries far greater significance. "Any law of physics dealing with gravity must also deal with acceleration." Einstein's reasoning again is lucid: If an ordinary observer cannot tell the effects of gravity (e.g. falling) from the effects of acceleration (e.g. inertia), then gravitation is a question of acceleration. Of course acceleration is a form of motion, and motion distorts space and time. Does this mean that gravity is a question of distorted space-time? Finally let us add the suggestion we made in the previous chapter: The presence of mass bends space and time.

By now the reader may glimpse a profound insight: There should be a firm chain of logic between the presence of mass, the relativity of space and time, the unity of four-dimensional space-time, and the nature of gravity. Let us therefore delve into the second main principle of general relativity, the one that is most elusive and challenging, and at the same time is another revolutionary concept arising from relativity. The essence of this principle is the title of the next chapter.

CHAPTER 10: MATTER DISTORTS SPACE-TIME

We recall that the first postulate of general relativity is the equivalence principle. We now turn to the second postulate which states that all matter distorts nearby space-time. That is to say, *an inherent property of any mass is its ability to curve four-dimensional space-time in its vicinity.* The curvature (or distortion or indentation or alteration or deformation as it is also called) of surrounding space-time governs the behavior of other nearby masses so that they exhibit gravitation.

The notion that matter alters space-time and that altered space-time determines gravitation is foreign to our common sense and to our every-day experience, but we have a good analogy. Space-time can be pictured as an elastic and stretchable surface, like that of a trampoline. We also imagine a graph-paper-like grid imprinted on the surface of the trampoline so that changes in the shape of the surface are visible and so that motion of something rolling or sliding across that surface is measurable. The edges of the trampoline represent time and space coordinate axes, as in our space-time diagrams, and we assume that they can be bent.

The trampoline analogy is designed to aid in the study of general relativity, but this analogy will mean more if we first apply it to special relativity. For instance we mentioned that natural muons formed in the high atmosphere can reach the Earth's surface despite their very short life span. Special relativity explains that their life has been extended by relative motion. Let us now imagine that these muons are moving across the trampoline, causing *a dragging of the flat surface in the direction of motion*, so as to *stretch* the edge of the trampoline that represents time. (We must imagine that dragging can stretch the trampoline without indenting its surface. The trampoline only becomes longer.) This stretching represents the effect of uniform motion on measured time. It allows the muons more time to cross the trampoline. In like manner *space-time is "stretched"* in the direction of the muons' motion—enough so that the muons do not "run out of time" before reaching the ground.

We now shall look at the trampoline analogy applied to general relativity: "Matter" is envisioned *indenting* the otherwise flat surface of the trampoline, as your weight indents a trampoline when you stand on it. The indentation distorts the squares of the grid printed in the surface, and matter can also bend the edges of the trampoline. Greater mass causes more indentation, and the amount of indentation decreases farther from the mass. *The indentation represents a gravitational field.* Another nearby mass in this gravitational field is represented as a small freely rolling marble on the surface of the trampoline close to where your feet indent that surface. The motion of this marble is determined by the shape of your indentation. *Allowing the marble to roll into the indentation represents falling. Letting the marble spiral around the indentation represents orbiting. If the marble is blocked from moving, its tendency to roll into the indentation represents weight.*

We emphasize how the shape of the trampoline helps us envision the key feature of general relativity: Indentation mimics the effects of accelerated motion, notably that associated with gravity. Since mass is associated with gravity, mass is associated with a complicated non-flat "curved" shape of space-time. Thus when we indent our trampoline's surface, we are analogically reproducing *the curvature of space-time caused by a mass*; we are depicting the second postulate of general relativity.

At the same time, the trampoline analogy has some flaws. First, the indentation of a real trampoline by an object is obviously the result of gravity, while we interpret the indentation to represent the *cause* of gravity. Second, the trampoline provides a spatial analogy, yet it is not just space but space-*time* which is indented by matter, and it is difficult to envision "indented" time. Third, anything rolling on a real trampoline encounters friction, so that the behavior of a marble moving in a circle on a trampoline is different from, say, the moon orbiting the Earth. Fourth, if a beam of light skims along the surface of the trampoline, the analogy requires that the beam dip into any indentation, which of course does not happen on a real trampoline.* Finally, this analogy tempts us to think of space-time as having an actual firm surface, but space-time itself in a mathematical entity whose "surface" exists only in the abstract.

We must also assume that the surface of our trampoline is perfectly smooth and still. It cannot have the grain of a fabric, nor can it have small-scale vibrations or fluctuations. In other words in this analogy we can stipulate that the surface has no texture and that no one bumps into or jiggles the trampoline. These details seem insignificant when we imagine a ball or other object placed on a trampoline, but if this surface is to represent space-time as envisioned in general relativity, microscopic smoothness and immobility are essential. We will see why in our chapter covering quantum mechanics, but suffice it for now that relativity does not allow points of space-time to undulate, whereas quantum mechanics may do so.

We may now acknowledge an inaccuracy in our assertion that gravity is a property of mass (page 58). In the framework of general relativity, this is true only indirectly. A better statement is that curved space-time is a property of nearby mass, while gravity is a consequence of the shape of space-time. In short, *gravity exists only where matter curves space-time.*

◆

In theory every mass can curve space-time. This inference raises an additional point: If we bring several masses close to each other, each of which engenders its own indentation in space-time, any mathematical analysis becomes very complicated. In mathematics this is called a "many bodies" problem, and a complete solution with many nearby masses indenting a surface defies computation. To avoid this we ignore the effect of very small masses. For example when a marble falls towards the Earth, in theory there is also very slight motion of the Earth towards the marble, since the latter also indents space-time to some extent. When we neglect this mutual effect, we call the smaller falling or orbiting object a "test object" to indicate that its own contribution to gravitation is negligible and that it "tests" the gravitational field around the larger object.

However our reliance on test objects is more than just for convenience. It illustrates the change in paradigm we described on page 66, in which gravitation is ascribed to properties of a field rather than to force acting on the mass of an object under the influence of gravity. Since the mass of such an object is not essential in this paradigm, we treat the object as a test object whose gravitational behavior is determined only by the field. This is why we will find that "*m*" (for mass) does not appear in the basic equations of general relativity.

---

* In theory there is some gravitational interaction between a photon and the mass of the trampoline, but it is negligible.

117

✦ ✦ ✦

We now introduce the role of Euclidean and non-Euclidean geometry in relativity. Euclid's principles are quite valid when applied to certain shapes and surfaces, and Euclidean geometers were certainly sophisticated enough to consider curved surfaces. However the founders of geometry labored under that assumption that flatness is a basic shape of nature. Euclid's geometry teaches that on a flat surface the shortest distance between two points is a straight line, and that the sum of a triangle's angles is 180 degrees. Likewise, as is clearly very important to us, on a flat surface the pythagorean theorem applies. (We include the pythagorean theorem in Euclidean geometry, as this theorem was known to Euclid and can be derived from his axioms.) For us *the most significant case of a flat surface is space-time with no nearby masses and hence no gravity.* Special relativity covers these conditions, as these allow uniform motion.

Among Euclid's conclusions, one stands out as a clue that there is room for refinement: his axiom, also know as the "fifth postulate," that parallel lines never cross. He himself suspected that this premise is problematic because it required knowing whether or not something (the crossing of lines) occurs at infinity, so that it is not conclusively confirmed in human experience. Indeed later mathematicians, geometers, and even philosophers tried in vain to find a definitive proof of this axiom. Their failure suggests that flatness may not be a ubiquitous attribute of our universe.

The modern "non-Euclidean" view is that the geometry of nature, specifically of space-time, is fundamentally curved. This means that Euclidean geometry does not suffice for the description of space-time in the presence of mass; we say that *space-time influenced by matter is non-Euclidean.* In particular, on indented space-time, parallel lines do meet; triangles do not contain 180 degrees; and *the pythagorean theorem does not apply.* The axiom that the shortest distance between two points is a straight line also does not hold. Rather, points on a curved surface are connected by *geodesics,* which are the straightest *possible* lines between such points. We recall that a curved world-line forms a geodesic, and thus *general relativity holds that only non-Euclidean space-time governs the world-lines of falling or orbiting objects.*

Two mathematicians who pioneered this topic interest us in particular. They are Gauss (1777-1855) and his student Riemann (1826-1866). The non-Euclidean geometry used by Einstein to study space-time is called Riemannian geometry, which can be applied to an infinite variety of shapes, only one of which is flatness (and which can accommodate four dimensions; more on this later). We can say that special relativity is a special case of general relativity as Euclidean geometry is a special case of Riemannian geometry. In this context, we think of a curved surface as one which is changing, and we will see that Riemann superimposed Newton's differential calculus upon Euclid's geometry so as to be able to study changing surfaces.[*]

---

[*]A more precise wording of Euclid's fifth postulate is that if a geometric point lies outside a straight line, then one and only one other straight line can be drawn through that point such that the two lines are parallel. Riemann's geometry revises this tenet to state that no parallel line can be drawn through any outside point, allowing for curved surfaces on which parallel lines eventually cross. Other systems of geometry, not relevant here, claim that an infinite number of such lines are parallel.

Under routine Earth-bound conditions, in which there is relatively little curvature of space-time, we are deceived into thinking that flatness prevails and that Euclidean geometry is adequate. The weakness of Euclidean geometry applied to space-time becomes manifest only when extremes are explored, as with very sensitive measuring devices or at very great distances or in places in the universe with detectably curved space-time, such as near the sun during the 1919 solar eclipse. This can be likened to Euclid having built a toy car which can only go on a flat trampoline; it might be perfect for flatness but it cannot go elsewhere. By this analogy Euclid built a geometry which cannot guide us wherever matter has deformed space-time. There non-Euclidean geometry is needed, which is why this kind of geometry is a major ingredient in the mathematics of general relativity.

◆　◆　◆

As an example of the role of space-time, let us consider an apple falling off a tree. The trampoline analogy illustrates how the unimpeded (free-fall) motion of the apple in four-dimensional space-time looks to us like falling: We imagine the Earth placed in the middle of the trampoline so that the surface of the trampoline is indented, and the apple is at the edge of the indentation. The trampoline is invisible (as is space-time anyway) and we are watching the apple roll into the indentation. The apple tends to approach the Earth faster and faster, suggesting an attractive force between the apple and the Earth. As long as the apple is allowed to move freely, we label this form of motion "acceleration due to gravity," which we interpret as "falling." This motion emerges in our perception as a trajectory through our familiar three-dimensional space. Of course once the ground prohibits further motion, we call the result "weight."

This scenario raises an obvious question: How does the apple "know" which way to fall? The answer is simpler to word than to imagine: Mass (in this example the Earth) distorts not just space, which is easily represented by our trampoline, but it also distorts time. As the apple glides into the indentation of space, it also glides into the indentation of time. *The apple seeks the easiest path, which is along a geodesic set by the shape of space as well as of time; the apple in free-fall obeys a four-dimensional geodesic.* In other words the dent in space-time governs all components of the apple's direction. In the case of space we call this direction "down." In the case of time we call it the future. Of course in theory all objects in undisturbed motion are always in free-fall, which means that all objects in the universe are always traveling on a geodesic toward their destination in space and toward their future in time.

Let us recall how Einstein equated gravity with inertia: Both are a consequence of acceleration. In the case of the falling apple, *it is space-time which provides the apple with its acceleration.* However this statement is somewhat misleading, for it implies that space-time moves, which is not what relativity argues. A more complex but more accurate statement is this: *The shape of space-time at each point along the apple's path determines a local "gravitational potential."* We will deal with gravitational potentials in mathematical detail later, but for now we can say that the gravitational potential expresses how much potential energy is available for gravitation at a given point in space-time. In other words the gravitational potential tells us how intensely this particular location generates gravitational effects such as falling, orbiting, weight, etc. In this paradigm—which is pivotal to general relativity—the basis of the progressively faster gravitational motion of our apple is really *the consequence of the progressively changing gravitational potential along the apple's world-line, while the gravitational potential is a consequence of local space-time curvature.* The apple obeys the shape of space-time by responding to local gravitational potentials, thus finding the easiest path.

We can raise even more basic questions: *Why* does the apple seek the easiest path? The reply will appear more completely in later chapters, but for now let us assume that an experiment with a suitable clock and ruler can be devised. (In fact Galileo did something like this.) We further assume that after falling for one second, a distant observer finds that the apple moves with a speed of 10 meters per second, and after another second of falling, the apple has gained speed to 20 meters per second. We are accustomed to saying that because of gravitational force, the apple accelerated to twice its speed in that last second. Relativity counters that the apple would have moved through space-time in an inertial manner (without changing speed) but that where the apple was after each second, *space-time was more deformed and the gravitational potential was stronger.*

Using the trampoline analogy, the dip in the surface is steeper where the apple was after two seconds. We can say that the apple "wants" to follow the easiest path by obeying inertia—by continuing in uniform motion—but the shape of space-time won't let it. We will revisit this notion in other contexts, both conceptual and mathematical, but we can say that *the law of inertia applies in relativity with the proviso that freely falling objects move along geodesics, the paths of which are dictated by the curved shape of space-time.*

Another question is whether the curvature of space-time is apparent to an observer who can accompany the freely falling apple. (We can imagine an observant ant on the apple.) For example, couldn't such an observer use a straight-edged ruler or a taut string to detect curvature? The answer is that any such device itself becomes curved by the very space-time being examined. Applying the trampoline analogy, a long ruler laid into an indentation will conform to that indentation. However a very short ruler seems straight to the observer, leading him to conclude that his space-time is flat. The observer will even be deceived using a beam of light as a straight-edge; to him, the beam appears straight even though it conforms to curved space-time. In short, any local means for detecting bending will cause space-time to appear locally "non-bent." This conclusion mirrors what Einstein noted in a freely falling frame of reference: An observer cannot detect his own accelerated motion and thus feels weightless.

Of course if the observer looks far enough away, he may detect acceleration after all, so we must stipulate that he is only a local observer, or, more precisely, that this observer surveys only local space-time—i.e. an area so small that it can be treated as a point. This notion will be important later, but we note an application of the term "local:" *In free-fall, local space-time appears flat.* This is a key concept we must accept: As far as a freely falling observer (say an ant) on the apple can tell in his frame of reference, there is no force, no acceleration, and no distortion of time or space. This observer feels weightless. From his local vantage he and the apple occupy a flat point in space-time and therefore exhibit uniform (inertial) motion.

Nevertheless if we consider this situation *in a more global wider frame of reference*, which implies a transformation to a different frame of reference, we find the apple to be in *non-uniform* space-time.[*] Hence as seen by a distant observer, the apple's speed increases. Likewise, if a freely falling

---

[*] In English translations, Einstein used the "finite" for the more global non-local case.

observer can track his progress on distant coordinate axes—for example if he can see the ground coming closer—his relative accelerated motion is revealed to him. As a drastic example, a freely falling observer feels weightless until his relative motion is "revealed" when he hits the ground. (Of course a sky diver in free-fall can detect relative motion with respect to the surrounding air, so for our purposes we must assume the absence of an atmosphere.)

We will reconsider this concept in detail, but since we can treat the world-line of a falling object as a series of points each of which is locally flat, we can say that the apparent change in the speed of falling results from *locally straight motion through globally curved space-time*. That is to say, *a falling object follows a geodesic which is as straight as possible in a non-straight environment*. (Later we will see how the concepts of space-time and gravitational field unite to characterize this "environment.") We can also anticipate the role of differential and integral calculus: The falling object's curved path can be analyzed as a series of small straight steps which, when chained together, form a curved line.

Before ending this section let us turn to another basic question: We asserted that as an apple falls, it "moves through time toward its future." Why should it do that? In reply, the relativistic principles of causality prevent objects from moving "backward" through time, but we can also state that *by virtue of their very existence*, all objects in the universe *always "move forward through time."* As we pointed out, we can stop objects in space, and we can alter the relative aging of objects (likewise, in theory, of people). However we cannot *stop* anything from aging. This point is noteworthy in general relativity, because in theory if an object stopped aging it would stop falling. In other words as long as an object moves through time, and as long as nearby space-time is curved, the object must try to accelerate through space and therefore must show the effects of gravity. (Later we will consider cases where space-time is extremely curved.) This means that wherever we find gravitation, objects move through space and time in an orderly fashion. However this notion raises the question whether such orderly behavior persists where the conditions for general relativity do not prevail. More on this in our chapter covering quantum mechanics, but for now we focus on the features of nature essential for the relativistic view of gravitation: Space-time is curved, and objects exist. Objects that exist must move forward through time. Space-time that is curved compels these objects to display gravitation.

✦

All forms of gravitation, not just falling and weightiness, come under these principles. The same logic relativistically explains astronomic orbits. To our perception, orbiting celestial bodies (moons, planets, comets, satellites) move in spherical or elliptical paths, whereas their four-dimensional world-lines are geodesics through space-time which is curved so as to guide these objects into apparent circles or ellipses. (In space-time diagrams an astronomic orbit forms a spiral; see page 71 or Penrose's Fig. 5.30.) For example the mass of our sun curves space-time, so that the Earth's straightest possible path in four dimensions becomes an orbit which we normally think of as existing in three dimensions. In other words orbiting is merely a kind of falling, and both are forms of free-fall. Of course if an object in a gravitational field cannot move freely because it rests on the ground, it merely feels heavy—it has weight—but the point is that in the eyes of general relativity there is no basic difference between a trajectory, an orbit, and the presence of weight.

Einstein cannot take credit for the discovery that the principles which explain falling also explain orbiting. Newton had already reached the analogous conclusion: the brilliant and revolutionary insight that planets orbit the sun for the same reason objects fall to Earth. In both cases the same laws of physics apply—though Einstein rewrote these laws. And even Newton cannot take full credit, because Galileo had already speculated that were it not for friction with the air, all falling objects would move together. We can now add that for the same reasons, all orbiting objects move together. This is why an astronaut can "walk" in space while the mother ship is traveling at thousands of miles per hour.

Here is an excellent example of how relativity, despite its seemingly bizarre notion of indented space-time, is a superior explanation of gravity. If a force were to govern falling as well as orbiting, what keeps the moon in its path? The Newtonian answer had to be that inertia balances the force of gravity just enough to prevent the moon from spiraling into the Earth but not enough to prevent its escape into outer space. As we already pointed out, this theory demands an exact balance between gravity and inertia but does not explain why such a balance should exist. Relativity paints a different and ultimately simpler picture: Passive motion along a geodesic through space-time indented by the Earth's mass accounts for the fall of an object as well as for the orbit of the moon.

We should add that even though falling and orbiting are familiar events, nature is filled with other manifestations of gravity. Examples: Stars, including our sun, shine because gravitational compression of their contents sustains thermonuclear reactions; stellar evolution is largely a gravitational process; and the only known impediment to the expansion of our universe is gravity. Incidentally, what if a nearby mass is so large that gravity is extremely strong? Then we may have a black hole. And if an array of beams passes near a large mass so that they are bent like a lens bends light rays, we have a gravitational lens (Schutz, p. 287). Both are of great interest in astronomy and cosmology, and we shall return to black holes later.

✦ ✦ ✦

Our space-time diagrams portrayed accelerated motion of an object as a curved world-line in straight coordinate axes or as a straight world-line in curved coordinate axes (page 93). Either way, space-time has no meaning without a system of coordinates. Indeed space and time axes are not just handy geometric tools; they represent *and define* the shape of space-time. Reichenbach (p. 283) is quite emphatic: "...everywhere and at all times there exists a *space-time coordinate system*...," and we can identify every point in space-time only by its location on three space coordinates and one time coordinate. We also pointed out that as long as motion is never absolute—and we include "motion" through time as well as space—a frame of reference is indispensable. The only way to quantify events in frames of reference is by means of systems of coordinates, and, in the context of relativity, the only difference between different observers of an event is their frames of reference.

Indeed whenever we measure an event with a ruler or a clock, we are using a system of coordinates in a frame of reference. However in the Newtonian view, rulers and clocks are autonomous tools by which to tell what is happening. In the relativistic view rulers and clocks do not reflect what is happening *unless we understand that they are a part of that event*; they are affected by it; they are included in its workings. Thus for example a ruler in curved space-time takes on that curvature. In other words, if a ruler measures a projection of space and a clock measures a projection of time on coordinate axes, then the coordinates become active participants in the event.

We note the principle, as it is at the heart of relativity: Space and time are not a rigid background upon which physical events occur; the events occur as they do because space and time are *non*-rigid. When we focus upon individual observations such as that gravity bends light, we "miss the point": First of all, it is the curvature of space-time that does the bending. But the more momentous concept is that the shape of space-time not only affects light; it affects objects, it affects ostensibly straight rulers, it affects pendulums of clocks, it affects the arms of a scale, it affects the natural path of objects, and it affects the coordinate axes along which we trace events. We touched upon this issue earlier, and now we can elaborate: We are very much in tune with relativity when we generalize that *all aspects of physical events unfold as they do because they participate with the deformation of space-time.* On their own, things do not shrink or expand, but the space-time in which we observe them does change. In short, the shape of space-time shapes everything.

The concept that all space-measuring and time-measuring (and mass-measuring) devices are themselves altered in gravitational events—and indeed during all relative motion—is not surprising once we ask the following questions: If such devices exist in space-time and every physical event is influenced by the shape of space-time, how could they *not* participate? Why should a clock or a ruler be an exception? In other words if time and space were absolute rather than relative, would it not mean that clocks and rules are somehow exempt from the laws of physics? After all, we pointed out that all time-consuming events consume energy (page 72), and of course "space-consuming" events do the same. Once it is stated, the answer is amazingly simple: We know that time and space are relative because clocks and rulers are a part of nature!

We can thus argue that a statement such as "gravity alters the measurement of time" is incomplete. Rather, the gravitational alteration of space-time embraces *each and every device by which we measure time.* This may be easy to imagine in the case of a clock based on a pendulum, but it also applies to time-measuring "devices" such as our own aging bodies. The strength of the gravitational field (at least in theory) affects how fast we age because we ourselves are altered along with space-time. In a sense we have come full circle. Lorentz was right, but for the wrong reason. Michelson's and Morley's apparatus did become deformed (it underwent apparent contraction), but not because of pressure from the ether. Rather, space-time was deformed, and the entire event, apparatus and all, participated in the deformation. From a philosophical point of view we can argue that Lorentz was right, but for the wrong *medium*: Space-time has taken the place of the ether. However from a physical and geometric point of view, space-time is not an equivalent nor a substitute for the ether. The ether was thought to be *where* gravitational events occur; space-time is *why* they occur.

The trampoline analogy is useful here: To fully discern the surface of the (initially flat) trampoline, we drew a grid of squares on it, which means we superimposed a system of coordinates on it. The analogy is apt in that this grid is not just laid down loosely but that it becomes part of the surface, so that if the surface is stretched or curved, the grid participates in the deformation. Similarly (but not easily imaginable) a four-dimensional system of coordinates can be drawn on the "surface" of space-time, and it too must participate in any deformation, notably the deformation associated with gravitation. The implication is that frames of reference containing gravitational events necessitate "deformed" systems of coordinates, and that adding gravitation to a physical event is tantamount to transforming space-time coordinates from straight to curved.

Although it may seem rather pedantic, we can even dispute that "space-time has curvature." A more meticulous statement is that space-time only has non-Euclidean *geometry* or that only *the geometry of space-time has curvature*. To heed this distinction we ought to say that nearby matter does not really change the shape of space-time; it changes the geometry of space-time. In the same context, since space-time itself is but a mathematical tool for describing quantitatively the kind of geometry envisioned by relativity, we might also dispute the assertion that space-time alters "everything": Only the geometry of space-time is altered, but "everything" is in that geometry. These refinements in terminology emphasize the role of geometry, but it is customary simply to say that space-time is curved.

Still we should know what is meant by "altered geometry." In effect we have identified two geometries. In the first of these the traditional pythagorean theorem holds and the shortest distance between two points is a straight line. The second geometry is such that the traditional pythagorean theorem fails, and the shortest distance between two points is a geodesic. *Relativity maintains that matter transforms the prevailing geometry from the former to the latter*. We can even say that *the content of space-time determines its geometric nature*. Since we equated matter with energy via $E = mc^2$, this idea invites us to ask what is meant by the "content of space-time." In fact eventually a detailed answer will occupy us. Let it suffice now that space-time may have one of two shapes, depending on its content: flatness in the absence of matter, for which Euclidean rules suffice; and matter-induced curvature, for which non-Euclidean rules are needed.

We see that systems of coordinates then play four distinct roles in relativity, and we should recount these explicitly. First, coordinates are the way we identify the locations and progress of physical events in space and time. Second, the shape of coordinate axes corresponds to the shape of space-time. Third, any system of coordinates participates in physical events. The fourth role, as we will detail later, is the most complex and the most critical: We maintain that the effect of all motion, uniform or accelerated, is equivalent to a transformation of the coordinates. Since gravity is a consequence of acceleration, *gravity is a question of transformation of space-time coordinates*. We quote Einstein: "...we are able to 'produce' a gravitational field by changing the system of coordinates." (The Principle of Relativity, p. 114.) This idea will help us write laws of physics valid in all systems of coordinates.

We should also be aware that the following entities, each of which can be described in systems of coordinates, are ultimately synonymous: What the physicist calls "prolonged events," "trajectories" or "orbits," the geometer calls "geodesics" and "world-lines," and the relativist calls "prolonged space-time intervals." Such notions bring us back to the idea that when we work with diagrams of curved space-time, we may consider world-lines to be straight and the coordinate axes to be curved. This routine concurs with the concept that moving objects tend to progress uniformly point-to-point along the straightest possible path through *non*-uniform space-time. We depicted this graphically in diagram L. The upper part of the diagram reflects our intuition: Space and time axes are straight; trajectories are accelerated. The lower part of diagram L reflects relativity: Falling through space-time is as straight as it can be; the space and time axes are bent; space-time is curved. Let us therefore restate: Objects fall or orbit in straightest paths through curved space-time. World-lines or geodesics run as straight as possible through non-straight systems of space-time coordinates. By nature, gravitational motion in space-time "tries" to be straight, but space-time is not straight. Individual points in space-time allow locally uniform motion, but objects in space-time move point to point; they do not stay local.

Lest the idea of straight world-lines in curved coordinates appear to be an idiosyncrasy of general relativity, we should consider a basic technique in terrestrial navigation: Given the curvature of the Earth, maps with curved coordinates (like those drawn by Mercator) are routinely designed so that routes appear as straight lines. In fact we will make use of analogies based on this method.

✦ ✦ ✦

We indicated that a freely falling observer finds only flat local space-time. But we recall that if two strings are hung from a ceiling on Earth, their lower ends are slightly closer because each string points to the Earth's center—i.e. they converge. This convergence in a gravitational field is an example of tidal change-of-shape effects found in natural gravity, and it means that in a freely falling frame of reference, space-time need not be 100% flat. We suggested that we may avoid this complication by considering only a small region of space, so small that the structure of the mass responsible for gravity does not matter. For example we can limit the space to that occupied by only one infinitely thin string.

In the terminology we are using in this chapter, tidal change-of-shape effects are consequences of the curvature of space-time, but they are *non-local* in that they involve two separated points, such as the ends of the two suspended strings. Considering only one thin string is equivalent to considering only *local* space-time. In the case of our falling apple, in theory the apple undergoes a tidal change-of-shape, and an observer might be able to detect this deformation. For instance an ant on the apple could find that two points on opposite sides of the apple are drawn together, like the ends of two suspended strings (or like two distant points on the Earth's surface during a high or low tide). Therefore we should consider only an infinitesimally small apple, *one that occupies only local space-time.* The usefulness and details of this procedure will engage us in chapter 15.

We mentioned Einstein's prediction that light from a distant star is deflected by the sun's gravity during a solar eclipse. Now we can enlarge upon this observation. First, we ask why the experiment required such unusual conditions: Because nothing on Earth is massive enough to make the bending of light by gravity discernible. However even with the sun's gravity put to use, the light beam must be very long to make a bend in it measurable, so that light from a distant star was needed. But sunlight obscures starlight, so that only during a solar eclipse could the displacement of the image of the star be observed.

As relativists we say that this displacement is the result of the shape of space-time near the sun as the beam of light from the distant star passes through an indentation along a geodesic. In order to accurately predict the displacement of the image of the star, *two* components must be considered. First, the beam from the star is bent toward the sun. Second, as the beam "dips" into the "indentation" of time near the sun, it is delayed, just as any time-consuming event near a mass takes longer than it would without the mass. Therefore to a distant observer such as one on Earth watching the star, the path of the beam is altered in two ways, and each provides part of the total effect: The beam is displaced, and it is slowed. We note that the displacement of its path has to do with space, and its slowing is a question of time. An observer of this event thus witnesses the effects of *altered space-time.*

In historical perspective this experiment really began when Newton suggested that light consists of particles ("corpuscles"). If these particles have any mass, some bending of a beam by the sun's gravity must occur solely on the basis of Newton's laws. The mathematics had been worked out as far back as in 1801; the magnitude of bending was reckoned be about 0.875 arc seconds (less than 1/10,000 of one degree). Einstein took a novel approach by using relativistic equations. Initially in 1911 he recognized the relativity of time but still assumed that space was flat. In other words he did not consider curved space-time. He calculated a result numerically similar to that of 1801 (about 0.83 arc seconds; there was also an arithmetical error). Had he stopped there, the conclusion might have been that Newton's laws are correct, rendering relativity superfluous. Thus if the solar-eclipse experiment had been carried out in 1914 as originally planned, the results could have severely discredited Einstein.

As it turned out, World War I forced a delay in the experiment until the next solar eclipse five years later. During that postponement, and after obtaining help with the mathematics, Einstein corrected himself. With curved space and curved time figured in, he recalculated that the amount of bending should be about 1.75 arc seconds (twice 0.875). The findings in 1919 confirmed that the bending indeed was greater than what could be anticipated by Newtonian calculations, and the *excess* closely matched Einstein's prediction. The news dramatically bolstered Einstein's reputation, and practically overnight relativity became a household word. (In the term "excess" we are sampling a theme found in any quantitative study of relativity. Under many circumstances the effects of relativity appear as an excess beyond the effects calculated from Newton's laws. We will use this term in this sense again.)

◆

In interpreting the solar eclipse observations, we must appreciate that the beam of light appears curved only to a distant (a very "non-local") observer. His frame of reference is large enough to reveal the effects of space-time curvature; he can see the geodesic. If this observer is ignorant of relativity, he may conclude that this beam violates Euclidean geometry, as the shortest distance between two points (the Earth and a distant star) does not appear to be a straight line. Meanwhile a local observer, traveling with the photon if he could, would report that the beam faithfully obeys the rules of Euclidean geometry about straightness. We know why: A photon "falling" to Earth, just like our falling apple, illustrates free-fall. In a freely falling frame of reference, local space-time appears flat, a geodesic looks straight, and the observer's proper data are unaffected.

If the local observer understands relativity, he explains that what his distant colleague sees is not a violation of geometric law but evidence of the curvature of space-time! In fact if we think about it carefully, for the distant observer light appears to be bent by gravity precisely *because* a beam seeks straightness but it is traveling through bent space-time. In this context we can recast earlier statements that the experiment of 1919 proved that gravity bends light. What it really proved is that *curved space-time makes light appear bent.* (What it really *really* proved is that the geometry of space-time makes light appear bent, but we need not be that meticulous.) In short, light is straight but space-time is not straight.

This idea can be restated as another contrast between Newton and Einstein. Newton in effect envisioned curved motion (e.g. a trajectory or orbit) in flat space through even time. Einstein envisioned *straight motion through curved space-time.* In this sense Einstein extended Newton's law of inertia (objects seek uniform motion) to realms even Newton did not anticipate; his law endures, whether space-time is flat or curved. Since gravitational behavior is straightest possible motion through curved space-time, and since straight motion is uniform, *gravitational behavior is merely inertial motion curved by space-time.* The bending of a light beam passing by a mass is therefore not because the beam does something exceptional or unnatural. It is because the beam faithfully obeys the laws of physics and geometry *while it responds to space-time.* As we said before, relativity shifts our attention away from the properties of the moving object—here represented by apples and photons—and toward the properties of the surrounding environment, specifically space-time.

We may well ask how the speed of a light can be decreased by passing close to the sun if photons move uniformly through space-time and if c, the speed of light, is constant. Here we recall that special relativity limits the postulate about the constancy of c to uniform motion, whereas gravity represents accelerated motion. Thus only to an observer in uniform motion, and in the absence of gravity, does the speed of light appear constant. (Any deviation in speed and/or direction from uniform motion, including bending, is considered to be "acceleration." Technically, even slowing is called acceleration.) If an observer is *accelerating* toward a source of light, the speed of light will appear greater; if an observer is accelerating away from a source, the speed will appear reduced; if an observer is accelerating across a beam of light, it will appear bent, as we described earlier aboard an imaginary accelerating elevator; and *if a beam passes through a gravitational field, i.e. through a distortion of space-time near a mass, a distant observer will indeed detect an apparent slowing as well as a bending*; he will perceive "acceleration" of the beam of light.

This concept leads us out of the apparent contradiction that gravity slows light but c should be constant. With our trampoline analogy let us have the beam of light conform to the trampoline: As the beam passes close to a large mass, it is "dipping" into an indentation in which space-*time* is stretched. Because it has more "distance" to cross, the beam takes a longer path to reach an observer. Its local speed is unchanged, but the beam is detoured; it arrives at an observer's location later than it would have without passing by a mass. If we think about it carefully, light appears to be slowed by gravity precisely *because* its speed is c but it is passing through distorted space-time. In short, c is constant but the shape of space-time is not constant; light travels uniformly but space-time is not uniform.

This is also why, when we mention the constancy of the speed of light, we include words like "appears" or "measurement of." The speed of light is constant as it moves through any one medium, but if an observer is in accelerated motion, or if the beam traverses a gravitational field, the speed of light *appears* to change. Relativity claims that this is artificial. Locally the speed remains constant, but because of space-time curvature the ordinary observer's perception of length and duration changes. As we said, we do not directly measure the invariant reality of an event in four-dimensional space-time. We measure *projections* of an event on our own space- and time-axes, and these projections change with relative motion, be it uniform or accelerated.

The slowing (delay) and displacement (bending) of a beam of light or of other electromagnetic energy can be detected more easily today than in 1919. For example it is possible to aim a radar signal from Earth to one of the other planets and to detect a delay in the echo, depending how close the beam comes to the edge of the sun. This effect is still very subtle, but its magnitude agrees with predictions of general relativity. Such radar may be used with space vehicles, so that anticipation of these effects has practical significance. For example if a long mission relies on radar ranging which is not adjusted for relativity, errors in navigation are unavoidable.

✦ ✦ ✦

We note that the phenomenon of gravity bending light draws upon special as well as general relativity. Let us envision a star one light year away from two observers, A and B, who are also far apart from each other and who can communicate with each other. We use diagram M. Suddenly the star explodes. The light from the star reaches observer A directly, but to get to observer B it glances by some massive celestial object. Of course this event (the explosion) is not known to the observers for at least one year. Then the flash reaches observer A, who immediately reports this to observer B. Some time later, observer B calls A with a similar report. We know why the two reports are not simultaneous: Light reaching observer B is delayed by an indentation of space-time near the intervening object. As a result, when A and B compare their observations, *the two observers disagree on the time of one event.* They may even conclude that there were two explosions one after the other.

This discrepancy is a consequence of *general* relativity. Observers may differ on the time and simultaneity of an event as a result of bent space-time. We also recall that *special* relativity showed how uniform motion alters measurements not only of time but of space as well as of mass. If the two observers plot the location of the star by sighting along the beam from the star, they will also reach conflicting results, namely that two separate stars exploded. Furthermore, if the observers calculate the mass of the star (which can be done from afar), their results will not agree; the "two" stars appear to have different mass. In short, *the observers disagree on the measurement of time, space, as well as mass.*

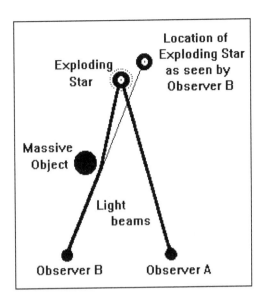

Observer A sees the exploding star in its
diagrammed location. Observer B sees
a delayed and displaced image of the same
event.

Diagram M. Two observers perceive one event, an exploding star. As an effect of general relativity, these observers conclude that two events occurred some distance and some time apart. (This is a space diagram, not a space-time diagram.)

Other concepts need attention here: In any scenario entailing general relativity the same rules apply (e.g. the speed of light is constant, though space-time is not constant). Indeed all valid laws of physics should appear the same to all observers in any kind of motion, uniform or accelerated. This statement can be expanded into a sweeping generalization which, like other aspects of relativity, has a philosophical and even a theological ring to it, and which accounts for some of the popular appeal and mystique of relativity: The laws of nature must be the same for everyone everywhere. The laws of the universe really are universal.

In special relativity we introduced $E = mc^2$ as such a law. But general relativity shows us that everything, mass as well as energy, is affected by curvature of space-time. Thus we detect the effects of gravity on objects as well as on electromagnetic radiation such as light and radar beams. There is no need to separate mass from energy. Both are affected by the properties of space-time. There is no need to separate special from general relativity. Both respect the laws of physics. There is no need to separate space from time. Both participate in physical events. There is no need to separate uniform from accelerated motion. Both alter space-time and both generate discrepancies in the measurements of space, time, and mass.

General relativity also cogently explains why different falling objects accelerate equally: They are influenced by the same distortion of space-time caused by nearby mass. There is no need to ascribe gravity and matching inertia to each object as several different objects fall together. There is no need to apply different laws of physics. It is necessary only to describe unimpeded motion through space-time and to translate this motion into what we can detect in our ordinary perception. To relativity, it does not matter whether the several objects "fall" together toward a mass or the mass moves toward the objects; space-time is distorted in either case, and the laws of physics are the same in either case. In other words the similar motion of falling or orbiting objects exists because they share a gravitational field. As long as we know what mathematics to use, that field appears the same from the point of view—from within the system of coordinates—of either object, or, for that matter, from the vantage of any observer. No particular frame of reference is favored, and switching frames of reference does not change the laws of physics.

Still, nature does not make it easy for us to verify the claims of general relativity, even when we accept its tight and elegant logic. We cannot feel space-time curving, nor can we observe events by traveling at the speed of light. The experiment of 1919 illustrates that enormous amounts of gravity are needed to make general relativity even barely noticeable. Thus all of relativity is shrouded by human and physical limitations which hide its effects from us, and the basic reason is that, in terms of the universe, we are very small, very slow, and very light-weight.

◆

Where else can we study the alteration of space-time by matter? The solar eclipse of 1919 helped us, and modern radar is even more direct, but there are other ways. We know that because of all the sources of gravity in our solar system, the orbit of the planet Mercury shifts each time this planet encircles the sun. (The orbit of Mercury is an ellipse, the long axis of which slowly turns like the hand on a clock.) Some shifting can be predicted using Newton's laws, but our actual measurements, if done with sufficient precision, show an *excess* shift. Only after the effects of relativity were incorporated into the calculations, which was accomplished by Einstein in 1915, did the mathematical predictions match the astronomic observations.

In fact the peculiar behavior of Mercury was the first main clue encountered by astronomers that Newtonian physics might be inadequate, and Einstein's ability to account for the discrepancy was a major impetus for relativity. However skeptics argued that while relativity was a novel explanation, there might still be others which fall back solely on Newton's equations. Moreover the bending of starlight (revealed later in 1919) had in part already been anticipated by Newtonian ideas. Therefore Einstein's feats regarding the orbit of Mercury and the bending of starlight were not as historically pivotal as the demonstration of the einstein red-shift (fully successful in 1960). The latter effect was more convincing because it was entirely unexpected before relativity entered upon the scientific stage, and it was very difficult to explain by any other means *except* relativity

We will return to the mathematics of Mercury's orbit, but we should note here that the improvement relativity offers over Newtonian predictions resides in several areas. Foremost is the consideration of the curvature of nearby space-time induced by the sun, which alters the path of the planet in a manner not explainable with Newton's formulæ. Furthermore special relativity tells us that Mercury's rapid motion increases its mass and augments the deformation of space-time. These effects are also foreign to Newtonian physics. In addition, precise calculations accommodate the observation that the sun and Mercury are not a perfect spheres and that they have internal irregularities which affect nearby space-time.

Incidentally, the relativistic explanation of the path of Mercury also compels us to revise the laws of Kepler, who, using instruments available at the time, noted that astronomic orbits form ellipses. Today we recognize slight deviations from fully elliptical paths on the basis of the shape of space-time.

✦ ✦ ✦

We note that general relativity includes some rather striking assumptions: Space and time have shape. Furthermore their shape can be altered. More revolutionary still is the concept that matter alters, and even determines, the shape of space-time. Finally, curved space-time is responsible for gravitational acceleration. Of course it is easy to say that matter shapes space-time, but how can matter "do" this? Matter appears passive and lifeless, yet general relativity endows it with immense capabilities. We even say that the curvature of space-time is an essential attribute of matter and that space-time is curved because matter exists. Yet where does matter get the energy to bend space and time? Basically relativity's answer is this:

$$E = mc^2$$

Clearly this equation, based in special relativity, has profound significance in the framework of general relativity: The energy needed to deform space-time, that is to say the energy-source for gravity, stems from the equivalence of matter and energy. In its capacity to curve space-time, matter and energy are only different aspects of the same phenomenon of nature. Incidentally, the ultimate origin of all energy and matter in the universe is a separate topic in cosmology—i.e. the "big bang," which also invokes quantum mechanics. Nevertheless we can envision that at the instant of the big bang, only energy existed, and that all the matter we find in the universe today was converted from that energy via the above equation. In that sense $E = mc^2$ was the first law of nature!

Nevertheless we can argue that matter can be responsible for the *existence* of space-time without necessarily being responsible for its *curvature*. That is to say, since matter appears everywhere in the universe, why is space-time not regular and why is gravity not even? Why, for example, is gravity stronger on the sun than on Earth? Because the intensity of gravity depends on the *distribution* or the *concentration* or the *density* of nearby matter and energy. The idea of distribution of matter and energy seems abstruse, but it is actually inherent in Newton's law of gravitation: The total amount of matter encompassed by this law is "distributed" between m_1 and m_2. (Although "mass" and "matter" can be used as loose synonyms, wherever matter appears in the laws of physics and in the formulæ which express these laws, the word mass and the symbol m are preferred. Of course our intuitive definition of mass is something that has weight.) The idea of distribution of matter and energy is also inherent in relativity through the notion—embodied in $E = mc^2$—that matter is merely concentrated energy.

131

Combining the notion of distribution or density with the equivalence of mass and energy allows us to restate the second postulate of general relativity: *It is the distribution of matter and energy in the universe that determines where and how much space-time is curved, and gravitation is the result of where and how much space-time is curved.* In areas of high density, the local alteration of the shape of space-time is more prominent and the effects of gravity are more pronounced, whereas in the absence of matter-and-energy, space-time is flat and there is no local gravity. (Of course if we measure accurately enough, there must be some gravity everywhere, because there is always matter somewhere, but we can ignore the very weak gravity in "outer space.")

Nonetheless it is not enough to say that the distribution, concentration or the density of matter and energy "determines" the curvature of space-time. We must express this statement in mathematical terms, or else we cannot call it a law of physics, and it can neither be used nor proven. In particular, given a link between space-time and gravity, any relativistic law of gravitation must basically be a mathematical equation linking the curvature of space-time and the density of matter and energy. By now the reader can tell where we are heading: Toward a quantification of the relationship between space-time and matter/energy, and we shall close this chapter with a rough formula that expresses the core of general relativity in more or less mathematical terms:

$$\boxed{\text{Curvature of space-time} = \text{Density of matter and energy}}$$

Our next goal is to recast this assertion into explicit equations. The steps are intricate and require us to continue switching back and forth among various topics. In the next chapter we will consider tensor calculus, which is the mathematical language of general relativity. The subsequent chapter is on the metrical tensor, which is the critical element in the relativistic description of the curvature of space-time, the left half of the above formula. Then we will consider the density of matter and energy, the right half of the above formula. Finally we will tie the two halves together. The end product will be a family of tensor equations called the field equations of general relativity, and we will be able to solve them.

✦

At the close of this chapter we pause to note that we have been charting of a major scientific and intellectual revolution: Before the twentieth century we naturally accepted a sanctity of space and time, as well as the security of our privileged role as observers of physical events. However special relativity shows us that our measurements of time and space for an event depend on—are relative to—uniform motion. General relativity shows us that time and space are also relative to accelerated motion, which means that time and space can be curved, and that gravitation is crucial to the character of our universe Most importantly, curved time and space provide a superior explanation of gravitation and a new framework for the laws of physics, one that is valid for all observers and all places in that universe, but one which undermines our time-honored assumptions.

CHAPTER 12: TENSORS AND GENERAL COVARIANCE

Even Einstein was surprised by how difficult it is to apply general relativity to an actual situation, such as the orbit of Mercury or the bending of starlight. Not only is traditional geometry insufficient, but a branch of mathematics called tensor calculus is needed. Tensor mathematics is another facet of relativity reputed to be incomprehensible, and many dissertations simply avoid it. While a comprehensive presentation would require a huge tome, we still can clarify the subject for our purposes. The natural opening question is why "non-tensor" mathematics is insufficient for general relativity. The answer reiterates one of the most dramatic discoveries in science: Laws of physics dealing with time, space, and mass may not hold during relative motion; something happens between moving frames of reference which invalidates these laws, unless they are reformulated in a certain way, specifically as tensor equations. Also, tensor math can deal with the complicated components of general relativity such as curved space-time.

As a preview of the problem, Newton's laws are usually very good approximations, but that does not qualify them as laws, because circumstances exist where the equations for these laws give conflicting results. Of course what Newton accomplished represents an unprecedented intellectual breakthrough exhibiting keen logic: Without force, objects never accelerate. With force, they do. Objects accelerate under the influence of gravity. Therefore, Newton reasoned, gravity is a force, and a force-law should always be valid. Nevertheless we are justified to heed the following argument: Newton's laws yield false predictions at high velocities and are more precise at low velocities. In fact they are perfect only at zero motion. But when there is no motion, there cannot be accelerated motion, which then means there is no gravity. In other words these laws are incontrovertible only in frames of reference in which masses or observers are at rest, and the equation for these laws are different for other frames of reference. The conclusion is that technically Newton's law of gravitation is a fully valid law only in the absence of gravitation, and that his laws about motion are undermined by relative motion!

Where shall we find the remedy? We have a clue: Lorentz showed us how some laws of physics can be legitimatized: To ensure that all *uniformly* moving observers perceive the same physical event as identical—or to ensure that all inertial frames are equally valid—we apply the special relativity factor or the Lorentz factor, and we can appreciate what it accomplishes: It can generalize laws so that they retain their validity, albeit only while at rest or during uniform motion—only under the conditions of special relativity. In fact in a limited sense the special relativity factor gives an equation a key attribute of a tensor equation. That attribute is called *covariance*. "Covariance" in itself is not a difficult concept. A mathematical quantity has covariance if it varies with respect to another quantity in a fixed way. For example the area of a flat circle (A) has covariance with its radius (r), meaning that any observer who knows the radius can find the area of a flat circle ($A = \pi r^2$). A "covariant" relationship exists between A and r, and it will do so in any system of coordinates. Moreover, as we will detail shortly, a covariant relationship is "general" if it survives transformation to any system of coordinates. (Einstein in his early notebooks used the term "invariant," where contemporary scientists might prefer "covariant.")

In the sense used later by Einstein, the "principle of general covariance" states that a law of physics formulated mathematically *must hold for any observer*, even if there is *relative motion* between observers, even when such motion is *accelerated*, and in diverse frames of reference, even those *with or without gravitation*. In practice, general covariance means that *equations expressing the laws of physics are form-invariant or simply "invariant."* There is no need to alter them for universal use. Lieber (p. 52 and p. 95) says it well: General covariance requires us to "...formulate the laws of the universe with equal right and equal success [for all observers in all frames of reference]...." In a narrower mathematical sense, as we will detail soon, Einstein's equations for general relativity are covariant because they are based in *tensors* and have components which transform accordingly. The Riemann tensor in a good example; please see page 265.

In terms of coordinates, laws of physics which have covariance do not fail, and their equations are invariant, *despite transformations of space-time coordinates*. In other words equations for covariant laws are independent of whatever system of coordinates is selected for their application. We already met the idea in special relativity that there should be no privileged system of coordinates. The principle of general covariance is the extension of that idea: All systems of coordinates should enjoy equal worthiness. Einstein raised this issue (as in The Meaning of Relativity, pp. 139-140) by pointing out that pre-relativity laws of physics hold only in inertial frames of reference. Since Euclidean systems can describe *only flat* space-time, this means that in Newtonian physics, Euclidean systems of coordinates are necessarily "singled out" (Einstein's words).

Einstein also emphasized (Relativity, pp. 61-62) that without general covariance, which means without invariance of certain equations in inertial as well as in accelerated frames of reference, absolute accelerated motion would be detectable. That is to say, covariance must be general because motion, uniform or accelerated, can only be relative. We anticipated this idea on page 62: Covariant laws of physics are such that an observer cannot determine his state of motion by any experiment done in any frame of reference. This concept can even be appreciated in a wider philosophical context: If general relativity is truly a law of nature, its form should not depend on any human-made and therefore artificial system of coordinates. I stress that the applications of tensors was the break-through that allowed Einstein to write equations satisfying this requirement.

I warn readers consulting various sources that confusion arises between "invariance/invariant" and "covariance/covariant." The exact meanings may depend on the context, notably physics vs. mathematics (and Einstein was not always clear; see Isaacs, pp. 195-198). In strict terminology an equation is invariant when its form as well as its other content do not change; all remains same in any system of coordinates. Covariant equations change their coefficients together when systems of coordinates are changed in any legal way, but they remain valid and true. As I said, in Einstein's context a covariant equation retains its form and function for all observers, and for the tensors he employed, these tensors retain the relationships between their components despite transformations of space-time coordinates. In the invariant case everything about the equation should be unaltered; in the covariant case the form of the equation must be unaltered. A covariant tensor equation retains its form though this tensor's components change upon ("under") a transformation of coordinates. Even Einstein overlooked this point, as I will detail later when discussing a certain key tensor.

Thus a law whose equations have invariance can also be covariant, and here we may encounter the term general invariance. E.g., if the equations X=0 or X=Y are true and are covariant, then X=0 or

X=Y should be true in any system of coordinates. Most (but not all) covariant equations are invariant. And as a preview for why I stress and dwell on this issue, this example holds in relativity math when X and Y are tensors and the equations are correct tensor equations. Thus if such an equation—being covariant—predicts that an object here weighs 1 gram, any user of that equation in any state of motion should concur. The confusion is compounded by the fact that tensors can be subdivided into co-variant or contra-variant types, which means something different, having to do with certain mathematical-geometric properties we will discuss later (page 274 if curious).

More discussion of the distinction between invariant and covariant equations is not critical here, but the property of invariance itself is important to relativity. First, we recall the constancy of the speed of light, and we find c to be an invariant entity during uniform motion. Second, we showed that a space-time interval is invariant under a transformation of coordinates. Third, we shall soon deal with vectors which we will also find to be invariant under a transformation of coordinates. Fourth, we note that covariant laws of physics can be formulated to be mathematically invariant in all frames of reference. No wonder that Einstein himself initially favored the word "invariance" over "relativity" for his discovery. But he realized early on that if he were to author a new universal law of gravitation–which in essence is general relativity–his equations had to be universal–which in essence means "covariant"–and it turned out that tensor equations have that feature.

Because the special relativity factor is derived using the pythagorean theorem on a flat "Euclidean" surface and without regard for accelerated motion, we can say that special relativity grants "insufficient covariance" or "restricted covariance," suitable only for frames of reference without curved space-time. In short, the covariance allowed in special relativity is not general enough. Despite the confusing terminology—it is acceptable to say "Lorentz-covariant"—laws outfitted with the special relativity factor are said to be invariant under the Lorentz transformation or "*Lorentz-invariant*." Einstein realized that the special relativity factor bestows Lorentz-invariance but not general covariance, and that Lorentz-invariance alone was ill-suited for gravitational fields.

Nonetheless we should be able to build upon the attributes of Lorentz-invariance in flat space-time, and we should be able to link these to the concept of a field. Where in nature, except in remote "outer space," can we find flat space-time? The very important answer is, *during free-fall*, i.e. in a freely falling frame of reference. Indeed if we start with any gravitational field and transform to a sufficiently small region of this field, notably a region in free-fall, we find that gravity disappears but that Lorentz-invariance survives. An effective way of wording this is that by a proper selection of the frame of reference, gravity and the curvature of space-time can be "transformed away" locally. However Lorentz-invariance survives locally.

We have several ways to "transform gravitation away." A mathematical way which we already mentioned is to consider only local regions of gravitational fields. We also create a zone of free-fall in the Earth's gravitational field by envisioning a small loose elevator. We can even achieve free-fall in an airplane as it loses altitude at a certain rate. A more modern way is to "go into outer space," which is why we use space vehicles to study weightlessness. (Orbital motion is a form of free-fall.)

Yet while a small frame is falling, it is enveloped in a wider more global area of indented space-time, so that upon a transformation from the local to the global frame of reference, the flatness of space-time is supplanted by curvature. Gravity is "transformed back," or (as we quoted Einstein on page 124) a gravitational field is produced by changing the system of coordinates. Such a transformation is tantamount to a change in the frame of reference from one in uniform motion (no acceleration) to one in accelerated motion, while acceleration provides a gravitational field. We conclude that *free-fall provides a locally flat sample of space-time, wherein the laws of physics are Lorentz-invariant*. Of course special relativity applies where space-time is flat.

This is why Einstein found the conditions during free-fall to be so useful. After all, he insisted that every frame of reference be equally valid, including ones in which we cause local gravitation to vanish. We can say that the inertial observer inspired Lorentz to fix the laws of physics for uniform motion, and the freely falling observer inspired Einstein to rewrite these laws for accelerated motion and, ergo, for gravitation. Kenyon (pp. 63 and 46) summarizes these issues as follows: "The principle of general covariance...postulated by Einstein has two components. Firstly, physical laws must be expressible...so that they remain valid under transformations to any accelerated frame. Secondly, when specialized to a frame in free-fall the physical laws should reproduce the established laws consistent with [special relativity]."

In this context Einstein's task had two parts: The first was to find laws which hold in flat space-time—which are Lorentz-invariant—and, for reasons which will be clear shortly, to express them in vector and tensor form. In the process Einstein envisioned a global mosaic of small local pieces of space-time, each of which hides gravity but each of which accepts special relativity. Hence we subdivide curved space-time into points, we give each point coordinates of its own, and we treat each point-and-its-coordinates as freely falling frame of reference. An observer in a such a frame can describe his experience concisely: "*If a law works for me here and it works for you elsewhere, then it works for anyone anywhere.*" In other words Einstein composed his equations so that they remain unchanged when moved from "here" to "elsewhere" and hence to "anywhere." This concept lets us reword the goal of general covariance: "The equations for the laws of physics should retain their form under general transformations." Here, in effect, "general" means "all."

The second part required that a law for special relativity should be legitimate for general relativity when curved space-time can be made to appear flat, notably during free-fall. Conversely the laws which apply in the presence of a gravitational field and accelerated motion should envelop and agree with the case with no gravitational field and no accelerated motion. Using the standard terminology, the equations of special relativity must be a "limiting case" of the equations of general relativity. Here "limiting" can be interpreted as "restricted to uniform speeds and negligible gravitational fields." We will demonstrate these notions mathematically later.

The inference that the equations of general relativity must be consistent with special relativity is essential for the integrity of causality in general relativity. We pointed out that special relativity leads to orderly causality and bars reversed causality. The underlying stipulation is that because they cannot reach speed c, "signals" and ordinary objects can only move on time-like geodesics. This means that ordinary observers can only move on time-like geodesics, and it means that we expect freely falling observers to experience orderly causality. For instance freely falling observers cannot find an instance in which an effect antedates its cause. But if freely falling observers, whose local space-time is flat, do not detect reversed causality in any physical event, then *causality can be no different once the same observers become aware of globally curved space-time*. Therefore *all observers must experience the same orderly causality*, even under the conditions for general relativity. This concept is a pivotal insight on the part of Einstein, but once thought through is rather self-evident: Nature makes no exceptions for observers in free-fall, given that these observers differ from others only in the choice of frame of reference.

This brings us back to the issue at hand: How should science explain gravitational behavior if all observers experience the same laws of physics, including the same rules of causality? We also recall our earlier passage (on page 56): If the laws of physics which apply during uniform motion need to be restated to make them independent of uniform motion, how ought the laws which apply during accelerated motion be restated? In other words how should these laws be formulated so as to survive transformations between frames of references which are in various states of motion? In particular, if the Lorentz transformation provides some covariance in the *absence* of gravity, what equations provide covariance in the *presence* of gravity?

At first glance we might consider simply amending the special relativity factor (valid in special relativity) for accelerated motion as used in general relativity. However with "*v*" replaced by "*a*,"

$$\sqrt{1-a^2/c^2}$$

is *not* valid. This is exemplified for the law of conservation of momentum, $\Delta \sum mv = 0$. To give it more general covariance, it cannot just be converted to $\Delta \sum ma = 0$. As a preview, it must be written as an equation whose terms are tensors; it must be written as a *valid tensor equation*, such as

$$T_{ik};k = 0.$$

This "tensorial" form of this law is not merely Lorentz-invariant. It enjoys unrestricted covariance, which, as I already explained, is "*general covariance*" in Einstein's terminology. In this context we can rephrase our earlier statement: "If a physical law in tensor-equation form works for me here during free fall, then it works for anyone anywhere; this law in this form is generally covariant." In this context I believe that a quote out of Lieber (p. 149) states this idea well enough: "...since General Relativity is concerned with finding the laws [of physics] which hold for ALL observers, and since various observers differ from each other...only in that they use different coordinate systems...Relativity is concerned with finding out those things which remain INVARIANT under transformations of coordinate systems. Now...a vector is such an INVARIANT, and, similarly, tensors are such INVARIANTS, so that the business of the physicist really becomes to find which physical quantities are tensors...since they hold good for all observers." The capitalized words are Lieber's, though again the idea that tensor components should change when coordinates are changed is not revealed explicitly.

We now are in a position to anticipate how the two tasks cited earlier can take on mathematical shape. We will select an equation valid under the conditions of special relativity—which means an equation that is Lorentz-invariant—and we will recast this equation into tensor form. Where do we find conditions of special relativity? In free-fall. Which equation should we work on? One that will allow us to describe the effects of curved space-time. Why not select Newton's law of gravitation? Because it is not valid under the conditions of special relativity; it is not Lorentz-invariant. What is the remedy, at least in principle? *To describe nature by means of tensors and to recast the laws of physics, specifically those dealing with gravitation, as tensor equations.*

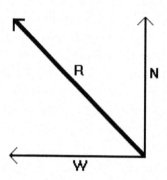

A vector is formed from two components. One component is labelled N (north) and the other component is labelled W (west). The resultant vector is labelled R.

The length of R can be calculated with the pythagorean theorem

$$R^2 = W^2 + N^2.$$

Diagram N. The formation of a vector from its components.

✦ ✦ ✦

We see that in order to understand the mathematics used in relativity we must study tensors in detail, but to do so we first consider *vectors*. A vector tells us how much of something there is at a given point, and in what direction it is going. A vector is thus a "directed quantity," and *motion* is the natural prototype. One good example is a moving boat: "North-west at 14 miles per hour" expresses motion as a vector, specifically a "resultant vector," which succinctly describes the behavior of the boat.

Resultant vectors have *components*, each of which is itself a vector, and the number of component-vectors depends on the number of dimensions. Thus an engine may direct our boat in one direction, say north at 10 miles per hour—one component-vector—but the wind pushes it in another direction, say west at 10 miles per hour—another component-vector. Starting with the component-vectors, "vector analysis" tells us how fast and in what direction the boat moves. This composite quantity is the resultant vector. In two dimensions, north and west in this example, a resultant vector has two component-vectors. To be concise we say "*vector*" for the resultant vector and "*components*" for the component-vectors.

A crucial fact for us is that the basis of vector analysis is the *pythagorean theorem*, as in diagram N. We can show the north-at-10-miles-per-hour component by an arrow 10 millimeters long pointing north; we label this arrow *N*. Likewise the west-at-10-miles-per-hour component can be an arrow 10 millimeters long pointing west and labeled *W*. The vector (a resultant vector) is then an arrow which we label *R* pointing north-west, and its length is calculated as the hypotenuse of a right triangle:

$$R^2 = W^2 + N^2$$

The length of *R* (in this example $\sqrt{200}$ or about 14) corresponds to the net speed of the boat. This routine applies even if the arrows are of unequal length. For example the engine may generate 20 miles per hour, the new direction of *R* is approximately north by north-west, and the new magnitude is $\sqrt{500}$. The calculation must be further modified if the directions of the components are not perpendicular to each other, but this is not pertinent now. However we note that expressing *R* *requires two numbers (W and N)*; two numbers carry the information contained in this vector, while *R* resides in two dimension.

This example can be augmented with additional dimensions. We imagine an airplane. The engine aims it one way, the wind another, and gravity a third, so that we have three-dimensional motion. In three dimensions the vector has three components and entails three numbers, but the calculation still hinges on the pythagorean theorem. Moreover, just as vectors can describe quantities with magnitude and direction in two or three dimensions, they can do so in four dimensions. Thus, looking ahead to the needs of relativity, *a vector in four dimensions has four components*. Expressing it calls for four numbers; four numbers carry the information.

We should comment that the use of arrows as symbols for vectors may imply that a vector acts along the length of the arrow. Actually the vector acts only at the one point illustrated by the beginning of the arrow. This is significant because we will assign vectors (and tensors) to individual points in space-time.

✦

To explain key features of vectors, we focus on the vectorial aspect of gravitational motion—an object falling in a certain direction at a certain speed. We neglected the distinction between speed and velocity till now. To be precise "speed" fits whenever direction does not matter, and "velocity" fits when direction is significant. Speed, or any quantity without a specific direction, is called a "scalar quantity" or a "scalar" for short, and it is one single number. Velocity is an indication of how quickly an object moves in a certain direction; hence velocity, including that of a falling object, is a "vector quantity" or simply a vector.

Furthermore in the traditional three dimensions or coordinates (X, Y, and Z axes), falling motion is a vectorial event with three components, but to handle four dimensions we add T. Designating the "velocity vector" of a falling object as VECTOR and its components as V_X, V_Y, V_Z, and V_T, the VECTOR is linked to its components *in a four-dimensional pythagorean equation*,

$$(VECTOR)^2 = V_X^2 + V_Y^2 + V_Z^2 - V_T^2.$$

CROSS-SECTIONAL DIAGRAM OF SPACE-TIME

(The "surface" is facing up and the "inside" is down.)

Flat space-time — No gravity

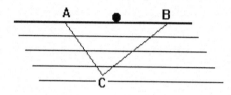

Curved space-time —
Gravity is present

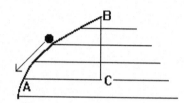

An object shows no effects of gravity when local space-time is flat.
An object shows the effects of gravity when local space-time is curved.

Diagram O. Flat versus curved space-time.

Each right-side term is a component that contains two pieces of information, a magnitude and a direction. V_X is that component of VECTOR that has a magnitude of V in the direction of the X axis. Similarly, V_Y is the component along Y, V_Z is the component along Z, and V_T is the component along T. Each component is a number, and four numbers characterize VECTOR. (For now we disregard that the T-term is negative.)

Incidentally, terms such as vector space, unit vector, and vector basis arise in the literature of physics. If a vector is assigned to every point in a space, the space is a vector space. For example if the direction and speed of the wind can be measured at any location in a wind tunnel, this interior is a vector space. A unit vector is a hypothetical vector whose magnitude is one unit (whatever unit is used—inch, meter, etc.). If all vectors in a vector space are assigned a magnitude of one—if they are all unit vectors—this set of vectors is the vector basis of this space. A vector basis is equivalent to a system of Cartesian (orthogonal) coordinates. Thus a basis for a three-dimensional vector space defines a three-dimensional system of orthogonal coordinates. For instance a point in the wind tunnel can be the origin of three unit vectors u_1, u_2, and u_3 which form the basis of the three-dimensional space in the tunnel. Any vector, such as for the wind, whose orthogonal components are c_1, c_2, and c_3 can be expressed as $c_1 u_1 + c_2 u_2 + c_3 u_3$. This system facilitates vector analysis of physical events and can be extended to four dimensions and to spaces with tensors at each point (tensor spaces).

Let us turn to more geometry. In the left half of diagram O, a horizontal line represents the flat "surface" of space-time. On this surface there is an object in uniform motion between A and B. We create a right triangle ABC, and AB is its hypotenuse. (Of course an object always "moves" through time as long as it exists, but here we ignore this point.) This "flat" triangle has two pertinent features: First, line AB is straight (no curve) and has no slope (no slant). Second, according to the familiar pythagorean theorem, AB is such that

$$(AB)^2 = (AC)^2 + (BC)^2.$$

If we quantify this object's motion, the velocity vector would simply be AB. Under these conditions, elementary algebra and Euclidean geometry suffice for mathematical analysis

However the object might fall. Hence in the right half of diagram O, the object is depicted following a geodesic in *curved* space-time, and we show a "cross-section" of curved space-time. Several differences appear. First, the object is "rolling" into an indentation, and it has a velocity vector shown by an arrow whose length represents that speed and whose direction represents that direction. Second, the triangle ABC no longer has a straight hypotenuse, so that $(AB)^2 = (AC)^2 + (BC)^2$ does not hold, indicating that we have transcended Euclidean geometry. Third, line AB has a slope which changes along the B-to-A path (it becomes progressively steeper) which means that the velocity vector changes continually during this event. Therefore to know the direction of the falling object's motion at any specific point along AB, we use differential calculus which allows us to treat the path from A to B as a chain of points. We then determine the slope of the line AB at any such point. (In the language of differential calculus the slope of a line is its first derivative.) Diagram O shows only two dimensions, and the vector in the right half of this diagram has two components, one along B-to-C and another along C-to-A, but in four dimensions more components would be needed. We see that *to analyze gravity relativistically we must employ non-Euclidean geometry, differential calculus, and vector analysis with four components.*

✦

With vectors in mind, diagram P details a prolonged event in space-time. For simplicity the diagram contains only the coordinate axes for time, but it shows the path of a freely falling apple as a curved geodesic. (The diagram could have shown the event as a straight line and the axes curved, but that makes it harder to interpret.) At point A we draw a vectorial arrow representing the apple's velocity at that point. The arrow's direction indicates where the apple is heading, and its length shows how fast the apple is falling. In fact as the apple falls from one point in space-time (the tree) through a second point (point A) to a third point (the ground), we can treat each point as a momentary event which has a magnitude and a direction. That is to say, the motion of the apple can be regarded as a chain of very short vectorial events.

As diagram P also shows, a vector can be projected onto coordinate axes, and a projection is a component of the vector. If we could portray all four axes, we would see that when ordinary observers study a gravitational event, such as how much time it takes or how much space it occupies, they are only measuring *components of a space-time velocity vector*. Thus, in terms of vectors, *our perception of this physical event is actually an evaluation of its vector components.* And the main activity in this event is motion.

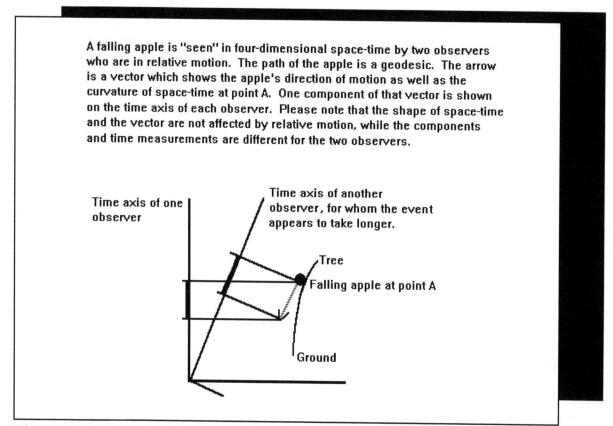

A falling apple is "seen" in four-dimensional space-time by two observers who are in relative motion. The path of the apple is a geodesic. The arrow is a vector which shows the apple's direction of motion as well as the curvature of space-time at point A. One component of that vector is shown on the time axis of each observer. Please note that the shape of space-time and the vector are not affected by relative motion, while the components and time measurements are different for the two observers.

Time axis of one observer

Time axis of another observer, for whom the event appears to take longer.

Tree

Falling apple at point A

Ground

Diagram P. Two relatively moving observers measure the time for an event, the fall of an apple. The apple's motion is a vector. One component is the measured time for the event, which appears different to the two observers.

In other words *events entailing motion, which covers practically all of physics, can be treated as vectors*, and they are *space-time intervals*. In equation form, where *s* is a space-time interval,

$$s^2 = (EVENT)^2 = (VECTOR)^2.$$

Accordingly, we equate our ordinary measurements of events with components and with projections of space-time intervals on coordinate axes. If an observer finds that an event with motion covers X amount of space by Y amount of space by Z amount of space in T amount of time, *its vector covers a certain amount of space-time, which is an interval s.* Thus the four-dimensional equation about a prolonged event in space-time,

$$s^2 = (EVENT)^2 = X^2 + Y^2 + Z^2 - (cT)^2,$$

corresponds to the four-dimensional vector-equation

$$(VECTOR)^2 = V_X^2 + V_Y^2 + V_Z^2 - V_T^2.$$

$$(VECTOR)^2 = V_X^2 + V_Y^2 + V_Z^2 - V_T^2.$$

Therefore, upon conversion of time units to distance units, the equation

$$(VECTOR)^2 = X^2 + Y^2 + Z^2 - (cT)^2,$$

signifies that *our measurements X, Y, Z and T of prolonged events are equivalent to vector components V_X, V_Y, V_Z, and V_T.*

We further recall that a space-time interval does not change despite a rotation of the coordinate axes, and as diagram P depicts, *the same applies to vectors.* Just as projections on the tilted axes emanate from the same interval, *new components still refer to the same vector.* Just as space-time intervals survive transformations but ordinary measurements do not, vectors survive transformations but their components do not. Thus when two observers move relative to each other while they study a vectorial event, *ordinary measurements are components about which the observers disagree, but the vector itself—both its magnitude and its direction—is invariant.* In diagram P, the time axis for each observer is included, and the projections of the length of the apple's invariant vectorial arrow on these axes are unequal. Likewise, (though not included in diagram P) the projections of the apple's invariant vectorial arrow are unequal on the space axes of the two observers.

In short, *a vector is invariant under a transformation of space-time coordinates.* The mathematical counterpart of this statement is that for relatively moving observers A and B studying the same event,

$$(VECTOR_A)^2 = (VECTOR_B)^2,$$

and if we consider the X's, Y's, Z's and T's to be components,

$$X_A^2 + Y_A^2 + Z_A^2 - (cT_A)^2 = X_B^2 + Y_B^2 + Z_B^2 - (cT_B)^2.$$

We emphasize three distinct features of this equation: One, measurements of components are taken along all four coordinate axes. Two, there are disparities between corresponding components measured by relatively moving observers, so that X_A^2 may differ from X_B^2, Y_A^2 may differ from Y_B^2, etc; components are relative. Finally, the sums of the squared components are the same for the observers. We may also now amend our statement on page 124: What the geometer calls "geodesics" and the relativist calls "events," the physicist/mathematician calls "vectors." Moreover what we ordinarily measure are "components." We can say that *events are invariant vectors in space-time, but we live in a world of varying components.*

As useful as it is, the notion that events correspond to vectors poses a problem. A vector acts at one point in a system of coordinates (even though we can represent a vector by an arrow), whereas an event can be prolonged; i.e. a space-time interval can be long and curved. What guarantees that the direction of a vector matches the direction of a space-time interval? The reply calls for differential calculus. As we said, a prolonged interval may be treated as a chain of infinitely small intervals, each of which we label as *ds*. The same applies to vectors. With sufficient miniaturization, the *ds*, the event, and the vector will have the same direction. Furthermore, we say that the *ds*, the event, and the vector occupy one point in space-time.

In this system, as the falling apple's location along the X axis changes by an infinitely small distance dX, the apple covers an infinitely small space-time interval *ds*. The same holds along the Y, Z, and T axes, and the previous equations for observer A can be written as

$$ds^2 = d(EVENT)^2 = d(VECTOR)^2 = dX_A^2 + dY_A^2 + dZ_A^2 - (cdT_A)^2.$$

Then, for two observers A and B,

$$dX_A^2 + dY_A^2 + dZ_A^2 - (cdT_A)^2 = dX_B^2 + dY_B^2 + dZ_B^2 - (cdT_B)^2.$$

Again, in four dimensions dX_A^2 may not be the same as dX_B^2, dY_A^2 may not equal dY_B^2, etc., but we know that in a rotated systems of coordinate axes, *ds* is invariant:

$$ds_A^2 = ds_B^2$$

This concept also applies to differential vectors. If $d(VECTOR)_A$ is a vector as it appears in system "A" while $d(VECTOR)_B$ is that vector as it appears in system "B," then

$$d(VECTOR)_A^2 = d(VECTOR)_B^2,$$

which declares that $d(VECTOR)^2$ is invariant. However the components of d(VECTOR) are relative. Suppose for instance that plotting a falling apple's trajectory shows that the wind pushes the apple in a certain direction. If we draw arrows for the two vectors on a diagram, their relationship (for example the angle between them) is invariant *even if we rotate the map—which is equivalent to rotating the axes*. The *projections* of the vectors are not invariant.

This means that not only are the vectors invariant, but *so is their equation*, which is a crucial point. We recall that the Lorentz transformation accommodates uniform relative motion and that laws of physics can be invariant under a Lorentz transformation. We see that when d(VECTOR) is a *ds*, *vector equations are also invariant under a Lorentz transformation.* Moreover *if a law asserts a relationship between vectors—i.e., if it is expressed by a vector equation in one frame of reference—then that equation is invariant upon a Lorentz transformation to any other frame of reference.* In short, our vector equations are Lorentz-invariant.

◆

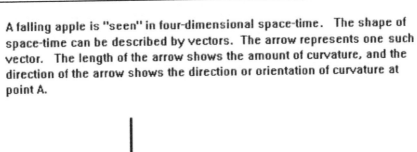

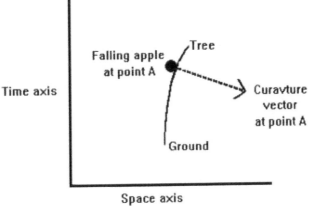

Diagram Q. A curvature vector for space-time.

Diagram Q omits some features of diagram P, but it emphasizes another concept: The *curvature of space-time is also a vectorial quantity*. It has a magnitude, which is the radius at point A. It has a direction, which is its orientation at point A. It has components which can be projected on coordinate axes, and it is invariant if the axes are rotated. As we shall see, space-time may have *several* curvatures, each a vector with magnitude and direction, but let us note an implication which is unique to relativity: Time can have vectorial .curvatures; i.e. the "shape of time" can be specified by vectors. Since a vector has relative (varying) components, time is relative, even in the language of vectors

Now we turn to the "tangent vector" which is linked to the curvature of space-time in a useful way. In diagram P the arrow depicting the fall of the apple is "tangent" to the "surface" of space-time at point A. Readers unfamiliar with the notion of a tangent should hold a flat object, such as a book, against a round one, such as an egg. The book is tangent to the egg at the point where they touch. (In this setting, a tangent is not a trigonometric function.) Please compare diagrams P and Q. P includes a velocity vector for an object in a gravitational field. Q includes the radius of curvature for space-time. If the two diagrams are superimposed, the vector and radius will be at right angles to each other at point A. We therefore say that the velocity vector is tangent to the curvature of space-time and perpendicular to its radius at that point. Since A is on the geodesic of the apple, the velocity of the apple at A is a "tangent vector." In other words the tangent vector depends on local curvature but it is also a local description of whereto and how fast the apple is falling. The tangent vector thus links the behavior of an object in a gravitational field with the curvature of space-time. This means that *as an object falls or orbits, its velocity vector—how it moves—is a tangent vector—how it follows the shape of space-time.* For example where the tangent vector becomes more vertical from steeper curvature, the path of the object also becomes steeper. In short, *the shape of space-time is linked to the effects of gravity.* This of course is just another way of saying that space-time sets the gravitational path of an object, but now we articulate this idea in the language of vectors.

We note three ways of expressing gravitational behavior: We can say that *ds*, which is a vector quantity, is the space-time accounting of falling or orbiting. The *tangent vector* is a geometric accounting of the same event. The *velocity vector* is a physical accounting, in as much as velocity is a tangible and observable quantity. We should add that since we are dealing with objects which do not exceed velocity c, we are limited to time-like geodesics and hence to time-like tangent vectors. Also, the exact calculation of gravitational motion involves an "equations of motion" which entails vectors and which we will use later.

Incidentally, we recall that we can define velocity in differential terms as $\dfrac{dx}{dt}$; an interval of distance per interval of time. We can augment this definition by using the vector concept and by considering four dimensions. This step gives us the velocity vector defined as $\dfrac{dx^a}{d\tau}$, where *a* runs from 1 to 4 and where τ stands for proper time. (The term $\dfrac{dx^a}{d\tau}$ is also called the four-velocity.) The tangent vector then is $\dfrac{dx^a}{ds}$, which is a definition we will use again later.

◆

We emphasize for future reference that physical events, including gravitational ones, are considered to be composed of infinitesimally small space-time intervals in four dimensions. These intervals are endowed with an "immunity" against the effects of transformation of coordinate axes. Intervals in space-time, as well as the shape of space-time itself, can also be treated as vectors which are similarly "immune." Since an observer relies on projections of events on his time and space axes, he in effect tries to devise laws of physics based on components. These laws may yield conflicting results for observers in other frames of reference. In order to delineate physical events so as not to depend on the relative motion of observers or on the relative motion of the events themselves, the strategy should hinge upon the invariance of vectors.

From this juncture our direction is clear. Vectors provide the link between the concept that laws of physics are universally successful and the mathematical necessity that these laws be expressed in universally applicable equations. Covariance in physical law stems from invariance of intervals treated as vectors in space-time. Since we can handle vectors in small "differential" increments, we should be able to recast equations about events in space-time from Lorentz-invariant equations to generally covariant equations. This means that *formulating laws of physics, including those for gravitation, through differential vectors in four-dimensional space-time should endow the mathematical expressions of these laws with general covariance.*

✦ ✦ ✦

But where are our tensors? Earlier we suggested that the principle of general covariance can be accommodated by the use of tensors. (Tensor analysis, today called tensor calculus, was developed at the end of 19th century. The basic derivations are credited to Ricci [1853-1925] who initially was not concerned with relativity. See Borisenko and Tarapov, pp. 66-68, or Lieber, pp. 134-139. Einstein reapplied Ricci's work.) Clearly the next step is to connect vectors with tensors, and for this purpose please imagine an event that might be described by a vector, something with direction and magnitude. Now *if that event is so complicated that a vector will not do, we use a tensor to describe that event.* I liken this situation to the problem of describing an oddly-shaped object in two words. Impossible; more words are needed, and if I describe this object with just two items of information, you will not know that I'm talking about. This analogy is quite apt, in that a vector can hold two numbers, but a tensor consists of more than two numbers. In short, a tensor has room for more information.

A common way to approach this issue, seen in almost any scientific explanation of tensors, is to try to describe the physical stress on a malleable or elastic object. Please picture a heavy wooden chest of drawers. To open and close a drawer we must pull or push on the knob. Doing either of these, in the eyes of physics, can exert a "stress" on the face of the drawer where the knob is attached. An "event" might be opening the drawer, and a vector can be the physical description of that event in terms of magnitude and direction: how hard we pull, and which way we pull. We could call this description a "stress vector," and an important point is that this vector has two components, basically because it can be depicted in two dimensions.

Now let's make the scenario more complicated. The drawer is stuck, and we pull so hard that we deform (e.g. we bend) the entire chest. In fact we can push hard on the drawer to try to loosen it, or lose our temper and hit its top, or push on the sides, or even bang against the bottom from bellow. Let's also assume that we want to describe the stresses we can place on this malfunctioning chest, but we are physicists and mathematicians, and we seek a scientific description. (Saying that we "push hard" is not scientific.) Enter the "stress tensor."

We envision that a point inside the chest has a infinitesimally small cube (a purely fictitious mathematical construct). Each side of this cube faces the same way as the walls of the whole chest of drawers. In other words the cube–which is a "volume element"–has "faces" or planes, each with a direction. One face faces front, one faces up, one faces right, etc., so obviously the cube has six faces. But to be scientific, we envision a three-dimensional system of coordinates, so that the front and back face the "x" direction, the sides (left or right) face the "y" direction, and top and bottom face the "z" direction.

If we must identify the member of any pair, e.g. front rather than back, we can give each number a + or − sign. However the signs are often unnecessary; in this case opposing faces are mathematically equivalent (because the cube is so small), so we need to specify only three directions, one for each of the three dimensions. Thus the cube has only three faces that interest us.

The next step is the most intricate. To describe that happens to the chest, we must consider that each face of the tiny cube experiences an amount stress–a magnitude–and it does so in one of three directions. Here a diagram [courtesy Google Images] is helpful.

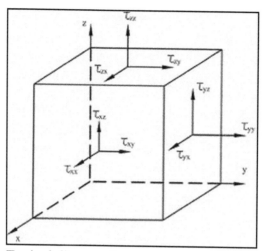

Typical diagram for a stress tensor. The stresses on each face are labeled with a Greek letter τ. The text details the stress on the front face of the cube.

We see in the diagram that stress can act in three ways (x, y or z) with respect to each of the three faces of a tiny cube. However, we should also be aware that this stress is a certain amount of deforming force; the amount of push, for instance, at this point (this cube) in the chest of drawers. Let's focus on the front face of the cube, which is facing in direction x. The stress "felt" by this face can exert itself toward the front or back of the cube (xx), or it can stress [a verb] the cube along its surface left or right (xy), or it can stress the cube up or down along its surface (xz).

Stress perpendicular to the face of a cube is called "normal stress," since "normal" is a geometric synonym for "perpendicular." Stress along the surface (parallel to the face) of the cube is called "shear stress." If necessary, a pulling normal stress can be identified, which is called "tension stress," and its value gets a + sign, whereas "compression stress" may be indicated by a negative value. These terms may important, for example, where beams and cables are stressed and possibly deformed, and in general stress has a magnitude measured as the force per area. The deformation, e.g. sagging of a weight-bearing beam, is called strain; force causes stress which results in strain. Hence tensors loom large in structural engineering.

What happens at a point (at a cube) in the chest is describable by means of three items of information, one that specifies which face is stressed, and two that reveal which way a stress–a magnitude of deforming force–is exerted. Let us use "S" to stand for the cube's stress. Let us further give each S two subscripts to stand for the face's direction and the stress's force in its direction. The first subscript will be a number represented by x, y or z. The second will also be an x, y or z.

The obvious question is how many different S's can apply at any cube. Easy, and we can list them:

$$S_{xx}, S_{xy}, S_{xz}, S_{yx}, S_{yy}, S_{yz}, S_{zx}, S_{zy}, \text{ and } S_{zz}.$$

Each of the nine S-terms is a component, and there are nine of these components in the case of stress. Let's focus on, say, S_{yx}. The y, the first subscript, identifies the face being stressed. The x, the second subscript, tells us which way a certain amount of force stresses this surface of the cube. (That amount of force is expressed as the force per unit area, here acting in the x direction.) Obviously we must agree beforehand which datum will be reported by which subscript. One way that is easy to remember is "face first, stress second," but the reverse convention is also legitimate, and readers consulting various sources should not be confused.

.

No matter which is listed first, the orientation of the face or the stress on that face, let us emphasize that the above list keeps track of three items of information for the cube: the face being stressed, the magnitude of that stress, and its direction with respect to that face. *That entire list of information constitutes a tensor,* in particular a stress tensor. A vector clearly cannot keep track of all the data; as I said, a tensor has more room for information.

A list is a clumsy was to organize the tensor's data. A much better way is in a table or a grid called a *matrix*, and we can give this matrix a label S_{uv}. The first subscript "u" can be x or y or z (depending of which face) and so can the "v" (depending on the direction of the stress). Here is this tensor, expressed as a matrix with its label and with all nine components; it is a matrix of components, each describing one stress:

$$S_{uv} =$$

S_{xx}	S_{xy}	S_{xz}
S_{yx}	S_{yy}	S_{yz}
S_{zx}	S_{zy}	S_{zz}

(In this format no commas are needed between the letters.) One more step will be helpful to drive home the difference between tensors and vectors. Let us assume that the chest is repaired, that the drawers now open with a slight amount of force, and that this force represents too little stress to deform the chest. Now opening a drawer can be described simply by one direction and one magnitude in two dimensions; the event can be described with a two-component vector. That vector can be labeled as V_j, where j is x or y. Now

$$V_j =$$

V_x
V_y

Two components suffice, but the whole matrix is still simple if we wish to consider all three dimensions. (Please recall the path of an airplane.) Our j can be x or y or z, and the matrix is

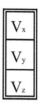

We have three components, of course, and this can be a "velocity vector" for an airplane. This vector can thus be categorized as a "column vector," but in general if a column is inadequate and a matrix must be used, we no longer call this a vector but a tensor. In short, when you see a matrix, think tensor. In this case a column suffices, and even in three dimensions, the vector holds two items of information, a magnitude and a direction. Looking back at the stress tensor, also in three dimensions, we noted that it holds three items, one magnitude and two unrelated directions, so that a vector would not have been sufficient. We need columns and rows.

We can visualize the relationship between vectors and tensors by means of the symbol for a vector, an arrow (whose length shows magnitude and whose direction shows direction). A vector is a mathematical entity which can be completely portrayed by *one* arrow. When a mathematical entity *cannot be fully revealed with just one arrow*, as in the case of stress, that entity is a *tensor*. If we call a vector "one-arrowed," a tensor is "multi-arrowed." A tensor is therefore more inclusive than a vector. Conversely a vector is a limited type of tensor. We can also say that *a tensor is a vector-like descriptive mathematical entity with more components than dimensions*. However, it helps to keep a key concept in mind, lest we forget why all this ado: *A tensor does not change if we change the coordinates, but its many components do change; they are relative or variable.*

Furthermore, looking ahead, a vector may reside in four dimensions, making it a "four-vector" and a tall column vector. However, not every four-vector deals with the four dimensions of space-time. For example in the math of quantum particle physics (notably in Dirac's equation; page 474), an electron can be described by a vector wherein two components reveal whether its mass is positive or negative and another two whether its spin is one way or another.

We also may note what happens at the site of our cube in the chest: Each of the three significant faces of the cube has a very small three-dimensional stress vector sitting on its surface, so we really have three magnitudes-and-directions-of stress in three directions; hence nine components and two subscripts. This scheme, breaking down an external physical stress into nine internal stresses, was worked out by Cauchy in the 1820's. Hence we see the name "Cauchy stress tensor."

In fact the concept can be organized into a hierarchy: A scalar, which entails just one number and no subscript, can be called a zero rank tensor. A vector is then a rank-one tensor requiring one subscript, and a the stress tensor is a rank-two tensor, needing two subscripts. (The word "order" can be used instead of "rank." To figure the number of components c from the number of directions d and the number of dimensions D, we solve $D^d = c$.) But things may get even more complicated: An even more general entity than a tensor is a "spinor," which can describe rotations in space, and which plays a role in modern particle physics. The details regarding spinors are complex, so I defer these for the chapter on how relativity helps in particle physics.

We discuss stress here because it exemplifies a physical quantity *whose mathematical description requires a tensor rather than a vector*. However the stress tensor is based on two dissimilar vectors (for force and surface). We will see a simpler analogy, one based on two vectors that are alike, and later we will find that the metrical tensor, which is vital to general relativity, is easier to envision than the stress tensor.

As we saw, when the three dimensions of space are included, the stress tensor is still of rank two—it still calls upon two vectors—but it has 9 tensor components; 9 numbers carry the information. *Now, if we include a fourth dimension, a tensor of rank two entails 16 components, and the information resides in 16 numbers.* In fact we should keep two combinations in mind because they recur in general relativity: A tensor which depends on two vectors in four dimensions (rank 2, 16 possible components), and a tensor which depends on four vectors in four dimensions (rank 4, 256 possible components, and, in theory, 4 arrows). As we will see shortly, we can often get by with fewer than the maximum number of components (less than 4, less than 9, less than 16, or less than 256 respectively) because of duplication—because two components may carry the same information.

To see how we symbolize tensors, let us devise an analogy for a tensor *which depends on two similar vectors in four dimensions*. We let an automobile dealer order model T cars, while each car comes in a color composed of two tints. We imagine four possible tints, labeled

a, b, c, or d.

Tint "a" can be green, black may be "b," white "c," and blue "d." In this analogy each tint is a vector quantity. Thus the direction might be the hue, such as green, while the magnitude might be "how green." (We could give each tint a number [1, 2, etc.] but for the present purpose, letters are better.) The dealer can pick one of 16 possible colors for any car, as there are 16 possible pairs of tints. The entire population of cars is described by the tensor "T," and the color of each car is given by a component of this tensor.

In particular, each automobile is described according to the two tints that distinguish it, such as tints "a,b" for a green-black color, "b,b" for black-black, "b,d" for black-blue, etc. The individual cars can be labeled T_{ab}, T_{bb}, T_{bd}, etc. The letter "a" in *ab* means the effect of tint "a" on the color of T_{ab}. Likewise "d" in *bd* means the effect of "d" on T_{bd}, etc. We call the effects of each tint a "partial" effect, in the same sense of the word we used earlier and in the same sense we will apply later in tensor equations. The use of colors to explain tensors suggests that tensor components can be likened to adjectives, and we will use this idea in another more complicated analogy. However we note that here we are dealing with *four* "dimensions"—greenness, blackness, whiteness, and blueness—but we need *more than four* "components"—green-black, black-black, etc. According to our practical definition, the cars cannot be described by a vector but only by a tensor.

How should all the possible cars be described? We could write something like

$$T_{aa} + T_{ab} + T_{ac} + T_{ad} + T_{ba} + T_{bb}...etc.$$

with a total of 16 terms, each for one color, and *each term is one component*. This could be shortened to

$$T_{(aa + ab + ac + ad + ba + bb...)}$$

but it is still unwieldy. To avoid long strings of symbols, more abbreviation is needed. An efficient method of notation is to represent the a, b, c, and d in the first position by *one subscript* (u for this example) and likewise the a, b, c, and d in the second position by *another subscript* (v for this example). We then designate the entire population of cars by the *tensor*

$$T_{uv}$$

where *u* and *v* each stand for tint a, b, c, or d. In other words "T_{uv}" means "$T_{tint\ a,\ b,\ c,\ or\ d\ with\ tint\ a,\ b,\ c,\ or\ d}$" and the tensor T_{uv} has 16 components. Eventually we will use subscripts as well as superscripts; both are also called "indices." As we will discuss later, the distinction between subscripts and superscripts is related to the fact that tensors as well as vectors can appear in "co-variant" or "contra-variant" forms.

Another way of listing all possible cars is in the following 16-box matrix of T_{uv}:

T_{aa}	T_{ab}	T_{ac}	T_{ad}
T_{ba}	T_{bb}	T_{bc}	T_{bd}
T_{ca}	T_{cb}	T_{cc}	T_{cd}
T_{da}	T_{db}	T_{dc}	T_{dd}

This matrix is an explicit version of the tensor T_{uv}. Each box holds one component of this tensor and provides one color, while all 16 components describe all possible T-cars. To say something about *all* these cars in tensor terminology, we use T_{uv}. To say something about *only* the T_{bd}-colored car, we use T_{bd}. This symbol implies that none of the other tints come into play, which we can show by using zeros (or, for simplicity, by blanks), and we say that T_{bd} is the only "non-zero" component of T_{uv}. In that case, in equation form,

$$T'_{uv} = 0 \ except \ T_{bd},$$

which for example may mean that out of the 16 possible cars the dealer picks none other than the black-blue one. The matrix of T_{uv} for this particular car then contains only one component, which we can present as

0	0	0	0
0	0	0	T_{bd}
0	0	0	0
0	0	0	0

In tensor terminology, T_{bd} is the "*bd* component of T_{uv}." At the same time we must appreciate that T_{bd} is *not* merely the resultant of two other vectors, *b* and *d*. We must assume that this car's appearance cannot be described by means of only one vector. Rather, the color of this T can only be portrayed by two vectorial tints. No shorter term can specify it as black-blue. In more technical wording, the 16 components of a rank-two tensor are *not* the resolution of 16 vectors into one. In terms of vectorial arrows, such a tensor is a combination of *two* four-component vectors, corresponding to two distinct arrows. This was the case for a stress tensor, but here both arrows refer to one property, color, a circumstance similar to the main tensor relativity uses to describe space-time .

Not only do symbols like "T_{uv}" and "T_{bd}" streamline the expression of tensors but they ease writing tensor equations. If we wish to equate two tensors (T and S) having the same components (u and v), we can write "$T_{uv} = S_{uv}$," and we have conveyed the information efficiently and compactly ("every model T and every model S car has the same color"). This notation also simplifies vector and tensor arithmetic. For example an equation for the addition of tensors is

$$T_{uv} + U_{uv} = V_{uv}.$$

(Adding doors "U" of the same color does not change the color of the car, even though now the car is model "V".) An equation for tensor multiplication is

$$T_{uv} U_{xy} = V_{uvxy}$$

but the color analogy does not help here. In mathematical terms tensors are built out of vectors by a kind of multiplication. For example if x^u is a vector, $x^u x^v$ forms a tensor. (Why we prefer superscripts will be clear later. Details appear in Lawden, p. 92, or any text on tensor algebra.)

Tensors cannot be divided by each other, but we can execute other mathematical operations on tensors, such as summation, substitution, contraction, differentiation, and covariant differentiation. We will encounter these later because they are significant for the tensor equations of general relativity, and we will see how our abbreviations simplify such operations.

We can also deduce two notable properties of our analogical tensor. First, if it does not matter whether we write T_{uv} or T_{vu} (the subscripts are reversed) then this tensor is said to be "symmetrical." I.e. tensor symmetry means $T_{uv} = T_{vu}$. In our analogy a car is the same color whether it is a,b or b,a, and this holds for all other pairs of tints. (Again, this is why the stress tensor is not an ideal analogy for us here.) If the order of indices does make a difference (if for example a green-black car is not identical to a black-green car), the tensor is said to be "antisymmetric" or "skew-symmetric." Rules for handling symmetrical and antisymmetric tensors will interest us later.

Second, even though the above table contains 16 combinations, there is duplication. For example car a,b is the same as b,a; component a,d is the same as component d,a; etc. (Here green-black is the same as black-green, etc. This concept would not apply for stress.) In fact only 10 combinations are unique, which means only 10 numbers carry all the information given by T_{uv}. Or as we said earlier, two components can carry the same information, which in this case makes 6 components superfluous. In tensor terminology we say that only 10 components are "independent." Symmetry governs which components are independent, and again this will interest us greatly.

Other examples of vectors are momentum and force, and we already illustrated that the curvature of a surface can be treated as a vector quantity. However the following fact about curvatures will be very important to us: If a surface is so complicated (i.e. so unevenly curved) that at any point on this surface *one vector is not enough*, we must describe it by a *tensor with rank greater than one*. As we mentioned, a location in space-time may have *more than one* curvature, each of which has magnitude (radius) and direction, suggesting that *we shall be describing space-time using tensors which have a rank of at least two*.

We should be aware that theoretically any physical quantity can be classed as a scalar, a vector, or a tensor, although we rarely bother calling scalars rank-zero tensors nor vectors rank-one tensors. Nevertheless the pattern has significant implications: Scalars, vectors, and tensors share some mathematical properties which are important for relativity. We will find that an important property is how scalars, vectors, and tensors behave under transformations.

In any case when we use indices, a scalar requires no index at all. A vector needs one index because it has one set of components; for example V_u. In two dimensions a vector has two components, so that its single index runs from 1 through 2 (u = 1 and u = 2 in this case). In three dimensions a vector still calls for one index, but that index runs from 1 to 3. In four dimensions a vector has one index which runs from 1 to 4. Since symmetry is expressed by the order of more-than-one indices, scalars and vectors are automatically symmetrical.

154

Usually each index letter stands for a number (rather than for another letter, as in our automobile analogy). Many mathematical discussions of relativity use Latin letters (a, b, c, etc.) when the value of the index can only be 1, 2 and 3, as when only three dimensions are considered. Then when the value of the index can be 1, 2, 3 and 4—as when dealing with four dimensions—Greek letters are often used (α, β, γ, etc.). In this convention V_u implies that u runs from 1 to 3 and that the vector V is in a three-dimensional system of coordinates. Then V_μ implies that μ runs from 1 to 4, and the vector is in a four-dimensional system of coordinates. I will be more explicit and indicate the number of dimensions when the information is critical.

We saw that if a tensor depends on two vectors, it is written with two indices such as T_{uv}, and thus we know its rank is two; three indices are needed for rank-three tensors such as T_{uvw}, etc. The rule about number of dimensions applies as it does to vectors, so that for example a tensor in four dimensions involving four vectors (rank four) has four indices each running from 1 to 4. Such a tensor might be written as T_{uvwx}. It has 256 components, but, as in our example with colors, we saw that not all are necessarily independent. The number of independent components and the symmetry will be an issue when we use tensors with more than one rank.

✦

A critical advantage of vectors we already cited is that while the components of vectors change under space-time coordinate transformations, the vectors themselves do not. *The same applies to tensors.* This should not surprise us. Since vectors are one kind of tensors, the invariance of vectors should hold for tensors. Moreover since we can treat relative motion—which is a vectorial quantity—as a transformation of coordinates, *tensors, like vectors, survive transformation between moving frames of reference.* And, again like vectors, *tensors have components which differ (vary) under transformations* and which may correspond to ordinary measurements of time and space. In geometric terms, components change when the axes are altered but tensors themselves do not. We also say that the altered components represent the same field. The point is that tensors enjoy immunity despite the alteration, and if a quantity needs a tensor of rank greater than one for its description, it still shows immunity. As we will discuss shortly, this implies that tensor *equations* remain invariant under transformations and that they remain invariant during relative motion.

But first we should reconsider scalars. A scalar quantity does not change upon transformation to any other frame of reference. In other words a scalar has no components which can appear different on coordinate axes. Another example of a scalar is temperature; all observers agree on a measurement of heat. Electric charge is also a scalar quantity; if it appears to be 1 volt to one observer, it will be 1 volt to any other, even during relative motion. A scalar of interest to us is the gravitational potential, which is the intensity of local gravity, and because the gravitational potential is a scalar, we will be forced to supplant it by a tensor for the purposes of general relativity. In fact, unlike Newtonian physics, relativity accepts neither time nor space nor mass as scalar quantities. Instead, as we know, measurements of time, size, and mass may vary among observers—they can appear different on different coordinate axes—which means that these behave like tensors, even if the rank is one.

However we reiterate that vectors are insufficient for many descriptions, such as space-time curvature. Hence the main rationale for using tensors to describe physical events relativistically is two-fold: As we said, *tensors can describe complicated things*, but moreover *tensor equations remain invariant during relative motion*. That is not to say that scalar and vector equations cannot be invariant, but many—though not all—crucial physical relationships of interest in relativity are expressed as tensor equations.

We can restate this idea overtly: If "$M = S$" is an equation in which mass and space-time are treated as scalars, it is of no use in relativity because mass and space-time are not scalar quantities. If "$M_a = S_a$" is a vector equation, it is inadequate because space-time may be too complicated to be delineated by a vector. Now if "$M_{ab} = S_{ab}$" is an equation between tensors which holds in the system of coordinates of one observer, and if the tensors survive transformation, then this equation holds in any other system of coordinates. (It should embrace four dimensions, but that is another issue.) In mathematical terms we have form-invariance, as the form of the equation needs no alteration despite the change of coordinates. This, very importantly, means that *a tensor equation successful in one frame of reference succeeds in all, while tensors can describe the complicated things we deal with in general relativity.*

Therefore equations between tensors possess *general covariance*, which means that *tensors allow us to write the laws of physics, including those espoused by general relativity, so that any observer in any state of motion, slow or rapid, uniform or accelerated, can rely on the same mathematical formulation of these laws.* This feature, namely general covariance, gives such laws enormous vitality. Clearly, Einstein adopted tensor-based mathematics precisely because it allows equations to be generally covariant, and indeed we study this form of mathematics because upon it rests the power of general relativity. Yet because of a famous error in his (and his math advisor Grossmann's) tensor algebra, the feature that *tensors survive coordinate transformations even though their components do change* eluded Einstein for about two years (Holt, p. 101-103) while he believed that a tensor itself changes when its components change. This frustrated him so deeply that he temporarily abandoned his principle of general covariance "only with a heavy heart."

Nonetheless we should point out that general covariance is not inherent in relativity, nor for that matter in any other system of physical laws. As we shall see, covariant equations need not have tensor form, and indeed any equation subjected to enough algebra can be made generally covariant. However that does not mean that it lends itself to relativity.

But which tensors are pertinent in relativity? We can already guess: In view of the link between matter and space-time, the tensors of concern to us should fall into one of two categories, *tensors for the shape of space-time*, and *tensors for the distribution of matter and energy*. We defer the problem of how matter and energy can be described by tensors, and we turn next to the basic tensor for the shape of space-time, the metrical tensor.

Though we have used some mathematics to "look into" space-time, two critical issues arise: One, how do we determine whether and to what extent space-time is curved; and, two, how do we use this determination to formulate laws of physics in tensor form? First, if space-time is curved, can we measure its radius? Assuming space-time has a visible surface, we could rise above that surface and look down, like observing the roundness of the Earth from a spaceship observing the roundness of a ball from afar. We could then compare the surface to other curved shapes and determine the radius. However there are obstacles. This method, called "extrinsic," requires us *to leave the surface in question and separate ourselves from it.* We would also have to assume that what we can compare our observations of shape against absolute and invariant standards.

In other words an extrinsic method for ascertaining curvature requires *an additional dimension.* For instance when we detect that a two-dimensional surface is curved by analyzing it from afar, we are looking at three dimensions. This means that determining the shape of four-dimensional space-time extrinsically requires a fifth dimension. First of all there is no physical evidence for the existence of more than four dimensions, and in fact the use of exactly four dimensions also has some theoretical advantages. (See also Barrow and Tipler, pp. 258-276, though there are systems based on more than four dimensions. These are concerned with encompassing relativity into a wider analysis of nature, and we will cite such a system later.) Likewise, if we assume five dimensions, why not six, and then seven, etc? The way to preclude such a nightmare of mathematics and logic is to accept four dimensions for space-time and to gauge the shape of space-time using only these four.

Of course even if we accept four dimensions, we are inside space-time and *there is no visible surface.* We can observe the results of space-time curvature, and in order to interpret these observations we *postulate* a surface for use in computations and in analogies (like our trampoline). That surface still is imaginary, but we treat gravitational motion as occurring exclusively on this surface. In the language of general relativity, the geodesic traced by a falling or orbiting object conforms to the shape of space-time, which means that we think of the behavior of the object as relying solely on the dimensions of the surface of space-time at the object's location. No additional dimensions are invoked, and we assume that an object has no other way and needs no other way of "knowing" space-time. Its behavior is geodesical, but it is not extrinsic.

In this context we will encounter the idea that an imaginary surface is "embedded" in space-time. (For example see Bishop and Goldberg, p. 40.) Imagine a taunt trampoline, particularly its surface, submerged in a large vat of Jell-o that has (of course only in theory) four dimensions. We need not take the idea of embedding any further but we accept that if the Jell-o is deformed, so is the trampoline's surface, which portends that sooner or later we will have to consider all four dimensions. Of course our analogies fail us in four dimensions, forcing us to simply trust four-dimensional mathematics, but for now we can continue to think of space-time as having a surface on which gravitational events occur, even while that surface is fictitious.

Then how do we determine without extrinsic clues whether that surface is curved? Magellan of all people gave us a hint: If the Earth is round, it can be circumnavigated, and doing so confirms the roundness without leaving the surface. This method does not rely on anything extrinsic. It is an "*intrinsic*" method, exploiting what we call an "intrinsic" property of the Earth, namely its roundness; soon we will define "intrinsic" further. But can we really circumnavigate space-time?

Euclid and Pythagoras give us a more practical approach. A surface demonstrates various geometric axioms and theorems (some of which we mentioned already). We can mark out points on a surface and arrange these points into groups of three to form right triangles. We can then test these triangles for their compliance with the pythagorean theorem. Such a method was devised by Gauss, *allowing the curvature of a surface to be assessed solely from that surface*, without external comparisons, without leaving the surface, and without help from additional dimensions. The underlying principle is simple: If the pythagorean theorem works, the surface it's on is flat; if the theorem fails, that surface is curved. Thus two-dimensional clues, such as compliance with the pythagorean theorem, can be used to detect three-dimensional features, such as roundness. This method is *intrinsic*, and we say that is reveals "intrinsic curvature." It is also an exercise in "metrical" geometry.* The pythagorean theorem is a metrical equation because the right triangle can be used to measure and evaluate things. Readers versed in geometry will know that we could also test triangles for whether each contains 180 degrees; they do so only on flat surfaces. Another such test is to see whether a circle of radius r has a circumference of $2\pi r$; it does so only when the circle is drawn on a flat surface. However, it is easier for us to detect intrinsic curvature by testing for compliance with the pythagorean theorem. As we mentioned earlier, such compliance is a sign that Euclidean geometry suffices, and a failure to comply with this theorem is a sign that non-Euclidean geometry is needed; we will reapply this concept later.

The term "intrinsic" in this context means "belonging only to the surface." A method for detecting curvature on the basis of only surface clues is intrinsic; and if curvature is found in this manner, we call it intrinsic curvature. As an important corollary, a geodesic is intrinsic; it exists strictly on the surface (page 104). In his original derivation Gauss restricted himself to intrinsic methods and surfaces with two dimensions, but Riemann later discovered that the same approach works in any number of dimensions, including the four that still later interested Minkowski and Einstein. The fact that these dimensions include time and allude to space-time poses no mathematical barrier. Indeed Riemann's work hints at curved space-time, black holes, tensor calculus applied to gravitation, etc., but he died at age 39 in 1866, and we can conjecture whether he might have pre-empted Einstein had he enjoyed better health.

Let us illustrate the intrinsic method of detecting curvature in a simple case, essentially retracing Gauss' reasoning. We perform a two-part experiment. First, on the surface of a basketball or similar sphere, we draw a line as straight as possible, making it as close as possible to 28.284 cm. long. We let this line be the hypotenuse of a right triangle by adding two other lines, forming the sides of this triangle. We can use diagram C (page 11) by assuming that cT_B is 28.284 cm. and that sides D and vT_B are made equal; we create an isosceles triangle. On the basis of the pythagorean theorem the sides of the triangle should each be 20 cm. long, since $20^2 + 20^2 = 28.284^2$.

---

*The term "metrical," which also appears as "metric," stems from the same root as "measuring," as in the word geometry, which literally means "Earth-measurement."

Yet when we actually measure our right triangle on the basketball, are the sides really 20 cm. each? No, they are longer *because the surface is convex*, and on this scale the curvature is quite apparent. But what if the hypotenuse of our triangle is only 14.142 cm. long? It may be harder to tell with ordinary tools, but the result is proportionately the same; the sides are not exactly 10 cm. long. However the curvature is less obvious. And what if the triangle had a 1.4142-cm.-long hypotenuse? The sides are not exactly 1 cm. long, and with the triangle so small, the curvature is hardly noticeable. Finally we imagine the smallest possible triangle, a minuscule one which we call a "differential" triangle. The curvature is concealed by the diminutive size of the triangle, but if the measurements are sufficiently precise, *the triangle still will not comply with the pythagorean theorem*. Locally the surface seems flat, but *its non-pythagorean property remains mathematically manifest, revealing that the surface actually is rounded.*

Now the second part of our experiment, one that is less mathematical and more intuitive: Out of a flat sheet of stiff cardboard we cut out several right triangles. For convenience the triangles have two sides of 1 cm. each, and if we are meticulous, each hypotenuse will be 1.4142 cm. long. Assuming we cannot bend the cardboard, we will find it impossible to fit several triangles on the surface of the basketball in mosaic-like fashion without some overlapping. (A mason has a similar problem fitting tiles on a spherical surface.) The way several triangles can be made to conform to this surface is by trimming them to that they are no longer right triangles—i.e. *they are no longer pythagorean*. Of course there are gaps between the flat pieces of cardboard and the curved surface of the basketball, but if the triangles are small enough, these gaps are negligible. Nevertheless the need to make each triangle non-pythagorean *discloses underlying curvature*. (The orientation of each triangle is also related to the shape of the ball, but we will defer this aspect of the analogy.)

Recalling our discussion of non-Euclidean geometry, we can now draft a practical general definition of intrinsic curvature: Such curvature exists when a surface does not accept Euclidean-pythagorean geometry. How do we know that a surface does not accept Euclidean-pythagorean geometry? Either because right triangles cannot be pythagorean or because pythagorean triangles do not fit. Which surfaces have intrinsic curvature? A basketball, the Earth, an egg, an indented trampoline, and, as is obviously critical for us, space-time with matter nearby. Extrapolating from an ordinary surface to four-dimensional space-time is intellectually difficult, but nevertheless *wherever space-time has intrinsic curvature, it is mathematically non-Euclidean and non-pythagorean*. Conversely which surfaces have no intrinsic curvature? A flat surface, including that of space-time far from matter (and also a cylinder and a cone, which are just flat surfaces rolled up; the pythagorean theorem holds on such surfaces).

Now we add an important point: When we measure distances and angles among the coordinate points laid out as right triangles, we are in effect conducting *vector analysis*. The sides of these right triangles are vector components, and the hypotenuses are vectors themselves. Moreover Riemann was able to include cases which require more than one vector, allowing us to use *tensors*. As we will detail soon, this line of reasoning leads to the *metrical tensor*, which has a rank of two and which is the mathematical device used to express *the intrinsically determined shape of space-time*. Or, said in reverse, the metrical tensor gives the shape of space-time its metrical property—the property of being measurable by intrinsic means. Given that the shape of space-time is responsible for gravitation, we can expect the metrical tensor to be vital to the laws of physics compatible with general relativity. Indeed this tensor is crucial to the main equations of general relativity.

Accordingly, let us assemble five concepts for use in this chapter: First, four-dimensional space-time rather than separate space and time holds the key to deriving certain correct laws of physics. Second, geometric/mathematical behavior can manifest intrinsic space-time curvature, and this approach calls upon the pythagorean theorem. Third, simple curvatures can be expressed as vectors, and we already hinted that complicated curvatures can be expressed by tensors. Fourth, vectors in space-time are represented by *ds*, while their components are our ordinary measurements of space (dX, dY, and dZ) and time (dT). We recall that transformations of space-time coordinate systems do not alter vectors but do change their components, and this concept extends to tensors. Hence to respect general covariance, we should express gravitational laws of physics by tensor equations. Finally, a geodesic is both an intrinsic feature of a surface and a basic aspect of gravitational motion in space-time.

Since we will work with *ds*, we should mention its several meanings. Depending on the context, *ds* is the smallest possible segment of a world-line, it can occupy a single world-point, it can be a minuscule part of a geodesic, an infinitesimally short interval between momentary events in space-time, an infinitesimal sample of a prolonged event, an invariant vector in a transformed system of coordinates, and the hypotenuse of a possibly pythagorean triangle. We also recall that when we ask "how does a falling object behave here," we seek the solution of equations for *ds*. Bringing these concepts together, *we can expect to derive some form of the pythagorean theorem solved for ds.*

◆

We have just grouped two kinds of intrinsic curvatures together. For instance a basketball (ideally) is a sphere with only one radius and one curvature, but an egg is more complicated. Indeed this is a vital distinction: The shape of indented space-time is also complicated, implying, as we noted earlier, that space-time can have more than one radius and more than one curvature. Let us elaborate by means of another example of a complicated curved object, an American-style or a rugby football. *A point on the surface of this object has two different radii.* The radius along the length is greater than the radius of a cross-section. (The same idea is often depicted by the shape of a saddle. One radius is that of a circle straddled by the rider's legs. The other radius is that of a circle in line with the horse's spine.)

Gauss also pioneered this concept. "K" represents "Gaussian curvature" or "average curvature," and we can designate the two radii of a complicated curved surface as r_i and r_k. Here

$$K = \frac{1}{r_i r_k}.$$

One advantage in expressing curvature as a reciprocal (1/r) is that a greater K is then associated with greater curvature, and it is more intuitive to associate a flat surface with zero curvature than with infinite radii. However the main attribute of K is that it is a convenient measure of *intrinsic but complicated shape.* (In the terms of calculus, the two radii are the minimal and maximal "principal" curvatures at a point of a surface.)

What about surfaces that have two radii but four dimensions? It helps to think in terms of the number of coordinate planes on which we can draw a geodesic, and, in the terminology applied to Gauss' concept, each geodesic has one Gaussian curvature. If we have X, Y, Z, and T Cartesian coordinates, there are 12 coordinate planes on which to draw a geodesic. One plane is bounded by X and Y, another by X and Z, another by X and T, etc. Altogether the possible planes are

XY XZ XT YX YZ YT ZX ZY ZT TX TY TZ

Deleting duplications—XY eliminates YX, XZ eliminates ZX, etc.—leaves 6 unique planes, 6 geodesics, and 6 Gaussian curvatures. In general the number of Gaussian curvatures in terms of n dimensions is calculated as

$$(n - 1)n/2,$$

which is 6 in four dimensions. This fact will raise the question why the metrical tensor has 10 independent components whereas in Gaussian terms space-time has only 6 unique intrinsic curvatures. We will see later that four of the 10 components of the metrical tensor are affected by the choice of coordinates. (A geometric reason for 4 fewer Gaussian curvatures than metrical components is simple: Gaussian planes do not include XX, YY, ZZ, and TT, which are meaningless in this context.) In any case, despite its historical and crucial importance in how to handle complicated intrinsic curvatures, Gauss' method applied to space-time is unwieldy. Multiple curvatures in four dimensions can be treated more efficiently in terms of vectors and tensors than in terms of Gaussian curvatures, especially when performing transformations of coordinates (as will do later). Riemann is credited with introducing most of these improvements.

The concept of Gaussian curvature is also open to challenge on the following grounds: Gauss dealt with the curvature of a sphere in two dimensions, yet a sphere has *three* dimensions. Wouldn't any surface be three-dimensional if it is curved? Yes, the two-dimensional surface of an object which is deformed does gain a dimension, but the *intrinsic* Gaussian curvature of such a surface still has only two dimensions. As we pointed out, an observer need not leave the surface to detect its deformation. We can think of him as an "intrinsic observer" who, using Gauss's method, can *gauge three-dimensional features solely from two-dimensional clues*. In particular, the observer can remain intrinsic while testing for pythagorean compliance, and this insight, though not overt in Einstein's definitive equations, was indispensable to the development of general relativity.

Indeed Riemann extended Gauss's derivations to more than two dimensions, but this is not the same as a three-dimensional departure from a surface. As difficult it is to envision, relativity treats space-time as a four-dimensional "Riemannian space," which can have an (embedded) intrinsically curved two-dimensional surface. True, four-dimensional space-time can be flat or it can have a complicated curvature, but the number of dimensions and the number of radii are separate features.

Incidentally, some significance resides in whether K is positive, zero, or negative (another reason I stress it). This touches on a question in cosmology as to what is the shape of all of space-time or what is the shape of the universe in terms of space-time. (When K is positive, right triangles have overly long sides. When K is negative, triangles will have shorter sides. See Penrose, p. 207.) If space-time's K is positive, our universe may be a "closed" and finite sphere. If K is zero, our universe may be flat and endless. If K is negative, our universe may be saddle-shaped and endless. Most cosmologists favor the idea of a universe with positive K, though the issue is not settled.

◆

We saw that any point on our football (we ignore the tips) has two radii. Let us apply this fact to space-time. Diagram Q depicts the complicated shape of space-time. Any point on this surface has two radii, each of which can be treated as a vector that has a direction and a magnitude. One radius is shown by an arrow pointing toward the right and down. The second radius for the curvature at point A is difficult to show in a flat diagram, and its arrow is not included in diagram Q. However, the magnitude of this second radius is related to the size of the indentation at the level of point A. Since an indentation becomes progressively steeper and narrower farther down the object's path, both radii change and both arrows are different at each successive point. Their difference resides in their lengths; they can share the same direction, but in any case the important lesson for us is that *any one point of curved space-time has two radii that are independent of each other.*

For now let us set aside the issue that the lengths (and directions) may differ at some other point. The pertinent implication is this: We treat radii as vectors. If a vector, which is an example of a rank-one tensor, is insufficient for assessing space-time, we need a tensor of rank two, namely the metrical tensor. *The metrical tensor depends on two vectors because the shape of space-time depends on two radii.* In other words *the two curvatures of space-time ordain the two ranks of the metrical tensor*, enabling the metrical tensor to define the intrinsic and complicated curvature of space-time at any one point.

We should note that having two curvatures is not a peculiarity of space-time or of relativity but a basic geometric feature of a complicated surface. *Any* intrinsically curved object, except a perfect sphere, has two different radii at any one point on its surface. For instance we can describe our football by means of the metrical tensor, and later we will see how to determine the metrical tensor for another common object.

◆

Let us recall that events in four-dimensional space-time are revealed to us through ordinary measurements along four coordinate axes X, Y, Z, and T. We derived an equation,

$$s^2 = X^2 + Y^2 + Z^2 - T^2,$$

in which *X*, *Y*, *Z*, and *T* are also the sides of an abstract four-dimensional right triangle, and in which *s* is the interval or vector that represents an event. Even if a triangle in four dimensions is impossible to visualize, this equation adapts the pythagorean theorem to a fourth dimension (though it assumes flatness). However we are faced with two immediate obstacles. First, *X*, *Y*, and *Z* are in ordinary units of distance, such as centimeters or meters, whereas *T* expresses how much time an event consumes, such as in seconds or minutes. It is desirable for all terms to be in the same units.

Furthermore the above equation has an arithmetically negative T-term, so that when we apply the equation under ordinary conditions (as on page 105) we obtain a negative s^2, which in turn means that s is an imaginary number. To avoid a negative s^2 and an imaginary s, the above equation can be written as

$$s^2 = -X^2 - Y^2 - Z^2 + T^2,$$

but now space is considered to be arithmetically negative. There is no physical difference between these two versions, and depending on the mathematical needs, one or the other version is used, but for present purposes it is better if all terms were positive.

The issue then is to convert the value of T to the same sign and the same units as the other coordinates. This conversion was proposed by Minkowski and Poincaré in the following manner. First, T is again multiplied by the speed of light, c, so that it becomes a measure of distance. This gives us

$$s^2 = X^2 + Y^2 + Z^2 - (cT)^2$$

or, as is sometimes preferable,

$$s^2 = -X^2 - Y^2 - Z^2 + (cT)^2.$$

We should be aware that $(cT)^2$ may be written as c^2T^2.

Second, cT is multiplied by an imaginary number called I (explicitly, it is $\sqrt{-1}$, the square root of -1) so that a negative time-term becomes positive. We use the Greek letter tau to represent this positive-time-which-acts-like-a-distance. In that case tau or $\tau = icT$, and

$$s^2 = X^2 + Y^2 + Z^2 + (icT)^2$$

or

$$s^2 = X^2 + Y^2 + Z^2 + \tau^2.$$

These two equations avert an imaginary s in their solution, and all four coordinates are expressed in positive and compatible units. Incidentally, using icT in $X_B = X_A cosR + T_A sinR$ and solving for $cosR$ leads us to our first equation, $cosR = 1/\sqrt{1-v^2/c^2}$, but then R is an "imaginary angle."

We may question the prudence of using an imaginary number in equations for physics, but because the i in icT always ends up squared ($i^2 = -1$), the solutions do not contain i.

We may also question the selection of c for cT, but c has a practical virtue: The analysis of general relativity begins with the conditions of special relativity, namely uniform motion. Under these conditions we expect "$RT = D$" as well as R itself to apply for all observers. The rate which meets these requirement is c. We can also derive cT on the basis of the light-like interval.

However the reliance on $\tau = icT$ has a notable arithmetic consequence: Any term related to τ must be a very large number, since c is a very large number. (One second of T becomes 300,000 kilometers of cT.) That is to say, unless X^2, Y^2, and/or Z^2 are large, τ^2 is the dominant factor determining s^2. Therefore the leap from three dimensions in

$$s^2 = X^2 + Y^2 + Z^2$$

to four dimensions in

$$s^2 = X^2 + Y^2 + Z^2 - (icT)^2$$

usually generates a very large numerical difference.

This comes as no surprise: Motion along coordinates X, Y, and/or Z over time implies speed, and when speeds are low compared with c, the effects of relativity over and above Newtonian physics are negligible. We can write that if v<<c (v is much less than c), then

$$X^2 + Y^2 + Z^2 \ll \tau^2,$$

which means that in four-dimensional geometry, slowly falling objects consume much time to cross little space.

Indeed, even though in relativity "time is added to space," it is rational to reverse this statement. We could say that under every-day conditions relativity adds *space to time*, and we could call the result "time-space" instead. In other words relativists are often accused of appending an exotic fourth dimension to the other three familiar dimensions, yet arithmetically the situation is the opposite: During low speeds the time dimension is the major ingredient, and the three spatial dimensions are minutiae.

The notion that time can stand as a mathematical peer next to height, width, and depth is a fundamental aspect of relativity. For example it is the basis for drawing space-time diagrams and for writing equations in which a value for X, Y, or Z can be easily added to a value for τ. More importantly we recall (page 101) that the incorporation of time within space-time ensures the invariant nature of the space-time interval.

Solving the issue of how to handle T still leaves a critical problem: How do we handle a curved s? We invoke differential calculus, permitting us to treat any s as a series of infinitesimally small intervals, ds's. Similarly, switching as we must to four dimensions, X, Y, Z, and T can be subdivided. Thus we write that

$$ds^2 \approx dX^2 + dY^2 + dZ^2 - dT^2,$$

and after converting T to τ, we obtain a four-dimensional differential pythagorean equation,

$$ds^2 \approx dX^2 + dY^2 + dZ^2 + d\tau^2.$$

(The "$\approx$" used here means "is approximately equal to." We will explain in the next paragraph.) The differential form of the pythagorean equation "straightens out" the sides of our triangles in the sense that if the triangles can be made sufficiently small, their own curvature is concealed. Of course it is not feasible to take actual measurements on such miniaturized triangles, but, mathematically speaking, that is not an objection, especially when we apply this equation to space-time.

Let us go back to why we just used $\approx$ rather than =, and at the same time let us use equations to express compliance and non-compliance with the pythagorean theorem. Our basketball analogy helps here. If we deflate and flatten the ball, in theory we can draw a two-dimensional differential right triangle on its surface which complies with the pythagorean equation

$$ds^2 = dX^2 + dY^2.$$

Once we re-inflate the ball *so that its surface has intrinsic curvature*, the pythagorean theorem ceases to hold. That is to say, the surface has become non-pythagorean and non-Euclidean. Now, in equation form,

$$ds^2 \text{ does NOT } = dX^2 + dY^2.$$

Likewise *even in differential form and in four dimensions*, as long as we have curvature of space-time (and the term "*does NOT =*" is replaced by the corresponding mathematical symbol $\neq$),

$$ds^2 \neq dX^2 + dY^2 + dZ^2 + d\tau^2.$$

This point is important enough to repeat: Use of the differential form conceals curvature of a triangle, but it does *not* conceal non-compliance with the pythagorean theorem. The word "intrinsic" is understood.

◆

We now turn to the mathematical basis of the metrical tensor, as we recall our overall motivation: How do we quantify the curvature of space-time? Non-compliance with the pythagorean equation informs us that space-time is not flat, but how can we attach a numerical value to the amount of non-flatness? Can the above equations be adapted to *quantify* curvature? Clearly *the extent of non-compliance* can provide the answers.

To use this notion we must appreciate that the above equations are actually abbreviated forms in which certain parts are omitted because their existence is implied. (We are also assuming that we are using a Cartesian system of coordinates.) Here, limited to two dimensions for simplicity, is an un-abbreviated form:

$$ds^2 = (A)dX^2 + (B)dXdY + (C)dYdX + (D)dY^2$$

This is called a "generalized" form of the pythagorean equation because it encompasses the quantities A, B, C, and D, each of which may vary. (The above equation is based on the algebraic rule that

$$(X + Y)^2 = (1)X^2 + (1)XY + (1)YX + (1)Y^2.)$$

Here A, B, C, and D are *coefficients*, meaning that each is multiplied by its corresponding quantity to give the proper value. (dX^2 is multiplied by coefficient A, $dXdY$ is multiplied by coefficient B, etc.) In the context of pythagorean equations we may call A, B, C, and D "metrical coefficients." We can think of coefficients as adjectives. An adjective modifies a noun as a coefficient modifies a mathematical term. For example if "A" is a large number, and "dX^2" is a value, then "$(A)dX^2$" means "large value."

A critical detail is that the pythagorean theorem tacitly assumes that on a *flat* surface

$$A = 1, \; B \text{ and } C = 0, \; \text{and } D = 1.$$

This means that we can rephrase the question posed earlier in this chapter, "how do we determine whether a surface is curved using intrinsic clues?" The answer is *we apply the generalized form of pythagorean theorem to a right triangle on that surface. If the coefficients A, B, C, and D are not 1, 0, 0, and 1 respectively, then there is curvature.* Conversely if *A, B, C,* and *D* are 1, 0, 0, and 1 respectively, then there is flatness. Form here we can make the obvious leap: *How far A, B, C, and D are from 1, 0, 0, 1 can tell us how much curvature* there is, which of course is our current goal. In other words *the numerical values of A, B, C, and D quantify curvature.* Clearly this principle will be very useful to us, but for now we must continue covering certain basics for the metrical tensor.

Another important feature of the generalized form of the pythagorean theorem is that dXdY is the same as dYdX, so that the use of four coefficients is unnecessary; only three are needed. A better way of saying this is that only three coefficients are independent; *C* is not independent of *D*, as we saw in our analogy with colored automobiles, where certain combinations of tints were redundant. We can write that

$$ds^2 = (A)dX^2 + 2(B)dXdY + (D)dY^2.$$

Now *on a flat surface A = 1, B = 0, and D = 1,* in which case

$$ds^2 = (1)dX^2 + 2(0)dXdY + (1)dY^2,$$

which reduces to the familiar equation we started with,

$$s^2 = X^2 + Y^2.$$

The foregoing considers only two dimensions, and eventually we must deal with four, so clearly we will end up with a bewildering number of symbols in a long equation. A more efficient notation is based on the idea that mathematically all dimensions can be treated the same. We use indices for this purpose; *dX* becomes dx_1 and *dY* becomes dx_2. Using *dx*'s,

$$ds^2 = (A)(dx_1)^2 + (B)(dx_1)(dx_2) + (C)(dx_2)(dx_1) + (D)(dx_2)^2,$$

and since a separate *C* is unnecessary,

$$ds^2 = (A)(dx_1)^2 + 2(B)(dx_1)(dx_2) + (D)(dx_2)^2.$$

Even though our indices here are *subscripts*, there is a rationale for *superscripts*. Einstein used subscripts in this original derivation, but in modern texts the above equation is written as

$$ds^2 = (A)(dx^1)^2 + 2(B)(dx^1)(dx^2) + (D)(dx^2)^2,$$

which is mathematically more precise but which is harder to interpret. For example in $(dx^2)^2$ the first $...^2$ identifies the dx (as dY) and second squares it, which is clearer written as $(dx_2)^2$. Since many terms in our equations will be squared or will have other exponents, subscripts avoid confusion. We will explain the difference and use superscripts when the distinction is significant. (Even Einstein's biographers switch to superscripts. Compare p. 64 of The Meaning of Relativity with Pais' p. 219.)

In any case the next step is to associate each coefficient with its corresponding dx's, or, to use the grammatical analogy, to link each adjective with its noun. For this we label all the coefficients with the same letter, by tradition "g," and we give the g's *subscripts which match the subscripts of the dx's*:

The coefficient A becomes g_{11} to show that it is to be multiplied by $(dx_1)(dx_1)$ or $(dx_1)^2$,

the coefficient B becomes g_{12} to show that it is to be multiplied by $(dx_1)(dx_2)$,

the coefficient C becomes g_{21} to show that it is to be multiplied by $(dx_2)(dx_1)$, and

the coefficient D becomes g_{22} to show that it is to be multiplied by $(dx_2)(dx2_2)$ or $(dx_2)^2$.

This allows the following explicit form (omitting the non-essential parentheses):

$$ds^2 = g_{11}dx_1dx_1 + g_{12}dx_1dx_2 + g_{21}dx_2dx_1 + g_{22}dx_2dx_2.$$

However just as $dXdY$ is the same as $dYdX$, so $g_{12}dx_1dx_2$ is the same as $g_{21}dx_2dx_1$, in which case

$$g_{12} = g_{21}$$

and, absorbing the unnecessary g_{21},

$$ds^2 = g_{11}(dx_1)^2 + 2g_{12}dx_1dx_2 + g_{22}(dx_2)^2.$$

Several points should be emphasized about the above equations:

First, the g's are coefficients which adapt the pythagorean theorem to a curved surface. We will elaborate on this later.

Second, we can arrange the coefficients in a table or matrix,

g_{11}	g_{12}
g_{21}	g_{22}

By crossing out the duplicated g (g_{21}) we can show that only three g's are independent. For two dimensions all we need is:

g_{11}	g_{12}
	g_{22}

Third, the term ds^2 does not need a coefficient. This reiterates that if one set of g's and dx's is changed (transformed) to another in a correct manner, the space-time interval ds does not change.

Fourth, it is no accident that this system is also used to label vectors and tensors where the subscripted terms refer to *components*.

Finally, it is also no accident that tensors may have only a limited number of independent components, while the above equations have only a limited number of non-duplicated g's. These two ideas are identical.

With subscripted g's to streamline the notation, we can readily consider three dimensions by writing a generalized differential pythagorean equation which includes Z. All that is required is subscripts which include the number 3 so that dx_3 replaces dZ, but it is still a rather formidable equation:

$$ds^2 = g_{11}(dx_1)^2 + g_{12}dx_1dx_2 + g_{13}dx_1dx_3$$

$$+ g_{21}dx_2dx_1 + g_{22}dx_2dx_2 + g_{23}dx_2dx_3$$

$$+ g_{31}dx_3dx_1 + g_{32}dx_3dx_2 + g_{33}(dx_3)^2$$

In a matrix form we could have 9 g's, but the same pattern holds true: Some of the coefficients share the same value, they are not independent, and therefore they can be eliminated. For example dx_1dx_3 is the same as dx_3dx_1, so that $g_{13} = g_{31}$. All in all

$$g_{12} = g_{21}, \quad g_{13} = g_{31}, \quad and \quad g_{23} = g_{32}.$$

This gives the matrix only 6 non-zero g's:

g_{11}	g_{12}	g_{13}
	g_{22}	g_{23}
		g_{33}

We note that the leap from two to three dimensions entails additional terms but still calls upon the basic pattern of the pythagorean theorem; new terms are squared and added in. One might suppose that with three dimensions the terms ought to be cubed (raised to the third power), which is not the case.[*]

What about all four dimensions? Yes, we can write a generalized four-dimensional differential pythagorean equation which includes a term for time, and we know how to convert dT to dτ. *The subscripts include the number 4, and dτ is represented by* dx_4. This equation is even more forbidding, involving 16 g-containing terms, and we need not present it all. It begins with

$$ds^2 = g_{11}(dx_1)^2 + g_{12}dx_1dx_2 + g_{13}dx_1dx_3 + g_{14}dx_1dx_4 + g_{21}dx_2dx_1...$$

and it ends with

$$...g_{43}dx_4dx_3 + g_{44}(dx_4)^2.$$

Please consider what we just accomplished: In admitting dx_4 we have fully assimilated *time* into our pythagorean equation, and we have done so "democratically." Subsequent computations will treat dx_4 the same as the other dx's, and in fact there will be no distinction between any of the dx's. For instance a 10% change in time (after multiplying by c and i) has the same relativistic significance as a 10% change in distance; space-time is just that: not separate space and separate time but space-time. This is not to say that the values of the dx's are all the same, only that mathematically each can be handled the same as the others. For example we already indicated that dx_4 is often the largest of the dx's, which means that in an equation, the term associated with g_{44}, such as $g_{44}(dx_4)^2$, is often the largest.

---

[*] This point touches upon a famous riddle in mathematics, Fermat's last theorem. Before he died (in 1665) Fermat claimed that this equation

$$X^n + Y^n = Z^n$$

does not hold if n is any whole number over 2. Although Fermat's assertion seems correct by trial-and-error, proving has been very difficult, yet it implies that even with more than two dimensions, our equation can not have cubed terms.

Again the pattern holds true for four dimensions as it did for two and three: Some of the coefficients have the same value, they are not independent, and therefore they can be eliminated:

$$g_{12} = g_{21}, \quad g_{13} = g_{31}, \quad g_{14} = g_{41}, \quad g_{23} = g_{32}, \quad g_{24} = g_{42}, \quad g_{34} = g_{43}.$$

The matrix without the duplicated *g*'s follows a similar pattern:

g_{11}	g_{12}	g_{13}	g_{14}
	g_{22}	g_{23}	g_{24}
		g_{33}	g_{34}
			g_{44}

This leaves 10 independent coefficients, and we must make note of this number. It means that in the equation for the generalized four-dimensional pythagorean theorem only 10 *g*'s are needed, just as on page 154 only 10 out of 16 possible colors are unique. Like "a,b" = "b,a" so $g_{12} = g_{21}$; in effect "b,a" and g_{21} are redundant.

(To predict how many coefficients are essential all we need to know is the number of dimensions; for 2 dimensions, $1 + 2 = 3$ coefficients; for 3 dimensions, $1 + 2 + 3 = 4$ coefficients; and for 4 dimensions, $1 + 2 + 3 + 4 = 10$ coefficients. The same short-cut applies to the number of components of a symmetrical rank-two tensor. But if the tensor is antisymmetrical, the number of coefficients is less.)

Even with only 10 *g*'s, an equation for ds^2 is unwieldy. We need two additional abbreviations. First we see that the above equations have + signs, which of course means that parts of the equation are added or summated. We could use the symbol Σ to show that these parts are summated, but we can apply the "summation convention," devised by Einstein, which allows the omission of this symbol and the + signs when indices appear twice in one term. (The summation convention is discussed in Bishop and Goldberg, p. 65, and we will illustrate it later.)

Second, instead of labeling each term with its own numbered subscript, we can indicate the subscripts by letters, for example i and k. In the equation for four dimensions we let the subscript

$$i = 1 \text{ through } 4$$

and we let the subscript

$$k = 1 \text{ through } 4.$$

In other words we hold i at 1 while we let k = 1, then k = 2, then k = 3, then k = 4. Next we hold i at 2 and again let k be 1, then 2, then 3, etc. This is called summing over i and k. There are 16 i-k combinations, 1-1, 1-2, 1-3, 1-4, 2-1, 2-2, 2-3, 2-4, 3-1, 3-2, 3-3, 3-4, 4-1, 4-2, 4-3, and 4-4. (By eliminating duplications such as in the case of 2-1 and 1-2, 10 remain independent.) Now the four-by-four matrix of g_{ik} can be expressed much more economically:

$$g_{ik(i\,=\,1,2,3,4)(k\,=\,1,2,3,4)}$$

In this manner, instead of writing g_{11} and later in the equation g_{12}, we can write g_{ik} wherever a g appears. Thus the term

$$g_{ik}$$

is a collection of all the coefficients for $dx_i dx_k$. With these steps combined, the entire equation is written

$$ds^2 = g_{ik} dx_i dx_k.$$

(Without the summation convention this is written

$$ds^2 = \sum_i^{1,...,4} \sum_k^{1,...,4} g_{ik}\, dx_i\, dx_k.$$

Although less compact, this format reveals that we are dealing with a sum of many multiplied terms, and that i and k equal 1 through 4.) We already mentioned that later on we will prefer a form with superscripts,

$$ds^2 = g_{ik}\, dx^i dx^k.$$

As we may have guessed, g_{ik} *is the metrical tensor.* The obvious question is how a particular set of coefficients, which is what g_{ik} stands for, can be a tensor.

The answer lies in two areas:

One, these coefficients can quantify curvature, and curvature can be treated as a vector quantity. However if the curvature is sufficiently complicated, its description requires a *tensor* rather than a vector, and the coefficients are its components. Two, g_{ik} shows mathematical behavior, notably how it transforms, that is common to tensors. Eventually we will elaborate on both areas. Suffice it for now that the two subscripts confirm that the metrical tensor has *a rank of two*, and we just saw that it has exactly *10 independent components*; it carries its information in 10 numbers. We also note that the g's in the previous matrix are those 10 components.

An important characteristic of the metrical tensor is that it is *symmetrical;* its indices, i and k, may be reversed because the components represented by each subscript carry equal weight and similar meaning, just as on page 154 the order of our cars' characteristics made no difference. Later we will show that this symmetry can be corroborated. The choice of letters such as "ik" is arbitrary. However let us reserve h, i, j, k, l and m for components of the main tensors we will deal with, and let us use other letters for demonstrating the mechanics of tensor algebra.

We can call the equation $ds^2 = g_{ik}dx_i dx_k$ (and $ds^2 = g_{ik}dx^i dx^k$) "the metrical equation for the generalized differential pythagorean theorem." It is generalized by the g's so as to admit other-than-flat surfaces. It is made differential by the d's. It is pythagorean in its basic form, which is solvable for ds. And it is metrical by what it does, which is measure surfaces, specifically their curvature. For brevity the term "*metrical equation*" suffices, and we use "g's" as a synonym for the metrical tensor g_{ik} or for its components. Many authors call the "metrical equation" simply "the metric."

We should note that any equation solved for ds can be called a "metrical equation" when it permits the determination of an interval between two points on a geometrical surface. Since our "surface" and points are in space-time, "$ds^2 = g_{ik} dx_i dx_k$" is a space-time metrical equation (even though the surface is fictitious and, as we will note later, even though the points are infinitely close together). While this is the most important and the most general metrical equation, it is not the only one we need. For example intervals on a sphere will be found more conveniently via the metrical equation "$ds^2 = g_{11}dr^2 + g_{22}d\theta^2 + g_{33}d\varphi^2 - g_{44}c^2dt^2$." Metrical equations in various forms are pivotal in general relativity; we will use them repeatedly.

Before going further, we must appreciate the relationship between tensors and tensor components. We already used colors in an analogy for tensors, but that scheme is too simple for our present purpose, except to show that a tensor is a complicated mathematical "object" characterized by many components, much like a complicated physical object that can be characterized by many adjectives. A tensor of rank two then is a mathematical object requiring two sets of adjectives; to identify it, either set alone does not suffice. We assume now that the object in question, labeled as "g," is a "table top" which is characterized by "g_{ik}." In this analogy, a simpler object might need a vector, which means it would need only one set of adjectives.

We imagine that an observer must identify this object by its adjectives. First adjective: "Square-shaped." Second adjective: "Flat." Third adjective: "Wooden." Fourth adjective: "Heavy." Fifth adjective: "Man-made." Sixth adjective: "Furniture-like." Etc. We assume that altogether 16 adjectives exist for "table top," but that six are redundant. In fact 10 adjectives organized into two sets suffice, and once all 10 are enumerated, the observer knows it is a table top. (We could even limit each adjective to 4 choices [4 shapes, 4 kinds of wood, etc.] so as to represent 4 dimensions, but this detail is not essential here.)

The point of this analogy is that tensor components are mathematical descriptions which identify something complicated, much like adjectives are linguistic descriptions which identify something complicated. Adjectives let us recognize a certain object. In the context of relativity, components of a tensor let us recognize the shape of space-time. The analogy also assumes that we have no single word for "table top" and that we know it only by its adjectives. Of course we normally do not say "square-shaped flat wooden heavy man-made furniture..." for a table top, but to name a tensor we enumerate its components. The enumeration may be symbolic and abbreviated, and its individual members may appear as a list or a matrix, but without the components, the tensor has no identity. Thus "$..._{ik}$" represents a collection of "adjectives" which are grouped into two sets, labeled i and k, and which taken together give "g_{ik}" its entire meaning.

A weakness of this analogy is that in speech the order of adjectives is rarely critical, nor is their grouping. A table top is flat and heavy or heavy and flat. In tensor calculus the order of "adjectives" may matter, and we assign a group for each. For example, the g_{43} component has a specified site in g_{ik}, and if we place it elsewhere (e.g. at g_{33}) or if we form more groups (e.g. g_{331}), we alter its meaning.

On the other hand the analogy illustrates *the concept of transformations* in that the adjectives effectively characterize an object even if it is observed form another location, from another vantage, or from another state of motion. For example our table top on Earth may be observed from the moon. In this new frame of reference the adjectives are not all the same. "Wooden" is still fitting, but the table top may no longer appear "square-shaped" (since relative motion affects the appearance of shape). Nevertheless the observer can tell that a table top is a table top because he heeds certain rules in the use of adjectives; for example he knows how squares change during relative motion. (We shall use other analogies for tensor components to make this point.)

Thus, even if we switch form one frame of reference to another, the rules—as long as they are applied correctly—are such that a certain collection of adjectives still reveals the identity of this object. In theory *all observers anywhere in the universe should agree that the object is a table top.* This idea is not unique to relativity: The appearance of objects—hence the adjectives we give them—depends on how we examine them, but this does not prevent us from agreeing with other observers about the identity of what we see. We encountered the same concept at the end of the previous chapter, where we identified a vector or tensor despite different components on differently-tilted coordinate axes. In the case of the metrical tensor, its 10 components, condensed as g_{ik}, allow other observers to discern invariant features of space-time, even when components change under a transformation.

Clearly, to write a reliable law of physics dealing with table tops, as for example "matter bends table tops", we should incorporate a description of table tops based on tensors, so that the form of this law can be invariant on Earth, on the moon, or anywhere else in the universe. Likewise *to write a reliable law of physics dealing with the shape of space-time ("matter bends space-time"), describing*

173

that shape with a tensor allows us to design a generally covariant equation, one that holds unchanged upon transformation to any frame of reference. Indeed in later chapters we shall write generally covariant tensor equations that tells us to what extent matter bends space-time.

◆

We underscore that the metrical tensor can be a *real description of the shape of a real object*, and we can perform a simple experiment to find a set of *g*'s: Ideally each point on a table top has the same shape, flatness. We draw Cartesian coordinates on our table top, and we select two nearby points at which to sample the surface; one point is at "X by Y" or, as we prefer to label it for two dimensions,

$$x_1 \quad by \quad x_2;$$

the other point is at

$$(x_1 + dx_1) \quad by \quad (x_2 + dx_2).$$

The distance between the two points is *ds*. In other words in shifting from location "x_1 by x_2" to location "$(x_1 + dx_1)$ by $(x_2 + dx_2)$," *ds* is how far we have gone. In diagram N if we change location so that one component of our motion is N and the other is W, R is how far we have gone. Of course calculating R from N and W necessitates the pythagorean theorem, as does figuring *ds* from dx_1 and dx_2.

In theory, and in keeping with the principles of differential calculus, we should visit and examine every possible point on out table top (which, again in theory, entails an infinite number of points) but to be practical we assume that the location of x_1 by x_2 is representative. We also assume that we can measure the *dx*'s and *ds* with a ruler. We enter our results into an explicit version of the metrical equation for two dimensions,

$$ds^2 = g_{11}(dx_1)^2 + 2g_{12}dx_1dx_2 + g_{22}(dx_2)^2.$$

For example, we find *ds* to be 5 inches, dx_1 to be 3 inches and dx_2 to be 4 inches. *We then try out different combinations of g_{11}, g_{12}, and g_{22} until we satisfy this equation.* If our table is flat we find that

$$g_{11} = 1, \quad g_{12} = 0, \quad and \quad g_{22} = 1$$

because

$$5^2 = (1)(3)^2 + (2)(0)(3)(4) + (1)(4)^2$$

and because this surface accepts the pythagorean theorem. Please compare these g's with the A, B, and D on page 166. In matrix form, the metrical equation is satisfied if the metrical tensor, g_{ik}, has these components:

$g_{11} = 1$	$g_{12} = 0$
	$g_{22} = 1$

Since we have determined all three components, we have succeeded in determining the metrical tensor. The result tells us that this set of values for the g's in this system of coordinates everywhere on the surface indicates that this table top has zero intrinsic curvature. Stated conversely, flatness gives this surface certain g's. Again we can use our grammatical analogy: We might say that we see a "flat" surface. In tensor mathematics we say that we see a "$g_{11} = 1$, $g_{12} = 0$, $g_{22} = 1$" surface. In English just one word, flat, carries the information about shape. In tensor mathematics three numbers, 1, 0, 1, carry the information given by g_{ik}, which in this case has three independent components.

As we already hinted, the obvious possibility to consider is that g_{11} *may not be 1*, g_{12} *may not be 0, and/or* g_{22} *may not be 1*, and we can guess that this means that *the table is not flat.* We can even focus on just one g, say g_{11}. Since "$g_{11} = 1$" is associated with flatness, the implication is that *how far* g_{11} *deviates from 1 is an indication of how far the shape of a surface deviates from flatness.* We will soon use another analogy to demonstrate how g_{11} does not equal 1 on a curved surface, and will also consider other complications, namely two curvatures at one point and point-to-point variation in curvature. Suffice it for now that if our table is badly warped, a speaker might say that its surface is "very non-flat," just as the mathematician can say that the metrical tensor for this surface is "very non-$g_{11} = 1$," all because a triangle drawn on this surface is "very non-pythagorean."

Of course determining the metrical tensor for four-dimensional space-time is not as simple as it is for a table top, but the concepts are just as basic. Flatness of a four-dimensional surface is indicated when, at all points (using Cartesian coordinates) the 10 independent components of the metrical tensor in $ds^2 = g_{ik} dx_i dx_k$ are

$$g_{11} = g_{22} = g_{33} = g_{44} = 1$$

and

$$g_{12} = g_{13} = g_{14} = g_{23} = g_{24} = g_{34} = 0.$$

This case is obviously important in relativity, since space-time is flat in the absence of a gravitational field. In English the one adjective "flat" might suffice to identify the shape of space-time that is free of gravitation, but tensor mathematics is more exacting: The values for the 10 *g*'s of the metrical tensor are needed, each *g* has a certain value (1 or 0), and these values are the same—they are constant—at every point in question. (We still ignore the details about how we subdivide space-time into a myriad of points, but later we will find that our calculus allows us to do so and to compare points.) With only the non-zero *g*'s identified, the corresponding explicit four-dimensional equation is

$$ds^2 = g_{11}(dx_1)^2 + g_{22}(dx_2)^2 + g_{33}(dx_3)^2 + g_{44}(dx_4)^2.$$

Thus we have fulfilled the expectation to outfit

$$ds^2 \approx dX^2 + dY^2 + dZ^2 + d\tau^2$$

with tensors, and we have begun to answer the question of how to quantify the intrinsic curvature of space-time. The answer lies in *how different a set of g's is* from the "flat" values listed in the next matrix. Obviously this notion will be important to us since space-time is never flat in the proximity of matter, except locally within free-fall.

◆

We should comment on how the components of the metrical tensor can be displayed. On flat space-time and in ordinary Cartesian coordinates, we saw that

$g_{ik} =$

1	0	0	0
	1	0	0
		1	0
			1

We recall that the upper-left box holds g_{ik} when $i = 1$ and $k = 1$; the second box in the top row holds g_{ik} when $i = 1$ and $k - 2$, etc. Dependent values are omitted, and here the "diagonal" set of non-zero *g*'s, g_{11} g_{22} g_{33} g_{44}, contains 1's. (A set of *g*'s, such as 1, 1, 1, 1, is called the "signature" of a tensor, but we avoid the term because it can also mean the sum of these *g*'s.) We will find that four critical non-zero *g*'s often occupy the diagonal in a matrix. However as already suggested, where space-time is not flat, the *g*'s in the diagonal need not be 1's, and those off the diagonal need not be 0's.

There are other ways of labeling the *g*'s. One possibility is to consider the *g*'s associated with space coordinates to be positive and those associated with the time coordinate to be negative. This convention has certain advantages we mentioned, such as avoiding negative space, though it usually means that ds^2 is negative. The matrix then appears thus:

g_{ik} =

1	0	0	0
	1	0	0
		1	0
			- 1

Another version reverses the signs so that a negative ds^2 can be avoided:

g_{ik} =

- 1	0	0	0
	- 1	0	0
		- 1	0
			1

We will find that for many purposes the lower-right corner value, g_{44}, also holds c^2 as a conversion factor to accommodate the fact that *this g is associated with time*. I.e. g_{44} = $(1)c^2$ or simply c^2. This again means that g_{44} usually is a very large number.

177

We note that flat space-time, represented in each of the above matrices, is operative in special relativity, or to quote Einstein, "...the special theory of relativity as a special case of the general theory is characterized by the g_{ik} having the constant values [as above]." (The Principle of Relativity, p. 157. In that work the matrix on p. 120 has negative spatial coordinates and a positive time coordinate.)

Let also us digress briefly on the format of the underlying equation. When each g on the diagonal has a value of 1, we can write

$$ds^2 = -(dx_1)^2 - (dx_2)^2 - (dx_3)^2 + c^2(dt)^2.$$

This is a format common in the literature, and we note that time (t) is not converted to τ. Hence c^2 appears in the last term, and its sign is opposite from that of the other terms. However this format allows us to rearrange the order of the terms so that the time-term stands out

$$ds^2 = c^2(dt)^2 - (dx_1^2 + dx_2^2 + dx_3^2).$$

Upon solving for *ds* (assuming that the three *dx*'s are equal to each other, defining *v* by $v = dx/dt$) we can derive the equation

$$ds = \frac{dt}{\sqrt{1 - v^2/c^2}}.$$

We can treat this equation as another version of the Lorentz transformation of special relativity, and it will be useful to us later. This version also avoids a negative ds^2.

◆

We have ascertained the effect of flatness on the components of the metrical tensor, and now we are ready to consider pertinent details about the effect of *curvature*. In particular, since an intrinsically curved surface has at least one radius, *we will uncover the mathematical relationship between a radius of curvature and the g's*. Our basketball (better than our table-top) provides an illustration if we start with the ball deflated and flattened, so that its surface accepts Cartesian coordinates. At some point x_1 by x_2 (at X by Y) on the ball we draw a very small right triangle with dx_1 and dx_2 as its sides and with *ds* as its hypotenuse. For simplicity we let $dx_1 = dx_2$ so that our triangle is isosceles. We know that in order to satisfy the metrical equation

$$ds^2 = g_{ik}\, dx_i\, dx_k$$

on this flat surface (using Cartesian coordinates), our g's are

$$g_{11} = 1, \quad g_{12} = 0, \quad and \quad g_{22} = 1.$$

178

We next imagine that the ball is re-inflated and again spherical. Ideally we should no longer use ordinary Cartesian coordinates but some system meant for spheres. We will use such a system later, but switching coordinates now would introduce complications which obscure our issue. Let us just note that dx_1 and dx_2 *change upon re-inflating the ball*; they acquire an arc and, when measured carefully, each is longer than it was. That is to say, once we endow the surface with intrinsic curvature, the values 1, 0, 1 for g_{11}, g_{12}, and g_{22} respectively no longer allow ds^2 to equal $g_{ik}dx_i dx_k$. We discern the curvature by the need for *new g's*, and, more pertinently now, *the extent of the change from flatness to curvature corresponds to how different from 1, 0, 1 the new g's must be* in order to satisfy the metrical equation. Since we are dealing with two dimensions, this difference is easier to find using the more explicit but abbreviated metrical equation

$$ds^2 = g_{11}(dx_1)^2 + 2g_{12}dx_1 dx_2 + g_{22}(dx_2)^2.$$

We can assume that $g_{12} = 0$. (Later we will see when this is not so.) Our $dx_1 = dx_2$ and our $g_{11} = g_{22}$, so that we need to find only one new g, either g_{11} or g_{22}. This simplifies the issue to finding what g satisfies

$$ds^2 = g(2)(dx_1)^2$$

on the surface of this object at point x_1 by x_2.

We label each side of our triangle as it appears on the flattened ball as dx_{flat}. On this surface $g = 1$, so that

$$ds^2 = (1)(2)(dx_{flat})^2.$$

On the round (inflated) ball both dx's are elongated into dx_{curved}, so that we need a *different g*:

$$ds^2 = g(2)(dx_{curved})^2.$$

However *ds need not change* (it is invariant), in which case

$$(1)(2)(dx_{flat})^2 = g(2)(dx_{curved})^2,$$

or more simply $(dx_{flat})^2 = g(dx_{curved})^2$. Solving for g gives

$$g = \left(\frac{dx_{flat}}{dx_{curved}} \right)^2.$$

Since dx_{curved} is longer than dx_{flat}, *this g must be less than 1*.

If for simplicity we let dx_{flat} equal 1, the g can be expressed as

$$\left(\frac{1}{1 + excess}\right)^2,$$

where the term "excess" is used in the same sense as on page 126. The g we seek will conform to this pattern. We emphasize that the "excess" signifies the difference made by curvature, so that if the surface at this location is flat, this "excess" is zero and the g is 1.

We now invoke two equations of basic solid geometry: As the ball is inflated (as dx_{flat} becomes dx_{curved}), a triangle on that surface gains in area, and because

$$area_{sphere} = 4\pi r^2,$$

the area of the triangle is proportional to $4r^2$. This means that the "excess" should be inversely proportional to $4r^2$. Meanwhile, for a sphere with radius r drawn in Cartesian coordinates,

$$r^2 = (x_1)^2 + (x_2)^2$$

or

$$1 = \frac{(x_1)^2 + (x_2)^2}{r^2}.$$

Therefore the above "*excess*" equals

$$\frac{(x_1)^2 + (x_2)^2}{4r^2},$$

and again it is inversely proportional to $4r^2$. By fitting this "excess" term into the pattern $\left(\frac{1}{1 + excess}\right)^2$, we determine the sought-after g: At point x_1 by x_2,

$$g_{11} = g_{22} = \left(\frac{1}{1 + \dfrac{(x_1)^2 + (x_2)^2}{4r^2}}\right)^2.$$

This equation is still based in Cartesian coordinates which limit its usefulness on the steep curvature of our basketball, but as we said, switching coordinates now needlessly complicates the issue. Nevertheless we see explicitly that the *g's needed for compliance with the metrical equation are governed by the amount of curvature*. In this instance *the radius (r) determines components of the metrical tensor* (g_{11} and g_{22}). (Rucker, pp. 46-49, and Feynman, vol. II, pp. 42-1 to 42-6, show more details.)

In matrix form the components of g_{ik} on our basketball are

$\left(\dfrac{1}{1 + \dfrac{(x_1)^2 + (x_2)^2}{4r^2}} \right)^2$	0
0	$\left(\dfrac{1}{1 + \dfrac{(x_1)^2 + (x_2)^2}{4r^2}} \right)^2$

As we expect, if the ball stayed flat its r would be infinite and the "*excess*" would be zero, in which case

$$g_{11} = g_{22} = \left(\frac{1}{1} \right)^2 = 1.$$

Thus "$g_{11} = 1$" means zero curvature, and the same for "$g_{22} = 1$." Once the ball is inflated so that it has curvature, these g's are less than 1, and we now have an equation to show that the greater the curvature—the shorter the radius—the smaller these g's. Earlier we had equated "very non-flat" with "very non-g_{11}-equals-1." Here we can be more specific since we have found that "non-flat" means "g_{11} less than 1." (Let us not delve now into the case where g_{11} is not equal to g_{22}, but it would mean that the two radii at this point are unequal and that our basketball is not a perfect sphere.)

We can attach a real number to g_{11} experimentally. On page 159 we drew a right triangle with a hypotenuse of 14.142 cm. on the (inflated) basketball, and the sides of the triangle were longer than the expected 10 cm. In fact on a ball of radius 12 cm. the sides are about 10.3 cm. each, which means that g_{11} is $\dfrac{10^2}{10.3^2}$ or about 0.94. Thus in this case "$g_{11} = 0.94$" means "12 cm. radius."

We reemphasize that we relied only on intrinsic data to quantify curvature by way of g_{11}. We evaluated the surface of the basketball without observations from afar, without measuring the depth of a hole to the center, and without invoking additional dimensions. Of course this is an analogy, but that does not preclude applying the relationship between the g's and the radius to the "surface" of space-time. That is to say, the g's reflect shape in four dimensions just as they do in two dimensions. We can show this by accommodating four dimensions mathematically, even if we cannot imagine a four-dimensional surface. Let us assume that an observer on such a surface is located at x_1, x_2, x_3, x_4 (at X by Y by Z by T in Cartesian coordinates), where he "draws" a four-dimensional right triangle of size dx_1, dx_2, dx_3, dx_4. If the surface at this location is flat, we expect the hypotenuse ds to conform to

$$ds^2 = (dx_1)^2 + (dx_2)^2 + (dx_3)^2 + (dx_4)^2.$$

(The last term can be negative; let us ignore that detail.) But if the surface is intrinsically curved, we must include the radius r, and because location makes a difference, we must spell out the x's. The equation follows the same pattern we saw for two dimensions, and it becomes

$$ds^2 = \frac{(dx_1)^2 + dx_2)^2 + (dx_3)^2 + (dx_4)^2}{1 + \dfrac{(x_1)^2 + (x_2)^2 + (x_3)^2 + (x_4)^2}{4r^2}}.$$

As we saw for two dimensions, the denominator can be converted to a g such that *in four dimensions*, the r explicitly determines the g:

$$g = \left(\frac{1}{1 + \dfrac{(x_1)^2 + (x_2)^2 + (x_3)^2 + (x_4)^2}{4r^2}} \right)^2$$

Nevertheless there remains a problem. Though these two equations handle four dimensions, *they do not provided for two r's at any point.* We could try to get around this by recalling the definition of the Gaussian curvature,

$$K = \frac{1}{r_i r_k}.$$

This equation suggests that it might be feasible to express intrinsic curvature at each point of space-time by means of the Gaussian curvature K rather than g's. Why not replace g by a term with K, such as

$$\left(\frac{1}{1 + \dfrac{K\left((x_1)^2 + (x_2)^2 + (x_3)^2 + (x_4)^2\right)}{4}} \right)^2 ?$$

Because doing so only compounds the problem. On a surface which has two curvatures (on our basketball g_{11} does not equal g_{22}), circumventing the g creates confusion when we wish to spell out which radius are we identifying. We really are better off with g's and two indices, such that the index in the first position goes with one curvature (the i-curvature for x_i and dx_i) while the index in the second position goes with the other curvature (the k-curvature for x_k and dx_k). In fact this method gives the term "partial" (page 92) a more specific interpretation: In the equation

$$ds^2 = g_{11}(dx_1)^2 + g_{22}(dx_2)^2 + g_{33}(dx_3)^2 + g_{44}(dx_4)^2,$$

each measurement along an axis (dx_1 along X, dx_2 along Y, etc.) is modified by a g (g_{11}, g_{22}, etc.), but each g acquires a *partial contribution* from *both* curvatures. For instance g_{11} is one member of g_{ik}; it is not a member of only g_i or only g_k. This is a subtle but soon significant detail. Let us also recall that the g's with the two indices, i and k, are components, which means they tell us about magnitude and direction. That is to say, g_{ik} expresses curvature in the language of vectors and hence in the language of tensors.

We can turn this around and assume for a moment that the "surface" of space-time is either flat or perfectly spherical and hence never entails more than one radius of curvature at any point. In these cases we might call the collection of g's not the metrical tensor but simply the "metrical vector," and even in four dimensions a maximum of four components would suffice. Such a vector might be called a "four-vector" and corresponding the metrical equation might read

$$ds^2 = g_i \, dx_i \, dx_i$$

where i runs from 1 to 4. (Also, in two dimensions, $g_1 = g_2$.) Likewise K alone, with no specification of more than one radius, might suffice. But, alas, space-time is not that simple. K is not an efficient vehicle, and the metrical tensor for space-time must have a rank of not one but two.

However there is another complication which the above derivations ignore: We must allow for the possibility that a surface, notably that of space-time, may vary place to place. Or, by analogy, our basketball may be lumpy or egg-shaped. We then need to sample the curvature at many locations—we must compare the g's at many points—to discern the overall shape. Such point-to-point comparisons implies differentiation of the metrical tensor. Later we will verify that one component of a tensor dealing with this differential indeed is linked to Gauss' K as it should be, but one component is not enough for the case of curved space-time.

All this means that the mathematical manipulations entailed by general relativity are much simpler when performed on components of tensors. In summary, we prefer g_{ik} over K because of *three features of space-time*. First, a point in curved space-time has two different curvatures; hence we need a tensor, g_{ik}, with two ranks. Second, space-time is four-dimensional; hence i and k in g_{ik} have one of four values. Third, the radii can vary among points; hence we will need to expose g_{ik} to differentiation.

Let us not lose sight of why the equation

$$g_{11} = g_{22} = \left(\cfrac{1}{1 + \cfrac{(x_1)^2 + (x_2)^2}{4r^2}} \right)^2$$

is of interest to us. It is because of the relationship between the *g*'s and the radius r. Said explicitly, since r is a measure of intrinsic curvature, and since r is linked to the *g*'s (g_{11} and g_{22} here) of the metrical tensor, this tensor must be pivotal in dealing with the shape of space-time in the context of general relativity. Of course we need to include the other 8 *g*'s, and we shall. Meanwhile we recall that mathematically speaking *the motion of an object in a gravitational field forms a tangent vector* (page 145) *on the surface of space-time which is determined by the curved shape of that surface.* We can say that gravitational motion is intrinsic, or that the geodesics traced by falling or orbiting objects conform to the intrinsic shape of space-time. Quantifying the radii of space-time curvature via the *g*'s will allow us to predict the gravitational behavior of objects, because by enumerating the components of g_{ik}, we can solve $ds^2 = g_{ik}\, dx_i\, dx_k$ for *ds* in any frame of reference. This allows us to compute the path for falling or orbiting objects, which in turn means that we can understand gravitational events, while gravitational events are ubiquitous in our universe.

✦ ✦ ✦

We must now separate and clearly define two issues. One is that the *g*'s depend on the curvature of the surface in question. The second issue, which we glossed over so far, is that the *g*'s also depend on the system of coordinates. In other words we need to distinguish between the effect of changing the *shape* and the effect of changing the *coordinate axes* in which we study that shape. By analogy, "adjectives" for objects are changed when we alter the shape of the objects as well as when we alter the vantage from which we observe them. Let us explore the effects of transformations of coordinate axes on the *g*'s.

It helps to go back to a simple case: The horizontal or X axis of a Cartesian system is altered while the vertical or Y axis is undisturbed. The reader is invited to participate: On ordinary flat graph paper we draw an oblique line in the same way the EVENT-line is drawn in diagram I (page 86). We call this line *ds*, and for now it represents the shape (the length) of an object. We label the lower and left edges of the graph paper with one consecutive number for each box, creating X and Y coordinate axes (like X-A and T-A in diagram I). Let us assume that the oblique line is 5 units long (ds = 5), that it measures 4 units projected onto the lower edge of the paper (dX = 4), and that it measures 3 units (dY ⁻ 3) along the left edge; of course according to the pythagorean theorem $5^2 = 4^2 + 3^2$.

But what if we re-label the bottom edge of our graph paper with one consecutive number for every two boxes? There will be twice as many units of Y per inch than units of X per inch, *as if the X axis had been stretched horizontally by a factor of two.* Now the projections of the oblique line cannot comply with the equation

$$ds^2 = dX^2 + dY^2.$$

In effect we have selected a new system of coordinates, one with a new X axis, and as a result this equation has not withstood the transformation. Indeed we know we have transformed the coordinates *because they no longer comply with the traditional pythagorean theorem,* simply because 5^2 does not equal $2^2 + 3^2$; the new system of coordinates disobeys Euclidean geometry.

Of course compressing the graph paper would give the same effect as stretching an axis, but we are deferring this point for now. Instead we bring the following important point into the picture: When we "transform" our paper (along the X-direction by a factor of 2 in this case) the oblique *ds*-line neither shrinks nor grows. It is still 5 units long—*it is invariant.*

If we insist upon using a pythagorean equation, the dX^2 must be multiplied by 4, so that $5^2 = (4)(2^2) + 3^2$, but then the equation no longer represents the standard theorem. It becomes a "non-pythagorean non-Euclidean theorem,"

$$ds^2 = 4dX^2 + dY^2.$$

Like an adjective, the 4 is a coefficient which changes the meaning of X. We can consider "4 " (which is 2^2) to mean "axis stretched-by-a-factor-of-two." That is to say, *this coefficient tells us which way (along X) and to what extent ($\sqrt{4}$) the system of coordinates is non-pythagorean or non-Euclidean.*

We can recast this exercise in terms of the *g*'s of the metrical tensor. If we rename X and Y as *x*'s with subscripts and specify that x_1 has been stretched, then

$$ds^2 = (dx_1)^2 + (dx_2)^2$$

does not hold unless $(dx_1)^2$ is modified by a coefficient, namely *g*. The adjusted equation is

$$ds^2 = (g)(dx_1)^2 + (dx_2)^2.$$

Meanwhile we have available our generalized metrical equation

$$ds^2 = g_{ik}\, dx_i\, dx_k,$$

185

as well as the more explicit two-dimensional version,

$$ds^2 = g_{11}(dx_1)^2 + 2g_{12}dx_1dx_2 + g_{22}(dx_2)^2.$$

The coefficient for $(dx_1)^2$ is now the g_{11}-*component of the metrical tensor*. For this transformation $(dx_1)^2$ is adjusted by having g_{11} equal 2^2 or 4.

We consider what has just occurred: A two-dimensional system has been transformed such that *the components of the metrical tensor* g_{ik} *are changed* from

$$g_{11} = 1, \quad g_{12} = 0, \quad and \quad g_{22} = 1$$

to

$$g_{11} = 4, \quad g_{12} = 0, \quad and \quad g_{22} = 1.$$

We went from the "pre-stretched" (pythagorean) equation

$$ds^2 = (1)(dx_1)^2 + (0)dx_1dx_2 + (1)(dx_2)^2$$

solved as

$$ds^2 = (1)(16) + 0 + (1)(9) = 25$$

to the "x_1-stretched-by-2" (non-pythagorean non-Euclidean) equation

$$ds^2 = (2^2)(dx_1)^2 + (0)dx_1dx_2 + (1)(dx_2)^2$$

solved as

$$ds^2 = (4)(4) + 0 + (1)(9) = 25.$$

However this equation

$$ds^2 = g_{ik}\, dx_i dx_k$$

retains its form under this transformation, signifying covariance.

186

From these data, namely *a change in the value of the g_{11}-component of the metrical tensor from 1 to 4*, we infer a two-fold stretching of the axis, and the arithmetic is no more complicated than the fact that 4^2 is the same as $(2 \times 2)^2$. However we note that in either case, before or after stretching, *the metrical equation yields the same ds^2, 25.* I.e., the oblique line is still 5 units long, signifying no change in the shape (length) of the object.

What about a vertical deformation of coordinates? If we compress the Y axis, the generalized pythagorean theorem applies but now g_{22} no longer equals 1. Thus upon deformation in both directions, say a three-fold horizontal stretching together with a two-fold vertical compression of axes, we need our equation to read

$$ds^2 = (9)(dx_1)^2 + (0)dx_1 dx_2 + (\tfrac{1}{4})(dx_2)^2.$$

The applicable metrical tensor g_{ik} has components $g_{11} = 9$ and $g_{22} = \tfrac{1}{4}$. In a matrix,

9	0
0	¼

Again ds and ds^2 are invariant; we did not disturb the shape of the object.

The significance of our exercise is this: *Changing to another system of coordinates gives us new components of the metrical tensor, but these still refer to the same ds.* That is to say, the transformation is such that shape can remain invariant but new g's arise.

Of course we should consider the case of a complicated object describable by a *rank-two tensor* projected onto four-dimensional graph paper, but this is too difficult to imagine, and we will use a better analogy shortly. Nevertheless we must be prepared to use the two separate indices so that we can tell which line is projected onto which axis. If "*ik*" are the projections before the deformation of axes, if "*ab*" are the projections afterwards, and if the g's are the coefficients, then

$$ds^2 = g_{ik}\, dx_i dx_k = g_{ab}\, dx_a dx_b.$$

This means that the new values for the g's allow an invariant ds. Again the form of the metrical equation is likewise invariant.

While stretching or compressing axes is instructive, these operations are of little use on intrinsically curved surfaces. A more pertinent step is to change the X-Y crossing angle so it is no longer 90 degrees. I.e., we cause the axes to be no longer "orthogonal" (perpendicular). They become "oblique" but they remain "rectilinear" (straight). Thus we transform the axes into an "oblique rectilinear system" of coordinates. Such a system can be used on a flat surface *or on the surface of a sphere—which of course means on an intrinsically curved surface.* (We do not rotate both axes together; that is also a transformation but it would complicate our current issue.)

We touched upon why an oblique rectilinear system lends itself to a curved surface by way of the analogy about stiff triangles fitted into the surface of a basketball (page 159). To cover the ball, the triangles had to be trimmed so that they were no longer orthogonal, or as we said, so they were no longer pythagorean. In this sense, the standard pythagorean theorem, which presumes a orthogonality, must be "trimmed." The equation for a system in two dimensions adjusted for oblique axes (trigonometry's "law of cosines") is

$$s^2 = X^2 - 2cosR\,XY + Y^2,$$

where the angle between X and Y is R, and where "- *2 cosR XY*" represents what we "trimmed." If R is 90 degrees, cosR is zero, which means X and Y are orthogonal. Then the term - *2 cosR XY* drops out, and we are left with the plain right-triangle pythagorean theorem. If R is, say, 80 degrees, cosR is about 0.17. In general, cosR increases with greater obliquity.

The above equation can also be written in a general differential form with g's,

$$ds^2 = g_{11}(dx_1)^2 + 2g_{12}dx_1dx_2 + g_{22}(dx_2)^2,$$

or its compact version

$$ds^2 = g_{ik}\,dx_i\,dx_k.$$

Now g_{11} and g_{22} each still equal 1, but

$$g_{12}\ (\text{as well as } g_{21}) = -\ cosR.$$

In matrix form we have g_{ik} =

1	- cosR
- cosR	1

This set of g's, i.e. this particular metrical tensor, reveals the transformation to oblique coordinates by way of g_{12} and g_{21}, unlike the stretched and compressed case wherein g_{11} and g_{22} are affected; we should compare this matrix with the previous. Again ds^2 remains unchanged, as does the form of the metrical equation. (If an oblique rectilinear system is placed on a sphere, it is possible to calculate the radius from the g's, but this is rather complicated. As we shall see shortly, we have a better way.)

Let us digress to touch upon a concept that will be useful when we analyze the shape of space-time in more depth: We can adapt an oblique system of coordinates to curved surfaces that are more complicated than spheres. This is accomplished by allowing the angle of obliquity to vary between different points on the surface. We can even imagine that an oblique system can be miniaturized and set up at every possible point of a surface. We recall that such a surface is a *field*. We assume that we can gauge the angle of obliquity at any point on the field and that we can quantify *exactly how the obliquity varies point-to-point across that field*. This means that the coordinate axes can be curved and that the amount of curvature can vary. In more technical terms we can transform from a rectilinear to a *curvilinear* or "Gaussian" system of coordinates. We already introduced this type of system graphically in our Minkowski-like diagrams by drawing curved axes.

We need to look at one more kind of transformation of coordinates which invokes new *g*'s. Let us start with the "bearing and range" system used on a practically flat surface such as a small region of the Earth. Bearing is the direction, possibly a compass reading such as φ number of degrees, and range is the distance, such as *r* number of yards. Thus if an object is located 100 yards to our right of us, we might say that its bearing is 90° and its range is 100 yards. This system is used at sea: "Ship at bearing 60°, range 500 yards." If the ship moves to a new bearing $\varphi + d\varphi$ and a new range $r + dr$, we can say that it moves through a distance *ds*, similarly to the way we did on page 174 using a Cartesian system. The equation for *ds* in terms of changes in bearing and range is

$$ds^2 = dr^2 + r^2 d\varphi^2,$$

and we note that instead of X and Y or x_1 and x_2, we see a distance and an angular direction.

Because the bearing-and-range system uses a center that can be the north or south pole of the Earth, it is a "polar" system—which may be convenient for Arctic and Antarctic navigation—but it is inadequate when the overall roundness of the Earth needs to be considered. However we can take advantage of the features of this system by extending it into a more comprehensive system called *spherical-polar coordinates*. Here the equation written for two dimensions is as follows:

$$ds^2 = (g_{11})(d\theta^2) + (g_{22})(d\varphi^2)$$

Geography again provides a suitable scenario for spherical-polar coordinates. The above equation is used to compute the distance *ds* between two points anywhere on the Earth's surface, but now we implicate the radius of the Earth (*r*), and instead of distance on the surface, the right side of the equation uses degrees (or radians) of latitude and longitude. The θ represents the latitude and φ is the longitude. If we changed our geographic location by $d\theta$ and $d\varphi$, solving this equation for *ds* tells us how far we have traveled. Thus by moving 2° west and 1° south across the Earth, we might have gone 100 miles. (The exact distance depends on where we began, because the lines of longitude are not parallel. Incidentally, in this context range and radius are different measures. The center of range may be a geographic pole or other location on a surface, but in a spherical-polar system the radius originates at the center of a sphere and extends to the surface of that sphere. Nevertheless we will see later that in an astronomical setting the distinction vanishes, since we will define range as the distance from the center of an imaginary sphere to the surface of that sphere)

We can set up this equation so it can be solved for the radius r. In other words if a 2°-by-1° shift moved us 100 miles, we can calculate the radius of the Earth (but again we must also know where we began). In fact in spherical-polar coordinates the relationship between one of the *g*'s and the curvature is surprisingly simple:

$$g_{11} = r^2$$

The other *g* under consideration is more complicated but also involves the radius: $g_{22} = r^2 \sin^2 \theta$.

Of course the equation for a spherical-polar system is another special case of the generalized metrical equation, but the matrix here (for two dimensions) is

r^2	0
0	$r^2 \sin^2 \theta$

During the exercise in which we found the *g*'s for a spherical basketball, we used orthogonal Cartesian coordinates. This is why g_{11} and g_{22} turned out to be so complicated on page 180. Had we converted to spherical-polar coordinates, solving for the radius would have been easier. However changing shape from flat to spherical affected the *g*'s, and transforming coordinates from Cartesian to spherical-polar also affects the *g*'s, so that it would have been awkward to do both together. Nevertheless in chapter 17, where we solve equations in curved space-time, we will prefer spherical-polar coordinates, even though it means that we will think of the "surface" of space-time as having latitude and longitude.

◆

We have seen that the components of the metrical tensor, the *g*'s, depend on the system of coordinates in use as well as on the shape under consideration. Since use of any system should be allowed—none is absolute—the implication is that we can never be sure of the shape of a surface. For instance if we place graph paper on our table top and later find that we need different *g*'s, how do we know whether the surface of the table became warped or whether the axes of the graph were deformed? Likewise if we find new *g*'s on our trampoline on which we painted a grid, how do we know whether we are witnessing an altered surface or altered coordinates? More pertinently, how can we ascertain whether a set of *g*'s reflects the choice of system of coordinates or the shape of space-time?

To settle this critical issue, let us consider the following: We can assume that the stretching of the X axis occurs at the rate of, say, 100% per week (i.e. the X axis is doubly stretched every week). Does it matter whether we start with ordinary graph paper on which

$$ds^2 = dX^2 + dY^2$$

or whether we start a week later with stretched graph paper on which

$$ds^2 = 4dX^2 + dY^2?$$

(After one week the coefficient for dX^2 is 2^2.) In other words whereas our choice of coordinates affects the g's, does it affect *the rate at which the g's change?*

In actuality it does not. In the first week the g in question increases four-fold, 1 to 4. In the second week the stretching-by-a-factor-of-two again results in a new coefficient such that the g changes from 4 to 16, still a four-fold increase. That is to say, *the manner in which this g varies under the transformation does not depend on the selected system of coordinates.* No matter which we start with, $ds^2 = dX^2 + dY^2$ or $ds^2 = 4dX^2 + dY^2$, a two-fold stretching of the X-axis is associated with a four-fold change in a particular component of the metrical tensor.

More explicitly, and using subscripts, the latter transformation is a switch from the "x_1-already-stretched-by-2" equation

$$ds^2 = (2^2)(dx_1)^2 + (0)dx_1dx_2 + (1)(dx_2)^2$$

to the "x_1-stretched-again-by-2" equation

$$ds^2 = (4^2)(dx_1)^2 + (0)dx_1dx_2 + (1)(dx_2)^2.$$

We also note the invariant result; either way $ds = 5$. (In the first case

$$ds^2 = (4)(4) + 0 + (1)(9) = 25$$

and in the second case

$$ds^2 = (16)(1) + 0 + (1)(9) = 25.)$$

The same notion applies for more elaborate transformations, *as well as to transformations between points in space-time.* For instance let us assume that we have determined the g's for two points, A and B, in a region of space-time, and these two points have different curvatures. First we used a Cartesian system of coordinates, and then we repeated the process in a spherical-polar system. Let us also assume that in the Cartesian case, the value of g_A is double that of g_B. Thus $g_{A\text{-}Cartesian} = 2g_{B\text{-}Cartesian}$. In effect a transformation exists between A and B; a coefficient is doubled. Of course the g's also changed once we switched to the spherical-polar system, so that $g_{A\text{-}Cartesian}$ is not the same as $g_{A\text{-}spherical\text{-}polar}$, and the same holds at point B. However—and this is the crucial observation—$g_{A\text{-}spherical\text{-}polar} = 2g_{B\text{-}spherical\text{-}polar}$! That is to say, *the way in which the g's vary point A to point B was determined only by the shape and not by the choice of coordinate systems.* This means that transformations to any system will *not* hide the true point-to-point shape of space-time as reflected in the *variation in the g's.*

The idea here is that we must not only consider the transformation of coordinates itself but also the *rate* of the transformation. Of course the rate at which coordinates axes undergo simple stretching or compression is of little interest in general relativity, but let us keep in mind that relative motion may be equated with a *rotation* of axes. In particular we pointed out (page 93) that while the angle R quantifies a rotational transformation, we are not limited to an invariable angle R. The implication then is that how angle R varies—its *rate* of change and hence its differential (its derivative)—will demand our attention. This concept applied to space-time and to natural gravitational fields is very important: The rate at which the *g's* change from point to point on the one hand reveals the regional curvature of space-time and on the other hand is *independent of the choice of the system of coordinates.* This duality parallels and coincides with the idea that general relativity implicates curved space-time—which means variable *g's*—while it demands general covariance—which means all systems of coordinates should be equivalent.

✦ ✦ ✦

Let us compile five statements we have made about the components of the metrical tensor:

Statement 1: Each *g* requires 2 indices; i.e. the metrical tensor is rank-two.

Statement 2: The *g's* depend upon the shape of the surface they are describing.

Statement 3: The *g's* also depend upon the selected system of coordinates.

Statement 4: How the *g's* vary across a surface is not influenced by the selected system of coordinates.

Statement 5: In space-time the metrical tensor has 10 *g's*; i.e. it has 10 independent components.

The analogy useful here is our basketball fully covered by a mosaic of very small cardboard triangles. With the ball at rest, each triangle faces a certain direction, and each triangle is non-pythagorean by virtue of the roundness of the ball. Therefore any point on the surface of the ball can be characterized by which way a triangle at this location faces (a direction) and by the extent to which the triangle is non-pythagorean as determined by the radius of curvature at this location (a magnitude). Hence each point possesses a vector.

This vector can be projected on three coordinate axes; for example the basketball may be inside a cubical box, and the vectorial arrow may be visible through the top, front, and side of the box (just how does not matter—the box might be translucent and the arrow might cast shadows). *By studying this vector as it appears projected on the walls of the box—these walls represent a system of coordinates—we can determine the shape of the ball inside that box.*

However what if the object in the box is somehow transformed from a basketball into a football? Now each point on surface of the ball has *two* radii and hence *two* vectors. This means that the components of a rank-two tensor are needed to describe this shape. Hence statement 1: Each g requires 2 indices.

Furthermore once we convert the basketball to a football and ascertain the components of the radii for many points on the football, we find *different projections*. The projections are the components of g_{ik}, and the set of g's depends on the shape in question. Hence statement 2: Different shape means different g's.

Of course just as various systems of coordinates are available, we are not bound to use only a cubical box. A differently-shaped container can be chosen. However then the projections of the vector on the walls of this new container appear *different*. This can be envisioned as placing a ball into a, say, cylindrical hat-box. Without changing the shape in question (same ball), the set of observed g's hinges on the selected system of coordinates (different box). Hence statement 3: Different coordinates mean different g's.

Unfortunately the football may be lumpy—i.e. it may not have the same two radii at other points on its surface. Still, *the variation in the g's from one point to the next is the same* whether the ball is in a cubical box, in a cylindrical box, or in any other box. This means that even if the amount of curvature changes point-to-point, we can still ascertain the shape of the entire ball in any packaging—in any system of coordinates—by noting the diversity in the g's point-to-point. Hence statement 4: How the g's vary across a surface is not influenced by the selected system of coordinates.

A corollary to the observation that new coordinates mean new g's is that if we search hard enough, we can find some box in which *one point of any ball looks flat*. For example we will always be able to select a system of coordinates such that locally the g's are

$$g_{11} = g_{22} = g_{33} = g_{44} = 1$$

and

$$g_{12} = g_{13} = g_{14} = g_{23} = g_{24} = g_{34} = 0.$$

This set of g's is associated with flatness. In other words curvature can always be "transformed away" locally (see page 135). But if the object has intrinsic curvature, *another point cannot have the same g's*. In other words if this ball in this box at point A appears to have a particular shape, even flatness, then at point B of this ball in this box, the g's will not be identical; the g's must vary point-A-to-point-B. However no matter what box we use—no matter which system we select—the *same A-to-B variation* will appear.

Here is the point of statement 4 applied to general relativity: If we subdivide *curved space-time* into points, if we "place" each point into a "box" of its own, and if we treat each point-and-its-box as a freely falling frame of reference, the result is the same: The shape of space-time corresponds to how the *g*'s vary point-to-point, regardless of the choice of "boxes"—regardless of the system of coordinates. If two observers use different systems—which means the observers are in relative motion or one observer is global and another is local—they disagree on the *g*'s but they agree on how the *g*'s vary point-to-point. Therefore *they agree on the shape of space-time.*

Eventually we will find that the *g*'s are linked to the gravitational potential and hence to the gravitational field. This means that the components of the metrical tensor by themselves describe both the gravitational field *and* the system of coordinates. Surely the reader can sense the predicament: We need a way to describe the gravitational field without worrying about the choice of the system of coordinates. We already foresaw the solution: We will be compelled to rely on the *differential variation of the g's.*

To visualize statement 5, which states that the curvature of space-time entails 10 components, let us reconsider a basketball, but we must assume it is in a four-dimensional box whose edges are X, Y, Z, and T. The ball has one radius, r. To an observer able to see in, r is projected on X, on Y, on Z, and on T. Obviously the observer is measuring four components of r on a four-dimensional system of coordinates.

In the case of a football in a four-dimensional box, which represents space-time, we must consider two radii at each point, r_i and r_k. Now there are more components to contend with. For instance the r_i from one point on the ball's surface is projected along the X-edge of the box as X_i while r_k from that point is projected as X_k. The r_i from that point is projected along the Y-edge Y_i while r_k from that point is projected as Y_k; etc. Altogether 16 combinations reveal the radii of the football at any point:

$$X_i \text{ with } X_k, \quad Y_i \text{ with } X_k, \quad Z_i \text{ with } X_k, \quad T_i \text{ with } X_k,$$

$$X_i \text{ with } Y_k, \quad Y_i \text{ with } Y_k, \quad Z_i \text{ with } Y_k, \quad T_i \text{ with } Y_k,$$

$$X_i \text{ with } Z_k, \quad Y_i \text{ with } Z_k, \quad Z_i \text{ with } Z_k, \quad T_i \text{ with } Z_k,$$

$$X_i \text{ with } T_k, \quad Y_i \text{ with } T_k, \quad Z_i \text{ with } T_k, \quad T_i \text{ with } T_k,$$

but the observer does not require, say, both "Y_i with X_k and X_i with Y_k," nor "Z_i with X_k and X_i with Z_k," etc. Six combinations are redundant, and the observer needs *10 items of information*—10 numbers—corresponding to the 10 independent components and the 10 *g*'s of the metrical tensor that quantify two curvatures at one 4-dimensional point. (We again note more components than dimensions, which brings us back to statement 1: We need a tensor, not just a vector.)

Here we emphasize an issue sometimes blurred in the literature. We will find statements to the effect that "each point of curved space-time has 10 curvatures." Such wording reflects the Gaussian system. We do better with the terminology of vectors and tensors by envisioning that each point in curved space-time has two "arrowed" vectors. When two arrows are projected onto four coordinate axes, we find 16 projections, 6 of which are duplicate. We therefore prefer to say that "each point of curved space-time has *two* curvatures which project as *10 independent components*."

Another issue is why the observer of the ball-in-a-box couldn't just solve the equation

$$r^2 = X^2 + Y^2 + Z^2 + T^2$$

for each radius. Because so doing does not reckon the radius by intrinsic means—i.e. from data obtained only at the surface of the ball. We therefore resort to the less direct method of first using the 10 *g*'s of the metrical tensor to ascertain to what extent the pythagorean theorem fails and then calculating the radius from the *g*'s at that point. This approach allows us to determine how the *g*'s vary point to point, independent of the system of coordinates. We will pick this thread up again later, but now let us further examine the metrical tensor itself.

As we stated earlier, the difference between a Euclidean and a non-Euclidean two-dimensional surface is that on the former this equation will hold,

$$ds^2 = dX^2 + dY^2,$$

whereas on the latter we need

$$ds^2 = (A)dX^2 + (B)dXdY + (C)dYdX + (D)dY^2.$$

We can say that the equation for the pythagorean theorem is to Euclidean geometry what the metrical equation is to non-Euclidean geometry.[*] Hence we likened Euclid's geometry to an automobile built to go on a flat road; there are many places it cannot take us. The traditional pythagorean theorem is like that automobile; it can only guide us on a Euclidean surface. The first step to non-Euclidean geometry is summoning coefficients (A, B, C, and D above) and calling them *g*'s. By furnishing the pythagorean theorem with *g*'s, we adapt it for non-Euclidean surfaces. In other words we free the pythagorean theorem from confinement to traditional Cartesian coordinates, and we let it to operate in curvilinear (Gaussian) coordinates. This analogy also suggests that our derivation allows intrinsic assessment: The metrical tensor g_{ik} improves Euclid's "pythagorean automobile" so that an airplane is not needed to detect and measure the curvature of a surface.

---

[*] Technically both equations are "metrical," but we reserve the term for the kind outfitted with coefficients.

The role of the metrical tensor in general relativity is explicitly revealed if we temporarily omit it: Under conditions of special relativity, *ds* is invariant—i.e., a space-time interval is immune to the effects of *uniform* relative motion—if, in four dimensions,

$$ds^2 = (dx_1)^2 + (dx_2)^2 + (dx_3)^2 + (dx_4)^2.$$

A more general form of this equation, which explicitly accommodates uniform relative motion, is

$$ds^2 = \left(\frac{dx_1 + vdx_4}{\sqrt{1-v^2/c^2}} \right)^2 + dx_2^2 + dx_3^2 - c^2 \left(\frac{dx_4 + vdx_1/c^2}{\sqrt{1-v^2/c^2}} \right)^2.$$

(See pages 99-100.) These forms of the pythagorean equation might be used to answer the following question: In the *absence* of gravitation, how does an event in space-time, sampled by *ds*, appear when measured along an axis of an inertial, uniformly moving observer?

Under conditions of general relativity, an event in space-time appears the same—*ds* is immune to the effects of *any* relative motion, notably including accelerated motion—if

$$ds^2 = g_{ik}\, dx_i\, dx_k.$$

Here the *g*'s are the coefficients for the *dx*'s, so that this (non-Euclidean) metrical equation can be used to answer the following question: In the *presence* of gravitation, what is the geodesic, as sampled at the *ds*? As we mentioned, when we have ascertained the 10 values for the *g*'s needed to solve this equation, *we can gain full understanding of gravitational events.* Why? Because the *g*'s quantify intrinsic space-time curvature, while the (non-Euclidean) shape of space-time directs falling or orbiting objects along geodesics. (We will devote chapter 17 to calculating a set of *g*'s and then calculating an orbit from these *g*'s.)

At the same time the *g*'s are the interface between (1) events in space-time described as intervals and (2) events in our physical world described as components on coordinate axes. Indeed when we examine the metrical tensor we find ourselves at the crossroads of general relativity, as is evident simply by inspecting the metrical equation: To the left of the equal sign is *ds*, to the right are the *dx*'s, and g_{ik} links the two. If we know from ordinary observations how much distance a physical event covers (dx_1, dx_2, and dx_3) and how long it takes (dx_4), and if we ascertain the values of the *g*'s—which means if we know the shape of local space-time—then we can calculate what the event (*ds*) is like in space-time. *The metrical tensor allows us to convert relative time intervals and distances into invariant intervals in curved four-dimensional space-time.* Hence the metrical tensor is also our doorway into space-time; it overcomes our natural inability to "see" events directly in four dimensions; it gives us a clear and precise "look" at the events in the elusive interior of our analogical curved house.

Furthermore, because *ds* is invariant and does not depend on the choice of coordinates, the metrical equation ensures that the connection between the shape of space-time and gravitation is the same in all frames of reference. That is to say, an event in four-dimensional space-time appears identical even if observers are in relative motion and even if that relative motion is accelerated rather than uniform. This means that *all physical events in the universe appear such that locally* $ds^2 = g_{ik}\, dx_i\, dx_k$. Hence the metrical equation illustrates the powerful but reassuring doctrine (page 129) that the laws of nature are invariant and consistent no matter where we go or what we do. We can even argue that if this were not so, all scientific study of the universe would be futile. Of course we must write the equations for these laws in a certain way; general relativity shows us how, but the metrical tensor is a fundamental ingredient.

✦

The metrical tensor embodies the ideas of many diverse contributors: Euclid and Pythagoras taught that space can be measured. In particular Pythagoras devised his geometric theorem on flat surfaces. Gauss, in seeking to survey a curved surface without leaving that surface, taught that non-compliance with that theorem signals intrinsic two-dimensional curvature. Riemann taught that the same method can be adapted to determine the intrinsic curvature of an i-and-k-dimensional surface. (The choice of letters is arbitrary; the point is that Riemann's method allows multi-dimensionality and complexity.) Minkowski taught that physical events are ordinarily observed in space and time but occur in four-dimensional space-time wherein time can be treated as another dimension. Lorentz taught that laws of physics should be portable (transformable) between uniformly moving coordinate systems. Einstein taught that there is a link between gravitation and the curvature of space-time, and he realized that Riemann's "i-and-k-dimensional surface" can be that of space-time.

✦ ✦ ✦

With the metrical tensor in mind we can refine our thinking about fields and in particular about the relationship between the gravitational field, the gravitational potential, curved space-time, and the *g*'s. The importance of fields in general relativity will become clear as we change our focus from the notion of a local set of *g*'s to the notion of the *point-to-point variability of the g's*. Surely the latter entails more complex mathematics, especially aspects of differential calculus applied to tensors, but as we already intimated, it is crucial.

Novices in relativity are tempted to claim that Einstein replaced "gravitational force" with "gravitational field." Is this is merely a question of terminology—new words for an old concept? In truth the distinction is substantial and profound, and we will say more about it shortly. Nor should students infer that Newton's system necessarily denies the possibility of a gravitational field. What it denies is anything but an inertial frame of reference. By Newton's conception, any gravitational event occurs in an inertial frame of reference—in what we now call flat space-time—in which a local gravitational potential appears as a force. Please see the detour on page 59.

Our trampoline analogy is useful in explaining the previous paragraph: If a marble rolling toward a large object represents an object falling to Earth, the Newtonian interpretation is that the marble and the large object must be attracted to each other by virtue of their masses. Since the surface of the trampoline is assumed to be flat, no other explanation suffices. Einstein's interpretation holds that every nearby point on the surface of the trampoline is tilted and curved toward the large object. This curvature gives every such point a gravitational potential, but of course in this analogy "point" and "curvature" denote those of space-time. If we examine the shape of one point in space-time, we know the gravitational potential at that point. If we compare this data with *neighboring points*, we know the gravitational *field*. As we already saw, the mass of the marble is irrelevant in this paradigm, which is part of the reason we treat it as a test object.

In this way we can think of a gravitational field as the collection of adjacent varying gravitational potentials. We will apply this idea later to gravitational equations, but we can already say that *the point-to-point variability of the g's in a region of space-time around a mass is a measure of the variability of the gravitational potential, while the variability of the gravitational potential propagates the gravitational field.* Conversely if a gravitational field is not variable, it is not gravitational—objects will not fall or orbit. (Some authors say that the gravitational field is "inhomogeneous," but the term is confusing because, as we shall see, the universe as a whole appears homogeneous. [The terms have yet another meaning in algebra.])

Einstein concluded (The Meaning of Relativity, p. 140) that general relativity must rely on the field concept, and that it is space-time that gives a gravitational field its structure. Since space-time is altered by mass and described by the metrical tensor, a gravitational field can be seen as an alteration of space-time induced by mass and measurable via the components of the metrical tensor. Thus the working definition of the gravitational field is *"variably curved space-time,"* each point of which has a gravitational potential that can be expressed in terms of the components of the metrical tensor, the *g*'s. (The gravitational potential is not a tensor but a scalar, which is an issue we will discuss shortly.)

We see that Newton and Einstein differ widely on their interpretation of the gravitational potential. Newton linked it to a force between two objects. Einstein linked it to the shape of local space-time. According to the latter concept, the apparent "force" exhibited by a falling or orbiting object at any point along its path is actually *the response of that object to conditions at that point of a field.* These "conditions" can be characterized by the gravitational potential, but they arise from the shape of space-time given by the *g*'s. The practical import of this notion, contrasted with the Newtonian rendition, is that different mathematics and geometry are needed, that the pertinent laws of physics appear different, and most importantly that the results agree with all known observations of gravitational behavior.

In this paradigm, the troublesome Newtonian idea of action at a distance is supplanted by the idea of local interaction of objects (or particles) with the field. Likewise we can settle another problem with Newton's law of gravitation: This law obviously predicts that gravitational attraction vanishes *in the absence of mass* (e.g. in outer space), but then how does it explain the absence of gravitation *during free-fall?* After all, getting on a freely falling elevator on Earth does not erase the mass of the Earth. Newton had to assume that during free-fall there is just enough inertia to erase the attractive force. Relativity's reply is that there is no such force. Instead there is global curvature of space-time which is not erased. The gravitational field still exists, but space-time appears locally flat; its curvature, as we said earlier, is "transformed away."

◆

The role of the gravitational potential, which is a scalar quantity, suggests that the gravitational field is a *scalar field*. This is partially correct; as we said, the gravitational potential reflects the variable nature of the gravitational field. Nevertheless, for the purposes of general relativity, the scalar gravitational potential is not reliable. To describe a gravitational field in terms of space-time we use *a rank-two tensor*, namely the metrical tensor. A collection of its components for all points in a region of space-time makes up a field of metrical tensors—a "*tensor field*"—which, as we shall find, is not as forbidding as it sounds. Of course there still must be a link between the gravitational potential and the metrical tensor, and we will examine this link later. For now we reiterate that *the variability of the g's in a region of space-time propagates a field of rank-two metrical tensors which describe the complicated curvature shape of space-time point-by-adjacent-point.* To be precise, gravity exists because of this variation, though we commonly say that gravity exists because space-time is curved.

Since we treat gravity mathematically as the variation between adjacent points of space-time, a generally covariant law of gravitation must first be legitimate under conditions of free-fall, wherein space-time appears flat. Our approach in studying the gravitational field therefore employs the concept that, given a sufficiently local area of space and a sufficiently brief duration of time, a frame of reference appears inertial. (Pages 135-137.) Indeed by analyzing space-time in small enough increments, we can fall back upon the rules of special relativity and yet we can apply the invariance of the interval (*ds*) in space-time. Under the conditions of general relativity we envision curved space-time, but because the metrical tensor and the metrical equation are derived in differential form, this curvature is not a problem, including under transformation of coordinates.

Since the metrical equation survives such transformations, it has the critical property required by relativity: *When the laws of physics are written respecting the metrical equation, they will retain their form and validity under any transformation of system of space-time coordinates; they will not be undermined by relative motion, uniform or accelerated; and they will apply with equal success for any observer of physical events.* The goal of formulating generally covariant universally valid laws of physics thus devolves upon the mathematical behavior of g_{ik} in variable space-time.

We still have other topics in this chapter, namely integration, parallel transport, and the equations of motion. These will be our tools with which to proceed.

◆ ◆ ◆

We have labeled the path of an apple falling off a tree as *s* and we subdivided *s* into *ds*'s. The issue to broach is how to make use of the *ds*'s. We can say that we fragmented this prolonged event into many momentary events, each of which is like one frame of a motion picture, and now we wish to "play back" the entire movie so as to view the event as one continuous entity. In other terms we broke the world-line of the apple into a myriad of world-points, and now we will string together the world-points to draw the world-line. In yet other words, we separated the geodesic into many ultra-small intervals, and now we will reconstruct it.

We know that the metrical equation applies for each momentary event, each world-point, or each interval (depending on how we think of points in space-time). For our present objective this equation is clearer written as

$$ds = \sqrt{g_{ik}dx_i dx_k}.$$

The aim is to *tally all the ds's* so as to calculate the entire *s*. The mathematical tool for this purpose is called integration, which is a part of mathematics called integral calculus. For instance we "integrate" *ds* between two points (the tree and the ground). The basic formula contains terms which look like this:

$$\int ds$$

or

$$\int \sqrt{g_{ik}dx_i dx_k}$$

The symbol $\int$ (a.k.a. "summa") stands for integration, and in a sense it is the opposite of the *d*'s which stand for differentiation; $\int$ is therefore also called the antiderivative. A synonym for integration might be "adding together," in the sense that by differentiation we separate something into infinitely small parts, and by integration we add them together.

Since *the entire geodesic is symbolized by s*, and since *the metrical equation gives us ds*, here is the equation for the whole process; *Tr* is the-apple-at-the-tree and *Gr* is the-apple-at-the-ground:

$$s = \int_{Tr}^{Gr} ds$$

By solving for *s* we can predict the progress of the apple in the Earth's gravitational field.

This example is trivial, but the same ideas can be used in more meaningful scenarios. For example in another chapter we shall solve the metrical equation and apply integration to calculate the gravitational motion of an orbiting celestial object. In theory, solving the metrical equation for *ds* and then solving the above equation for *s* should allow us to describe gravitational motion, but in addition to integration, the actual computation will call for an indirect method, one that necessitates what are called the equations of motion and other mathematics we have not covered so far.

✦ ✦ ✦

It is intuitively obvious that if we let an apple roll freely into an indentation in a trampoline, it will move according to the shape of the surface. However this is only an analogy, which is why we raised questions as to how an apple "knows" which way to fall and how a gravitating object "knows" its geodesic. Our erstwhile answer hinged upon several somewhat glib generalizations, such as "light is straight but space-time is not," but this question can be restated more precisely: Why does an apple move from the tree to the ground in the particular path that is straightest in space-time, rather than in any one of a theoretically infinite number of other possible paths? How does any falling object or particle find *the* path in curved space-time? These are important questions, because unsatisfactory replies can destroy the basic premise that all gravitational events occur along geodesics. Accordingly, we will seek answers in several areas of physics and mathematics.

One place to look is in the law of conservation of energy. If an apple "tried" to fall "up," additional energy would be required. Conversely, the most energy-efficient path for a freely falling object is what we call "down." In more general terms any course except along the straightest possible path entails the introduction of some outside influence or force, something that does not occur spontaneously (e.g. a push, a propeller, a sail). In theory this means that when an object falls, it consumes a certain minimum amount of energy. Under experimental conditions we see this as the einstein red-shift wherein the depletion of energy from the act of moving provokes a decrease in frequency. One implication, to which we will return in a more fitting context, is that as Einstein derived the relativistic equations of gravitation, he had to ensure that they do not violate the law of conservation of energy.

Let us take another approach, one based on certain laws of Newton and Euclid having to do with motion. Of course these laws are insufficient to explain general relativity as a whole, but they exemplify the concept of *parallel transport*. Newton's first law requires undisturbed motion to be *straight*, and Euclid's notion of *straightness* can be expressed as follows: We imagine many streets meeting one long highway, and all the streets are parallel to each other: _/_/_/_/_/_/_. If all such streets intersect the highway at the same angle, then the highway must be straight. This geometric principle is called "parallel transport" though we see synonyms like "parallel displacement" or, recalling that we can "slide" coordinates, "parallel translation." Parallel transport is a round-about way of evaluating straightness, but we will find it to be very useful. (The concept of parallel transport was developed by Levi-Civita shortly after Einstein published his first paper on general relativity. There is no requisite that every case of transport be parallel, but when it isn't, it has no meaning for us.)

To observe parallel transport as it applies to relativity, we need three props: a globe of the Earth, a piece of string, and a small protractor. We stretch the string between two distant points on the globe, say New York and Moscow. *On the curved surface of the globe, the path of the string is a geodesic.* Were it not for the sphericity of the globe, the path would be straight, but since there is curvature, the path is "straight" along a curved surface; hence a geodesic is the "straightest possible line" or the "shortest curve" which can connect two points on a curved surface. We can say that a straight line is a special type of geodesic, one that happens to exist on a flat surface, and in this sense a geodesic is a generalized straight line.

We note—and the reader is encouraged to confirm this by experimenting on a globe—that the direction of the string does not matter. *If the string is taut between any two points, which means if it is as straight as possible so that it represents the shortest distance between its ends, the string forms a geodesic.* The implication is that there is something distinctive about geodesics—that the shape of any geodesic connecting two points possesses a naturally favored inherent efficiency. We will return to this notion is a mathematical setting, but obviously it bolsters the validity of geodesics as a basic ingredient of general relativity.

Let us get back to our exercise with parallel transport. We slide the protractor along the string as if traveling from New York to Moscow. Any angle selected on the protractor will point in the same direction anywhere along the string. (In this exercise we are using only intrinsic methods. We must not confuse this with navigation by use of a magnetic compass, which is an external tool.) Wording this in reverse is more significant: If we slide an intersecting line along the string and maintain the same angle, we will mark out a geodesic. This means that parallel transport on a flat surfaced draws a straight line, and *parallel transport on a curved surfaced draws a geodesic.*

Essentially we are subdividing the string into a number of small steps or points. As we progress from point to point heeding parallel transport—as we keep our protractor parallel form step to step—we automatically generate a geodesical line. In fact it is not even necessary for the protractor to mark an angle; the angle can be zero degrees, which means in the direction we are heading. Then all we are doing is "following our nose," with the proviso that we do not turn our head and that we take small steps. In short, *parallel transport keeps us on geodesics.*

Einstein said it succinctly on p. 77 of The Meaning of Relativity: "A [geodesical] line may be constructed in such a way that its successive elements arise from each other by parallel displacements." Of course he applied this concept to space-time. In the case of flat space-time, a geodesic appears as a straight line between any pair of adjacent points. In the case of curved space-time, *parallel transport generates a geodesic between any pair of adjacent points.* We note that in theory, parallel transport is a point-to-point process which can be applied repeatedly, including of all over a gravitational field.

We can now reconsider how an object "knows" which way to fall: Our freely falling apple need not "see" between the tree and the ground. All it needs to "know" is the next infinitely close point in space-time, and all it needs to do is keep moving at the same angle, i.e. to simply "follow its nose," which guarantees a geodesic. We therefore say that *the path for falling or orbiting is the result of point-to-point parallel transport in space-time.* In short, parallel transport makes falling geodesical.

Clearly, parallel transport aids in the relativistic explanation of gravitational behavior, but is it valid? Why should an object keep moving at the same angle? The answer is surprisingly familiar: *Because it obeys Newton's first law.* (Recall page 127.) This law seeks motion ("inertial" motion) in the same direction and the same speed. The Newtonian term for steady motion parallel to an apparently straight line is *uniform motion*: An object's motion tends to remain uniform. A falling object likewise tends to remain in a uniform and straight path in space-time, *even when space-time is neither uniform nor straight but curved.*

Another analogy will help us here. Let us consider a pilot flying from New York to Moscow, and the smallest increment of his progress is one inch. He is told to set his course parallel to some direction, but he is very forgetful. After traveling one inch he reads his instructions, and these give him his course. When he reaches two inches, he again reads his instructions, and again he is told to set his course parallel to his given course. When he reaches 3 inches.... There is no need for the pilot to know his distant destination; he need not "see" from New York to Moscow. In effect all he needs to know is parallel transport, and he can rest assured that he is on the correct path.

Likewise let us consider an object moving into an indentation in space-time in the *a*-direction of a system of coordinates labeled by *x*'s. (The object moves along the *x*-axis in the *a*-direction, which might be eastwardly. We prefer superscripts; why so later.) The smallest increment of progress is dx^a. The object is very "forgetful" about what is straight and uniform motion. After traversing the first dx^a, the object finds itself at a point where space-time appears (locally) flat. The object "reads its directions" and accordingly continues along the *x*-axis in the *a*-direction. After traversing the second dx^a, the object repeats the process and again continues along the *x*-axis in the *a*-direction. After traversing the third dx^a... As in the case of the pilot, there is no need for the object to "know" anything but parallel transport. The correct path follows automatically.

In this scenario the object does not "know" that *the contour of space-time in a gravitational field changes point to point.* In fact the entire system of space-time coordinates is tilted differently at the site of each new dx^a, so that the *a*-direction is different at each position. Fortunately, however, differential calculus allows us to subdivide the measurement of gravitational motion into dx^a's that are so small that *each dx^a (and each ds) behaves like a straight line, even if it is on a geodesic.* Therefore as far as the object is concerned, all it "knows" is locally straight and uniform parallel progress in ultra-small steps. (As far as a distant observer is concerned, all he sees is accelerated motion, but the object does not "know" this. Its proper motion is straight and uniform.)

Naturally the laws of physics are the same at every point in space-time, regardless of its overall shape: Each point appears locally Euclidean (flat), and at each such point Newton's first law suffices to guide the object "straight down" in space and—given four dimensions—"toward the future" in time. In this sense parallel transport is a partner of Newton's first law for motion and of Euclid's criteria for straightness. In fact as far as parallel transport is concerned, the curvature of four-dimensional space-time is locally indistinguishable from flatness and straightness. We can say that like geodesics are as straight as possible, parallel transport is as parallel as possible.

Einstein and Infeld asserted this conclusion (The Evolution of Physics, p. 147). They were describing Maxwell's theory, but their statement applies to gravitation: "The field here and now depends on the field in the immediate neighborhood at a time just past. The equations allow us to predict what will happen a little further in space and a little later in time, if we know what happens here and now. They allow us to increase our knowledge of the [gravitational] field by small steps. We can deduce what happens...by the summation of these very small steps." The metrical tensor tells us about "here and now," the "small steps" are the local space-time intervals, parallel transport tells us how to proceed, and "summation" is integration.

◆

We may now ask by what right we invoke Newton's first law to justify parallel transport. After all, this law excludes acceleration, yet accelerated motion is the prominent feature of the world-line of any falling or orbiting object. We will tackle this issue in the next few pages, using both physical and mathematical approaches.

In Newton's science, gravitating objects attract by a force that is proportional to their mass. According to relativity mass curves space-time which guides objects into gravitational motion. How can such apparently dissimilar explanations apply to one phenomenon? In practical terms the answer is obvious—the difference is usually imperceptible. But in principle how can *curvature* mimic *force*, and how can a law of gravitation arise from the confusion? We already saw that gravitation can mimic force, but in preparation for further details, let us focus on the notion that *the curvature of space-time* can mimic force.

The assumptions we must make for the moment are these: The Earth's surface still represents the "surface" of space-time, and there are two imaginary pilots who forgot that the Earth is round. We again look at a globe, preferably one with lines of longitude and latitude clearly labeled. Let us have the two pilots fly their respective airplanes to the north.

One pilot starts where the prime meridian crosses the equator, a point in the Atlantic Ocean near Ghana. He sets his course perpendicular (90 degrees) to the equator. The other pilot starts where the west-30-degree line of longitude crosses the equator, a point off the coast of Brazil. He too sets his course perpendicular to the equator. Both pilots resolve to fly north as straight as possible, which to us means that they will move geodesically along the Earth's surface.* Of course they both must fly at the same surface-hugging altitude; if they go high enough to see the Earth's roundness, their method ceases to be intrinsic and defeats our analogy. Each pilot believes that their courses are parallel because each flies straight and perpendicular to the equator. Applying basic Euclidean geometry and believing that the Earth is flat, neither pilot sees any risk of collision. In fact they begin their trips about 2,000 nautical miles apart, and they expect to maintain that separation.

Of course we know what happens. Because the Earth is round, the paths actually converge, and the planes meet at the north pole. How do the pilots interpret this outcome, still assuming that they are ignorant of the Earth's shape? As long as it is intuitively apparent to them, as it was to Euclid, that parallel lines should not cross, they surmise that the non-Euclidean behavior of their paths is

---

* In navigation such paths are called great circle routes, and of course they need not follow any particular meridian. The equator and all lines of longitude are geodesical great circle routes. Each is called a great circle because its radius is that of the entire sphere.

evidence of a *force* which pulled them toward each other. Likewise, if it is apparent, as it was to Newton, that undisturbed objects should move in straight lines, the pilots interpret their convergence as evidence of an attractive force. In other terms, because space-time, here represented by the Earth's surface, is slightly but definitely curved, objects behave as if there were a force of gravity.

Though the analogy is less dramatic, we can also envision the two pilots on our trampoline. Starting out next to each other at one edge, each pilot tries to walk straight across to the opposite side. Of course their weights indent the surface, and they end up hitting each other. Moreover heavier pilots induce more indentation, especially when they get close to each other. If the pilots are unaware of the indentation, they seem to "attract" each other, and their weights, m_1 and m_2, as well as their separation, D, seem to govern their attractive "force" according to the ratio in Newton's law of gravitation,

$$force = \frac{(m_1)(m_2)}{D^2}.$$

How logical the basic assumptions can be a to arrive at such a profound misinterpretation of nature: Undisturbed motion is uniform, and parallel lines do not cross. We also note that the pilots are misled by their own common sense even while they follow geodesics.

The modern view is that $\frac{(m_1)(m_2)}{D^2}$ in the above schema is not evidence of a gravitational force but the consequence of the shape of space-time. We say that Newton was obliged to conceive this force so as to preserve both his own first law (of inertia) and Euclid's geometry. Indeed once we apply modern instruments, we find that the above ratio is not correct, and since Einstein managed to replace force by four-dimensional non-Euclidean space-time, we can rewrite the applicable laws so as to predict gravitational behavior to the highest degrees of accuracy.

We emphasize that this new interpretation *reinforces Newton's first law* and even calls upon his second laws on a local basis: Since there is no gravitational force, there is no acceleration (*a*) as a result of that force, and *motion is straight and uniform*; $F = ma$, but if $F = 0$ then $a = 0$. Then why do falling objects accelerate? Why is their "*a*" not zero? Because globally, *space-time is not straight or uniform*. In other words the only difference between the world-line of a uniformly moving object and that of a freely falling object is the non-local shape of the geodesic in space-time. All objects *do* tend to move in a straight line, whether they are in a gravitational field or not, but the geometry of space-time defines straightness. Geodesics, like our two pilots' paths which start out parallel, indeed do not converge *except where space-time is curved*. Thus in the 1919 observations, as the photons from the distant star passed near the sun, the law of inertia was not repealed. In fact on a local basis it was substantiated. Only the flatness of space-time near the sun was "repealed." We can speculate that if the 1919 data had been available in 1684, Newton would have been doubly delighted with his first law but would have discarded his law of gravitation.

205

We can summarize this important inference: *Gravitational acceleration results from local compliance with Newton's first law during compliance with the overall curved shape of space-time.* When we fall or orbit, we drift uniformly and straight-as-possible through space-time which, except locally, is neither uniform nor straight. Parallel transport applies, and all laws of physics hold true, but the effects of non-local space-time curvatures—i.e. the effects of point-to-point variation—are superimposed; we call these the effects of gravity.

✦

Before proceeding we need to link parallel transport and geodesics with vectors. Since a vector can be depicted as an arrow, let us imagine that a *vectorial arrow* is moved through space-time from point to adjacent point so that this vector stays parallel—i.e. so that it traces a geodesic. In this scenario, two features of a local vector remain the same. One is the direction of the vector, so that parallel transport can be represented by a movable arrow maintaining the same angle to the geodesic on the "surface" of local space-time; parallel transport maintains direction, which gives parallel transport its name. The other feature of a vector which remains the same is the length of the arrow; parallel transport maintains magnitude. In summary, *a vector's parallel transport preserves its local length and angle.* (Three comments: First, some authors say that in parallel transport these two features are "conserved," indicating that laws of conservation apply. Second, we assume the vector does not spin on its axis, though this detail is rarely stated overtly. Finally, tensors can also be parallel-transported, though this detail is rarely essential.)

However, a vectorial arrow in space-time not only symbolizes a certain direction in space but also a certain direction in time. Theoretically, it points to where as well as to "when," so that during parallel transport a vector maintains its posture in space as well as in time. Hence we could replace the term "parallel transport" by the less efficient but more descriptive term "transport which maintains a vector's features relative to the local four-dimensional surface." The vector here can be a tangent vector, one whose direction matches the local curvature, since the tangent vector is in the same plane as velocity. This detail reflects the idea that gravitational motion is tangent to the surface of space-time.

Any vector can be parallel-transported along a geodesic, but—as we shall emphasize later—it must be a local vector, and clearly the most convenient local vector in the realm of general relativity is the tangent vector. The role of the tangent vector in relativity, together with the concept of parallel transport, allow us to better define geodesics. *A geodesic is a curve on a surface on which the tangent vector is always parallel-transported.* In other words a curve is a geodesic if it is created by parallel transport of its tangent vector, and (as we saw in our string-on-globe experiment) when we need to create a geodesic, we can call upon parallel transport.

This function of parallel transport has a practical application which will interest us extensively: Since gravitational motion occurs along geodesics in curved space-time while parallel transport (of a tangent vector) on a curved surfaced generates a geodesic, the mathematical analysis of gravitational motion embraces parallel transport, and it treats parallel transport as a vectorial process. This concept makes parallel transport a powerful tool—albeit an abstract mathematical and intangible tool—because it ensures that we comply with geodesics while we derive basic equations of general relativity.

We will use this equation later, but if we label our tangent vector on the surface of space-time by the letter *U*, we can write

$$U^a = \frac{dx^a}{ds},$$

where the direction of this vector's hypothetical arrow is *a* (which could be "down"). The geodesic generated by parallel transport of U^a heads in the *a*-direction, but again we note that the specific direction (*a*) does not matter; any one direction of parallel transport creates one geodesic.

Using an arrow to represent a vector, once we "aim" our tangent vector and continue parallel transport, the resulting path will be geodesical. As we pointed out for our experiment we need not have picked Moscow as our goal. The only prerequisite was that the string be as straight as the surface of the globe allows. As we said, there is something distinctive about geodesics, which means there is something distinctive about parallel transport, particularly when applied to a tangent vector.

The notion that any direction for parallel transport generates a geodesic is particularly conspicuous if we start at the north pole and stretch our string to the south pole. Any direction we pick (and meridian of longitude) gives yields a geodesic, and therefore the path from the north pole to any spot on the Earth also yields a geodesic—as long as the string is taunt.

Moreover there is nothing imperative about starting at the north pole; any spot on the Earth can also serve as the start. The conclusion is inescapable: *Any two points on curved space-time are connected by a geodesic.* And lest we forget, any two points on *flat* space-time are connected by a special kind of geodesic—a straight line.

Meanwhile we recall that we view vectors by their components reflected on coordinate axes (page 143). As long as space-time is only flat, the direction, magnitude, and components of a vector do not change when parallel-transported point to point. However when a vector is transported through *curved* space-time, its direction and magnitude with respect to a geodesic still does not change, but *its components, as reflected on curved coordinate axes, do change.* That is because the curvature of space-time transforms the angle of the tilt of the coordinate axes. Therefore the parallelism of the vector is not absolute; *it is relative to the shape of space-time along which the vector is being transported.* In effect space-time regulates the result of parallel transport of a vector.

The concept (stated in the previous paragraph) that "components of a vector do not change when parallel-transported point to point" needs amplification. First we should restate the concept in more formal language: "We can find a system of coordinates such that the changes in the components of a parallel-transported vector vanish—they are zero—for transport in *infinitesimal increments* along any path that originates from some point on the surface of space-time." Second, in equation form, we say that

$$\frac{dV^i}{ds} = 0,$$

where V is the vector. This time we let i be the direction of the path of parallel transport from some point, and ds is an infinitesimal segment of that path. Of course the surface may be curved, so that any such path is a geodesic, but a geodesic is *locally* straight. This equation, though it defines parallel transport in vectorial language, at first glance appears to contradict the notion that components of a parallel-transported vector may change, but we will say more about this equation later.

Let us show some features of parallel transport graphically. In the left half of diagram R a segment of a geodesic appears as a curved line, while a vector, depicted by arrows, is parallel-transported A to B. This particular vector is not tangent, but that makes it easier to diagram. At both points the angle between the vector and the geodesic is the *same*, say, 30 degrees, but because of the difference in curvature between A and B, the arrows do *not* point in the same direction with respect to the coordinate axes.

Parallel transport is more evident in the right half of diagram R, where the geodesic is drawn as a straight line, in keeping with Newton's first law and with Euclid's definition of straightness. The vectors at A and B are clearly parallel, each 30 degrees to the geodesic, but then the axes do not appear straight. However the axes, particularly the vertical one, will appear straight if A and B are very close together, which means that the curvature of the vertical axis is indiscernible in some system of coordinates. Points A and B may inhabit the trajectory of our apple falling to Earth.

Point A is "higher," where space-time is less curved, and point B is closer to the Earth, where space-time is more curved; viewing diagram R upside down makes this clearer. The projection of the vectorial arrow on the space axis is longer at point B than at A, while the projection of the vectorial arrow on the time axis is shorter at point B than at A. Hence *the components of the vector at A differ from the components of the parallel vector at B, but this is only true if the path from A to B encounters curvature.*

This concept is important and useful enough to reiterate: The orientation of a parallel-transported vector relative to a geodesic stays constant, but if a vector is transported to a different location in curved space-time, it acquires changed components. We can envision this in a more realistic setting: New Yorkers are certain that a dropped apple falls straight down to the ground—we can say that this apple has a vertical velocity vector. If this event were duplicated in Moscow, Muscovites would also claim that the apple falls straight to the ground—again it forms a vertical vector. However based on the sphericity of the Earth, the two vectors point in very different directions (as a globe of the Earth would reveal). In fact the vectors could only point in the same direction if the Earth were flat. In other words each vector complies with local conditions, but these conditions are globally different by virtue of curvature. Each vector has the same relationship to the ground, but the ground is not the same. If the two apples were dropped side by side, their velocity vectors would be

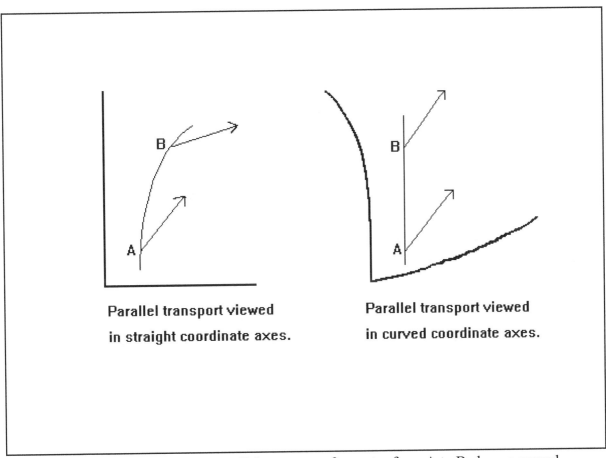

Parallel transport viewed
in straight coordinate axes.

Parallel transport viewed
in curved coordinate axes.

Diagram R. The left half shows parallel transport of a vector from A to B along a curved geodesic in straight coordinate axes. The right half shows parallel transport of the same vector, but the geodesic is drawn straight and the coordinate axes are curved. On an infinitesimal scale, with A and B infinitely close together, the components of the vector would show no change going from A to B.

parallel, but that it is besides the point here. Of course this scenario is difficult to visualize in the relativistic case, since we must accept that the "conditions" include the "shape" of time. In relativity we summon the metrical tensor, in which we link the amount of curvature of space-time (and hence the gravitational potential at each point) with ordinary physical measurements that reveal what we call the effects of gravity. The metrical tensor, by representing the shape of space-time, defines the path of a falling object. Parallel transport, by representing Newton's first law, shows how a falling object follows that path. In short, vector transport can be parallel, but in curved space-time the components of the metrical tensor are not constant; transport can be parallel but geodesics are curved; motion prefers uniformity but space-time near mass does not.

◆

Parallel transport seems quite logical, but does it apply in the real world? Does parallel transport occur, at least in a mathematical sense, in gravitational events? And does parallel transport help us work with the curvature of space-time? Obviously we do not claim that at each microscopic point in space-time a falling object produces a protractor and checks whether the vector of its motion is still parallel. However there is a device which can demonstrate parallel transport and which provides a sound physical analogy for parallel transport. It is no more complicated than a spinning top, but its sophisticated version is the gyroscope. A well-made gyroscope, particularly in a gyrocompass (a compass based in a gyroscope rather than on a magnetized needle) maintains its orientation even when it is moved from place to place; it keeps pointing in the same direction in space. This feature can be couched in terms of vectors. The behavior of a gyroscope is like that of a vector which maintains the same angle. We then simply imagine that when a gyroscope is moved or "transported," it maintains its "parallel" posture. With this in mind we should once more look at a globe of the Earth.

Let us have one pilot fly his airplane north, again starting where the prime meridian crosses the equator and still hugging the surface. Using a gyrocompass he sets his course perpendicular to the equator. For simplicity the gyroscope is tilted north, so that the vector of his course (specifically his direction; speed is immaterial here) is identical to the vector of his gyroscope. All the pilot has to do is note the tilt of the gyroscope, and as long as he keeps his airplane at the same angle—zero degrees—to the gyroscope at each point along this part of his trip, he will in effect heed parallel transport while moving north along a geodesic.

Our pilot's intent is to trace a large triangle. When he reaches the north pole he turns left and heads along the west-30-degree line of longitude back to the equator. The pilot now has a new course and a new angle to the gyroscope, but again he keeps his airplane at a constant angle to the gyroscope, and he will head toward the equator near Brazil along another geodesic. Finally at the equator he turns east back to his starting point near Ghana, keeping his airplane at the new angle to the gyroscope. Obviously he is on a third geodesic, which is the equator.

We can do a rudimentary calculation on this triangular course. The angle between each "leg" and the equator is 90 degrees, and the "legs" meet at the north pole at an angle of 30 degrees. Therefore we have a triangle of 90 + 90 + 30, which is not 180 but 210 degrees. If the pilot starts off knowing he is 90 degrees from the equator and at the north pole he changes course to head south at 30 degrees, he expects (à la Euclid) to intersect the equator again at an angle of 60, not 90, degrees. In this triangular course *there is an excess of 30 degrees.*

The pilot notes the key point: When he started out, he aimed his airplane zero degrees to the tilt of the gyroscope; he essentially followed its tilt. But when he returns to the starting point, his airplane is no longer at zero degrees to the gyroscope. *The final angle differs from the original one,* but *only because the path spanned a curved surface*, here the surface of the Earth. This is called the geodetic effect, though to make the analogy effective we allow this surface to represents the surface of space-time. With respect to the course of the airplane, the direction of the gyroscope has undergone angular rotation of 30 degrees in this example. The difference is called the "excess" angle. The best way to visualize this is on a globe, but diagram R may help if we imagine three curved geodesics forming a triangle; Kenyon on p. 33, Fig. 3.8, adds more details and presents a diagram.

The finding of an excess angle suggests a useful mathematical-geometric maneuver: *parallel transport around a closed loop of space-time.* The loop itself can be any shape; in this analogy we used a triangle, though later we will find a square to be advantageous. We also recall that when a vector is transported through curved space-time, its components change. This means that just as a pilot can gauge the Earth's roundness by the change in the angle of a real gyroscope as it is carried around a triangle on the Earth's surface, *so we can gauge the curvature of a region of space-time by the change of the components of a vector after it has been parallel-transported around a loop.* The concept is very important: By analogy, the behavior of a gyroscope can reveal the curvature of the Earth's surface, like the behavior of a parallel-transported vector can reveal the curvature of space-time's surface. We stress that these methods are intrinsic. For example our pilot gauged the curvature of the Earth only from surface clues, and we must assume that he did not notice the Earth's roundness by looking at the horizon from a high altitude.[*]

The observation that a vector has new components after completing a loop on a curved surface is called "path-dependence." This term can be explained as follows: On a flat surface, the final components of a vector upon parallel transport in a loop depend only on the initial orientation; a flat surface alters nothing, no matter what path is followed. In contrast, on a *curved* surface, the final components depend on the initial orientation *as well as on the selected path*; such a surface alters the orientation, so the path makes a difference in the components. We see that path-dependence is a sign of intrinsic curvature.

―――――――――――

[*] While a gyroscope provides a useful analogy for the study of geodesics and parallel transport, we should be aware that the physical details of how a gyroscope works are quite complicated. In particular, a gyroscope shows a gradual slight deviation from perfect alignment. This deviation is called precession, and most of it is not a consequence of relativity but of the spin of the Earth. (In fact Newton had worked out some of the technicalities.) Relativity explains an additional source of precession which we call "excess" precession and which includes the geodetic effect. We will examine this excess precession later, including as it is revealed in the precession of the orbit of the planet Mercury and in experiments designed to confirm relativity.

Therefore we can quantify the shape of a region of space-time *by the extent of path-dependence, which means by the alteration of the components of a vector that is transported around this region.* We also know that the metrical tensor is a measure of the intrinsic curvature at any point in space-time and that this curvature may vary point-to-point. These concepts taken together give parallel transport another prominent role in general relativity: We can achieve a comprehensive assessment of the intrinsic shape of space-time *by monitoring how the g's vary during parallel-transport of a vector around small closed loops.* We will elaborate and apply this method later.

But why should we use such a complicated method to assess shape? We already know how to quantify curvature by testing for compliance with the pythagorean theorem, so why bother with testing for path-dependence during parallel transport? The answer is that in order to study the *variability of the g's in a region of space-time,* we must be able to quantify curvature at more than just one point, and our method must be intrinsic. Parallel transport around a loop will enable us to meet these needs. At the same time parallel transport ensures that the curvatures in question are geodesics.

◆

Although we did not identify it as such, the observation that moving clocks lose time is an example of parallel transport. On page 74 we "transported" a clock around the Earth. Once the clock completed the "loop," it no longer agreed with a stationary clock because any transported time-measuring device conforms to the shape of space-time. Just as transporting a vector tilts direction, transporting a clock "tilts time." Hence we can equate the aberration in the clock with the change in a vector's components upon parallel transport. Moreover, just as an observer traveling with the clock senses no discrepancy—each "proper" tick appears "parallel" to the previous—so an observer traveling with the vector senses no tilt. The direction at each point appears parallel to the previous.

Earlier we generalized that time and space are relative because clocks and rulers are a part of nature, and now we can advance the reasoning: The components of a transported vector, and for that matter the orientation of a gyroscope, are relative to space-time because vectors and gyroscopes are a part of nature. In fact the behavior of a gyroscope, namely that it keeps pointing in the same direction in *space and time,* is quite useful in the experimental study of space-time, as we shall see later.

◆   ◆   ◆

We have been using the Earth as an analogic model for space-time. Let us detour to emphasize that in reality the curvature of space-time is quite different from that of the Earth. Of course the only reason the shape of the Earth is not readily apparent to us is that we normally view it "locally." However its curvature can be revealed with only a few external aids, such as a telescope and a tall distant mast, and the calculation of its radius is only a matter of simple geometry. (In the third century B.C. Eratosthenes gauged the Earth's radius by noting that at any one time the lengths of shadows are different in different parts of the world.)

The reason the shape of space-time is not apparent to us is more intricate. First of all, space-time is a mathematical abstraction, one which even ascribes a "shape" to time. Moreover no matter how powerful our telescope, a beam of light entering it has conformed imperceptibly to the shape of space-time, obscuring any curvature (the beam seems perfectly straight). Only under exceptional conditions, as during the eclipse of 1919, do we find a measurable radius for space-time, and even this tells us nothing about the space-time near our Earth. Measuring the radius of space-time near the Earth calls for four-dimensional equations applied in difficult experiments. For example since the orientation of a gyroscope conforms to the shape of space-time, a very precise gyroscope can assess the curvature of space-time; such experiments are being performed in space-vehicles in orbits near the Earth. Actually this effect must be added to the effect of the motion of the space-vehicle itself as it "drags" space-time. Furthermore, the orientation of a gyroscope is altered by the spinning of the Earth, which also "drags" space-time. Therefore such experiments, to which we will return, are very intricate. However it is easy to *calculate* the radius of space-time on the surface of the Earth. Without going through the actual numbers, we can apply parallel transport by essentially considering what happens as a vector is moved around a "loop." A space-time diagram can elucidate the germane relativistic principles.

Diagram S represents a closed loop A-B-C-D around which a vector is parallel-transported from point A to B to C to D and back to A. We imagine that the loop is in the Earth's gravitational field and that point A is closest to the center of the Earth. Although the loop covers a space on this page, it portrays an event in space-*time*, so that, for example, a shift from A to B denotes a change through time, not space.

We begin our imaginary trip at point A by simply waiting a moment with the vector at rest. This advances the vector through time from A to B without a change along the space axis. Next we turn the vector 90 degrees and move through space from B to C very quickly so that practically no time passes. In similar fashion we attempt to complete the loop. We now summon the idea that according to general relativity, measured time depends on location; this is why a clock slows when brought down from a mountain. (A prolonged event, even the tick of a clock, needs more time at a lower altitude.) In diagram S the location of line A-B is closer to the center of gravity than line C-D, which means that the interval C-to-D is slightly shorter than the interval A-to-B. Therefore the *rate of time* is different going from A to B than going from C to D. Moreover the linkage between space and time means that the horizontal segments B-C and D-A are also not the same length (D-A is somewhat longer). That is to say, not only is time different in different spatial locations in a gravitational field, but space is different in different times! This is another reason why simultaneity is not absolute.

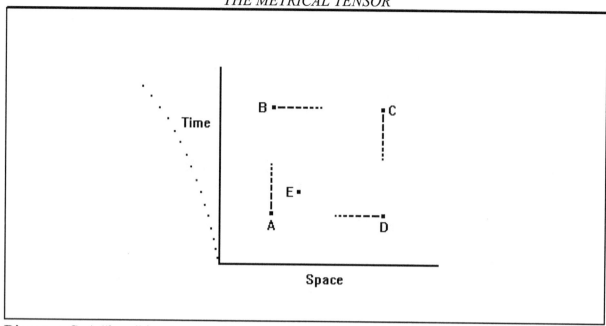

Diagram S. A "loop" is a traversed in space-time. The goal is to transport a vector from A to B to C to D and back to A.

The result of our trip is predictable: If the rate of time for A-B is unlike what it is for C-D, and if the space in D-A does not equal B-C, then we cannot return to our starting point at A; we will not close the loop. While heeding parallel transport, we have drifted away from our goal. In diagram S the actual end of our trip is represented by point E. The geometric principle is simple: A perfect square—like a "pythagorean" triangle—cannot be fitted onto a curved surface. How then can we close our square? Diagram L gives us a hint: To make the points A-B-C-D in diagram S mark out a square, either the legs A-to-B, B-to-C, etc. must be bent, or, as is equally valid, *the time and space coordinate axes must be curved*. In the latter case the square can be completed, but then the axes are *no longer Euclidean*. (To keep diagram S uncluttered, this is shown only for the time axis.) Of course this is another case where space-time is deformed and no longer accepts Euclidean geometry, but we add that *if parallel transport around a loop disrupts the components of the vector, the traversed space-time cannot be flat*. We will use the same logic when we discuss the behavior of the planet Mercury: If a point in Mercury's loop-like orbit ends in an unexpected location—just like point E in diagram S is in an unexpected location—then space-time near Mercury must be curved.

We extract two conclusions from the above procedure: First, by giving a numerical value to the gap between A and E at the surface of the Earth in terms of distance (space) per time, we can compute that an ordinary falling object should accelerate by about 10 meters per second every second, which is what experiments disclose. Second, when we compare Newton's laws and relativity, we find that the latter predicts slightly greater acceleration. The *excess*—using this term again in the context of general relativity—reflects the curvature of space-time near the Earth.

We can equate that excess with the divergence of the curved dotted "relativistic" time axis from the straight "Newtonian" time axis in diagram S. (Only the bend in the time axis is shown, and the curvature of the relativistic [dotted] axis is greatly exaggerated.) If bending the axis allows points A and E to coincide, then *the radius of that bend at A-and-E is the radius of space-time.* By calculation it turns out that the radius of space-time at the Earth's surface is about 240 million miles, vastly longer than the radius of the Earth itself, which is less than 4,000 miles. In terms of the universe, we make barely a dimple is space-time! (To be thorough, we should compute two radii here, signifying that space-time is not perfectly spherical.)

The size of the number 240,000,000 reflects how time is assimilated into the equations of relativity; *T* is multiplied by c, which greatly increases its numerical influence. Also we recall the analogy in which an observer attaches a ball to a string and spins it around over his head while he is inside a moving elevator; the elevator's motion denotes motion through time. To the observer inside the elevator the path of the ball looks like a circle, but to a distant observer it looks like a spiral or a coil. According to this analogy, the former observer experiences this event in space; the latter in space-time, where the ball is moving in a spiral. However we must imagine that this spiral has been pulled and stretched so that the radius of its coils is very long—about 240,000,000 miles.

Of course this means that the Gaussian curvature of our nearby space-time is very slight; K is only about $\dfrac{1}{240,000,000^2}$, which indicates that our Earth induces comparatively little deformation of space-time. Therefore Newtonian calculations for gravitational acceleration yield about 10 meters per second every second, while relativistic calculations yield a very slightly greater result. That is to say, "Einstein's Earth" is a bit stronger gravitationally than "Newton's Earth." The excess stems from the curvature of space-time and is not addressed in Newton's laws. As we shall discuss later, the universe does contain objects which affect their surrounding space-time to a more dramatic extent.

The concept of excess curvature has another implication we already touched upon: On a spherical surface the excess in the number of degrees beyond 180 of a triangle is a gauge of the intrinsic curvature. However we must consider that this excess need not be positive; on a saddle-shaped surface, some of the excess may be negative, similar to the cosmological implication of a negative Gaussian curvature K; see page 161. However modern cosmology holds that the Gaussian curvature of space-time is naturally positive because gravity always appears as an attraction (never as repelling) and because mass and energy are positive.

◆

We have digressed. We should confirm the correspondence between gyroscopes and parallel transport in a setting that reveals curved space-time. Fortunately an opportunity to do so exists: The planet Mercury behaves as a gigantic gyroscope, and we can treat the orbit of Mercury as a case of parallel transport around a closed loop. Let us assume that we can use the distant stars as a system of coordinates by which we track Mercury's orientation and orbit, so that we can tell which way its axis points and where it is at any moment. If we also assume that space-time curvature does not exist (and that the other planets are too far to influence the situation and that the sun and Mercury are perfect spheres), then no matter how many orbits Mercury completes, the orientation of its axis and its orbit should show no unexpected changes.

But we know from decades of observation that after every orbital "loop" the axis of Mercury drifts by a small amount, so that there is a gradual shift in Mercury's position and orientation. This drift can be detected from the Earth as a failure of Mercury to return to where it should after each orbit. Relating this finding to diagram S, Mercury fails to return to point A. This discrepancy is known as Mercury's "precession," and it includes an "excess" beyond Newtonian predictions. That excess measures altogether about 43 seconds of arc per century. After we have acquired the necessary mathematical tools we will see the applicable equations, but we already know what this "excess" tells us: The gyroscopic behavior of Mercury *conforms to the shape of space-time.* In other words Mercury, by virtue of its rapid spinning through curved space-time, reveals an excess to the geodetic effect mentioned on page 211. (We note that most of the amount of precession of a gyroscope on Earth as well as of a planet is not a relativistic effect but is attributable to the motion of the Earth or planet. Newton's equations alone account for this source of precession. That is to say, if space-time were not curved, Mercury's precession would not show an excess.)

The inference to be gleaned from the activity of Mercury is this: *As with parallel transport of an imaginary vector, the shape of local space-time determines the orientation of a real gyroscope.* In particular, if that shape is flat, a transported gyroscope retains its original orientation. *It stays parallel.* If the shape has curvature, the transported gyroscope *still* stays parallel, but the effects of its parallelism are altered by the curvature. In other words a gyroscope does maintain a constant angle to its geodesic on the surface of space-time, but *the surface of space-time is not constant.* Moreover, if parallel transport occurs in a closed loop, this curvature is revealed in the overall change of the angle of the gyroscope.

As mentioned earlier, Newton's paradigm, despite its flaws, constituted an enormous leap in our understanding of nature, for it meant that the laws for objects on Earth are identical to the laws for objects in the sky. Einstein's coup was to demonstrate that space and time are relative and that the inaccuracies in Newtonian laws on Earth are identical to their inaccuracies in the sky. Just as Newton realized that planets stay in orbit the same way apples accelerate when they fall, Einstein realized that Mercury over-shifts its orbit the same way clocks slow down when they fall.

Incidentally, pre-relativity astronomers tried various ways of explaining Mercury's odd precession. One hypothesis postulated a hidden planet, dubbed Vulcan, which was thought to generate enough gravitational force to disturb Mercury's orbit. Of course no such planet was found, and the ability of relativity to explain the discrepancy constituted one of Einstein's first triumphs.

✦ ✦ ✦

We may now enlarge upon the concept that general relativity requires the shape of space-time to mold gravitational motion into geodesics. This point raises the following challenge: While it is apparent that an object in motion tries to pursue a straight and uniform path—i.e. that it obeys Newton's laws—does this principle hold for a *geodesic*? Moreover if we know the geodesic and we know Newton's laws, do we necessarily know relativistic gravitational motion? In this context we can even question our analogy about traveling from New York to Moscow: Why must our trip conform to a particular arc along the surface? Why not follow any of an infinite number of alternative paths? The replies entail the "principle of least action" to show why an object moving in a gravitational field seeks one particular optimal path favored by nature, a path that indeed is geodesical.

To study this topic we should go back to a basic and self-evident point: A key element of physics is motion: *Objects move*, and much of our attention is devoted to *how they move.* (In the context of general relativity, falling and orbiting are prime instances of motion.) In general we describe motion mathematically by means of various *equations of motion*. The very familiar equation of motion, valid under ordinary conditions, is Newton's $F = ma$, as is clearer if we solve for acceleration:

$$a = \frac{F}{m}.$$

(To make this equation more useful, we can incorporate the direction of motion [by an arrow over the *a*, as for a vector] and we can use a differential form, but that is not essential here.)

Despite the scientific break-through it represents, Newton's equation of motion poses several problems: One is that force and mass (*F* and *m*) are difficult to define in a manner that allows the equation to apply under various circumstances. For example, motion is possible without mass, and—as is vital in Einstein's paradigm—gravitational motion occurs without classical force. Furthermore if try to solve complex physical problems using $F = ma$ with the notion of force intact, the calculations become prohibitively cumbersome because each force must be enumerated separately, usually by means of vectors . Finally we know that $F = ma$ itself is not generally covariant, in that this equation presumes only inertial frames of reference. This means that whenever we need to heed the effects of general relativity, we can use $F = ma$ only locally—i.e. at only one point in space-time at a time.

The implication is obvious: Can we find more widely applicable equations of motion? In particular *can we find an equation that describes the geodesical motion of objects moving freely*—"naturally"—*in a gravitational field?* The answer is "yes." In the process we will learn that it is easier to describe motion in terms of a *path* rather than acceleration, and that it is preferable to write equations of motion in terms of *energies* rather than force or mass. One advantage of this approach is purely practical: Energy is a scalar quantity and need not be described by vectors.

Another advantage is that we can design equations—particularly equations of motion—that reflect a very basic property of the universe, namely that nature is not wasteful. We will look at some of the intricate details, but the introductory principles are not obscure: When an object moves, it does so along some path, and an underlying ingredient in this event is the expenditure of energy.

This brings us to the *principle of least motion*. The concept predates Newton: Long before science knew much about the physics of light, Fermat (1601-1665) attained a surprisingly profound insight regarding the motion of light. To appreciate Fermat's conclusion, let us imagine that a stream of photons moves from a source at A to a receptor at B. We also imagine that we can track the path of the photons and that we can measure how long it takes for photons to get from A to B. A critical feature of this scenario is that in theory a beam of light can take any of an infinite variety of paths to get from A to B, *some of which take longer than others*. However *one path exists which requires less time than any other*. By intuition alone, and more importantly by application of mathematics, we expect that this path is favored in nature.

Of course a beam of light may encounter various obstacles in it path, such as lenses, mirrors, various transparent media (air, water, glass, etc.), and—most pertinently—various gravitational fields. We can say that light was instructed by nature to get from A to B as quickly as possible despite obstacles. Without such obstacles, the optimal path is readily evident: A straight line. But how does the path of light behave in the presence of such obstacles, notably including curvatures of space-time? What determines how non-straight it is when straightness is impossible?

In reply we enlarge on Fermat's conclusion: Even with obstacles, light naturally finds the optimal path, *the one which consumes the least amount of time*. I.e., *light moves along the fastest path*. Once we think it through, this is rather obvious. For example, light reaching us from the sun encounters obstacles (radiation belts, various layers of Earth's atmosphere, clouds, smog, the gravitational fields of the sun and Earth, et al.) and yet it apparently always crosses the sun-to-Earth distance as promptly as these obstacles allow. This is even why radar and GPS systems work; they would not if Fermat had been wrong. Yet general relativity teaches that the path of light—and indeed of anything moving freely through a gravitational field—is a geodesical path through curved space-time.

So much for Fermat's contribution, which is limited to the time-efficiency in paths of light. Several years later Maupertuis widened the concept. The principle of least action, as crystallized by Maupertuis around 1746, is that *natural motion is efficient and frugal*. In this context "action" is a quantity that reflects energy used over time. I think of "action" as a measure of the effort expended to move something which has mass. "Least action" implies moving that "something" with least effort. A simple mathematical term for action or "effort" is mx^2/t, where m is mass, x is distance moved, and t is the time for motion, but this expression is inadequate in our context.

"Natural" means undisturbed; without external forces, human or mechanical. "Efficient and frugal" means that *some quantity is kept at a minimum* when things move naturally, such as when they fall or flow freely. In short, Maupertuis surmised that nature seeks to expend minimal effort. For example water seeks the steepest descent, and dropped objects promptly fall straight down. The exact history of this idea, particularly the chronology of individual contributions, is unclear.

218

Leibnitz and Euler had worked on it earlier and Lagrange later. In any case this "quantity" became known as the "action." The key insight and its mathematical consequence is that in a physical system *natural motion minimizes the "quantity of action."*

A critical issue was the quantitative identification of this "action." Maupertuis thought that action consisted of a measure of kinetic energy over some period of time, suggesting that systems tend to minimize the energy of motion. Euler (around 1748) roughly equated action with potential energy, suggesting that systems seek a state of least potential energy. (Thus a pool of water has a flat surface even though the bottom may be uneven, presumably because in this state the potential energy of the system is minimized.) Meanwhile Euler began work on the "calculus of variations" (more on this later, but Euler coined the term) and derived an important step in this manner. Lagrange (around 1788) implicated *kinetic as well as potential energy in his conceptualization of action.*

In particular, Lagrange, working cooperatively with Euler, devised a quantity now called the Lagrange Function or *"the Lagrangian,"* usually labeled as "*L*." The value of *L* is defined as kinetic energy, often expressed as "*T*," *minus* potential energy, often "*V*." (Other symbols are used. In this context *V* refers to potential energy; do not confuse with *v* for velocity. Kinetic energy can written out as ½ mv^2, and potential energy in a gravitational field is mGh (product of mass, Newton's gravitational constant, and height). Thus, in words,

$$L = Kinetic\ Energy - Potential\ Energy$$

and in traditional symbols, the key equation that defines the Lagrangian is

$$L = T - V.$$

I prefer the more explicit version,

$$L = K.E. - P.E..$$

However, *L* is usually treated as a function of generalized coordinates, *q*, and of generalized velocities, $\dot{q}$. (Dot means the derivative with respect to time; velocity is the derivative of location with respect to time.) In short, the value of *L* depends on location and speed in a certain way. This means that *L* can exist in phase space, where the coordinates are position and momentum, rather than in ordinary two-dimensional (e.g., coordinates x and y) space. The idea is that action is determined by the difference between kinetic and potential energies of the moving mass for all possible values of position and momentum. The very important concept regarding L is that *any point along a path has an L; it has a Lagrangian.* Moreover, and again very importantly, *the sum of all L's (for all points) in a path is called the action, usually labeled as "S."*

I must point out that the Lagrangian is a kind of compromise. For example if an object is tossed up in a gravitational field and allowed to fall, we can imagine three kinds of trajectories or "paths."

Each path is parabolic and takes the same time, but *each path has a different (total) action S, once the L's in each path are integrated (summed) for all its points.*

A flat and low trajectory entails little potential energy (the object never gets very high) but it entails little kinetic energy (the object need not move rapidly). On the other hand, a high arching path includes much potential energy (much height is achieved) but also much kinetic energy (much speed must be achieved). The optimum in-between possibility represents the compromise between the two energies, and that is what actually happens naturally; it is the usually observed "classic" natural trajectory.

We also note that the Lagrangian is a parameter (a numerical modifier of a function) that describes the path of a moving object. That is to say, *a moving object heeds its L at every point along its trajectory so that the sum (by integration) of all the values for the entire path is a minimum.* That sum, said again, is *the action "S."* These concepts will also interest us in the chapter on quantum mechanics, particularly regarding the modern-contemporary version of quantum physics which exhibits close cooperation with special relativity; for the moment only the connection between these concepts and general relativity is under discussion.

Meanwhile we know that we can analyze a gravitational path from point A to point B as a series of small steps chained together into a continuous line. Now let us add the assumption that at each point along this path we can calculate the difference between the kinetic energy and the potential energy—we can find the local Lagrangian, the *L*. If we integrate (see "∫" on page 200 about integration; integration is a form of addition or summation) for the total length of time needed to get from A to B, we obtain a certain quantity which can be written out explicitly as

$$\int_{t_A}^{t_B} (Kinetic\ Energy\ -\ Potential\ Energy)\ dt.$$

This quantity is called the "action" or, to be more complete, the "action integral." It is the action abbreviated as *S* and summarized in by the equation

$$S = \int_{t_A}^{t_B} (K.E.\ -\ P.E.)\ dt.$$

The t_A is the starting time and t_B is the ending time. The *dt* appears in these equations because we integrate the value of *L* at successive points in time. In more general terms "action" is defined by an integral of energy through time ("time integral" for short) in an energy-conserving system. Incidentally, this "action" does not have the same meaning as in Newton's "action at a distance." Rather "action" in this context is an indication of the overall difference between the kinetic and potential energy of an object moving along a path. We can think of "action" as a assessment of how effectively energy is consumed, and in this context energy is more fundamental than force.

As I mentioned, a trajectory or path can be calculated with Newton's *F = ma*, but this "classical" method is very difficult in many cases. It is easier to consider energies, and it is also more elegant in terms of the basic concept that nature is frugal; nature seems to favor the best compromise, and in my opinion this contrast represents a profound concept worthy of emphasis: Newton's equation of motion, as well as his other "laws," are purely descriptive, in that they describe mathematically *what* Newton observed but offer no underlying reason. The Lagrangian approach does more. It implicates basic features of nature, namely energies, so as to offer a glimpse into *why and how.*

Let me touch on the later contribution of Hamilton (1805-1865), who continued Lagrange's work with the calculus of variations and succeed in proving that an unimpeded (free) physical trajectory represents a mathematically optimal path. In the English literature this conclusion is called *Hamilton's principle of least action.* It states that *a moving object seeks a path for which the action S is the least.* As a result, the observed motion—the kind we see routinely in our environment—does not vary from an apparently ideal path. In more formal terminology, the principle of least action anticipates that the action S be an "extremum" for the path of motion exhibited by a moving object. The term extremum comes from differential calculus, and in theory it can mean a maximum or a minimum—as we shall see later—but it is usually though of as a minimum. We reiterate Hamilton's conclusion in a form we will use shortly: *Of all possible paths, the one naturally followed by freely moving objects is the path which minimizes the time integral of the difference between kinetic and potential energy.* Of course we will first concentrate on the role of Hamilton's principle in relativity, though it also arises in quantum mechanics. Hamilton also devised a particular expression for total energy called the Hamiltonian that plays a major role in quantum mechanics, which will interest us later.

The point for now is this: The foregoing equations led Hamilton to show that *to comply with the principle of least action, the action, given by the value of S, must be "stationary."* We will see that this means that when the principle of least action is satisfied, a very small variation from the optimal path produces no change in the action; hence we write

$$\delta S = 0.$$

In other words the optimal motion of an object between time t_A and t_B is such that an infinitesimal variation in S is insignificant, whereas substantial deviations from the optimal path produce larger variations in S. Therefore the principle of least action applies a "variational principle."

In this context we should note the—often confused—relationship between three concepts: (1) Nature favors an optimum path. (2) That path is characterized by least action; hence the principal of least action. (3) A mathematical consequence of least action is the variational principle tersely embodied in the equation $\delta S = 0$ and in the clause that S is stationary.

Before we get engrossed in details, let me insert other—often under-emphasized— mathematical points: The mathematical analysis of least action applies differential calculus and any text on calculus will dwell on this issue). The procedure is pictured with a graph of a wavy line which has minimum and maximum points (troughs or dips, and peaks or crests). As a basic fact in calculus, these points turn out to have zero first derivatives, and they are called extrema. Thus a minimum is an extremum and a maximum is an extremum, both characterized and identified by zero first derivatives. (The wavy line may also have inflection points, where the line changes from concave to convex or vice versa and where the second derivative is zero.) A typical graph of the action S has these features, notably extrema, but *the general and broad mathematical property of the action is that when S is at an extremum, the action is stationary.* This assertion signifies that a very small variation—a "perturbation"—in S, labeled δ, makes no appreciable difference in S; hence the equation $\delta S = 0$, which compactly summarizes the variational principle. I will restate these notions from various angles, including by way of analogies, but it may be helpful to see them here in one place.*

We can link this idea with the Lagrangian L: The "action integral" *S,* where $S = \int_{t_A}^{t_B} L\ dt$ and L is the Lagrangian, has a stationary value when the path of motion is what we normally observe in nature; i.e. it is not disturbed by outside influences; it is "limited" and it obeys the conservation of energy. In other words in an ideal case, such as the natural path, a stationary *S* expected no matter what its numerical value. This is the variational principle applied to "action." Mathematically the infinitesimal variation (δ) in the value of *S* "vanishes," which is another was of verbalizing the key equation $\delta S = 0$. Incidentally, the variational principle applies as much as possible even in the face of disturbances. We can say that nature does "the best she can."

For an analogy, let us reconsider Fermat's principle of least time: Please picture a large group of soldiers who have split up into several smaller groups. All start at point A and eventually arrive at B. However most groups are on markedly incorrect paths—they waste time—while one group is on an optimal path—it uses time most frugally. In the case of the "wayward" groups, even the slightest further error will result in significantly greater tardiness. In the cases of *the groups that have found the much better path, a small error will make little difference in how long the groups take to reach their destination.* Indeed, the closer the path is to the optimum, the less time-difference a small error will make, and if a group has found *the* path with zero error, any very small deviation (a perturbation) will make *zero* difference in how long it takes to reach the destination. We then say that the elapsed time is stationary. If we allow the soldiers to represent photons, we can say that this group has complied with Fermat's principle and has demonstrated a variational principle.

To make the analogy more pertinent, we can consider not time but "action" by assuming that the soldiers have a certain supply of kinetic and potential energy for that action. Now their task is to be frugal and use a minimum amount of *action* in moving from point A to point B in a certain fixed

---

* I am aware that I repeat many ideas, some perhaps too often, but in my experience this is appreciated by novices when the topic is esoteric, intricate and/or important (and tends to appear on exams). E.g. Euler-Lagrange and the variational principle will interest students of math (calculus), relativity, classical mechanics, quantum mechanics, and more.

amount of time. They are to do this by ensuring that at each step along their path, *they use the optimum balance in the amount of kinetic energy and potential energy.* They are to calculate the difference (labeled as *L*) between the amount of kinetic energy and the amount of potential energy at each step. When the soldiers reach point B, they are to add up their findings. *We call this sum the "action," we label it as S, and we find that in the ideal case this sum-by-integration to be the least it can be.* In short, the steps of an optimum path generate least action.

Now let's bring the variational principle into the picture. We compare the results of different "varied" paths. When a group of soldiers succeeds in following the optimal path—the path which generates the least action—we find that compared with another *slightly* different path, the values for the actions are almost identical. Moreover *the better the soldiers succeed in following the optimal path, the less the difference in the action, until a very small deviation from the optimal path no longer alters the action.* In effect as the soldiers become more proficient, the variation in action becomes smaller. Hence we then say that the action *S* for an optimal path is mathematically stationary. We already expressed this inference in symbols of the calculus of variations as $\delta S = 0$. We expect that *when any group of soldiers is left alone to move in a natural manner, its actual path routinely turns out to have a stationary S.*

We need not belabor this concept by introducing other variables, but what if the soldiers are *not* "left alone to move in a natural manner?" For instance what if half way along their path they are forced to pause or change direction. In that case their path before and after the disturbance *still conforms to the principle of least action*; each segment of their motion still has a stationary *S*, but the final outcome is different. The practical application of this notion is that when we speak of motion, we exclude factors like friction, other forces (such aid by an engine), an addition or subtraction of energy (such as a change in temperature), etc.

Upon connecting these concepts and their historical developments, the *action* in a limited system, the "*S*," is now firmly defined as *the integral of L over time.* (As mentioned, an integral is a kind of sum in calculus.) Thus we now have two key equations, wherein $L = T - V$ or $L = K.E. - P.E.$ and wherein we commonly use generalized coordinates. First,

$$S = \int_{t_A}^{t_B} L(q,\dot{q})\,dt$$

Meanwhile the notion of "minimal action" was refined by the application of the *calculus of variations.* Accordingly, the action S can be at a maximum or minimum, *as long as its variation is zero.* The requirement for zero variation in *S* is written succinctly in the second key equation,

$$\delta S = 0$$

The leap between these equations contains complicated and tedious calculus, most of which I will omit, but several keys are so important and subtle that they can be stressed repeatedly. (1) When *S* is at a minimum, a small shift in location generates only a second-order (second derivative; quadratic; not first-order) change in the value of the action S. (2) A comparison can be set up between [a] the actual path and [b] the actual path altered by a small deviation or "perturbation." (3) This deviation allows the consideration of kinetic and potential energy, which determine S. To calculate S, the difference between kinetic and potential energies, the L, is integrated (summed) between the staring time and ending time of the path taken by a mass at some velocity. In effect, the action is proportional to the average difference between kinetic and potential energy for an event, such as the motion of an object between two location in space. The crucial concept is that *a moving object naturally moves in a way that minimizes S*; the action is as small as possible, though mathematically it may be a minimum or maximum. In any case natural the motion is optimally frugal, and the outcome reflects a variational principle wherein action is stationary. Meanwhile, the idea had evolved that motion, including undisturbed natural motion, can be described by "equations of motion." These equations of motion can be derived using complex math that includes the calculus of variations.

These two conditions—S is defined by the integral above, and the very small variation in S yields zero—give rise to the famous and central *Euler-Lagrange equations of motion*. We will see that for a system with generalized coordinates and generalized velocities, the master equation reads

$$\frac{\partial L}{\partial q} - \frac{d}{dt}\frac{\partial L}{\partial \dot{q}} = 0$$

(This is a partial differential equation; derivative of one variable is taken while other[s] are constant.) If a disturbing force with a non-conservative component were applied, the right side of this equation would equal that component rather than to zero. An alternate version often used in the literature is

$$\frac{d}{dt}\left(\frac{\partial L}{\partial \dot{q}_i}\right) - \frac{\partial L}{\partial q_i} = 0$$

which is the previous equation multiplied through by –1. I also show optional parentheses, and I include subscripts (*i*'s) to show the dimensions being considered; *i* can be 1, 2 or 3.

For readers or students interested a very detailed derivation of the Euler-Lagrange equations, I suggest the videos available (at least in June 2014) in the web site www.digital-university.org. The procedure is quite lengthy because the steps entail various rules of calculus, notably the power rule, the chain rule, and integration by parts. However, I present a briefer version in the next few pages.

At this juncture let us recall the dictum of general relativity that geodesics are the shortest possible paths between points in space-time and that freely falling or orbiting objects move along geodesics. Let us now see *how the geodesics of relativity comply with the principle of least action*. In other words let us confirm that *an equation for geodesical motion can satisfy the equation δ S = 0*. Our analogy again is not perfect—it would be difficult to determine whether the motion of soldiers traces out a geodesic—but in theory we should be able to confirm that *if S is stationary, the path must be a geodesic*. This means that we should find an equation of motion compatible with both general relativity and the principle of least action. In the process we will encounter two problems. First, the mathematics, save for some highlights, is beyond our scope; references are listed. Second, the equation of geodesical motion we will see shortly contains terms we have not yet described. Therefore we will merely introduce certain equations and say more about them later when we get to use them.

To meet the goal voiced in the previous paragraph, we turn to some preparatory steps. Let us consider the path, traditionally labeled as *s*, of an object moving along a geodesic in a gravitational field from point A to point B. Here I use a lower-case "s" to distinguish this path from its action "S," though later a link will appear. We know that a clock moving on a geodesic measures proper time, so that the proper time (t) elapsed while moving from A to B is

$$t_{A\,to\,B} = \int_A^B dt.$$

Hence to consider the entire path *s* by integration between points in space-time rather than between just points in time, we use the equation

$$s_{A\,to\,B} = \int_A^B ds.$$

We also know that along a geodesic our metrical equation applies, namely

$$ds^2 = g_{ik}\,dx^i dx^k \quad \text{or solved for } ds, \quad ds = \sqrt{g_{ik}\,dx^i dx^k}.$$

Therefore we substitute for *ds* and write an equation to describe the path *s* as a geodesic,

$$s = \int_A^B \sqrt{g_{ik}\,dx^i dx^k}.$$

The next step is called parametrization, where we express the right side of the equation by the parameter *s*. In this context (which is more specialized than the one on page 45) parametrization is a mathematical tool that allows us to associate each point along a path with a real number. This number is a parameter, and in effect every point on a parametrized path corresponds to some value of the parameter. Analogies help here: When we place markings (inches, centimeters, etc.) on a ruler, we parametrize the ruler. Similarly when we print degrees of latitude and longitude on a globe of the Earth, we parametrize the globe. We can think of parametrization as providing a line (or surface) with local co-ordinates which are functions of the parameter, so that we can describe the motion of objects along this line. In particular, we can quantify the length of the path traversed by an object moving along that path. In this case the parameter *s* describes a path in space, so that any

225

point along s can be described along x coordinates; the directions of the path are given by superscripts i and k. With this method, locations given by x^i and x^k are functions of this parameter. (A generic parameter is also used, traditionally labeled as λ. For convenience the starting point is then $\lambda = 0$ and the end is $\lambda = 1$. The parameter λ is needed for the equation of motion for light—i.e. for null geodesics—since in that case ds vanishes. We can ignore this detail here.)

With s as the parameter, the above equation $s = \int_A^B \sqrt{g_{ik}\, dx^i dx^k}$ becomes

$$s = \int_A^B \sqrt{g_{ik}\frac{dx^i}{ds}\frac{dx^k}{ds}}\, ds.$$

For instance if our soldiers follow a geodesic, they can calculate the length of their trip by means of this equation. This information is important because we know that if a group of soldiers has found the optimal path, a small error will not make a significant difference in how far they moved to arrive at their destination. That is to say, *if the length of the path is optimal, a small variation from a geodesic—here the phrase "deviation from a geodesic" or "excursion from a geodesic"or "perturbation of a geodesic" may used—results in no significant change in that length.* In short, *the length of a geodesical path should be stationary, according to the variational principle.* Hence we can write the equations

$$\delta \int_A^B \sqrt{g_{ik}\frac{dx^i}{ds}\frac{dx^k}{ds}}\, ds = 0$$

and since $s = \int_A^B \sqrt{g_{ik}\frac{dx^i}{ds}\frac{dx^k}{ds}}\, ds$, we obtain the variational equation the likes of which we saw earlier,

$$\delta s = 0.$$

We are ready to combine two quantities. One is the stationary *length* s of the optimal and actual path. The other is the stationary *action* S of the optimal and actual path. Now

$$S = \int_A^B \left(g_{ik}\frac{dx^i}{ds}\frac{dx^k}{ds} \right) ds$$

(and the square root sign can be omitted). Just as the equation $\delta s = 0$ holds for a geodesic, so

$$\delta S = 0$$

holds for a geodesic!

The implication, which indeed is our goal, is that an equation of motion *based on the principle of least action* should also be an equation of *geodesical* motion. We can also say that since an actual path shows no (zero) variation from a geodesic, and since δ represents the variation of a path from a geodesic, *the actual path of an object falling or orbiting in a gravitational field complies with the equation*

$$\delta \int_A^B ds = 0.$$

This basic equation of geodesical motion satisfies the physical principle of least action and the mathematical variational principle. However the form of this equation, though important in principle, is not very useful. We will now seek a better one, and let me pose an introductory question, whose answer will help in following the many steps: Why do we resort to the calculus of variations; why not use plain differential and integral calculus? Because here we do not seek the minimum point or maximum point or inflection point on a line that is described by a function (an equation), as can be done with ordinary calculus. Rather we seek a function out of a whole class of functions (equations) that has a minimum or maximum (extrema). Traditional calculus is an inadequate tool for this complicated task. We also should keep in mind that the lines we are considering can represent any motion, including gravitational, that is treated by general relativity.

A typical exercise applying the above paragraph is to consider a function that is depicted on an x-y graph by a line (a "path" on a surface) connecting two points. The first goal is to show that the shortest distance between two points is a straight line. In effect, the questions that are posed are (1) what is the minimum distance between these two points, and once this is found (2) what is the equation—the function—for the line that meets this requirement. I.e., what is the function that has this minimum value for its length. However, other functions exist—indeed *a whole class of functions*—so that there are many possible lines connecting these two points, and we seek the one function whose equation meets this requirement.

The way to deal with this situation analytically is to consider two separate lines (two separate functions) added together; one of these (1) is presumed to be the shortest, and the other (2) is an extra bit that makes the whole line longer. In that case, the whole line is also presumed to be the shortest, simply because we expect natural motion between two points in space to be represented by an equation for s straight line. (Here, to make this easier, we consider just one dimension.) After much ado it turns out—surprise!—that the sought-after equation for line (1) is f(x) = mx + b, which plots a straight line; the function of minimum length corresponds to a straight line.

This outcome is applied to the Euler-Lagrange equation of motion, valid for natural motion by virtue of containing the Lagrangian L. Upon more ado, it turns out that this equation of motion obeys the variational principle; a perturbation, represented by line (2) has zero effect on the action S. In other words a stationary value of S signals that the Euler-Lagrange equation describes motion with least S. Now I will show somewhat more thoroughly how this is done in our context. (All this is covered nicely in two "youtube's"aired without cost in the aforementioned web site for digital.university as Analytical Mechanics #1 and #2.)

For this purpose let us recall our analogy and assume that the wayward groups of soldiers, the ones that deviated from the optimal path, can determine (thanks to parametrization) where they are at any moment and how their position changes moment to moment. That is to say, the soldiers can plot their path along x coordinates and can also calculate the derivatives of x with respect to time. These two parameters can be labeled x^i and $,x^i$ while $,x^i = \dfrac{dx^i}{dt}$; the former is location; the latter is velocity.

(We recall the use of commas to label derivatives, as on page 103.) We know that in four dimensions i will run from 1 to 4, so for brevity we can omit the i's. The optimal path for these soldiers, which corresponds to the actual paths of freely moving objects, is then described by (x and $,x$), but a deviation from that path is described by $(x + \delta x)$ and $(,x + \delta,x)$. The parentheses reduce confusion among the uses of commas. We compose an equation for the L (the Lagrangian measure of energy) for the wayward group so that we can examine the local variation in L; we note the term δL, signifying the very small deviation of L from the optimal path:

$$\delta S = \int_A^B \delta L \, dt$$

In order to spell out δL, we let the L for the wayward group be a function of

$$(x + \delta x) \text{ and } (,x + \delta,x).$$

As noted earlier, the L for the group on the *optimal* path is a function of only $(x + \delta x)$, as there is no second derivative when the variation vanishes into zero. For the *wayward* group, since the variation does not vanish, the equation has more terms:

$$\delta S = \int_A^B [L(x + \delta x), (,x + \delta,x) - L(x + \delta x)] \, dt.$$

We can "expand" the terms inside the brackets by use of the general rule [*]

$$f(a + h), (b + k) = f(a,b) + \frac{\partial f}{\partial a} h + \frac{\partial f}{\partial b} k + etc.$$

If we let $a = x$, $b = ,x$, $h = \delta x$, and $k = \delta,x$, the equation for δS becomes

---

[*] This rule is a "Taylor expansion" which gives a "Taylor series." It is a method to find an approximate value of a function by finding a series, similar to approximating the value of a number (such as the gap between 1 and 3) by finding a series (1 + 1/2 + 1/4 + 1/8). The members of the series provide the approximation. In our case a part of a Taylor series simplifies an otherwise difficult integration of a function near one point (x).

$$\delta S = \int_A^B \left(\frac{\partial L}{\partial x} \delta x + \frac{\partial L}{\partial ,x} \delta ,x \right) dt.$$

We note the content of the parenthesis, which is a partial derivative. (The derivative of one variable is taken while the other[s] are constant.) We can think of the first term, $\frac{\partial L}{\partial x} \delta x$, as a measure of the changes in L along the x-axis of only the optimal path. The two terms together, $\frac{\partial L}{\partial x} \delta x + \frac{\partial L}{\partial ,x} \delta ,x$, represent a measure of the changes in L along the wayward path, and in effect they represent the Lagrangian value for the soldiers who are off course. That is to say, these two terms reflect to what extent these soldiers misuse their kinetic and potential energy. If we solve for δS using these terms, the result will not be zero; the action for this path will not be the least; the action will not be stationary. Again our analogy does not show this well (we could say that the soldiers disobeyed their orders) but such a path should not be found in nature—it is not an "actual" and optimal path.

The next step is "integration by parts" of the second term in $\frac{\partial L}{\partial x} \delta x + \frac{\partial L}{\partial ,x} \delta ,x$. This is a technique for integrating equations with products of different types of functions. The general rule here is

$$\int_A^B u\, dv = uv \Big|_A^B - \int_A^B v\, du.$$

In the case of our $\frac{\partial L}{\partial ,x} \delta ,x$ in which $,x = \frac{dx}{dt}$, we let $u = \frac{\partial L}{\partial ,x}$, we let $dv = \delta ,x$, we let $du = \frac{d(\partial L / \partial ,x)}{dt}$, and we let $v = \delta x$. Altogether the result of this step is

$$\Big| \frac{\partial L}{\partial ,x} \delta x \Big|_A^B - \int_A^B \delta x \left(\frac{d(\partial L / \partial ,x)}{dt} \right).$$

The $\Big| \frac{\partial L}{\partial ,x} \delta x \Big|_A^B$ is called a boundary term for the interval A to B, and we note that it contains δx. This detail simplifies the situation since there is no variation possible at the very beginning of the path (at point A from which all possible paths diverge), nor can there be variation at the end (at B where all paths, optimal or not, converge). Hence $\delta x = 0$ at A and B, which eliminates the boundary term.

Meanwhile we have ignored the first term in $\frac{\partial L}{\partial x}\,\delta x \;+\; \frac{\partial L}{\partial ,x}\,\delta ,x$. The result of incorporating $\frac{\partial L}{\partial x}\,\delta x$ is the equation

$$\delta S = \int_A^B \left(\frac{\partial L}{\partial x} - \frac{d(\partial L/\partial ,x)}{dt} \right) \delta x \; dt .$$

We still insist that optimally $\delta S = 0$, so that we integrate the right side and equate it to zero, which yields our erstwhile goal,

$$\frac{\partial L}{\partial x} - \frac{d(\partial L/\partial ,x)}{dt} = 0,$$

which can also be written

$$\frac{\partial L}{\partial x} - \frac{d}{dt}\frac{\partial L}{\partial ,x} = 0.$$

These are forms of the *Euler-Lagrange equation*, and we note that they are set in terms of L and that each includes partial differentials.

The fully detailed origin of this equation is a separate and complicated topic, but a notable feature is that it can be derived in two ways, via the calculus of variations as outlined above, and via a Newton's equation of motion as we will outline shortly for interested readers. The Euler-Lagrange equation is more sophisticated and more broadly based than Newton's, and it represents an intermediate equation of motion between such equations found in Newtonian physics and those used in general relativity. A key point is that the action S, as in $S = \int_{t_A}^{t_B} L\ dt$, has a stationary value, such that $\delta S = 0$, only if the Euler-Lagrange equation applies.

This insight has vital relevance for general relativity: The Euler-Lagrange equation can describe the actual paths seen in natural gravitational events, and since gravitational events can be framed as the Euler-Lagrange equations, *gravitational events must exhibit least action.* On the other hand the Euler-Lagrange equation does not explicitly address geodesics, which means we have more work to do on this topic.

But first, I will embark on three detours. First, we stated in several ways that the variational principle applies in natural motion, and yet the foregoing derivation does not reveal explicitly why or where $\delta S = 0$

is essential. Here is the answer in words: The derivation of the Euler-Lagrange equation *requires that* $\delta S = O$. This derivation typically considers just one dimension, so that the distance covered by a moving object indeed is the shortest, i.e. along a path that is a straight line. The math proof of this requirement is laborious and subtle, so I will continue verbally rather then symbolically. (It appears deep inside tube #2 in the aforementioned web site for digital.university, and it is more prominent in en.wikipedia.org/wiki/Euler-Lagrange_equation [2014].)

At a certain stage of the derivation that eventually provides a Euler-Lagrange equation, a crucial equation appears. In this equation, usually highlighted in texts, three function are integrated between two points, the start and end of a line representing motion. On page 228 in the abbreviated derivation, this equation is equivalent to

$$\delta S = \int_A^B [L(x + \delta x), (,x + \delta,x) - L(x + \delta x)]\, dt.$$

One of these functions (second in the above) represents the perturbation or variation on the line, the bit that makes the line longer rather than shortest. But if natural motion is to require the shortest path (a straight line), then this function drops out, signifying that *when the path is shortest, δS vanishes into zero*. In more formal words, a necessary condition for the above-mentioned crucial three-function equation to have extrema (including the minimum length of the line) is that a certain function vanish, the one representing perturbation. This means that the complete derivation yields the Euler-Lagrange equation

$$\frac{\partial L}{\partial x} - \frac{d}{dt}\frac{\partial L}{\partial,x} = 0$$

only if this function is absent. Otherwise the left side of this equation will not equal zero, and we do not achieve the equation for natural motion. Thus Euler-Lagrange being true relies on δS being zero. Please also envision that our soldiers must march in step so as to move most efficiently in a straight line, and therefore a necessary condition is the vanishing of any effect of perturbations on their gait. And on this note I end the first detour.

Next let me also detour here into the historically interesting and confirmatory derivation of the Euler-Lagrange equation from Newton's equation. This approach starts with $F = ma$ but expresses F in terms of potential energy; $F = -\dfrac{d(P.E.)}{dx}$. Potential energy in this case is strictly a question of position (x), as for instance raising an object off the ground gives that object potential energy. We assume that F is a "conservative force," which means that it depends only on the change in position. (Gravity in a Newtonian paradigm is a conservative force, whereas friction is an example of non-conservative force.) Furthermore the a (acceleration) in $F = ma$ is expressed as a derivative of velocity, $\dfrac{dv}{dt}$, while velocity v is expressed as a derivative of position, $\dfrac{dx}{dt}$ or $,x$. Hence $F = ma$ takes on a new and more generalized form, $-\dfrac{d(P.E.)}{dx} = \dfrac{d}{dt}(m,x)$. Finally we invoke the other part of the Lagrangian L, namely kinetic energy, which can be defined in non-relativistic terms by $K.E. = \dfrac{1}{2}mv^2$ or by $K.E. = \dfrac{1}{2}m,x^2$. When all these expressions are assembled, we end up with the Euler-Lagrange equation of motion, $\dfrac{\partial L}{\partial x} - \dfrac{d}{dt}\dfrac{\partial L}{\partial,x} = 0$, which as we said helps to bridge the gap between Newton's and Einstein's equations of motion. However this derivation hinges on certain assumptions, such as the absence of the effects of relativity, and therefore it is weaker than the derivation based on the principle of least action. This ends the second detour.

My third detour is brief: Readers may find that modern particle physics and particularly quantum field theory draw heavily on "gauge field theory" and "gauge symmetry," where equations describing fields with particles in them are invariant. (See Gribbin [Q is for Quantum], p. 155.) The details are very technical, but the point is that the Lagrangian L is invariant under these field transformations, attesting to its fundamentality in that each law of conservation (e.g. of energy) corresponds to a symmetry of L (Noether's theorem). In this context, Einstein's general covariance is related to gauge invariance, and general relativity becomes a gauge field theory, as is vital for merging general relativity into quantum mechanics and modern physics (page 462). This concludes the three detours.

We now bring together two mathematical ideas: The above equations, including the Euler-Lagrange equation, deal with the Lagrangian L rather than with geodesics, while the equation

$$S = \int_A^B \left(g_{ik} \frac{dx^i}{ds} \frac{dx^k}{ds} \right) ds$$

does deal with geodesics. The latter equation (on the basis of $S = \int_{t_A}^{t_B} L\, dt$) implies that

$$L = g_{ik} \frac{dx^i}{ds} \frac{dx^k}{ds}.$$

A further series of steps (see Kenyon p. 187) is needed to transform this equation for L into a partial differential form and to combine it with the previous equations for S, but this process involves mathematics we have not yet used, and as we mentioned, the resulting equation contains symbols (namely Γ's) we have not yet covered. Therefore we will examine and apply this equation later and simply show it now:

232

$$\frac{d^2x^j}{ds^2} + \Gamma^j_{ik}\frac{dx^i}{ds}\frac{dx^k}{ds} = 0$$

This is a useful version of an equation of geodesical motion, and its key feature is that it respects both the principle of least action and the needs of general relativity. It is the equation will use to describe the actual motion of an object in a gravitational field, and it applies to anything moving on a time-like geodesic in space-time. (This equation can be written with other parameters besides *s*.)

Let us pause to recap this complex topic. Earlier we saw that in the geometry of space-time, geodesics are the straightest paths for gravitational events. Now we add that freely falling objects seek *least action*, and least action ascribes a physical basis for geodesic motion in terms of kinetic and potential energy. We deduce that when the motion of a freely falling or orbiting object enacts zero variation from the geodesical path, its "action" is least and it is stationary. Euler's-Lagrange's and Hamilton's achievement was finding how much potential energy and how much kinetic energy ensure that these criteria are met. Clearly this principle provides a compelling reason why a falling or orbiting object naturally seeks the most efficient path, namely a geodesic.

In order to further blend the principle of least action with general relativity, we should reconsider the problem in terms of *world-lines* and *space-time intervals*. This objective is aided by reverting to Fermat's principle about the minimum time for an optimal path. Our analogy with soldiers is not effective here, but we can adapt our round-Earth analogy by assuming that we wish to determine the best world-line for moving a clock from New York to Moscow. We must grant that while moving across the Earth, the clock "falls" naturally along some optimal and favored path. In essence we are trying to predict *whether or not that path is a geodesic.* Let us assume that ideally an otherwise undisturbed clock needs 12 hours to cover this distance. Thus if we send off a clock at noon, it should arrive at midnight (ignoring time zones and the effects of relative motion).

We need to consider only two world-lines. (Drilling straight through the Earth from New York to Moscow is not allowed; if nothing else, it would waste energy.) One world-line is a long and undulating path far off the Earth's surface into "outer space" and back. The other is a geodesic; it is as straight as possible and it hugs the Earth's surface. In terms of space-time intervals, both paths are a series of *ds*'s that connects two events, the departure from New York and the arrival in Moscow. (In terms of parallel transport, only the latter satisfies the dictum that parallel transport generates geodesics, but we defer this issue.)

In the language of differential calculus, the space-time interval along a time-like geodesic is "maximal"—it is the *longest* possible interval, not the shortest. (In the context of differential calculus the path taken by a freely falling object is such that the "action" *S* is a maximum or minimum. But how can this be, if falling or orbiting objects find the *shortest* possible line? Let us remember that our clock is moving freely on or near a large mass, which means that the clock is traversing bent space-time. Since space-time is bent the most closest to a mass, time runs slower and distances appear shorter when a clock is closer to that mass. If we set up an equation so that ds^2 is positive, as in

$$ds^2 = (TIME)^2 - (SPACE)^2,$$

then the quantity *TIME - SPACE* is *maximal* if *TIME* is as long as possible and *SPACE* as short as possible. In that case the clock measures *most time* by giving the maximal number of ticks between events in space-time. However, even though a stronger gravitational field slows a clock, *slower motion allows it to speed up, and faster motion also slows it.*

This brings us to a key question: Mathematically speaking, which of our two representative world-lines has the maximal space-time interval? In effect we are asking *which world-line allows the "falling" clock to have the longest proper time* and still get to Moscow at midnight. Therefore our clock must seek a compromise between (1) moving in an arc into a weaker part of the gravitational field so as to generate a faster rate and (2) not moving too fast to generate a slower rate. Though the analogy is imperfect, we can picture throwing the clock and trying to find the best combination of angle and speed—how low or high to aim and how slow or fast to propel this object—so as to hit Moscow on time. This problem is equivalent to that of finding *the minimal difference between potential and kinetic energies*, but for the moment we seek *the maximal proper time*. If we also assume that a clock can move at the speed of light as can a photon, we should also be able to confirm Fermat's principle that the path of light uses least time.

It turns out that if we throw the clock so that it ticks as many times as possible while in transit, its path conforms to a geodesic, and it arrives when and where it should. Any other number of ticks causes it to deviate from a geodesic. This finding demonstrates the apparently paradoxical statement (paraphrasing Rucker, p. 110) that a geodesic is the maximally straight time-like path consisting of maximal space-time intervals, and *it signifies that in four-dimensional space-time, the world-line with the longest proper time between the two events is a geodesic.* In other words though he knew neither about longest proper time nor about photons, Fermat was right and, as we suggested, his principle is linked to geodesical motion. Once we say it, this concept is rather simple: Motion that takes *least time* also takes *least space.* However these quantities are equivalent to most proper time and most straightness. This inference corresponds to the conclusion that the path with the minimal difference between potential and kinetic energy—the path generating the least action (I)—is a geodesic. Hence *the actual world-line taken by a falling or orbiting object is a geodesic, which also means that it is the path with the least S.* As we said earlier, and as Einstein had surmised, there is something distinctive about geodesics.

Indeed once we say it all together, the concept is even more concise: *Motion that takes least action also takes least time and least space.* We can also enunciate three main "principles of leasts." The most basic one is the principle of least action, which describes all natural undisturbed motion. The second is the principle of least time specifically for the path of light. The third can be called the principle of "least distance," by which we mean most straightness between points. The consequence of "least distance" in curved space-time is the geodesic, which describes the world-line of freely falling objects.

However let us not be misled by the elegance of this concept, for we must be careful in its application. For example when an ordinary moving object—not a photon but, say, a falling apple—uses "least" time, it does *not* reach the fastest *possible* speed, and therefore it does not consume the least *possible* time. Attaining or exceeding the speed of light violates the principle of least action because ordinary objects lack the energy and have too much mass to reach such speeds. Similarly when a moving object crosses "least" space, it does not "drill through" the curvature of space-time like a tunnel through a mountain, lest it violate the principle of least action. In other words least action is primal, and least time and least space can only follow and must satisfy least action. Though the following wording spoils the terseness of the concept that "motion that takes least action also takes least time and least space," we could extend this statement to read that *motion that takes least action also takes least time allowed by least action, and such motion takes least space allowed by least action.* In short, *least space-time allowed by least action* confines motion to world-lines on *geodesics*.

Let us look briefly at a particular world-line that obeys this concept. We can treat proper time in ultra-small differential steps labeled "d(proper time)," and we can integrate our moving clock's proper time as it moves from point A (New York) to point B (Moscow) by either path we are considering. In short, we can spell out the equation $t_{A \, to \, B} = \int_{A}^{B} dt$, so that

$$Proper \; time_{A \, to \, B} = \int_{A}^{B} d(proper \; time).$$

Equivalently we can call the path s, so that *ds* stands for ultra-small intervals of s. Locally, our clock is always "here" in terms of space, so that distance (dx, dy, and dz) is zero and so that d(proper time) is proportional to *ds*. As we already noted, this object's motion is a series of *ds*'s integrated between A and B:

$$s = \int_{A}^{B} ds$$

Our actual world-line in space-time—the one with maximal proper time—is such that *the value of*

$$\int_{A}^{B} ds$$

is maximal. However because $ds = \sqrt{(TIME)^2 - (SPACE)^2}$ we can write the term

$$\int_A^B \sqrt{(TIME)^2 - (SPACE)^2},$$

which again reveals its correspondence with the key term in the principle of least action,

$$\int_{t_A}^{t_B} (Kinetic\ Energy - Potential\ Energy)\ dt.$$

In other words *a geodesic which consists of maximum space-time intervals is analogous to a path which obeys the principle of least action.*

As we noted, the mathematics behind this conclusion calls on a branch known as the calculus of variations, and one of its basic symbols is δ. In this setting the term "variations" means the deviations away from something standard. Thus a longer line shows variation from a shorter one. We note the distinction between a derivative, d, and δ; a step along a line is a "d" but the variation between lines is "δ." By intuition the word "deviation" may express δ better, but the word "variation" is traditional.

Since δ represents the variation of a path from maximal straightness even on a curved surface, the world-line we seek is such that

$$\delta \int_A^B ds = 0.$$

Therefore this equation, which we already acknowledged for its theoretical significance, is *the basic equation of motion for our clock in its actual geodesical world-line* (although more utile equations are available). We therefore reiterate that the space-time path of an object which varies the least from a geodesical trajectory—meaning it varies zero—conforms to the basic equation of relativistic gravitational behavior.

Readers versed in differential calculus know how to find the minimum or maximum point on a curve. For example this type of calculus can find the high point of a trajectory. The *calculus of variations* addresses more difficult problems, such as *finding the geometric line or curve,* including the trajectory itself, which has certain minimum or maximum features. That line or curve can be a function or an equation; see Appendix. Thus the latter type of calculus can find a geodesic on the basis of maximal proper time, because it is designed to find the path (or curve, surface, etc.) for which a given function is stationary. That stationary value is an extremum, which, as we saw, means it can be a maximum or minimum. The practical benefit of finding a stationary value is that it allows us to identify a function which, in our case, leads us to equations of geodesical motion.

Problems that lend themselves to the calculus of variations have been tackled for centuries. In fact ancient Greek mathematicians sought solutions, at times with some success. A famous seventeenth century example solved by this kind of calculus is the problem of the "brachistochrone" ("least time") which consisted of finding the trajectory of an object between two points while that object is moving

236

only under the influence of one force, namely gravity. This trajectory is called *the curve of quickest descent*, and obviously it is fundamental to gravitational motion, though we replace force by space-time curvature. The same mathematics is used to solve many other difficult and basic problems in other branches of physics including quantum mechanics, as well as in geometry, geography, business, economics, biology, etc. (For example in geometry the calculus of variations confirms that the surface of a limited size which encloses a maximal volume is a sphere, while spheres appear naturally in countless instances in the universe—everything from bubbles to stars. In geography the calculus of variations predicts that on a small scale quiescent bodies of water have flat surfaces and on a large scale they conform to the Earth's convexity, while the same mathematics predicts that rivers seek the steepest descent. In business the calculus of variations allows computation the profitability of a complicated enterprise such a fish hatchery, where we consider variables such as the rate of reproduction of the fish, the size of the harvest, the price of fish, the costs, the current stock, and even the level of effort. In fact an equation of "fish-stock motion" can be derived.)

Incidentally, when we use this kind of calculus and the principle of least action to find the shortest distance—we called it the "least distance"—between two points on a flat surface, the result is the equation for none other than *a straight line*. Of course this is intuitively and physically self-evident—it was suggested by Archimedes in about 250 B.C.—and such a calculation is a case of mathematical "overkill," but this outcome is essential if we are to rely on the calculus of variations. In particular we would not trust the Euler-Lagrange equation to lead us to a geodesical equation on a curved surface if it did not lead us to a geodesical equation on a plane, which is just an elaborate characterization of a straight line. (Again see the Appendix.)

The calculus of variations was developed mainly by Lagrange, Euler, and the Bernoulli brothers in the eighteenth and nineteenth centuries, though Galileo demonstrated one of its effects experimentally and Newton also made use of it. More than anyone else, Hamilton merged it into modern physics. (Besides the web site I mentioned above, Menzel details the mathematics on pp. 164-169, Feynman devotes a chapter in vol. II, pp. 19-1 to 19-14, and Lovelock and Rund devote much of their book to this topic. Kenyon, Møller, and Einstein placed the derivations into appendices in their respective books, demonstrating that the calculus of variations is a separate and lengthy subject, though Kenyon's treatment is particularly concise on pp. 66-68 and 186-188.) An interesting historical point about the calculus of variations and the principle of least action is that these concepts were discovered long before they could be put to their best use. In particular they are not only valuable in relativity but are also indispensable to quantum mechanics, both of which are twentieth century developments. We can say that we had acquired a set of powerful mathematical tools before we appreciated all we could accomplish with it. In fact Hamilton came close to discovering quantum mechanics, but he had no access to the observation that came to light between 1900 and 1920 which alerted physicists to the need for this branch of science.

Another historical point is that the acceptance of the principle of least action and its implications revealed a certain amount of nationalism. Until Hamilton, who was Irish, published his work in this topic, scientists in Great Britain tended to champion Newton's system, whereas their Continental counterparts promoted the Lagrangian-Eulerian paradigm. Once Hamilton's contribution was recognized, British science espoused it eagerly, and even today the designation favored in English writings, "Hamilton's principle of least action," is often criticized by other nationalities for obscuring its central European origins.

◆

Earlier we asked by what right we invoke Newton's first law when dealing with accelerated motion. Part of the rationale is that we subdivide accelerated motion into points with uniform motion through non-uniform space-time. This makes geodesical world-lines, composed of *ds*'s, appear locally straight. Hence motion, notably free-fall, is geodesical—as straight as possible—even in curved space-time. Objects indeed prefer uniform motion, but they do so because they prefer geodesics, and they prefer geodesics because they seek least action.

Then why not just call Newton's uniform motion "Newton's least action?" Because, as we noted earlier, an equation of motion based upon purely Newtonian physics is too limited and cannot be generally covariant. In contrast the equation of motion

$$\delta \int_A^B ds = 0,$$

and its more useful versions are generally covariant. The reason is that *ds* survives transformations of coordinates and that motion in a gravitational field complies with $ds^2 = g_{ik}\, dx_i\, dx_k$ for all observers.

In effect the principle of least action, particularly as formulated by Hamilton, allows Newton's laws to operate locally in general relativity—in space-time coordinates which accommodate geodesics—while it corroborates the legitimacy of geodesics in general relativity. Indeed Einstein used this approach to support other parts of his thesis; see his paper on pp. 165-173 in The Principle of Relativity and his pp. 154-156 in The Meaning of Relativity. In this setting, where the principle of least action and the Euler-Lagrange equation enjoy primacy, Newton's laws can be considered to be corollaries. In other words despite history and chronology, we can say—in the realm of the most basic laws of physics—that the principle of least action and the Euler-Lagrange equations are the parents, and Newton's laws, notably $F = ma$, are the offspring. As explained parenthetically on page 231, we can start with $F = ma$ to derive the Euler-Lagrange equation, but the reverse process—using Euler-Lagrange and the principle of least action to derive $F = ma$—is crucial since the Euler-Lagrange equation and the principle of least action are more fundamental and broader than Newton's laws.

For interested readers, here is an outline of that process: We label velocity ($\dfrac{dx}{dt}$) as ,*x* (where *x* is

location). We express force in terms of potential energy so that $F = -\dfrac{d(P.E.)}{dx}$. We express kinetic

238

energy so that $K.E. = \frac{1}{2}m(,x)^2$, which means we ignore relativity. The Lagrangian L is expressed as $K.E. - P.E.$, so that $L = \frac{1}{2}m(,x)^2 - P.E.(x)$. We expect the path of an object to obey least action, and we know that $S = \int L\,dt$ satisfies the Euler-Lagrange equation, $\frac{\partial L}{\partial x} - \frac{d}{dt}\frac{\partial L}{\partial ,x} = 0$. It turns out that when we insert L as defined above (still ignoring relativity) into the Euler-Lagrange equation and perform the pertinent calculus, we find $\frac{d(P.E.)}{dx} - \frac{d}{dt}(m,x) = 0$. This equation can be written as $\frac{d(P.E.)}{dx} = \frac{d}{dt}(m,x)$ *which is entirely equivalent to Newton's* $F = ma$. This ends the outline.

We conclude that the alliance of the principle of least action and the calculus of variations provides a secure foundation for the notion that a geodesic is the correct description of the actual world-line of an object in gravitational motion. This key inference, like others we encounter in relativity, is elegantly logical and nearly self-evident: From the vantage of energy physics, *motion entails as little "action" as possible.* From the vantage of space-time geometry, *motion is as straight as possible, which means it is geodesical.* Indeed nature routinely seeks optimal shapes and processes, and this observation inspires an attractive physical as well as philosophical doctrine: *Nature is efficient and not wasteful.*

◆

The concept of variation, again symbolized by δ, can also be applied to an equation specifically about parallel transport of vectors. We expect that the deviation from "parallel-ness" with respect to the shape of space-time at any point is zero, which is another way of reiterating that parallel transport keeps us on geodesics.

Hence we have the equation

$$\delta(g_{ik}V^iV^k) = 0,$$

in which V^i and V^k represent the magnitude of the transported vectors and g_{ik} brings in the shape of space-time at any point. The two sets of components reflect the complicated curvature for each point in space-time, as we must consider parallel transport along each of its curvatures. This equation will also be useful to us.

◆

239

One more issue can be raised here. What law of physics keeps a gyroscope parallel (likewise, what keeps a bicyclist erect)? Basically it is Newton's first law justified by the principle of least action, but in this situation a specific derivation of that law is identified, the law of conservation of angular momentum: If undisturbed, the angular momentum of a spinning object remains constant because its total energy, or, to be relativistically correct, its energy-plus-mass, is constant. This means that the condition for conservation of energy (and momentum) exists when motion entails no variation from the geodesical path, and that conservation of energy emerges as a natural consequence of the principal of least action. We touched on this earlier and we shall see it again later, but one corollary of this concept is that other relativistic equations must comply with conservation of angular momentum. This idea is efficiently expressed by differential equations (which we need not derive). The motion of a test object in free-fall can be described as

$$\frac{dx^j}{ds} = a\ constant.$$

(The summation convention applies, and $j = 1$ through 4.) This equation means that the object's motion along four coordinates (x's), split into infinitesimal local steps (ds's), is constant even if space-time is curved. The terms "constant" and "uniform" refer to the same idea. Thus this equation is a four-dimensional space-time interpretation of Newton's first law, and it reflects the conservation of momentum in four dimensions. The term dx/ds will reappear in other contexts.

For a test object in an orbit, which will interest us in detail in chapter 17, the equation is more complicated. We let φ be the angle the object sweeps out in its orbit, and readers familiar with Kepler's laws of planetary motion will at once make the connection: Kepler noted that a planet sweeps out equal areas in equal times. Had he known about calculus and about geodesics in four dimensions, he would have written

$$r^2\frac{d\varphi}{ds} = a\ constant,$$

which signifies the conservation of angular momentum in space-time.

At the close of this chapter we can appreciate how the various pieces—kinetic and potential energy, the calculus of variations, the principle of least action, the metrical tensor, geodesics, free-fall, motion through space-time, parallel transport, the conservation of angular momentum—all begin to fit together to create the unified and coherent structure that is general relativity. We can even compose tidy adages to characterize this structure: "Locally the motion of a falling object in space-time is Newtonian, its path is Euclidean, and its coordinates are Cartesian. However globally, its space-time is neither Newtonian nor Euclidean nor Cartesian; it is curved." Recalling that many of the concepts about curved space-time stem from Riemann's work, we can claim that "gravitational motion is locally Newtonian, its path is locally Euclidean, its space-time is Riemannian, its geodesic is Hamiltonian, and therefore its global behavior is Einsteinian." Similarly we summarize that "Euclidean and Cartesian space is three-dimensional and flat. Gaussian space is two-dimensional and curved. Minkowskian space is four-dimensional and flat. Riemannian space is any-dimensional (including four-) and curved. Einsteinian space is four-dimensional and curved by the presence of mass and energy."

Yet if Euclid were capable of accompanying an object, even a photon, as it traveled from a distant star to the Earth, he would arrive as convinced as ever that the shortest distance between two points is a straight line. If Newton came along, he would arrive reassured that an object in uniform motion tends to maintain that motion. And if Euclid and Newton could participate in the observations of starlight during the eclipse of 1919 or in the relativistic analysis of the orbit of Mercury, they would *not* be forced to change their minds. However they would have to admit that "straight" and "uniform" are defined by the shape of space-time, and they would realize that physics must be adapted to these observations. In this sense Euclid and Newton were not wrong; their space-time is "wrong," or to be exact, their space-time is not curved. Qualities such as straightness, uniformity, and parallelism are the same everywhere, but their manifestations in our experience are recast by the geometry in which we find them.

We now address the most intricate aspects of relativity, the mathematical counterpart of the assertion that

> Curvature of space-time = Density of matter and energy.

We proposed this strategy already, and at last we can act upon it. A logical approach is to consider *what changes in the shape of space-time can be attributed to nearby matter and energy*. The answer will tell us *how the curvature of space-time is determined by the density of matter and energy*.

Since the metrical tensor reflects the shape of space-time, we expect to obtain equations solvable for g_{ik}, and we expect these equations to permit reliable predictions about the effects of gravity.

Therefore it would seem sufficient to assume that the metrical tensor is proportional to a tensor for the density of nearby matter and energy. The temptation might then be to write one or more equations which state that

the tensor g_{ik} = the tensor for the density of matter and energy??

but we add "??" to indicate that this may not be correct. We shall call such "??" equations our "provisional" field equations. They are preliminary formats for the equations we seek. We also presume that the right side of the above provisional field equation can be represented by the tensor T_{ik}. Accordingly we pose the following possibility:

$$g_{ik} = T_{ik}??$$

From our discussions so far, we know that this equation must meet several criteria: (1) The left side, g_{ik}, must describe space-time even when curvature varies point-to-point. (2) That description must hold in any frame of reference—it must be generally covariant—which means it remains in tensor form, but (3) it must also hold during free-fall and be Lorentz-invariant. Finally (4) that description must be compatible with other laws of physics. Meanwhile the right side of our provisional equation should also meet some criteria: (1) T_{ik} must accommodate the equivalence between mass and energy. (2) T_{ik} must accommodate the various forms of energy, and (3) it must be a tensor. We also expect to treat both gravitation and matter-and-energy as field phenomena. Eventually we will address each issue.

We can now roughly trace the evolution of Einstein's thinking as he struggled to derive the field equation that meets these criteria. Because this equation is really a family of 10 equations, it is called by its plural, the field equations of general relativity. (The reader may soon find empathy with

Einstein, who lamented that this quest threatened his very sanity.) He realized that he could find these field equations by allowing the gravitational potential to represent g_{ik} and by expressing matter as density. Therefore he charted his strategy on the basis of the Poisson (1781-1840) equation,

$$\nabla^2\Phi = 4\pi K\rho.$$

The main element in the left side, Φ, is the gravitational potential, which we recall is an expression for the intensity of gravity; we will discuss Φ and ∇^2 in detail shortly. The right half of the equation characterizes matter as ρ, which is density; i.e. mass per volume (not to be confused with p which stands for momentum.). With the proper constants, namely $4\pi K$, Poisson's equation declares *how the intensity of gravity is determined by the density of matter*, and Einstein made this his starting point.

We do not need the entire derivation of Poisson's equation, but we will cover steps which reveal pertinent intermediary equations and which illustrate how the principle of equivalence enters into the picture. To begin, let us imagine a room illuminated by a single lightbulb at its center, but the room has the shape of a sphere. The bulb has a certain intensity as it emits energy in the form of light and heat. The room encloses this energy so that every point in the room and every point on its (curved) wall enjoys a certain brightness and temperature. In this analogy we imagine that *a piece of matter "emits" gravity* much like the lightbulb gives off energy, and we equate the density of that piece of matter with the brightness and hotness of the bulb; denser matter "emits" more gravity just as a stronger bulb emits more light and heat.

It is important that Poisson described matter by its *density* ρ—*i.e. by its distribution*—rather than by its weight. For this purpose matter is quantified by M. The volume of a sphere of radius r is $4\pi r^3/3$, so that the density of matter can be expressed by the equation

$$\rho = 3M/4\pi r^3$$

or, showing how M depends on ρ,

$$M = 4\pi r^3\rho/3.$$

Let us next examine the gravitational potential Φ, also known as the Newtonian potential. We already noted that it has significance in general relativity, but to understand the basis of Φ we invoke Newton's equations, including his law of gravitation between two masses,

$$G = K\,\frac{m_1\,m_2}{D^2}.$$

243

In physics "work" is defined as force (in this case the force of gravity, G) multiplied by distance (D), so that "gravitational work" is GD. Furthermore "work" requires the expenditure of energy (E), and here it is "negative work" because as D decreases E increases. (As a mass is raised against gravity it gains potential energy, and as a mass falls through distance D it gives up energy.) Hence

$$GD = -E.$$

Multiplying Newton's law by D, we derive

$$GD = K\frac{m_1\,m_2}{D^2}\,D.$$

Since $GD = -E$ and two of the D's cancel each other,

$$E = -K\frac{m_1\,m_2}{D}.$$

Now we imagine that the two masses come together at the center of our spherical room (which has the radius r), and r replaces D. If we express mass as M, then "$m_1\,m_2$" can be replaced by M^2. The potential energy of this system is given by

$$E = -K\frac{M^2}{r}.$$

Dividing both sides by M, we have

$$\frac{E}{M} = -K\frac{M^2}{rM}.$$

For the next step we can think of the gravitational potential as an indication of the amount of gravitational energy in a mass (potential energy per density-of-mass), in which case

$$\Phi = \frac{E}{M}$$

and therefore the gravitational potential is a function of (it depends on) M at distance r,

$$\Phi = -K\frac{M^2}{rM} = -K\frac{M}{r}.$$

(Again K is Newton's constant and M is $4\pi r^3\rho/3$.) In other words *the gravitational potential, Φ, is the intensity of local gravity at a certain distance, r, from a certain concentration of matter, M.* According to our lightbulb analogy, just as every point in the spherical room has a certain amount of brightness and temperature, *every point around matter possesses a gravitational potential.* In this setting the space in the room is a *field*, and since the gravitational potential is a scalar quantity, *M* appears to be surrounded by a *scalar gravitational field.* Of course we expect a tensor field, but that issue can wait.

Still we can ask why bother with the gravitational potential and with the equation

$$\Phi = -K\frac{M}{r},$$

in view of the fact that we can quantify the intensity of gravity in a manner which gives us accelerated gravitational motion,

$$a = K\frac{m_2}{r^2}.$$

The problem with the latter equation is that it does not consider the density of mass in a field. On the other hand the equation $\Phi = -KM/r$ does not consider motion, meaning that we need additional equations, namely the equations of motion we already introduced, to calculate how objects actually fall or orbit. Nevertheless since Φ is compatible with the concepts of density and field, we prefer to derive the field equations using Φ—which means using Poisson's equation—even if doing so complicates the calculation of motion. Still, the equation $a = Km_2/r^2$ tells us something else useful: Acceleration is the derivative (rate of change) of velocity, and velocity is the derivative of motion (rate of change of location), so that the similarity between $\Phi = -KM/r$ and $a = Km_2/r^2$ indicates that we will be dealing with a derivative of a derivative.

Our lightbulb analogy is not as far-fetched as it sounds. Many physical entities follow Poisson's pattern, including heat, light, and electricity. For instance Φ can refer to the electrical potential and ρ to the density of electrical charge. (See Menzel, p. 48.) Nonetheless we deserve to see a more direct role of the gravitational potential in relativity. At the same time we ought to confirm $\Phi = -KM/r$ using relativity.

For these purposes we use a different and better suited scenario, similar to Einstein's in <u>Relativity</u> (pp. 79-81 and 129-131). We imagine that an observer examines two identical clocks, both of which are placed on a rotating disk (say a carousel) so that one clock is close to the center of rotation and the other is close to the edge. In effect the clocks are in relative motion with respect to each other (one clock moves in a wider circle) which in itself is not a relativistic effect.

If w is the velocity of rotation (the "angular" velocity) and r is the distance between the clocks, the relative velocity between the clocks is v according to the equations

$$v = wr \quad and \quad v^2 = w^2r^2.$$

We let the observer sit next to the more centrally-placed clock. According to this observer the two clocks do not tick at the same rate, and this *is* a relativistic effect. We can say that each tick of the clock near the observer takes t_o time, and t_r is the duration of each tick on the other (the relatively moving) clock. The observer finds that

$$t_r = \frac{t_o}{\sqrt{1-v^2/c^2}}.$$

This discrepancy of course is the consequence of relative motion, whereby any event at the location of t_r appears to take longer than at t_o. Unless v is very large (unless the carousel is huge and is spinning very rapidly), we can approximate and rearrange the equation to read

$$t_o = t_r \left(1-\frac{v^2}{2c^2}\right).$$

(See page 31 about this approximation.) Since $v^2 = w^2 r^2$, this can be written as

$$t_o = t_r \left(1-\frac{w^2 r^2}{2c^2}\right).$$

Meanwhile we recall that clocks at different altitudes from the Earth's surface also run at different rates. This means that the observer on the disk may report one of two sensations, which, *by the principle of equivalence*, are interchangeable and indistinguishable: He may conclude that inertia, manifest during rotation as a centrifugal force, is greater at the more peripheral location of the clock. Or the observer may conclude with equal validity that there is a *gravitational field* which is greater at the more peripheral location. (This is why centrifuges are used to generate gravitational fields.) In other words the location of t_r on the carousel represents the Earth's surface, and the location of t_o on the carousel represents a higher altitude where gravity is weaker.

If the observer measures the intensity of gravity and thus ascribes a value for Φ to each location, he will conclude that the difference in the duration of the clocks' ticking ($t_r - t_o$) is linked to the difference in Φ's ($\Phi_r - \Phi_o$):

$$\frac{t_r-t_o}{t_r} \quad \textit{is proportional to} \quad \frac{\Phi_r-\Phi_o}{\Phi_r}.$$

We see that stronger gravity slows a clock, so that each t depends on r—i.e. the rate of time depends on location on the disk or, equivalently, on location in a gravitational field. In either view, $t_r - t_o$ or $\Phi_r - \Phi_o$ must be negative; i.e. if $t_r > t_o$ then $\Phi_r < \Phi_o$. The distance between t_r and

t_o can be very small, so we can replace Φ_r and Φ_o by a single local term, Φ. Therefore Φ *is inversely proportional to r* (but one of these must be negative).

Naturally if a piece of matter is responsible for the effects of gravitation, Φ is the gravitational potential at a given location in the surrounding gravitational field. The mass of that piece is embodied by *M*, and *M* is directly proportional to Φ via a constant, K. We can tie Φ, *M* and *r* together in

$$\Phi = - K\frac{M}{r},$$

confirming the conclusion reached on page 244 in a relativistic setting: Φ *is inversely proportional to r and directly proportional to M.*

Since "work" in the physical sense must be exerted to transport a mass from the periphery toward the center of our spinning disk (just as effort must be exerted to walk toward the center of a moving carousel or "negative" work must be "exerted" to climb down a mountain), we can write the equation

$$\Phi = - \frac{w^2 r^2}{2}.$$

We recall that $v = wr$ and $v^2 = w^2 r^2$. The terms are squared because as the disk rotates, say, twice as fast, the effect increases approximately four-fold.

We now invoke

$$t_o = t_r \left(1 - \frac{w^2 r^2}{2c^2} \right).$$

A simple substitution yields the next equation, which can be rearranged and written in differential terms:

$$\frac{dt_o}{dt_r} = \left(1 + \frac{\Phi}{c^2} \right).$$

The difference between conditions at the location of t_r *and conditions at the location of* t_o *defines the gravitational potential* in the term

$$1 + \frac{\Phi}{c^2},$$

and the more conditions in this field vary—the more t_r and t_o differ—the greater the Φ. We recall that t_r and t_o differ according to the special relativity factor,

$$\sqrt{1-v^2/c^2},$$

which means that *we have accounted for the gravitational potential on the basis of relativity.* Moreover the involvement of the special relativity factor means that Φ can appear in Lorentz-invariant equations compatible with special relativity.

We can carry the logic further: If t_r and t_o do not differ at all, then

$$\frac{dt_r}{dt_o} = 1 \quad and \quad \frac{\Phi}{c^2} = 0.$$

This is quite significant, for it implies that if conditions across a field do not vary, there can be no gravitational potential and obviously no point-to-point variability of the gravitational potential. What "conditions" are we talking about? The "shape" of time and space! This is why the special relativity factor applies to both space and time, and in fact if we could measure the *sizes* of the two clocks on our disk, these would also show a discrepancy. This means that when "conditions" do not vary, neither time nor space is curved across the field. Conversely *the gravitational potential exists when and where space-time is curved.*

Of course these "conditions" can be quantified by g_{ik}, and we will pick up this thread again shortly by connecting Φ with g_{ik} in an equation. For now we emphasize that the variability of the gravitational potential propagates the gravitational field. Therefore we need a more precise notion of the *variability* of the gravitational potential.

Accordingly let us continue to survey Poisson's equation $\triangledown^2\Phi = 4\pi K\rho$ by considering $\triangledown$, also called "nabla" or "del." This is better explained in the lightbulb analogy in which $\triangledown$ is a "gradient." For example we find a gradient in brightness as we walk away from a lightbulb; our locally available light fades from one step to the next. A gradient is even more obvious for heat, because heat clearly flows from an point of higher temperature to one of lower temperature (according to the second law of thermodynamics), and Φ can represent the local intensity of heat, i.e. the temperature. The advantage of this paradigm is that if we know the temperature at any point in space and we also know how the temperature is different at an adjacent point, we can calculate a *vector-like quantity* which describes how heat flows—how it *varies*—point-to-point. For example if it is 20 degrees in New York and 30 degrees in New Jersey, we can draw a vectorial arrow to show the direction (west) and amount (10) of the temperature gradient. However a gradient is not exactly the same as a vector, and we will cite the difference later, but for now let us call it a vector.

This concept was originally developed in three dimensions with three minuscule vectors of heat flow. Each vector has an amount and a direction oriented along one of the three coordinate axes of space.

The $\triangledown$-term holds this data as a gradient of Φ for any direction in a space, including in the spherical space surrounding our lightbulb. (The final direction of flow is the one in which the gradient is steepest. See Hopf, pp. 41-44.) In other words *we can use $\triangledown$ to quantify how the temperature varies*. We do this by letting the partial contribution of each minuscule vector of temperature be differentiated with respect to each axis. For example the change in temperature while moving along the x axis can be expressed as

$$\frac{\partial \Phi}{\partial x},$$

where Φ is the intensity of heat. The same applies along the y and z axes, and these partial contributions can be added together into a sum of the minuscule or differential changes in temperature. Thus $\triangledown$ is a composite vector differential or vector derivative which reveals *how Φ is changing from point to point in each direction in a space*. (Please glance at the next equation.)

The notion that the flow of heat can be analyzed as the addition of partial contributions brings us to the notion of partial derivatives and partial differential equations, and here we should pause to explain the distinction between partial and ordinary derivatives. An ordinary differential equation examines a change in some physical quantity, say velocity, with respect to one independent variable, say time. The key term of such an equation might be dv/dt, which is the change in velocity over time (dv/dt is acceleration). This is the type of equation used in Newtonian physics, wherein physical events are reduced to motion of particles; in theory each particle has a dv/dt.

Relativity adopts the concept that physical events can be reduced to quantities which may have changing values at points in four-dimensional space-time. This means that four independent variables are considered, and in order to analyze the effect of each of these four, partial differential equations are needed. In the case of Φ, this property is labeled $\triangledown\Phi$, which is the symbol for the gradient ("grad" for short) of Φ. Hence the basic equation for $\triangledown\Phi$ in four dimensions becomes

$$\nabla\Phi = \frac{\partial \Phi}{\partial x} + \frac{\partial \Phi}{\partial y} + \frac{\partial \Phi}{\partial z} + \frac{\partial \Phi}{\partial t}$$

or more compactly (i = 1 through 4)

$$\nabla\Phi = \frac{\partial \Phi}{\partial x^{i}}.$$

The pertinent significance of the $\triangledown\Phi$ is this: Using our lightbulb analogy, Φ is the "temperature" or "brightness" of gravity, and we can think of $\triangledown$ as a set of vectorial arrows showing which way and by what magnitude gravity "flows" from its source. In this way $\triangledown\Phi$ is a measure of the point-to-point variation in Φ across a *field*, and it allows us to handle the scalar Φ as vector quantity. We thus can conceive of gravitation as *a gradient in the local gravitational potential across the field that surrounds a mass*, and by means of partial differentiation we can accommodate *the four dimensions of space-time.*

This notion will be even more meaningful because we can connect the gravitational potential with density of mass. So far we expressed Φ as $\triangledown\Phi$, and we expressed mass as ρ. Meanwhile our lightbulb-in-a-spherical-room analogy suggests that a certain "amount of gravity" "illuminates" the wall. Since the surface of a sphere of radius r is $4\pi r^2$, and since Φ at distance r is KM/r, that "amount of gravity" is equal to 4πKM. We therefore need a mathematical link between ρ and Φ on that "gravity-illuminated" surface.

Finding this link involves integration, for instance by the use of Stokes' theorem. (See Kenyon, p. 78, and Menzel, pp. 132-135, but other theorems can be used; see also Borisenko and Tarapov, pp. 137-144.) Although Stokes' theorem is intricate—the integral of the derivative of a differentially-expressed vector quantity in a certain volume equals the integral of that quantity on the surface of that volume—we can think of it simply as a way to determine what happens on the wall of our spherical room as a consequence of an event inside the room. If we imagine that the room is as small as possible—i.e. that its radius is so short and its surface so small that we are left with a point—it turns out that what is happening inside can be represented by $\triangledown(4\pi$KM); the steps are not essential for us. Using Stokes' theorem and expressing *M* as density via $4\pi r^3\rho/3$ leads to Poisson's equation,

$$\triangledown^2\Phi \;=\; 4\pi K\rho.$$

Here we note an important detail: As shown by the exponent 2 in $\triangledown^2\Phi$ (by virtue of Stokes' theorem), we are dealing with a double $\triangledown$. This means that Poisson's equation is of the "second order"—it entails a change within a change or a gradient of a gradient or a derivative of a derivative, a.k.a. a *second derivative*, as suggested earlier by the fact that "a" is the derivative of velocity and velocity is the derivative of motion. The term $\triangledown^2\Phi$ reads "nabla square of the gravitational potential" which is *a summation of the vectorial second partial derivatives of the gravitational potential with respect to x, y, z and t:*

$$\triangledown^2\Phi \;=\; \frac{\partial^2\Phi}{\partial x^2} \;+\; \frac{\partial^2\Phi}{\partial y^2} \;+\; \frac{\partial^2\Phi}{\partial z^2} \;+\; \frac{\partial^2\Phi}{\partial t^2}$$

(∇^2 is also called the laplacian operator.) Poisson's equation states mathematically that this double-differential expression of Φ is proportional to the density of mass at any point in space.

We can rephrase this conclusion in terms of a field: Since gravitation is a consequence of point-to-point changes in the potential in a gravitational field, and since gravitational fields surround concentrations of matter, Poisson's equation discloses *how the point-to-point variation in the gravitational potential is linked to the density of matter.* In particular, "$4\pi K\rho$ of density of matter" is linked to "$\triangledown^2\Phi$ of intensity of gravity." Because this idea is applied in space-time, some authors (e.g. Menzel, p. 383) use a "four-dimensional nabla," in which case Poisson's equation becomes

$$\Box^2\Phi \ = \ 4\pi K\rho.$$

In the absence of matter, i.e. in empty space where $\rho = 0$,

$$\Box^2\Phi \ = \ 0 \quad or \quad \nabla^2\Phi \ = \ 0,$$

and the latter is known as Laplace's equation. It implies that zero density of matter means zero gravity, which seems to be a superfluous statement, but this concept is quite fundamental. For one thing, if Laplace's equation were wrong, Poisson's would also not hold. Moreover part of Newton's law of gravitation contains the quantity "m_1 multiplied by m_2" to obtain the force of gravity G. If one of these masses, say m_2, were zero, Newton's equation insinuates that there would be no gravity: $G = 0$. But why should gravity disappear when there is no object (no m_2) to "gravitate?"

Poisson's and Laplace's concepts create a different notion: Mass, or more precisely the density of m_1, is responsible for a gravitational field which exists with or without m_2. *The gravitational field exists when the gravitational potential shows a point-to-point gradient.* That is not to deny that both masses can contribute to the deformation of space-time, only that gravitation, as Einstein emphasized, is a field phenomenon which has existence independent of an object to "gravitate."

We re-emphasize that by analyzing gravity as a point-to-point gradient (a second vector derivative of the gravitational potential, to be exact) we are no longer treating it as a scalar field but as a "vector field." I.e., the gradient of a scalar field is a vector field. In this way Poisson's equation gives concrete mathematical expression of the concept that gravitation is a field phenomenon. In other words in seeking the tensorial relationship between the curvature of space-time and the density of matter and energy, we have found the relationship between the vectorial behavior of gravitational potentials and the density of matter, and we already know that vectors lead to tensors. Thus we regard Poisson's equation as a fledgling "tensor field equation" for gravitation, one which begins to translate gravity and mass into the language of general relativity.

On page 137 we promised to recast an equation valid for special relativity into tensor form. We have only begun—we are still in vectors—but the special relativity factor in our derivation (page 246) does make Poisson's equation *Lorentz-invariant*. In the process Newton's G has been supplanted by $\nabla^2\Phi$, Newton's $K\dfrac{m_1 m_2}{r^2}$ has been supplanted by $4\pi K\rho$, and Poisson connects these in a partial-differential equation while it bypasses the concept of a gravitational force. ($\nabla^2\Phi \ = \ 4\pi K\rho$ makes no mention of force.) Obviously these conversions will be of interest to us once we link gravitational potentials with curvatures of space-time and once we include energy.

Meanwhile we suggested that the gravitational potential is specified by the intrinsic shape of space-time by way of the metrical tensor g_{ik}, and we already considered the notion of a vector and tensor field in relativity. Clearly, there ought to be a firm connection between the left half of Poisson's equation and the metrical tensor, and this connection may take the form of an equation between Φ and the g's.

In deference to history we should add that the introduction of the field concept did not immediately and automatically lead to, or even suggest, general relativity. While it seemed reasonable to view the conduction of heat as a field phenomenon, the initial assumption was that this occurs only inside matter and that fields do not exist where there is no matter. Thus when we liken heat conduction (which can not occur in a vacuum) to what happens in a gravitational field (which can exist in a vacuum), we are making a substantial leap. The major impetus for that leap came from the study of the electromagnetic field (which also appears in a vacuum; more on this later) and from the finding that an ether is not essential for the movement of energy. Many years passed before science accepted the notions that a gravitational field can exist outside of matter and—more astonishingly—that it is determined by the presence of matter.

✦

We must now establish the aforementioned connection between the gravitational potential Φ and the metrical tensor g_{ik}, for without it we cannot justify relying of Poisson's equation. We assume that our gravitational field is weak and that the speeds in question are slow. Here it helps to think of the metrical tensor as consisting of two parts, one of which applies to flat and gravity-free space-time; we label this part as f_{ik}. The other part, h_{ik}, represents *the deviation from flatness*. Using a rule similar to the addition of tensors,

$$g_{ik} = f_{ik} + h_{ik}.$$

(Some authors prefer other labels, and depending on how the equation is set up, some terms may be negative.)

Let us consider g_{ik} as a whole first. We know that in an ordinary system of coordinates the non-zero components of g_{ik} under "flat" conditions are all equal to 1. We also expect g_{44} to be numerically the major component of space-time curvature near a small amount of slowly moving matter. This allows us temporarily to disregard any changes in the minor non-zero components of g_{ik}, namely g_{11}, g_{22} and g_{33}, and to focus on how g_{44} behaves. We thus assume that the four-dimensional metrical equation

$$ds^2 = g_{11}(dx_1)^2 + g_{22}(dx_2)^2 + g_{33}(dx_3)^2 + g_{44}(dx_4)^2$$

can be reduced

$$ds^2 = (1)(dx_1)^2 + (1)(dx_2)^2 + (1)(dx_3)^2 - (g_{44})(dx_4)^2.$$

The (1) in each of the three terms means that only the change in $(g_{44})(dx_4)^2$ is significant. We also note that these three terms represent spatial dimensions whereas the fourth term represents the time dimension. In other words under these conditions (small mass, slow motion) ds is determined mainly by the curvature of time, not by the curvature of space.

Now we turn to f_{ik}, which we are treating as that part of g_{ik} representing flat space-time. Therefore the components of f_{ik} also equal 1, which includes

$$f_{44} = 1.$$

As for h_{ik}, we let it represent the deviation from flatness. Thus we can replace

$$g_{ik} = f_{ik} + h_{ik},$$

with

$$g_{44} = 1 + h_{44}.$$

This equation says that practically all of the shape of space-time, expressed by g_{44}, consists of flatness, expressed by 1, plus superimposed departure from flatness, i.e. *curvature*, expressed by h_{44}. Our restriction to a weak gravitational field and slow speeds means that h_{44} must be a small number. Moreover h_{44} must decline to zero farther and farther from the mass that is responsible for the field. Thus, in this set-up, objects are "weightless" in outer space because there $h_{44} = 0$.

We also disregard "second-order" terms containing h_{44}, which means we consider only linear changes in h_{ik}. (A more complete equation is

$$g_{44} = f_{44} + Ah_{44} + A^2 h_{44} + ...etc.,$$

but we admit only the first two right-side terms and let $A = 1$. The other terms are negligible.) That is to say, we consider the gravitational field to be free of irregularities (no tidal effects) by exhibiting no changes within changes. Students of differential calculus will recognize that this step "linearizes" the equation, justified in part because the pattern of Poisson's equation is linear.

Given the above limitations—a weak and regular gravitational field through which objects move slowly—the appropriate metrical equation becomes

$$ds^2 = (1)(dx_1)^2 + (1)(dx_2)^2 + (1)(dx_3)^2 - (1 + h_{44})(dx_4)^2,$$

and it remains for us to identify h_{44}.

We note the pattern of reasoning: We know what space-time is like without gravity:

$$g_{44} = f_{44} = 1;$$

$$h_{44} = 0.$$

We wish to determine the difference that a mass induces, such that

$$g_{44} \text{ no longer equals } 1;$$

$$h_{44} \text{ no longer equals zero.}$$

In this case g_{44} is associated with general relativity, f_{44} with special relativity, and h_{44} with the difference manifest as gravitation. Said in words, *space-time is flat until it is affected by nearby matter, and gravity is absent until space-time is curved.* This is a commonly used approach in the analysis of general relativity. It asks, "what difference does the appearance of matter make?" For example, as we mentioned, relativity may appear as an "excess" in gravitational acceleration beyond a Newtonian prediction. The "difference" or "excess" we seek should be a change in g_{44} in terms of the gravitational potential. In other words h_{44} ought to depend upon Φ, and we should see an equation structured as

$$g_{44} = 1 + \text{ some value set by } \Phi.$$

Actually we find a parallel pattern in pre-relativity physics. Newton identified gravitational accelerated motion, he recognized uniform motion, and he proposed gravitational force. *Motion is uniform until affected by force, and gravity is absent until force acts on matter.* This pattern prompts us to rephrase the question, "what difference does the appearance of matter make?" Rather we can ask, "when is the gravitational potential not zero?" Newton's answer would be, "when the force of gravity is added to otherwise uniform motion." Einstein's answer is, "when curvature is added to otherwise flat space-time."

Let us now fill in

$$1 + \text{ some value set by } \Phi,$$

and here the right side of the equation

$$\frac{dt_o}{dt_r} = \left(1 + \frac{\Phi}{c^2} \right)$$

comes to mind. We also recall that the g for the curvature of our basketball was such that

$$g = \left(\frac{dx_{flat}}{dx_{curved}} \right)^2.$$

However on our rotating disc we focused on dt's, not dx's, so that the g in question is such that

$$g_{44} = \left(\frac{dt_o}{dt_r} \right)^2.$$

In other words $(\frac{dx_{flat}}{dx_{curved}})^2$ on our basketball represents $(\frac{dt_o}{dt_r})^2$ on the rotating disk, and by virtue of the principle of equivalence, the turning of the disk is equivalent to a gravitational field. Just as curving (inflating) the flat ball altered a g, so changing location on the disk alters g_{44}, and so transformation between points in a gravitational field alters g_{44}. The above equations imply that to solve for g_{44}, the term

$$\left(1 + \frac{\Phi}{c^2} \right)$$

must be squared. Explicitly

$$\left(1 + \frac{\Phi}{c^2} \right)^2 = 1 + \frac{\Phi}{c^2} + \frac{\Phi}{c^2} + \frac{\Phi^2}{c^4}.$$

Here $\frac{\Phi^2}{c^4}$ is a very small number which can be omitted, so that

$$\left(1 + \frac{\Phi}{c^2} \right)^2 \approx 1 + \frac{2\Phi}{c^2}.$$

(This again allows us to disregard "second-order" terms, as they are negligible.)

We conclude that, in terms of g_{44}, local gravitational conditions vary by $1 + \frac{2\Phi}{c^2}$, which brings us to our immediate target: h_{44} equals $\frac{2\Phi}{c^2}$, while the connection between g_{ik} and Φ is such that

$$g_{44} = 1 + \frac{2\Phi}{c^2}.$$

Not only will this equation be useful in solving the field equations, but it guides us toward our goal: to bring the shape of space-time, here represented by the g_{44}- component of the metrical tensor, into a relationship with gravitation, here represented by the gravitational potential. We already know that the gravitational potential is related to the density of nearby mass according to

$$\Phi = -\frac{KM}{r}.$$

These two equations together tell us that g_{44} *deviates from 1, i.e. that space-time deviates from flatness, where the density of nearby matter is not zero.*

We see that the deformation of space-time is associated with the variability of Φ while Φ depends upon nearby matter. Poisson showed that "$4\pi K\rho$ of matter" induces "$\nabla^2\Phi$ of gravity." Relativity shows that matter changes g_{44} by $\frac{2\Phi}{c^2}$. The solution of our metrical equation for weak gravity and low speeds becomes

$$ds^2 = (1)(dx_1)^2 + (1)(dx_2)^2 + (1)(dx_3)^2 - \left(1 + \frac{2\Phi}{c^2}\right)(dx_4)^2.$$

In reverse, the absence of nearby matter is signaled by a null gravitational potential and by no change in the g_{44}- component. That is to say, when M and the gravitational potential are zero, g_{44} is *not* zero; it is one (1), and "one-ness" of g_{44} is associated with flatness of space-time. This idea is clearer if the equation for the gravitational potential is written as

$$\Phi = \frac{1}{2}c^2 (g_{44} - 1),$$

which reiterates that only curved space-time, signaled by g_{44} greater than 1, permits a gravitational potential to exist. If g_{44} is just 1, then $\Phi = 0$; gravity vanishes.

We can combine the above equation with Poisson's equation to yield

$$\nabla^2 \left(\frac{1}{2}c^2 (g_{44} - 1)\right) = 4\pi K\rho.$$

In the absence of matter, when $M = 0$ and $4\pi K\rho = 0$, we find that $g_{44} = 1$. Here space-time is flat, corresponding to what we saw on our basketball in two dimensions, where $g_{11} = 1$ meant zero curvature.

Gravity can vanish under another circumstance, wherein we see the mathematical rationale for the term "nearby matter." $g_{44} = 1$ not only if $M = 0$ but also if r is infinite. ($M/\infty = 0$.) That is to say, when matter is *not nearby* but very far away, it generates *zero* local gravitational potential and cannot deform nearby space-time. We said the same thing by $h_{44} = 0$. (Gravity also vanishes during free-fall because local space-time appears flat, but this is not pertinent here.)

We note that Φ in "$\Phi = - KM/r$" is negative, and the relationship between Φ and r is linear. Numerically, $2\Phi/c^2$ is usually a very small quantity. (When $2\Phi/c^2$ equals a certain number, it has an special astrophysical meaning which we shall cite latter.)

Let us recall that in the metrical equation, dx_4 includes the constant c to make $(g_{44})(dx_4)^2$ compatible with the rest of the metrical equation; since this term is squared it includes c^2, as in $c^2(dt)^2$. The constant c also appears in the term $1 + 2\Phi/c^2$, where c^2 stems from the special relativity factor connecting t_r and t_o. How "neatly" the uses of c^2 converge here!

We stated earlier that unlike acceleration ("a" in $F = ma$ and in $a = Km_2/r^2$), the gravitational potential says nothing about motion. Nonetheless there must be a connection between Φ and gravitational acceleration, or else the former would be of no use. Indeed "a" is the result of the gradient of the gravitational potential, so that

$$a = \nabla\Phi,$$

or, depending on the definition, $a = -\nabla\Phi$; falling objects can be considered to accelerate "down" a gradient. (In terms of Newton's laws, $\nabla\Phi$ determines the gravitational force: $F = ma$ and $G = m\nabla\Phi$.) We can restate this for relativity, namely in a manner that allows specifying vector components:

$$\frac{d^2x^i}{dt^2} = \nabla\Phi = \frac{\partial\Phi}{\partial x^i}.$$

The left term is acceleration expressed explicitly as the second derivative of motion. The two right terms hold the gravitational potential as a gradient and explicitly as a partial derivative. The equation states that accelerated motion in the i-direction is linked to a change in the gravitational potential in the i-direction. This means that *gravitational behavior stems from the point-to-point variation of the gravitational potential* in each direction. Of course this may seem obvious—the intensity of gravity governs how objects fall—but this equation underscores that we can analyze gravitational behavior in terms of a field rather than in terms of force.

We should pause and consider: g_{44} is linked to gravitation by its relation to the gravitational potential, which in turn is linked to matter by its relation to mass-density and which also is linked to gravitational acceleration. Meanwhile g_{44} is linked to space-time by its relation to the shape thereof.

In short, *the shape of space-time is responsible for gravitation because* Φ *adds to* g_{ik}. We can say that even though Φ is a scalar and not a tensor quantity, and even though g_{44} is only one component of a tensor, g_{ik} contains at least one relativistic tensorial counterpart of the gravitational potential. As we pointed out, the metrical tensor g_{ik} finds itself at the crossroads of relativity.

With the information at hand we can refine our metrical equation by estimating how the spatial components, namely g_{11}, g_{22} and g_{33}, deviate from one (1) in response to nearby matter. We know from special relativity that the effect of relative motion is "more" time and "less" space, and we found that g_{11}, g_{22} and g_{33} may each be *less* that 1. (On our basketball, g_{11} had to be less that 1 because dx_{curved} is longer than dx_{flat}.) We can assume that these three g's vary equally and that we remain in ordinary Cartesian coordinates. Since the gravitational potential Φ increases the value of the time coordinate dx_4 by a factor of

$$1 + \frac{2\Phi}{c^2},$$

we can assume that Φ should decrease the value of the space coordinates dx_1, dx_2 and dx_3 by a factor of

$$1 - \frac{2\Phi}{c^2},$$

so that

$$g_{11} = g_{22} = g_{33} = 1 - \frac{2\Phi}{c^2}.$$

This yields a more thorough "refined" solution of our metrical equation for weak gravity and low speeds (it is still an approximation),

$$ds^2 = \left(1 - \frac{2\Phi}{c^2}\right)(dx_1)^2 + \left(1 - \frac{2\Phi}{c^2}\right)(dx_2)^2 + \left(1 - \frac{2\Phi}{c^2}\right)(dx_3)^2 - \left(1 + \frac{2\Phi}{c^2}\right)(dx_4)^2.$$

We know that on a flat surface the key g's equal 1 (or -1, depending on the set-up). Here, on the curved surface of curved space-time, we find that none of these g's equal 1. For instance "non-g_{11} = 1" means "non-flat," which is that same as saying that the value of g_{11} in the presence of curvature is less that 1. Of course usually the g_{44}- term remains by far the largest (since dx_4 includes c), but the entire "refined" equation signifies that *curvature of space, as disclosed by* g_{11}-g_{22}-g_{33}, *and curvature of time, as disclosed by* g_{44}, *determine gravity.*

We note that this "refined" metrical equation is still not "the whole story." First of all the equation for the most important component,

$$g_{44} = 1 + \frac{2\Phi}{c^2},$$

does not tell us enough because it lets Φ affect only one component of g_{ik}; this is because g_{44} by itself ignores 9 out of 10 pieces of information; it describes the situation using one "adjective" while more are available. In other words this equation links g_{ik} with one gravitational potential whereas general relativity in effect claims that any one point in a gravitational field has as many as 10 gravitational potentials.

However even when we bring Φ to bear the other three non-zero g's ($g_{11} = g_{22} = g_{33} = 1 - \frac{2\Phi}{c^2}$),

Φ is still a scalar, whereas the g's are tensor components, which (by virtue of relativity) may appear dissimilar to various observers. As a consequence, any approximations or omissions in equations containing g's preclude general covariance, and full unanimity among all observers requires unabridged representation of all components. This becomes significant under the conditions important in modern cosmology and astrophysics, such as in black holes, during gravitational collapse, and for very rapidly moving stars. We will find better equations for these cases later.

On the other hand the mathematical link between the g_{44}- component of the metrical tensor and Φ means that under conditions with weak gravitational fields and low speeds, the equations of general relativity yield the same results as Newton's equations. This is revealed by the "Newtonian limit" which shows that general relativity limited to g_{44} reduces to Newton's law of gravitation. We will elaborate when we apply general relativity to a specific situation.

◆

Meanwhile what about the right side of Poisson's equation, $4\pi K\rho$? Why cannot we take advantage of this term and simply replace

$$g_{ik} = T_{ik}??$$

by

$$g_{ik} = 4\pi K\rho??$$

Yes, ρ is a term for the distribution of matter, but just as g_{44} does not tell us everything about the shape of space-time, ρ *does not tell us everything about matter.* For instance ρ, like Φ, is a scalar quantity, whereas relativity demands tensors. Also ρ says nothing about relative motion, whereas we know that motion affects our measurement of mass. The means that $4\pi K\rho$ will have to be replaced by T_{ik}, because T_{ik} is a tensor with 10 components which embrace the effect of relative motion of ponderable (weigh-able) mass. Not surprisingly ρ will appear in only one of the components of T_{ik}. We can say that g_{44} has given us only a sampling of g_{ik}, and that ρ is only a sampling of T_{ik}. Nonetheless Poisson's equation does give us an idea how the two are related. (To be exact, density in this context should be written as ρ_o to signify the absence of motion, and we will heed this detail later.)

✦

Let us return to the point that Poisson's equation suggests that a derivative of a derivative, i.e. a second derivative, should be implicated. In particular the 2 in $\triangledown^2\Phi$ tells us that the left side of our provisional equation should be in the form of a *second differential, or second derivative, of the metrical tensor.* Hence we propose that, rather than $g_{ik} = T_{ik}??$,

$$a\ second\ derivative\ of\ g_{ik} = T_{ik}??$$

Calculating a derivative, we recall, is a procedure in differential calculus which tells us how changing quantities change. A second derivative of g_{ik} will allow us to deal with the complex point-to-point rate of change of the g's with respect to space and time. Though this approach introduces various mathematical complications, we shall find it to be quite logical, and it addresses the question as to what changes in the shape of space-time can be attributed to given nearby matter. We also know that the g's depend on the chosen system of coordinates, but that the way in which the g's vary across the surface hinges only on its shape.

Not only should we expect a second derivative of g_{ik}, but considering the linear structure of Poisson's equation, we anticipate two further particulars: First, the equation for "a second derivative of $g_{ik} = T_{ik}??$" should follow a certain mathematical configuration (including that it should be linear). Second, a minus sign must appear in the definitive version. Meanwhile we already learned that in g_{ik}, $g_{44} = 1 + \dfrac{2\Phi}{c^2}$, and we already saw that T_{ik} cannot be adequately represented by $4\pi K\rho$, though the latter embodies the density of matter.

Clearly Einstein's utilization of Poisson's concept was a critical step, and Einstein called Poisson's equation his "model" (as on p. 82, The Meaning of Relativity). Poincaré, Minkowski, and Lorentz were all working on relativity, and they were aware that Newton's law of gravitation as well as the Galilean transformation are defective. Minkowski incorporated a time interval into his theory, but he still treated gravity as a force rather than as potentials in a field. It fell to Einstein to grasp Poisson's hint. Nevertheless it was only a hint, and we shall see that Einstein was forced to rely on the difficult achievements of Riemann (and others) to move forward, particularly to ensure that his field equations exhibit general covariance.

To see the relevance of Riemann's work we augment our analogy of a house with shaded windows. The metrical tensor allows us to look inside, but our house needs a particular interior, named a Riemannian space or Riemannian manifold. Let us briefly define "manifold." Generally, a manifold is a space containing mathematically interesting objects that has certain mathematical features, the most basic of which is a system of coordinates. In topology each point of a manifold appears Eucilean; a 2-dimensional manifold is a surface. For geography the surface of Earth is the manifold, while latitude and longitude form the coordinates. For relativity *space-time is a manifold, while the space-time axes form the co-ordinates*. From page 105 on we discussed time-like, space-like and light-like intervals; to accommodate these, the time component of space-time equations for *ds* must be negative (page 96), as was not inherent in Riemann's original work. Ergo to be exact this makes manifolds in relativity "pseudo-Riemannian." I prefer to call this space simply a "Riemannian space," and in either case it has two principal but interrelated characteristics, as follows:.

First, it is a four-dimensional space with intrinsic curvature in all four dimensions. This peculiarity complicates our analogy, in that our "house" must have a bent interior and pliable curvilinear walls, within which Euclid's and Pythagoras' rules are inadequate. Moreover similar clocks may tick at different rates and similar rulers may have variable lengths in different parts of this house. Nevertheless the events inside such a house must ultimately appear the same no matter what the structure of the house is. If the walls are bent, or if the events are moved, these events—and the applicable laws—must remain identical.

The second characteristic of a Riemannian space is that it is a "metrical continuum." The formal mathematical definition of a continuum—and of a manifold for that matter—is quite complex (see above, Wald, pp. 12-14 and 423-427), but the general features of continua are pertinent. Continua are abstract concepts, yet several analogies can illuminate these features: If a ramp is a continuum, a staircase is not a continuum. The surface of an ice rink is a continuum, unless it has a hole. A continuum in one dimension could be a ruler with markings so that a real number can identify any point along its length. In theory these markings must be infinitesimally close together and must be numbered in sequence, and if the ruler has a gap which cannot be identified by a number, it ceases to be a continuum. A continuum in two dimensions could be a sheet of graph paper with markings along the edges so that any point on its surface can be given two numbers. A three-dimensional continuum could be a cube made of graph paper with markings for height, width and depth. More difficult to imagine is a "hyper-cube" with markings for height, width, depth, and time, but it is a four-dimensional continuum, and of course each series of markings is one axis of a four-dimensional system of coordinates. However we must also grant the possibility that a ruler may be bent, that graph paper may have non-perpendicular or even bent sides, and, most importantly, that a "hyper-cube" can be similarly deformed. We anticipate the implication: Events exist a curved four-dimensional space-time continuum.

We say that the set of points in continuum are "compact"—as closely packed as necessary—and "connected"—adjacent to each other continuously and contiguously. We can also say that a continuum is a set of elements, including of course points in space-time, such that between any two of them, there is always a third. Hence no part of a continuum is inaccessible to point-by-point analysis, and all points are indistinguishable except as they are reflected on coordinate axes. These stipulations may seem trivial and redundant, but they are necessary so that equations dealing with continua can be legitimately applied to relativity.

Each point in a continuum may have some measurable parameter, and this property defines a field. For example a continuum of magnetic potentials—each point has a measurable amount of magnetism—is a magnetic field. An ocean is a continuum which defines several fields, since each point within the water has a temperature, a density, a salinity, a current, etc. As is more relevant for us, each point of a continuum may have a gravitational potential, so that such a continuum forms a gravitational field. Indeed a mathematically valid gravitational field must be a continuum, though as we know, general relativity treats gravitational potentials as consequences of the local shape of space-time (as in $g_{44} = 1 + \dfrac{2\Phi}{c^2}$). Therefore *space-time is a metrical continuum*, which means that not only are its constituent points compact, connected, and freely accessible, but each point has measurable shape.

We emphasize that not only is the space-time continuum is metrical, but each of its point can be described in four dimensions, while each dimension, even the one representing time, is treated the same. This feature allows each point to hold an ultra-small four-dimensional triangle which can be tested for compliance with the pythagorean theorem by means of the metrical equation. The degree of compliance provides a measure of local shape, while shape correlates with how clocks and rulers behave. Each point is also capable of hosting a momentary event, while a prolonged event is a chain of momentary events occupying a group of points. To study a gravitational event, we place it into our analogic "house" so that the trajectory of a falling or orbiting object is a time-like geodesics in the interior space of the house. To know what is happening, we can observe the projections of the object onto four walls or shaded windows.

We note some mathematical details of the space in this house: In an ordinary three-dimensional "Euclidean house" we can set up three Cartesian coordinates (x_1, x_2, and x_3) and an interval (ds) at each point such that

$$ds^2 = dx_1^2 + dx_2^2 + dx_3^2.$$

At each point in our four-dimensional house the same equation applies except, as conceived by Minkowski, it must include a fourth term, dx_4^2, which makes ds itself four-dimensional. However this space and this equation still offer no provision for intrinsic curvature.

At each point in a "Riemannian house" we can also find an interval (ds) in four dimensions, but as we noted while studying the metrical tensor, the necessary equation at each point is

$$ds^2 = g_{11}dx_1^2 + g_{12}dx_1dx_2 + g_{22}dx_2^2 \ etc.,$$

or in compact form,

$$ds^2 = g_{ik}\,dx_i\,dx_k.$$

The g's are essential because unlike Euclidean space, each point of Riemannian space possesses a definite amount of intrinsic curvature. We can say that our Riemannian house is built by the application of the metrical equation at each point, and each point can be characterized by the components of the metrical tensor. Moreover, as the metrical tensor can tell us how fast clocks tick and how long rulers appear at each point in our house, the basis of all gravitational events is that clock-ticks (i.e. time) and ruler-lengths (i.e. space) at different points of a Riemannian space are unequal. Poisson's equation said the same thing but in terms of the gradient of the gravitational potential, and Einstein went on to realize that a Riemannian space can accommodate the complicated shape of space-time which determines gravity—and indeed that the curvature of space-time *is* gravity.

Under special circumstances, namely in the absence of any nearby matter and energy, that amount of curvature is zero, but a Euclidean space never has such curvature. In this sense Euclidean space is a special type of Riemannian space, and Euclidean geometry is a special case of Riemannian geometry. Said in reverse, Riemann generalized upon Euclid, and this fact is part of an overall trend we glean in studying relativity—greater generalization. For example earlier we called g_{44} a sample of g_{ik} and we called ρ a sample of T_{ik}, implying that g_{ik} and T_{ik} represent generalizations of narrower concepts. Since Newton's laws implicitly presume only Euclidean geometry, we can also maintain that Newton's laws are but special cases (samplings?) of relativity, and that relativity is a generalization of the former.

The compact and connected nature of points in a continuum leads to another mathematical property of a Riemannian space, one that will interest us soon: Such a space is "infinitely continuously differentiable." This means that we can compare any set of points in it by the use of differential calculus. For this reason the overall shape of a Riemannian space is described by another tensor (bearing Riemann's name), one that is based on *how the components of the metrical tensor vary from point to point.*

We also note that a Riemannian space can be described as a *tensor field*. In fact Riemann's breakthrough was showing how the variable shape of a multi-dimensional space lends itself to tensor analysis. The tensor quality of Riemann's mathematics means that the ensuing equations hold in any system of coordinates. In terms of our analogy, Riemann's tensor methods allow us to write the laws of physics so that they are independent of the choice of "house," which means so that they appear and work the same for all observers. However what cannot be easily displayed analogically, as we

263

shall see later, is how generally covariant laws demonstrable inside any such house also satisfy the conservation laws.

While the idea of treating gravitation as a tensor field is not difficult to accept, eventually we will find it necessary to treat matter likewise as a tensor "material field." For the moment it suffices to think of a Riemannian space as the field-like house in which the interplay between matter and space-time transpires, but what is remarkable about this house is that its shape—the "geometry" of its space-time to be more precise—responds to the presence of matter, and at the same time this shape governs how matter moves.

We commented earlier that relativity shifts the explanation of gravity away from a study of masses and towards a study of the intervening environment. What is that intervening environment? In a general sense, it is a field or it is space-time, and in a more rigorous mathematical-geometrical sense it is a Riemannian space. This is not merely a change in emphasis; it has practical physical significance. For example the basis of the principle of equivalence of inertia and gravity is that regardless of which we call it, both are one characteristic of one Riemannian space; i.e. of space-time that contains matter. Moreover, in the eyes of general relativity all events, including our existence, occur in a four-dimensional Riemannian space ("pseudo-Riemannian" to be exact, as I mentioned).

◆

The key product of Riemann's work is the aforementioned tensor which fully describes four-dimensional Riemannian space: *the Riemann curvature tensor*. This tensor had been derived independently by Christoffel (1829-1900), so it is also called the Riemann-Christoffel curvature tensor; for brevity we call it *the curvature tensor*. It can be written in several ways but can be abbreviated as R_{hikl}. (The selection of indices is arbitrary. We omit the j because it is too similar to an i. As we shall see, a "mixed" form of the curvature tensor, such as R^h_{ikl}, is important.) Other mathematicians in this arena were Ricci and Levi-Civita who derived the curvature tensor using different approaches. Their publications were also of great help to Einstein, and Levi-Civita was particularly supportive.

Students of the history of mathematics view this topic with interest in that otherwise inconspicuous and generally unknown accomplishments of Riemann, Christoffel, Ricci, and Levi-Civita eventually contributed to what became the most celebrated topic in modern science. None of these individuals set out with relativity in mind, but Einstein became familiar with their work through the assistance he requested and received from his colleague Pick and, more importantly, from his former fellow student Grossmann, who was an expert on tensor calculus and who tutored Einstein in 1912-1914. (See Pais, pp. 212-213, and Clark, p. 254.)

Let us preview some features of the curvature tensor. Like the metrical tensor, it expresses the intrinsic shape of any multi-dimensional space, including four-dimensional space-time. It has a rank of four because it is ultimately composed of four vectors. It has 20 independent components by which it describes space-time around any point, just like 20 adjectives may fully identify an object. If there is no intrinsic curvature of space-time (no nearby mass and hence no gravity), all components of the curvature tensor become zero. Or in mathematical terms, in a Euclidean space the curvature tensor "vanishes everywhere," which is the case limited to special relativity. We also say that in flat space-time this tensor "vanishes identically." Conversely, if these components are not all zero, then space-time is intrinsically curved.

The term "everywhere" is not always essential but it means that all components of a tensor are consistently zero. Hence we can say that in a Euclidean space, curvature as described by the Riemann tensor is zero "everywhere." This idea is based on the observation that a vector vanishes completely if and only if all its components vanish. The term "identically" means for all values of the variables, in this case all values of h, i, k, and l in R_{hikl} or R^h_{ikl}. (Each of these values can be 1, 2, 3, or 4 in four dimensions.)

The concept of a tensor vanishing "everywhere" and "identically" brings us to a sophisticated way in which the property of covariance may be defined: If a tensor vanishes in one system of coordinates, it likewise vanishes upon transformation to any other system of coordinates. In other words whether or not all components of a particular tensor equal zero is determined by the intrinsic property of whatever the tensor describes, independent of whatever system of coordinates is selected. (We must not assume that *all* tensors used to describe space-time vanish in the case of "flatness." For example, as we shall see, the metrical tensor does not; it only remains constant point-to-point.) Einstein expressed the relevance of this notion to general relativity as follows (in The Principle of Relativity, p. 121): "If...a law of nature is expressed by equating all the components of a tensor to zero, it is generally covariant." The implication of course is that we have here an essential criterion: *If we can express a law of gravitation by setting an appropriate tensor equal to zero in one frame of reference and it is zero in all frames of reference, then that law is generally covariant.* We will also find that if we know the components of a tensor in one frame of reference, we can calculate its components for any other frame of reference. Given the ability of the curvature tensor to describe space-time, we must decide whether it is that appropriate tensor, one which can help us formulate a relativistic law of gravitation.

Let us therefore recall that we are looking for the left half of the provisional field equation

$$a \ second \ derivative \ of \ g_{ik} \ = \ T_{ik}??$$

Perhaps the desired *second derivative of the metrical tensor g_{ik} is the curvature tensor*: Can we say that

$$R_{hikl} \ is \ a \ second \ derivative \ of \ g_{ik}?$$

and if so, is the following provisional equation valid?

$$R_{hikl} \ = \ T_{ik}??$$

265

Clearly we need to know how tensors R_{hikl} and g_{ik} are related, and for this we need to know the merits of the *derivative* of g_{ik} over g_{ik} itself.

Let us reconsider a freely falling frame of reference such as a loose elevator falling to Earth. Though the elevator is in a gravitational field, observers inside detect no signs of a gravitational potential; their space-time is flat. To appreciate the entire situation they must look outside this elevator, which in mathematical terms means that in effect they perform a transformation to a more global frame of reference, one in which curved space-time is evident.

We can think of the transformation from a freely falling frame of reference to the global frame of reference as "getting onto another nearby elevator." This means we envision—as ridiculous as it is—that a gravitational field is densely populated by many infinitesimally small freely falling elevators. (The analogy with many points in our curved house is comparable but it obscures the notion of free-fall.) An observer making the elevator-to-elevator transformation immediately encounters *the point-to-point variability in the g's*—i.e. the complex change in the components of the metrical tensor—which is characteristic of curved space-time. Now we recall that the variability of the g's propagates the gravitational field (as does the gradient of the gravitational potential in Poisson's equation). Thus gravitation is characterized by the *rate of change* of the metrical tensor, and, according to differential calculus, a rate of change summons a *derivative*. We defer the question of just how observers calculate the variability of the g's, though we anticipate that parallel transport is implicated.

We can connect the notion of point-to-point variability in the g's with two statements we have made: One, if a law works here (in one freely falling elevator) and it works elsewhere (in another such elevator), then it works anywhere. Two, the task is to find laws whose equations remain unchanged when moved from "here" to "elsewhere." Therefore in order to find generally covariant laws—in order to find equations which remain invariant "elevator-to-elevator"—we must find invariant equations for the variability of the g's. That is to say, *we must find the derivative—to be exact, the second derivative—of the tensor g_{ik} in a form that remains valid going from one freely falling frame of reference to another.* We already know that the same metrical equation and the same principles of special relativity apply locally in each elevator, which means that Lorentz-invariance will be a prerequisite.

Let us not forget, however, that the values of the components of g_{ik} are influenced by the system of coordinates we select. This means that we can always find a system such that the g's in one elevator indicate flatness, but as long as the elevators are in a detectable gravitational field, the components of g_{ik} in any other elevator must be different. Meanwhile *the pattern of point-to-point differences*—i.e. the "elevator-to-elevator" variation—does *not* depend on the choice of system. A point-to-point *variation* in the g's in one system appears *the same* in any system. This concept will obviously be important to us in mathematical terms: *The components of g_{ik} are affected by the choice of system of coordinates, but derivatives of g_{ik} are not.* We can now also begin to apply the concept of vanishing tensors: If indeed R_{hikl} is a tensor and a second derivative of g_{ik}, then the vanishing of R_{hikl} (written as $R_{hikl} = 0$) must be an essential feature of the mathematics of general relativity.

We further note that flatness is a special case for the shape of space-time, but that does not mean that g_{ik} vanishes, only that its components keep certain values; they fail to vary; their rate of change is nil; they have no derivative. In yet other words, the metrical tensor responds to zero curvature by *remaining constant* at all points. We found an instance of this while examining the gravitational potential Φ and its relation to one component of g_{ik}, namely g_{44}. Under conditions with no gravitation, Φ is zero; there is no gravitational potential, but there still is a g_{ik}, which (using Cartesian coordinates) has the value of 1, but then it stays 1 everywhere space-time is flat.

We must glance at the tensor T_{ik} with which the density of matter (and energy) is represented: When there is *no* nearby matter and energy, T_{ik} responds differently than does the metrical tensor g_{ik}. T_{ik} responds to flat space-time by *vanishing*, which is intuitively predictable: Without gravity, T_{ik} should have no components. This means that there exist frames of reference in which g_{ik} has non-zero components while T_{ik} has all zero components. The contrast is especially salient if the same conditions (no gravity and flat space-time) are studied upon transformation to another system of coordinates. The result of the metrical equation, ds, is invariant under the transformation, but the g's are not. In a new system of coordinates their values are different, but they do not all vanish. In contrast if T_{ik} vanishes in one system of coordinates, it will also vanish in any other, and, based on the nature of tensors, we expect the same from R_{hikl}. I.e., when R_{hikl} vanishes, so does T_{ik} but not g_{ik}. The implication is that R_{hikl} is more appropriate than g_{ik}.

Hence we can recapitulate four reasons to prefer

$$a\ second\ derivative\ of\ g_{ik} = T_{ik}??$$

over

$$g_{ik} = T_{ik}??$$

First, we seek equations for the variability in the g's, meaning that we seek its derivatives. Second, a derivative of g_{ik} can be independent of the choice of systems of coordinates, making a derivative a more reliable description of the deviation of space-time from flatness. Third, the behavior of T_{ik} does not match that of g_{ik}. Fourth, the term $\nabla^2 \Phi$ in Poisson's equation suggests a second derivative. These considerations empower the provisional field equation

$$R_{hikl} = T_{ik}??$$

The question then is how we derive the tensor R_{hikl} so that it is a second derivative of g_{ik}. There are two hidden issues here. First, it would be logical to obtain R_{hikl} solely from g_{ik} without the introduction of additional data. In other words since the g's determine the metrical properties of local space-time, the components of R_{hikl} should depend only on these g's. Indeed Einstein managed to show how to derive the curvature tensor solely from the metrical tensor.

The second issue is more subtle: The derivative of a tensor is not necessarily another tensor, and if the derivative of the metrical tensor were not a tensor, the advantages of composing laws of physics in tensor form would be lost. This case therefore calls for a special type of differentiation called *covariant differentiation*, i.e. the finding of *covariant derivatives*.

"Differentiation" of course means that a change is calculated in ultra-small steps. The term "covariant" has two possible meanings here. We shall find that by melding the concept of parallel transport with tensor differentiation, the principle of general covariance is respected, and it is in this context that "covariant" differentiation merits its name. The second meaning is purely geometric. For clarity we use the word "covariant" in the former sense and "co-variant" in the latter. We will say more about the distinction later.

◆

Next we embark upon an intricate mathematical excursion to bring together transformations, differentiation of vectors and tensors, partial differentiation, covariant differentiation, the metrical connections, and the curvature tensor. (Lieber's book provides basic details but his format is difficult to unite with other concepts in relativity. Kenyon takes a broader approach, though he presumes considerable sophistication in calculus. The two works complement each other but together they entail several hundred pages of text.)

Let us begin with the role of partial differentiation. As we noted, the ∇ in Poisson's equation admits multiple independent variables and allows us to consider the derivatives of each variable while all other variables are held constant—i.e. to assemble the partial effects of several derivatives (pages 183 and 250). These variables of course include the four dimensions of space-time, which means that the field equations we seek are four-dimensional partial-differential equations.

Meanwhile we indicated that transformation equations as well as tensors are vital to general relativity. *Let us therefore apply partial differentiation to convert transformation equations into tensor form.* We picture observers A and B aboard two of our freely falling elevators in a gravitational field. We let these observers use the rotational transformation equations to compare their measurements of some event described by a vector. X_A is A's measurement of a component along his X axis, while X_B is B's measurement of a component along B's X axis. We ignore the Z and T axes for now. The effect of relative motion between A and B can be depicted as a transformation during which we tilt one set of coordinate axes through angle R. With parentheses included for clarity, this is expressed as follows:

$$X_B = (cosR) (X_A) + (sinR) (Y_A)$$

$$Y_B = -(sinR) (X_A) + (cosR) (Y_A).$$

We note that X_B depends on X_A as well as on Y_A. Likewise Y_B depends on X_A as well as on Y_A. This means that a change in X_B is the result of the *partial contributions* of any change in X_A *and* any change in Y_A. The corresponding is true for a change in Y_B; it depends partially on a change in X_A and partially on a change in Y_A.

We also note that X_B and Y_B each depend on angle R. However these rotational equations are related to the Lorentz translation-type of transformation—the latter envisions a "sliding" of X_A-Y_A axes past the X_B-Y_B axes—so that sinR and cosR are functions of the special relativity factor. We recall that this factor contains "*v*" for uniform motion, *not* "*a*" for accelerated motion. For example

$$cosR = \frac{1}{\sqrt{1-v^2/c^2}}.$$

Here sinR and cosR have three roles which we can spell out and which make the equations for rotational transformations more useful to us. One, as this equation exemplifies, they *reflect the relative speed, v, between A and B*. Two, as the previous pair of rotational equations shows, they act as *coefficients* for X_A and Y_A. Three, they act as *partial derivatives*.

The third role of sinR and cosR requires the symbol ∂ for "partial derivative." We let x^1 replace X, so that x^1 is X in A's system of coordinates and x'^1 is X in B's system. Likewise x^2 replaces Y. We use the apostrophe to mark terms that apply to B's system, and for reasons we will cover shortly, we use superscripts. (Here x^2 does not mean *x* squared; it means axis Y.) Thus in the equation $X_B = (cosR) (X_A) + (sinR) (Y_A)$, the term $(cosR)$ portrays the change in x'^1 per change only in x^1, which we write as

$$\frac{\partial x'^1}{\partial x^1}.$$

Likewise in this equation sinR written as

$$\frac{\partial x'^{1}}{\partial x^{2}}$$

portrays the change in x'^{1} per change only in x^{2}.

Since both coefficients are functions of R, a change in each is linked to a change in R. To be specific, while we seek the partial derivative of R with respect to $\dfrac{\partial x'^{1}}{\partial x^{1}}$, we hold $\dfrac{\partial x'^{1}}{\partial x^{2}}$ constant. If then we seek the partial derivative of R with respect to $\dfrac{\partial x'^{1}}{\partial x^{2}}$, we hold $\dfrac{\partial x'^{1}}{\partial x^{1}}$ constant. This isolates the two effects so as to admit the independent but "partial" contribution of each to what is called the total differential.

We cannot proceed without reconsidering a technicality. Vectors and tensors come in co-variant and contra-variant types, as well as a mixed type. Again, "co-variant" is not the same as "covariant;" the latter means that the laws of physics appear alike in all frames.* For the moment we can treat co-variance and contra-variance as a geometric issue which stems from how we interpret components in diagrams. We can depict a vector or tensor in co-variant or contra-variant manner according to which is more serviceable, and we will use mathematics to switch back and forth, which does not change the physical meaning of a vector or tensor. We will say more about "co- to contra-" interconversion later.

To be precise, vectors and tensors themselves are not co-variant and contra-variant; they are invariant. Only their *components* can be one or the other, though we commonly overlook this semantic point. Furthermore if we wish to be even more meticulous, we can say that a vector or tensor has not one, but two sets of components, one co-variant and the other contra-variant. Thus for example the metrical tensor can be said to have not 10 but 20 independent components in two sets. However this detail is rarely heeded.

We raise the co-variant/contra-variant issue because from here on we will pay more attention to the use of superscripts vs. subscripts. So far we have avoided superscripts so as to minimize confusion with exponents, even though this policy sacrifices some mathematical precision. *Superscripts are used for contra-variant vectors and tensors, and subscripts are used for co-variant ones.* This is why modern authors favor $ds^{2} = g_{ik}\, dx^{i} dx^{k}$ (or $ds^{2} = g^{ik}\, dx_{i} dx_{k}$) over $ds^{2} = g_{ik}\, dx_{i} dx_{k}$. In common terminology regarding tensors and vectors, we call a superscript a "contra-variant index" and we call a subscript a "co-variant index."

---

* Some authors use the term "covector" or "one-form" for "co-variant."

For now we focus the more intuitive contra-variant case, and we institute two changes: First, in order to admit four dimensions and four coordinate axes on which to plot motion, we again use x with the superscript 1, 2, 3, or 4. As is customary, X becomes x^1, Y becomes x^2, Z becomes x^3, and T becomes x^4. This allows us to consider differential changes of the x's along the four coordinate axes: dx^1, dx^2, etc., and as we already saw for X and Y, we use the same system for partial differentials.

Second, we used subscripts A and B to refer to two systems of coordinates (rotated relative to each other), but using so many indices is awkward. Therefore we still use an apostrophe on terms belonging to the "B" system. Thus dx'^1 replaces a measurement along X_B. etc.

Now if we label that total differential as dx'^1, then (still in two dimensions) the equation

$$X_B = (cosR)\,(X_A) + (sinR)\,(Y_A)$$

becomes more complicated but more functional,

$$dx'^1 = \left(\frac{\partial x'^1}{\partial x^1}\right)(dx^1) + \left(\frac{\partial x'^1}{\partial x^2}\right)(dx^2).$$

We note how the indices prevent ambiguity: $\dfrac{\partial x'^1}{\partial x^1}$ is a coefficient linking dx'^1 with dx^1, and $\dfrac{\partial x'^1}{\partial x^2}$ is a coefficient linking dx'^1 with dx^2. The same pattern applies to Y_B (which is dx'^2) as well as to dx'^3 and dx'^4 as we include the remaining dimensions. Thus the right side of the equation has a total of 16 terms (parentheses omitted) because

$$\frac{\partial x'^u}{\partial x^1}\,dx^1 + \frac{\partial x'^u}{\partial x^2}\,dx^2 + \frac{\partial x'^u}{\partial x^3}\,dx^3 + \frac{\partial x'^u}{\partial x^4}\,dx^4$$

is repeated four times as each u is 1, 2, 3 or 4. We note that four sets of terms are added together (summated) and that each set of terms contains a dx along one of four axes. Furthermore *each dx is modified by a coefficient which is a partial derivative and which depends on angle R.*

The 16 coefficients can be written more compactly by replacing both indices by letters (u and v) and by using the summation symbol, $\sum$. In four dimensions v, like u, is 1, 2, 3 or 4:

$$\sum_v \frac{\partial x'^u}{\partial x^v}.$$

271

Of course each dx is a vector component. For example dx^1 can be the contra-variant component of ds projected on the x^1 axis in A's system. We can therefore express how ds is projected on every axis in B's system by means of one efficient equation:

$$dx'^u = \sum_v \frac{\partial x'^u}{\partial x^v} \, dx^v.$$

Since we summate all the partial derivatives, dx'^u is a set of contra-variant vector components, and we label its vector as V'. The components in the other system then belong to vector V. Upon a rotation of coordinate axes, we find that

$$V'^u = \sum_v \frac{\partial x'^u}{\partial x^v} \, V^v.$$

This is a generalized transformation equation, and *it is in vector form*. It tells us that how the vector looks in the new system (/) depends upon *a summation of the partial derivatives*. In other words the *entire* difference in how this vector appears to two separate observers is determined by all the *partial* differences.

We reiterate that authors differ on the placement of some indices (for example Lawden's equation 31.3, p. 90, and Einstein's equation 56, p. 65 in The Meaning of Relativity), but we all routinely invoke Einstein's summation convention for omitting the $\sum$, so that the previous equation appears as

$$V'^u = \frac{\partial x'^u}{\partial x^v} \, V^v.$$

Having "vectorized" a transformation, we can easily "tensorize" it. Higher-rank tensors are built from lower-ranked tensors (or vectors) by the rule for multiplying vectors,

$$T^a U^b = V^{ab}.$$

Thus the same pattern for transformation holds for the case of tensors, but an additional partial-differential vector-term $\dfrac{\partial x^w}{\partial x^y}$, *one per rank*, appears in the equation (reflecting that tensors are built from vectors). Hence in the case of a *rank-two contra-variant tensor*, its transformation equation is

$$V'^{uw} = \frac{\partial x'^u}{\partial x^v} \frac{\partial x'^w}{\partial x^y} \, V^{vy}.$$

Again the x's with the apostrophe belong to B's system, which we can call the "new" system, and the plain x's belong to A's system, which we can call the "old" system. We note that to calculate the "new" tensor V'^{uw}, we use the summation of partial changes along the "new" axis per unit change along the "old" axis. Again the process generates new components (*uw* do not equal *vy*) and this equation is in *tensor* form. We note three further details:

One, since some indices, namely *v* and *y*, appear twice in one term, Einstein's summation convention can be used. This means that in four dimensions we let *v* and *y* run form 1 to 4; we "sum over the *v*'s and *y*'s." (When indices are not repeated, we do not sum over these, or else we would be running the equation one way, A to B in this case, and then back the other way, B to A, which is not what we want.)

Two, for a scalar (a rank-zero tensor) the transformation equation is simply

$$V' = V.$$

Three, in the absence of relative motion, the $\dfrac{\partial x}{\partial x}$-terms become 1, which (for a rank-one contra-variant tensor) means that

$$V'^u = V^v.$$

The transformation equations clearly show that non-moving observers find no difference in the components of V, which we expect since with no relative motion, angle R = 0, sinR = 0, cosR = 1. Then the equation

$$X_B = cosR \ X_A + sinR \ Y_A$$

reduces to

$$X_B = X_A;$$

non-moving observers find no discrepancy in their measurements of an event. This point is important: *The transformation equations demonstrate that if a vector or tensor vanishes in one frame, it does so in all*:

$$\text{If } V'^u = 0, \text{ then } V^v = 0.$$

Transformation equations—we can call them transformation laws—also serve as an powerful mathematical test: If a quantity transforms according to the pattern in the above equations, we know that this quantity is a vector or a tensor. For instance the metrical tensor passes this test (V could be *g* in the transformation equation for rank-two tensors), which is the evidence mentioned on page 171 that *the g's indeed form a tensor*. (There are other ways to authenticate tensors. See Lawden, p. 94, or Synge and Schild, p. 18. The transformation equations are valid because they are linear and homogeneous. See Borisenko and Tarapov, p. 76, or Synge and Schild, p. 119.) It is even feasible to define a tensor as a mathematical entity that transforms according to one of the transformation equations; see Synge and Schild, pp. 10 and 11. This means that we encountered two main ways of defining tensors, by how many components they have relative to the number of dimensions—as on page ?—and by how they behave under a transformation. In short, if it looks like a tensor or acts like a tensor, it's a tensor.

The transformation laws reveal a principle of paramount relevance in general relativity: *Equations with tensors are invariant when moved from one frame to another.* We already reached the same inference by another approach: *A tensor equation successful in one frame of reference is successful in all.* For example let us assume that a law of physics expressed by the equation "$G^{ab} = M^{ab}$" is proven in some system of coordinates. If the components of "G" and "M" do not transform to another system according to the transformation equations, then this law may not hold everywhere. But if these components in one frame are related to corresponding components in another according to the transformation equations, we can be assured that "$G^{ab} = M^{ab}$" is valid in all frames, including accelerated ones. Similarly, as shown above and as we mentioned in words, if "$G^{ab} = 0$" holds in one frame (if G^{ab} vanishes in one frame), it does so in all.

We should add that valid tensor equations obey other mathematical rules. For instance if an equation is to hold in all frames of reference, all its tensors must be co-variant or contra-variant to the same extent, and if an equation sets two tensors equal to each other, both must have the same rank. I.e., neither $G^{ab} = M^a{}_b$ nor $G^{ab} = M^{abc}$ hold in all frames.

These principles apply to tensors of any rank, and they apply in four dimensions. As we noted, all this is pertinent because Einstein demanded that laws be universally successful—i.e. that they have general covariance; that they appear the same under any transformation of coordinates, for any observer, and during any kind of motion.

On the other hand equations can be generally covariant without being tensor equations. $E = mc^2$ is an example, and $ds^2 = g_{ik} dx_i dx_k$ is not a pure tensor equation. This also does not mean that any tensor equation is automatically a valid law of physics. For instance as Einstein developed his field equations, which in essence constitute a law of gravitation, he had to meet several requirements, only one of which is tensorial status. Another requirement which gave Einstein much trouble and which we shall discuss later is that of compliance with the conservation laws.

We now turn to vectors and tensors with *co-variant* components First let us examine the geometric distinction between contra-variance and co-variance using diagram K (recalling that this in unrelated to the principle of general covariance). Point T_A on axis T-A is a projection of point P, and if that projection is made *parallel to the horizontal axis X-A*, then we call the distance from T_A to the vertex (where the axes meet) a contra-variant component[*]. But we can also project P *perpendicular one of the other axes,* and we call the result of such a projection a *co-variant* component. Since T-A and X-A are perpendicular to each other—they are orthogonal Cartesian coordinates—the two methods generate the same component. However if the axes are not orthogonal but oblique, such as T-A and X-B, a difference arises. Now the projection from P to C forms a co-variant component—it is perpendicular to the X-B axis—and it is not the same as the distance from T_A to the vertex. (The distance from T_A to the vertex is the same as the distance from P to X_A which is shorter than P to C. Angle R determines the difference.) This means that wherever orthogonal coordinates are not used, such as with curvature, contra-variant and co-variant components of a point, of a vector, or of a tensor are not identical. Moreover, as we shall see shortly, the equations that describe transformations are different for the two types of components.

While the geometric distinction is arbitrary—a projection can be parallel to one axis or perpendicular to another—some applications favor contra-variance and others favor co-variance. Velocity is a prime example of a vector usually represented as contra-variant, while a gradient is a vectorial quantity usually displayed in co-variant manner. Let us see why: We envision an object moving ("falling") with some known velocity across the "surface" of space-time. The intuitive way to depict the object's velocity at any point along its path is by means of an arrow, so that we create a tangent vector (see page 145). If we place this vector into a system of coordinates and analyze its length and direction as projections onto the axes in logical fashion, we obtain contra-variant components. If our axes are labeled x^1, x^2, x^3, and x^4 or simply x^u, we can assign the partial-differential term $\dfrac{\partial x^u}{\partial t}$ (distance *x* per time *t*, which is velocity) to each point along the object's path. If we then analyze how the components appear to a relatively moving observer, we derive the contra-variant transformation equation we saw earlier. In general if a vector or tensor satisfies this equation during transformation, we say that "it transforms in contra-variant manner."

But we can also envision that the object "falls" by responding to the local gravitational potential, Φ. Unlike velocity, the gravitational potential by itself is not a vector quantity—it is a scalar—so that we cannot give it a vectorial arrow. Nonetheless Φ varies along the "surface" of space-time, so that we have the *gradient* of Φ in our system of coordinates (as for Poisson's equation on page 249). If our axes are labeled x^u, we can assign the term $\dfrac{\partial \Phi}{\partial x^u}$ (gravitational potential Φ per distance *x*, which is a gradient) to each point along the object's path, but we note that unlike in the case for velocity, x^u is in the denominator.

An arrow will not do for a gradient either (why so shortly), which means that a gradient is not exactly a vector. Nevertheless gradients can have magnitude and direction which can also be projected onto

---

[*]Calling components "contra-variant" or "co-variant" is counter-intuitive, but we are stuck with this convention, which is based on the transformation equations that apply for each type.

co-ordinate the axes, so that we treat them as vectors. That magnitude is the steepness of the gradient—like the steepness of a dip our trampoline—and the direction is the direction in which the gradient is steepest. (A gradient can be drawn as a series of concentric lines that surround this point, like a contour map of a valley. The closer the lines to each other, the steeper the gradient.) Here it is more natural to project the components perpendicular to the coordinate axes, and these projections form co-variant components. In other words the gradient vector is 90 degrees from the tangent vector at this point.

However the situation is more complicated: Without knowing the local shape of the surface, we lack sufficient information in a gradient to define a vectorial arrow, which means that *without the metrical tensor* (without g_{ik}) *the gradient of* Φ *cannot supply the velocity vector for a falling object.* This is an important notion for two reasons. First, we will see how the g's are needed us to convert co-variant to contra-variant components and vice versa. Secondly, it is another way of stating that the shape of space-time (given in g_{ik}) ties the magnitude and direction of gravitational motion to the variation of the gravitation field (the gradient of Φ). This comes as no surprise, as we already know that $g_{44} = 1 + \dfrac{2\Phi}{c^2}$, but now we can say it more strongly and in the full spirit of general relativity:

We need to know space-time to know gravitation.

We still have a loose end: Is the transformation equation different in co-variant form? Yes, because when $\dfrac{\partial \Phi}{\partial x^u}$ is placed into another system of coordinates—which means when viewed by a relatively moving observer—the ∂x's of the "new" vector appear in the denominator. This step invokes the chain rule of differential calculus that says that $\dfrac{dz}{dx} = \dfrac{dz}{dy}\dfrac{dy}{dx}$, which gives us

$$\frac{\partial \Phi}{\partial x^{\prime u}} = \frac{\partial \Phi}{\partial x^v}\frac{\partial x^v}{\partial x^{\prime u}}.$$

Replacing $\dfrac{\partial \Phi}{\partial x^{\prime u}}$ by V'_u and $\dfrac{\partial \Phi}{\partial x^v}$ by V_v for simplicity, we obtain *the transformation of a co-variant vector:*

$$V'_u = \frac{\partial x^v}{\partial x^{\prime u}} V_v.$$

(Co-variant components take subscripts, but the summation convention still applies.) We note that this equation has the $\dfrac{\partial x}{\partial x}$-term reversed compared with its contra-variant counterpart. In other words to calculate the "new" vector V' we use the summation of partial changes along the "old" axis per unit change along the "new" axis. In general if a vector or tensor satisfies this equation during transformation, we say that "it transforms in co-variant manner."

We mentioned that there are mathematical relationships between the co-variant and the contra-variant components of a vector or tensor. These relationships descend from the determinant of the metrical tensor (more on this shortly). The rules for the conversion between co-variant and contra-variant forms are

$$A^a = g^{ab}A_b \quad and \quad B_a = g_{ab}B^b.$$

We again note the role of the g's: We need the metrical tensor to turn co-variant components, such as those of a gradient, into contra-variant components, such as those of velocity. As we just said, we can interpret this observation as a mathematical counterpart of the concept that *we need to know space-time*—here represented by the metrical tensor—*in order to know gravitation*—here represented by the velocity of falling. If the g's = 1 because space-time is flat, in which case Cartesian coordinates can be orthogonal, the dissimilarity between the two types of components vanishes.

We also have the "see-saw rule" for vectors:

$$A_d B^d = A^d B_d.$$

These rules allow us to raise and lower the indices of a vector or tensor, but only as long as it is done on both sides of an equation. Furthermore $C_i C^i$ is a scalar quantity, the value of which is invariant under a transformation. One application of these rules which we will use later is that

$$C, \text{ which is a scalar,} = g_{ab}C^{ab} = g^{ab}C_{ab}.$$

We also have mixed tensors, such as $D^h{}_{ikl}$, which have both contra-variant and co-variant components. This tensor could be the product of, say, B^h and C_{ikl}. The mixed form is of particular interest for us because the curvature tensor can be expressed as $R^h{}_{ikl}$. Using the above inter-conversion rule,

$$R^h{}_{ikl} = g^{hm}R_{mikl} \quad and \quad R_{hikl} = g_{hm}R^m{}_{ikl}.$$

Older literature calls the purely co-variant form (e.g. R_{hikl}) the curvature tensor "of the first kind" and calls the mixed form (e.g. $R^m{}_{ikl}$) the curvature tensor "of the second kind." Incidentally, to be precise $U^h{}_i$ is not the same as $U_i{}^h$ unless U is symmetrical.

The rules linking contra-variant and co-variant components also apply to the metrical connections, which we shall examine shortly. Often whether a particular tensor is co-variant, counter-variant, or mixed is not crucial, and we saw that the three types are equivalent in ordinary Cartesian coordinates. Therefore when dealing with certain approximations wherein curvature is ignored, the three types may be interchangeable.

Let us briefly look into the co-variant/contra-variant aspects of the metrical tensor. The term g^{ik} is a contra-variant (and symmetrical) version of the metrical tensor, also called its reciprocal or its conjugate tensor. Its formation involves the determinant of g_{ik} and the cofactor of the determinant of g_{ik}:

$$g^{ik} = \frac{\textit{Cofactor of } g_{ik} \textit{ within its Matrix}}{\textit{Determinant of the Matrix } (g_{ik})}$$

A mixed form of the metrical tensor can also be generated.

Determinants and cofactors are functions of a matrix used to ease algebraic solution of several linear equations together; texts on advanced algebra define these, and Lieber (p. 193) summarizes the definition. In our context, we need only to know that cofactors and determinants are used to transform the metrical tensor from its co-variant to its contra-variant form. The determinant of the matrix of g_{ik} is written $|g_{ik}|$ (but this is often abbreviated as g). The corresponding determinant of g^{ik} is written $|g^{ik}|$, and we find that $|g_{ik}| \cdot |g^{ik}| = 1$. For us this means that

$$g^{ik} = \frac{1}{g_{ik}}.$$

When restricted to two dimensions, the metrical tensor has only four components, g_{11}, g_{12}, g_{21} and g_{22}. An equation we will use later (which is based upon the definition of determinants) has to do with the determinant of such a matrix:

$$|g_{ik}| = g_{11}g_{22} - g_{12}g_{21}.$$

We appreciate that according to our mathematical conception of space-time, every point is characterized by the metrical tensor, so that such a collection of such tensors forms a tensor field. As the metrical tensor can have contra-variant or co-variant or mixed components, the entire tensor field shares this property. For instance if the metrical tensor with co-variant components at every point of a field heeds the corresponding transformation equation, we have a gravitational tensor field in co-variant form.

Before leaving transformation equations we should preview just what these mean to us. We envision an object moving in a certain direction at a certain speed, so that its motion constitutes a vector, V, with contra-variant components. If another observer views the same event while there is relative motion between the observers, each has the right to consider himself to be at rest and to consider the other observer to be in motion. The geometric representation of this scenario is two systems of coordinates tilted relative to each other by angle R. If one observer "sees" a component of V as V^v, the other observer "sees" a different value for the component, say, V'^u. The relative speed of the two observers determines the angle of the tilt, R, which can be expressed as a partial derivative $\dfrac{\partial x'^u}{\partial x^v}$. The relationship between

V^v, the components as seen by one observer,

V'^u, the components as seen by the other observer, and

$\dfrac{\partial x'^u}{\partial x^v}$, the representative of their relative speed,

resides in the transformation equation

$$V'^u = \frac{\partial x'^u}{\partial x^v} V^v.$$

This equation ensures that both observers indirectly—which means by way of components—"see" the same vector. After all, V is the same as V', because new coordinates do not change the vector in question, nor do they change the transformation equation. However they change the vector's components. For instance we recall that any space-time interval (ds) can be treated as a vector which itself does not change when its coordinate axes are rotated, while its projected components (dx's) do change. In terms we already used, the vector and this equation are invariant, but the components are relative. The transformation equation quantifies *how* relative. By extrapolation from vectors to tensors, "transforming a tensor" means changing its components according to the above pattern.

We also recall the concept that gravity is a question of transformation of space-time coordinates. In this context the transformation equations accomplish two additional things: One, they confirm the vectorial or tensorial status of the terms we use to describe space-time, matter, and other features. Two, they guarantee that different sets of components describe the same vector, or the same tensor, or the same space-time interval *ds*, as the case may be. Clearly these attributes are central to general relativity in that *a gravitational field is characterized by point-to-point transformations of coordinate axes*, as for instance by analogy from one freely falling elevator to another.

We can think of the transformation equation $V'^u = \dfrac{\partial x'^u}{\partial x^v} V^v$ as simply equivalent to

$$V'^u = R \ V^v.$$

Though algebraically inaccurate, this version reiterates that vector components (*u* and *v*) appear different during relative motion when that motion is expressed by the tilt (angle R) of coordinate axes. As we said, the term $\dfrac{\partial x'^u}{\partial x^v}$ recasts R as a partial derivative. But why do we bother recasting R in this manner? This issue brings us to the fact that so far we have not supplanted the assumption on page 93 about R, nor have we specified the type of axes. We shall soon tackle the prospect that *R need not be constant and the axes need not be straight*, and we will find that *a partial-derivative expression for R facilitates solving this problem.*

A key point for now is that to master general relativity, we must be able to "transform tensors." Let us recall which tensor we should consider at this time. We have seen that the gravitational field can be viewed as a field of metrical tensors, which means that g_{ik} should be our focus. We further recall that we are deliberating the proposal that

$$R_{hikl} = a \ second \ derivative \ of \ g_{ik}.$$

We also know that the variability within this field accounts for gravitation and that quantifying variability entails differentiation. These concept suggest the next series of steps: Mating the process of *differentiation* with *transformation equations* in order to find *derivatives of the metrical tensor.*

✦

However our approach must be indirect. First we consider a contra-variant vector (a rank-one tensor with contra-variant components) in a vector field. Since a field is a mosaic of points, we imagine we can move this vector's arrow from point to point (or again by analogy, from elevator to elevator). Moving something, even a vector, by an infinitesimal amount constitutes *differentiation with respect to location,* and we need to learn what happens when a vector is moved to an adjacent point of the field. We can treat this process as moving our entire rule for (contra-variant) vector transformation,

$$V'^u = \frac{\partial x'^u}{\partial x^v} \ V^v,$$

along an axis from point *x* to point *x* + *dx* in the *w*-direction; *x* with indices represents coordinate axes X, Y, Z, and T, but we take advantage of partial derivatives. For clarity in this step,

$$V'^u = \text{a}, \qquad \frac{\partial x'^u}{\partial x^v} = \text{b}, \quad \text{and} \qquad V^v = \text{c},$$

so that our vector transformation equation becomes

$$a = bc.$$

Since b represents $\dfrac{\partial x^{/u}}{\partial x^{v}}$ which is a differential term, we expect that this term will be converted to a differential of a differential, a.k.a. a second-order differential. In basic differential calculus when "$a = bc$" is subjected to differentiation with respect to x, the result is

$$\frac{da}{dx} = b\frac{dc}{dx} + c\frac{db}{dx},$$

and this holds for partial derivatives. Applying this process to our vector transformation equation yields

$$\frac{\partial V^{/u}}{\partial x^{w}} = \frac{\partial x^{/u}}{\partial x^{v}}\frac{\partial V^{v}}{\partial x^{/}{}_{w}} + V^{V}\frac{\partial^2 x^{/u}}{\partial x^{v}\partial x^{/w}}.$$

For our purpose we have taken liberties regarding indices (for example see Einstein in <u>The Principle of Relativity</u>, pp. 133-136 or Lawden, pp. 95-97), but for identification let us call this the "full" equation. *It uses partial differentiation to reveal the variability of a vector field.* Its left side means "partial change in vector $V^{/}$ per infinitesimal change in x in the w-direction" or, in effect, "the vector is moved to the next point in this field." The right side has two pertinent features: One, the first term has gained one index, namely w, which we made into a subscript or a co-variant index. Two, the term $V^{V}\dfrac{\partial^2 x^{/u}}{\partial x^{v}\partial x^{/w}}$, which is the second-order differential we expected, spoils the "tensor purity" of this equation; it would fail to transform according to the transformation law, which is not a second-order equation. The "full" equation can comply with this law only when this "untensorizing" term vanishes. Let us therefore plan to temporarily erase this term.

We should first streamline the notation. If V^{u} is a vector in a coordinate system of x's, the derivative of V^{u} with respect to x^{w} is $V^{u}{}_{w}$. Therefore the term $V^{u}{}_{w}$ can replace $\dfrac{\partial x^{/u}}{\partial x^{v}}\dfrac{\partial V^{v}}{\partial x^{/}{}_{w}}$ in our "full" equation, and the apostrophe is no longer helpful. When we erase the "untensorizing" term, our vector transformation equation becomes

$$\frac{\partial V^{u}}{\partial x^{w}} = V^{u}{}_{w}.$$

We note the "extension" from one to two indices; *differentiation of the rank-one tensor V^{u} has formed a new tensor $V^{u}{}_{w}$ of rank two.* We can say that the differentiated item has acquired a vectorial arrow as well as a subscript (rather than a superscript), but this will be easier to explain later, using an analogy. Corresponding rules apply for differentiation of co-variant and mixed vectors and tensors, enabling us to build new tensors by "extension" of vectors or tensors. We will eventually implement this process to "extend" the rank-two metrical tensor g_{ik} into the rank-four curvature tensor R_{hikl}.

Let us designate this form of vector and tensor differentiation as "ordinary differentiation." There is a critical problem with ordinary differentiation: We recall that (in two dimensions) $\dfrac{\partial x'^1}{\partial x^1}$ and $\dfrac{\partial x'^1}{\partial x^2}$ are coefficients representing angle R, and R may or may not vary, depending curved or straight axes. However if R does not vary, the "untensorizing" term in the "full" equation vanishes (the second-order differential of a constant is zero), which means that the right side of the transformation equation

$$V'^u = \frac{\partial x'^u}{\partial x^v} V^v$$

cannot vary, and which means that *ordinary differentiation presumes a constant angle R*. In other words if we enforce tensor purity by allowing no "untensorizing" term, we in effect demand straight axes that allow no variation in R. Einstein approached this issue by way of the observation that in pre-relativity thinking the laws of physics were deemed valid only in inertial frames of reference (The Meaning of Relativity, pp. 139-142), so that rectilinear (straight) Euclidean coordinate systems were "singled out...among all conceivable ones." Under this restriction, $\dfrac{\partial x'^1}{\partial x^1}$ and $\dfrac{\partial x'^1}{\partial x^2}$ are independent of the x's, *which means that space-time must be flat.* I.e. if location does not matter, the surface cannot be intrinsically curved, and *angle R cannot vary.* This is an intolerable restriction, since space-time has curvature which is progressively steeper closer to a concentration of matter, or as depicted by an indentation of our trampoline, the coordinates tilt more and more at each successive point closer to a mass. Hence we must attune our equations *to allow angle R to vary.*

Said another way, ordinary differentiation holds only for *linear* transformations such as that of Lorentz, in which the two systems move past each other only at uniform rate. Geometrically speaking, a linear transformation can occur only between Euclidean systems of coordinates. In a gravitational field, the geometry of space-time (except on a local basis) is never Euclidean and the rate at which the two systems "slide" past each other is never uniform.

All this means that if we subject a tensor to ordinary differentiation, our result does not meet the requirement of general relativity; it excludes curved space-time; it may not apply during any state of motion; it lacks general covariance. To remedy this, somehow the metrical tensor g_{ik} must be inserted into the equation for ordinary differentiation,

$$\frac{\partial V^u}{\partial x^w} = V^u{}_w$$

(which is the "full" equation for our contra-variant rank-one tensor V^u, but without the "untensorizing" term). The necessary modification, consists of the insertion of a term containing the *metrical connections*, the Γ's, which, as we shall see, introduce g_{ik} into the equation in an indirect manner. We write a general equation,

$$\frac{\partial V^u}{\partial x^w} + \Gamma^u_{wy} V^y = V^u{}_w.$$

Both equations evaluate the partial change in *u* per partial change in the *w*-direction, *but only the latter has the metrical connections.*

The metrical connections may be labeled in several ways, one of which is called the Christoffel symbols in honor of their inventor. In the literature we will encounter Christoffel symbols "of the first kind" (e.g. "[ab,c]") and those "of the second kind" (e.g. "$\{^c_{ab}\}$"), but this terminology is somewhat archaic, and nowadays the Greek letter Γ is preferred. ($\Gamma^c{}_{ab}$ is equivalent to $\{^c_{ab}\}$ and to $g^{cd}[ab,d]$; see Lawden, p. 109.) The rules for raising, lowering, and extending these indices apply as they do for vectors and tensors.

The form of differentiation in which the metrical connections appear is called *covariant differentiation*. Conceptually,

ORDINARY DIFFERENTIATION + TERM(S) WITH METRICAL CONNECTIONS

= COVARIANT DIFFERENTIATION.

As in the case of ordinary differentiation, *a tensor subjected to covariant differentiation gains a rank.* The terms with the Γ's may be preceded by a + or - sign depending on whether the tensor being differentiated is contra-variant, co-variant, or mixed.[*] The number of Γ-containing terms in this scheme is set by the rank of the tensor being differentiated. For example covariant differentiation of a rank-two tensor V_{uv} yields a rank-three tensor V_{uvw}, as in

$$\frac{\partial V_{uv}}{\partial x^w} - \Gamma_{wu} V_v - \Gamma_{wv} V_u = V_{uvw},$$

with some indices lowered or omitted for clarity.

---

[*] "Covariant differentiation" does not mean differentiation of only co-variant vectors or tensors. Some authors avoid this issue by calling covariant differentiation "invariant differentiation."

We emphasize that the rank has been "extended," and the presence of the Γ's signifies covariant differentiation. We also note that the added index is a subscript, so that we can say that covariant differentiation adds a co-variant index. (Shortly we will learn that the mere presence of the Γ's does guarantee covariant differentiation, but for now the above scheme suffices as our guide.)

We can concisely symbolize ordinary or covariant tensor differentiation by using a comma or a semi-colon. As an example, when a co-variant tensor V_{uv} is differentiated with respect to y we can write its *ordinary* derivative as

$$V_{uv,y}$$

and its *covariant* derivative as

$$V_{uv;y}.$$

The (negative) Γ's are implied in the latter.

In the case of a contra-variant tensor V^{uv}, its ordinary derivative with respect to y can be written as

$$V^{uv}{}_{,y}$$

and its covariant derivative as

$$V^{uv}{}_{;y}.$$

Here the implied Γ's are positive rather than negative. Differentiation of mixed tensors follows a corresponding pattern with commas and semi-colons, but the added index is still a subscript.

Readers will encounter other conventions for labeling ordinary and covariant derivatives based on lower case vs. capital letters. For example $\dfrac{dV^i}{ds}$ can signify the ordinary derivative of V^i with respect to ds, while $\dfrac{DV^i}{ds}$ can indicate the covariant derivative. (The D here may be called an operator because it is a mathematical instruction for an operation which, in this case, produces a covariant derivative.)

✦

We will see that *the Γ's obviate the restriction to linear transformations*, so that covariant differentiation is differentiation of tensors which itself survives non-linear transformations and which yields a new (higher-ranked) tensor. Why do the Γ's have this effect? Because they accommodate variation in the coefficients ($\frac{\partial x'^1}{\partial x^1}$, $\frac{\partial x'^1}{\partial x^2}$, etc., which, we note, are partial derivatives). This means that they entail differentiation of the quantities which embody the angle (r) of rotation in a rotational transformation. As a result the Γ's cover the possibility that *R is not constant* but that it may vary at a certain rate. Thus they allow for the uneven curvature of space-time and for the variability of the gravitational field. In effect the metrical connections, and hence covariant differentiation, resolve the predicament that the special relativity factor $\sqrt{1-v^2/c^2}$ encompasses only uniform motion ("*v*") and not accelerated motion ("*a*"). We can think of the Γ's as liberating us at long last from the confines of special relativity and therefore from restricted covariance. Ultimately the Γ's guarantee general covariance as we use derivatives of tensors. The practical application of this idea is that *the derivation of equations for generally covariant laws of physics should entail covariant differentiation.*

A related principle we have emphasized is that space-time intervals (*ds*'s) are invariant under transformations in which the axes are tilted or rotated. Now by admitting non-linear transformations we can add that *space-time intervals are invariant under transformations in which the axes are bent.* This supports the concept that vectors—and therefore tensors—are invariant under *any* transformation. (Non-linear transformations are called "general" transformations, which is why we can define general covariance as form-invariance under general transformations.)

Before leaving this topic, we should point out that our statement that the presence of the Γ's signifies covariant differentiation is an oversimplification. As we shall see shortly in detail, parallel transport plays a key role, so that the scheme

ORDINARY DIFFERENTIATION + TERM(S) WITH METRICAL CONNECTIONS
+ *PARALLEL TRANSPORT* = COVARIANT DIFFERENTIATION

is more accurate. Why do we stipulate *PARALLEL TRANSPORT?* Because we insist that motion in space-time, including between points in space-time, occur on geodesics, and we recall that parallel transport generates a geodesic.

✦

We have deferred the question of just how observers *calculate the variability of the g's.* Clearly the Γ's must play a key role, simply because the metrical tensor assesses various points in space-time, and differentiation assesses point-to-point variations. We also recall that the result of parallel transport can quantify space-time curvature. Hence we expect to summons parallel transport, and we devise the following scheme:

A vector, *V*, exists in a vector field, while a collection of infinitesimally close points constitute its surface. For identification, we number adjacent points as P_1, P_2, P_3, P_4, etc., and even though two adjacent points, say P_1 and P_2, are close together, we can identify a direction from P_1 to P_2;

285

we say that there is a path from P_1 to P_2. We think of vector V at point P_1 as a vectorial arrow placed on the surface of the field at P_1. We then slide vector V along the path from P_1 to P_2 while the direction (and length) of the vector remain unchanged; of course this process is parallel transport, and the point-to-point paths constitute geodesics.

Using a system of coordinates, we note the components of vector V at point P_1 and again at point P_2, and we anticipate that there may be *a change in the components of V* as a consequence of transport from P_1 to P_2. We determine whether the components depend on the path from P_1 to P_2. If they do not—if parallel transport along this path induces no change in components—the surface here is flat. Conversely, if the path between the two points influences the consequence of parallel transport—if the components of V change—the surface here is curved, and *the shape of the surface determines the change in the components.* This procedure is most informative if we consider *four* points arranged in a "loop," though the loop must be ultra-small and its four points must be infinitesimally close.

This is a rather laborious way to assess curvature, but it has important advantages. First of all it takes into account how a field changes point-to-point. It also allows us to express curvature as a tensor, and it is independent of the system of coordinates. At the same time this procedure illustrates how a gyroscope points to a different direction after a trip around a "loop" in space-time, akin to how the planet Mercury shows a shift after each "loop" around the sun.

The loop can be any shape, such as a triangle or orbit, but a four-sided square-like figure is easier to work with, similar to what we imagined on page 213. We pretend that we slide V's vectorial arrow around a city block, visiting all four corners. This city is hilly, and we must imagine, granted with difficulty, that the surface of the hills has four dimensions and that time is one of those dimensions. Of course this surface represents the variable intrinsic shape of space-time, and the city represents a gravitational tensor field.

We may further imagine that we can use a four-dimensional coordinate system. We monitor our vectorial arrow at each corner by noting whether moving from one corner to another alters the vector's components as they are projected onto the four axes of this system. Flatness of the entire block would induce no changes in components upon moving our vector. Conversely, if one corner has different local curvature than another, the components show different values. (We could also use an analogy with four freely falling elevators, but visiting four corners is easier to envision. Likewise the four points could be imagined on a "hilly" floor in our house, but a hilly city is more familiar.) The critical observation is that *upon completion of a round trip while heeding parallel transport, the change in the components of the vector discloses the intrinsic curvature of this region.* In other words since the components vary from their initial values, we can quantify the shape of the surface we covered by the behavior of the components among a loop-like group of points. Of course the metrical tensor describes the curvatures at each point, so that in effect we are delineating how the g's vary across the surface—*we are differentiating* g_{ik}.

To explore our block-in-space-time while sliding V around, we let P_1 through P_4 be the four points which form our square-shaped loop. Our starting point is P_1, the next "corner" is P_2, then P_3, P_4, and back to P_1. We imagine that the P's are connected by four infinitely short geodesical paths, but in keeping with our analogy, we call these "streets." We ensure that the vector V maintain a constant angle to the surface of the streets, giving us parallel transport around the block. (Another proviso is that V not spin around the paths; the only activity of interest is direct-as-possible parallel transport corner-to-next-corner.) We expect that upon completion of a loop the components of the transported vector depend on two factors: the vector and the shape of space-time. Altogether this entails three pieces of information: the initial components of the vector, a designation of the space-time interval across which the vector is transported, and an indication of the change in the shape of space-time in this region. However, these data need not be placed in this order.

We could describe the streets by using X and Y axes. P_1-to-P_2 and P_3-to-P_4 might run east-west, parallel to the X axes. P_2-to-P_3 and P_4-to-P_1 might run north-south, parallel to Y. Please see diagram T, keeping in mind that it is a mathematical-geometric abstraction meant to depict transport along geodesical intervals. (We will use the R's in this diagram later.) However the X-Y method is cumbersome, so we use our shorthand, even if it is less intuitive: All streets become x^a, and the superscript a identifies the direction of a street. Using +/- signs, we can make the east-bound direction positive and the west-bound direction negative; the same with north (+) and south (-). Since we consider P_1, P_2, P_3 and P_4 as infinitesimally close points, the interval between any two successive corners is dx^a.

Let us concentrate on the first segment of the loop. We label the direction in which we move "1" (a in dx^a is 1; it might be east). Analogically, we drive one block from corner P_1, where the coordinate value is x, to corner P_2, covering distance dx^1. We let V^p be the vector V at point P_1. As we cover dx amount of space-time in direction 1 from P_1 to P_2 while heeding parallel transport, *the term δV^u discloses the variation in the components of vector V.* The coordinate at P_2 is $x + dx^1$, where the vector components of V are $V^p + \delta V^u$. We will explain all the indices in this scenario shortly, once we have found the mathematical relationship between the surface at P_1 and P_2, the δV^u, the V^p, and the dx^a. The surface of space-time is quantified locally through g_{ik}, but since we are studying a point-to-point variation, we anticipate that the Γ's can bring the g's into the picture; this will be a separate intricate step, but one which is essential for the validity of this entire scheme.

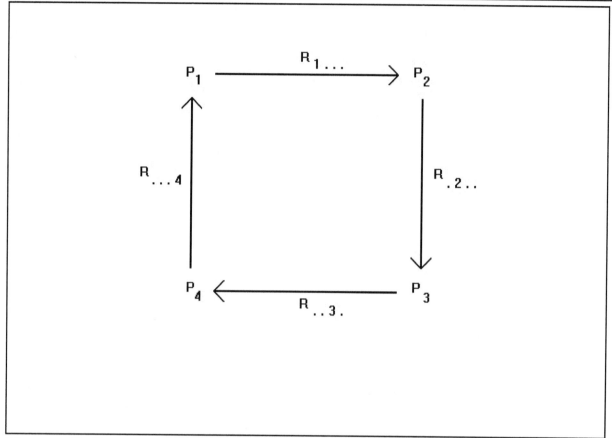

Diagram T. Transport around a "square" loop.

Although we in an analogy, this procedure can be an experiment which we can illustrate with numbers. We imagine that P_1 is 10 blocks east of the city's center ($x = 10$, east is "a"), and we wish to study space-time in the region of x. At x a gyroscope casts a shadow 2 meters long ($V^p = 2$). We move the gyroscope east one more block ($dx^a = 1$) to point P_2. There we find the gyroscope's shadow to be 0.1 meters longer; $\delta V^u = 0.1$ and $V^p + \delta V = 2.1$. Because δV is not zero, we know that space-time along this dx^a has curvature. Moreover there must be a mathematical relationship between that curvature and the 0.1-meter difference. The other parameters, namely the initial length 2 and the interval 1, should also appear in the equation.

From what we already know, it should not surprise us that the relationship between δV^u, V^p and dx^a invokes the Γ's, and that this relationship—like the transformation equation—is linear. Therefore

288

$$\delta V^u = -\Gamma^u_{ap} V^p dx^a.$$

The right side of this equation may have a + or - sign, depending on whether V is co-variant or contra-variant, and as we will see later, Γ^u_{pa} is also legitimate. In the left side the symbol δ stems from the calculus of variations, but later the term δV^u will be replaced by dV^u in the equivalent and more useful differential equation

$$dV^u = -\Gamma^u_{ap} V^p dx^a.$$

We can now positively identify the indices. The directions of parallel transport across intervals dx are a. The initial components of vector V are p. The final components that show change δV are u. In four dimensions, p and u each can be 1, 2, 3, or 4. The entire equation is the mathematical counterpart of the statement that the orientation of the parallel-transported vector, as revealed in its components, hinges on three factors: the initial orientation of the vector, given in V^p; the traversed space-time interval, given in dx^a; and the variation in the shape of space-time, represented by Γ^u_{ap}. This means that *the dependence of δV^u on V^p and on dx^a is governed by the metrical connections.* Again, the patten is linear. For instance if space-time is doubly curved as expressed in the Γ's, the δV's are doubled.

In other words the Γ's hold the consequence of transporting the vector. By "consequence" we again mean the effect upon the components of the vector, so that with the help of the Γ's, the equation solved for δV's shows that *the metrical connections determine how the components of this vector change upon parallel transport along a geodesic,* in this case from P_1 to P_2. If the Γ's = 0, i.e. if the metrical connections vanish, then parallel transport results in no change in V's components. This occurs when space-time is flat; the consequence of transporting the vector across this interval is such that $\delta V^u = 0$. In contrast the consequence of transporting the vector across *curved* space-time is such that the components change by the non-zero value of δV^u, which hinges on the Γ's. *The metrical connections reflect the rate of point-to-point variation in the shape of space-time,* and we can rephrase our earlier statement: The dependence of δV^u on V^p and dx^a is governed by the point-to-point variation in the g's. What remains to be stipulated is the mathematical link between Γ^u_{ap} and the g's, which, as we said, merits a separate step.

Again we note that in a gravitational field the components of the metrical tensor at P_1 cannot be the same as at P_2. This means that the gravitational potential at P_1 is not the same as at P_2. We already know the approximate relationship of one of these components with the gravitational potential,

$$g_{44} = 1 + \frac{2\Phi}{c^2},$$

and we cited how Einstein focused on this point (in <u>The</u> <u>Meaning</u> of <u>Relativity</u>, p. 142): The restriction to inertial frames of reference is associated with the independence of coefficients $\dfrac{\partial x'^1}{\partial x^1}$ and $\dfrac{\partial x'^1}{\partial x^2}$ on location. Conversely, the dependence of these coefficients on location frees us from that restriction, so that no specific frame is singled out. *The metrical connections define just how these coefficients depend on location.*

The benefit of the metrical connections is clear, for when we say that coefficients may depend on location, it means that the *g*'s can depend on location and that they can fluctuate across space-time as the curvature is different at one point than at an adjacent point. By analogy, space-time is different in two separate but adjacent freely falling elevators, and the Γ's quantify that difference. As we know, this also means that the tilt of coordinate axes need not be constant, and that under transformations among gravity-containing frames, angle R is not necessarily constant. And this is why we need covariant differentiation if we are to obtain a generally valid derivative of the metrical tensor in accordance to the proposal that

a second derivative of $g_{ik} = T_{ik}$??

Let us recap the tortuous road leading to transformation equations suitable for gravity-containing frames. While discussing special relativity, we rejected the Galilean transformation, replacing it with the "sliding" Lorentz transformation which accommodates the relativistic effects of uniform motion (*v*). We then replaced the "sliding" transformation with the more adaptable rotational transformation containing angle R, and we used the R-based partial-differential equations to derive transformation equations for vectors and tensors. Finally we applied the metrical connections to quantify changes in angle R. This step ensured that differentiation of tensors is covariant, and (aided by parallel transport) it allowed for the variability of the gravitational field as envisioned in general relativity.

✦

A note of explanation is needed regarding the mathematics of parallel transport: We have seen two equations which may appear to be contradictory: The equation $\dfrac{dV^i}{ds} = 0$, discussed on page 208, is a mathematical definition of parallel transport. It states that a system of coordinates exists in which *the changes in the components of a parallel-transported vector are zero* for transport in infinitesimal increments along any path from some point to another on the surface of space-time. The equation $\delta V^u = -\Gamma^u_{ap} V^p\, dx^a$, discussed on page 289, tells us how the components vary as a vector is parallel-transported across curved space-time, indicating that *the changes in the components are not zero*. The final components *u* are different from the initial components *p*. How can components of a parallel-transported vector be invariant according to one equation and differ according to another?

To solve this dilemma we first re-work each equation. Regarding $\frac{dV^i}{ds} = 0$, we note that parallel transport applies on the uneven curvature of space-time, which means that this equation should satisfy covariant differentiation. Following the convention whereby the comma indicates that $\frac{dV^i}{ds}$ is an ordinary derivative, we write $V^i{}_{,s}$ and we recast the equation allowing a semicolon to indicate a *covariant derivative*. This step, together with a change in the index for conformity, gives us an updated definition of parallel transport, one that includes covariant differentiation of the vector:

$$V^u{}_{;s} = 0$$

We also re-write $\delta V^u = -\Gamma^u_{ap} V^p dx^a$ solved for zero, $\delta V^u + \Gamma^u_{ap} V^p dx^a = 0$. More importantly we inject the parameter *s* (see page 225) and thereby we can use the differential form. Altogether $\frac{dV^u}{ds} + \Gamma^u_{ap} V^p \frac{dx^a}{ds} = 0$, and using the comma/semicolon convention on the first term,

$$V^u{}_{,s} + \Gamma^u_{ap} V^p \frac{dx^a}{ds} = 0.$$

Now we can apply the concept that the metrical connections (the Γ's) allow us to turn an ordinary differential into a covariant one. Granted, this argument entails some circular logic—we developed covariant differentiation using parallel transport, and now we clarify an issue regarding parallel transport using covariant differentiation—but it is valid, and it leads us to an "overall" equation for parallel transport,

$$V^u{}_{;s} = V^u{}_{,s} + \Gamma^u_{ap} V^p \frac{dx^a}{ds} = 0,$$

which holds much information: The covariant derivative of a parallel-transported vector ($V^u{}_{;s}$) is zero, while the ordinary derivative combined with the metrical connections (the parts between the equal sign) is also zero. In fact many systems of derivation accept the statement that a vector remains parallel over a curved surface only if its behavior agrees with this "overall" equation.

To apply the "overall" equation, we envision a scheme with infinitesimally close points in space-time (based on the one laid out by Kenyon on p. 59). We also recall our scenario on page 203 in which an object heeds parallel transport as it moves step by small step, but here we imagine a moving vector while we analyze its behavior in only one step or one increment of its motion. This vector, V, is to be transported in the a-direction between two points, P_1 and P_2. The gap between these points, which is one increment of the V's trip, is dx^a. At P_1 the components of V are determined in a system of curved coordinates (a non-Euclidean system), and these components are labeled as p. Now we introduce a stipulation: *The vector is transformed to a Cartesian system of coordinates, one with straight and orthogonal axes, and in this condition the vector is parallel-transported to* P_2. While it is in straight coordinate axes, the motion of the vector is equivalent to free-fall, which means it is equivalent to motion through locally flat space-time. Under these circumstances the equation

$$V^u_{;s} = 0$$

indeed applies (as does the simpler $V^u_{,s} = 0$) *since a flat surface induces no changes in the vector's components* (no changes in magnitude or direction). The vector is *transformed back to a non-Euclidean system* at P_2, where its components are again determined, labeled as u. When components p and u are compared, the equation $\dfrac{dV^u}{ds} + \Gamma^u_{ap} V^p \dfrac{dx^a}{ds} = 0$ applies, as does the somewhat abbreviated version

$$V^u_{,s} + \Gamma^u_{ap} V^p \frac{dx^a}{ds} = 0,$$

as well as the simpler $\delta V^u = -\Gamma^u_{ap} V^p dx^a$. *Again the surface that encloses* P_1 *and* P_2 *has curvature described by* Γ^u_{ap}. In effect this scheme melds two processes, a very local one that defines parallel transport by invariant components—the first equation under discussion—and a more global process that allows curvature to vary the components—the second equation under discussion. In the setup on page 203, each increment of motion represents an instance of parallel transport because each increment is locally flat (as it appears in each freely falling elevator), but the overall trip engenders a change in direction because of the presence of more global curvature. The paradox between the two equations is resolved. (Incidentally, the equation $\delta V^u = 0$ is a special case of $\delta V^u = -\Gamma^u_{ap} V^p dx^a$ for parallel transport but in the absence of curvature.)

◆

The term "affine" is used in this context. We say that a Riemannian space is an "affine space" in that its constituent points share mathematical "affinity," but the opposite may be easier to interpret: A space is not affine if vectors transported between its adjacent points fail to stay parallel, and in such a space the rules of vector/tensor transformation do not hold. Since the Γ's describe the consequence of parallel transport, they are also called coefficients of affine connections, or affine connections, or affine relationships. (Menzel, p. 119, and Weyl, p. 112, cover this point.)

We can safely skip the term "affine" and just ask what metrical connections connect. They connect points in space-time. More specifically they connect the components of vectors transported in parallel fashion between infinitesimally close points. Why are the metrical connections metrical? Because the geodesics between points have metrical properties, namely intrinsic curvature. Why is parallel transport implicated? Two reasons: Because the variations in the components of parallel-transported vectors reveal the metrical changes, and because parallel transport generates geodesics.

✦

The terms δ and Γ merit further comment. We encountered δ in

$$\delta \int_A^B ds = 0$$

applied to motion along a geodesic: We pointed out that the δ stands for the deviation or variation of the geodesic from straightest-ness, and if we add together (integrate) all the ds's from A to B, the deviation is zero. In our current setting A and B are P_1 and P_2, and transporting an abstract vector on a geodesic is akin to moving a physical object through a gravitational field.

Later we will use equations of motion in an astronomic setting, but we can make a prediction at this point: Since the Γ's describe the consequence of transporting the vector V^p from P_1 to P_2 along dx^a, and since the components of that vector change by δV^u if space-time between P_1 and P_2 is curved, then the Γ's must play a role in the equations of motion. In particular the Γ's should accommodate for the deviation or variation of motion away from straightness and uniformity. At the same time we already emphasized that the variability of the g's in a region of space-time propagates the local gravitational field. Therefore we expect that the g's enter into the equations of motion by way of the Γ's.

✦

We now turn to the actual equations linking the Γ's with the g's. What criteria must they meet? First, since all the data needed to describe the shape of space-time resides in the metrical tensor, we expect the Γ's to rely *only on the g's*. (We expect the same from the curvature tensor, R_{hikl}.) Second, such an equation ought to entail *differentiation of g_{ik}*, and we know that differentiation of tensors extends the rank by one. Therefore the Γ's should have a rank of three and should have three indices. However even though metrical connections behave like coefficients or components, they are not necessarily components of tensors. We may say that the Γ's have a "pseudo-rank" of three. (Authors vary widely on how they organize and label the pertinent derivations, making cross-comparisons difficult. For example compare Kenyon [pp. 61-62 and 194], Menzel [p. 120], Pauli [p. 38], and Wald [p. 35].)

A differential of g_{ik} along coordinate x in the w-direction might appear simply as

$$\frac{dg_{ik}}{dx^w},$$

but this represents full rather than partial derivatives. For our present purpose we call the direction j, and we need to consider the *partial differential of* g_{ik} per change in x^j:

$$\frac{\partial g_{ik}}{\partial x^j}$$

We recall that the equation for the covariant derivative of a tensor has one Γ-containing term per rank. *The same holds for the metrical tensor;* by virtue of its rank we expect two Γ-terms. Accordingly, if we recall the equation for covariant differentiation,

$$\frac{\partial V^u}{\partial x^w} + \Gamma^u_{wx} V^x = V^u_{\ w},$$

and apply it to g_{ik}, the covariant derivative of g_{ik} becomes

$$\frac{\partial g_{ik}}{\partial x^j} - \Gamma^b_{ij} g_{bk} - \Gamma^b_{kj} g_{ib}.$$

The Γ-terms are negative because g_{ik} is in its co-variant form.

In a freely falling frame of reference, where the g's do not vary, the ordinary derivative of the metrical tensor vanishes, so that

$$\frac{\partial g_{ik}}{\partial x^j} = 0.$$

In such a frame the covariant derivative of the metrical tensor must also vanish, so that

$$\frac{\partial g_{ik}}{\partial x^j} - \Gamma^b_{ij} g_{bk} - \Gamma^b_{kj} g_{ib} = 0$$

or

$$\frac{\partial g_{ik}}{\partial x^j} = \Gamma^b_{ij} g_{bk} + \Gamma^b_{kj} g_{ib}.$$

Now we draw upon the relationship between co-variant and mixed forms. It is governed by the rules used to raise and lower indices, $A^a = g^{ab}A_b$ *and* $B_a = g_{ab}B^b$. These apply to the Γ's with three indices as follows:

$$\Gamma_{ikj} = g_{ib}\Gamma^b_{kj}.$$

Upon lowering the b index (i.e. upon converting the Γ's into their co-variant form), and with rearrangement of terms, we reach an equation which, for reasons to be clear shortly, we call equation "A,"

$$\frac{\partial g_{ik}}{\partial x^j} = \Gamma_{ikj} + \Gamma_{kij}.$$

With three indices, the Γ's stand for 64 coefficients (4^3), but because of symmetry of the g's, as in

$$g_{ki} = g_{ik},$$

it does not matter which index, i or k, is considered first. Since Γ's with ik and ki are equivalent, the number of independent possibilities is less than 64, and the Γ's (though not tensors) show partial symmetry—only with respect to i and k—as shown by

$$\Gamma_{ikj} = \Gamma_{kij}.$$

(This symmetry of the Γ's is also a necessary condition for them to vanish in flat space-time.)

The next several equations will be more transparent if we arbitrarily assign numerals to the indices. Assume $i = 1$, $k = 2$ and $j = 3$. We rewrite the equation which we called equation "A," and in this format let us call it equation "B:"

$$\frac{\partial g_{12}}{\partial x^3} = \Gamma_{123} + \Gamma_{213}.$$

"B" tells us that the partial derivative of metrical tensor with respect to direction 3 is described by the metrical connections considering 1 followed by 2 as well as 2 followed by 1.

However we should also differentiate with respect to 1 and 2, leading to two additional possibilities which we call "C"

$$\frac{\partial g_{31}}{\partial x^2} = \Gamma_{312} + \Gamma_{132}$$

and "D"

$$\frac{\partial g_{23}}{\partial x^1} = \Gamma_{231} + \Gamma_{321}.$$

In effect "B," "C," and "D" explicitly consider the six possible sequences of the Γ's indices as each index is the one against which differentiation occurs. Ordinary algebra allows two of these equations to be added together and the third subtracted, producing a "composite" equation,

$$\Gamma_{123} + \Gamma_{213} + \Gamma_{312} + \Gamma_{132} - \Gamma_{231} - \Gamma_{321} = \frac{\partial g_{12}}{\partial x^3} + \frac{\partial g_{31}}{\partial x^2} - \frac{\partial g_{23}}{\partial x^1}.$$

The right side of this composite equation consists of the left side of "B" plus the left side of "C" minus the left side of "D."

The left side of the composite equation is analogous: The first two terms are the right side of "B," the next two terms are the right side of "C," and the last two terms, both negative, are the right side of "D."

Now more algebra: Because of partial symmetry (two adjacent indices of each Γ can be reversed), the second and fifth terms on the left cancel each other; Γ_{213} becomes Γ_{231} which cancels with $-\Gamma_{231}$. Similarly the third term on the left cancels out the last term because Γ_{312} becomes Γ_{321} which cancels with $-\Gamma_{321}$. These steps leave

$$\Gamma_{123} + \Gamma_{132} = 2\Gamma_{123} = \frac{\partial g_{12}}{\partial x^3} + \frac{\partial g_{31}}{\partial x^2} - \frac{\partial g_{23}}{\partial x^1}$$

which, using letter-indices again, is commonly written as

$$\Gamma_{ikj} = \frac{1}{2}\left(\frac{\partial g_{ik}}{\partial x^j} + \frac{\partial g_{ji}}{\partial x^k} - \frac{\partial g_{kj}}{\partial x^i} \right).$$

Based on the rule for lowering/raising indices, we can write the same equation in mixed form, compatible with how the Γ's are usually presented:

$$\Gamma_{ik}^b = \frac{1}{2}g^{bj}\left(\frac{\partial g_{ik}}{\partial x^j} + \frac{\partial g_{ji}}{\partial x^k} - \frac{\partial g_{kj}}{\partial x^i} \right).$$

Let us call these two the "Γ-equations." Clearly the Γ's *depend on the g's and their partial derivatives.*

We may also use commas to express more succinctly the partial differential change in the components of g_{ik} in the *j*-direction:

$$\frac{\partial g_{ik}}{\partial x^{j}} = g_{ik,j}.$$

The comma in *ik,j* means ordinary differentiation of *ik* with respect to *j*. Accordingly we can write

$$\Gamma_{ikj} = \frac{1}{2}\left(g_{ik,j} + g_{ji,k} - g_{kj,i}\right),$$

or

$$2\Gamma_{ikj} = g_{ik,j} + g_{ji,k} - g_{kj,i}.$$

This rendition shows succinctly how *the metrical connections stem from ordinary, though partial, derivatives of the components of the metrical tensor.* The relationship between the metrical connections and metrical tensor is called the fundamental theorem of Riemannian geometry. Verbally, *the metrical connections equal the rate of change of the metrical tensor.* We shall find the above equations quite useful.

◆

The metrical connections are abstract quantities; there is no direct way to view them. Nevertheless the Γ-equations entail the differentiation of a tensor, which as expected acquires one index—gains one rank—in the process. We can ask where this new index comes from. Since rank, even "pseudo-rank," is a matter of how many vectors are involved, the implication is that another vector has been introduced. In the case of the Γ's, by what right is this vector added? We touched upon this issue on page 281 (where differentiation added a vectorial arrow), but an analogy gives a more complete answer. Let us imagine that observers can "watch" parallel transport of a vector between two adjacent points, P_1 and P_2, in a "dip" in space-time. (This is not wholly theoretical, as real observers can watch a gyroscope moved between two locations.) Since the three indices of the Γ's admit 64 combinations, we can even imagine 64 observers, each of whom has a different vantage of this event.

Shifting anything from one point to another, even transporting a vector from P_1 and P_2, is in itself *a vectorial event*, one with direction and extent. Therefore from the observers' vantages, a *third* set of vector components has appeared, marking the shift from P_1 to P_2. In the above equation *j* may stand for the direction of this shift. By allowing each index to run 1 through 4, an observer from his vantage can assess all these components along four coordinate axes, one per dimension. Of course the results of such assessments are tied to the *shape of space-time*, which is quantified by the *g*'s.

For instance since the Γ's depend on derivatives of the g's, flatness is signaled by the vanishing of the Γ's. (This scenario becomes too cluttered if we consider that curved space-time has two local radii, so we ignore this issue. On the other hand by virtue of symmetry, many of the 64 observers would have the same views, so that fewer would be needed, but we defer this issue.)

Thus we can picture the metrical connections as expressing how the g's behave *between two points in space-time* from the vantages of various observers. This can be expressed in an equation with the grouping of the terms emphasized:

$$2\Gamma_{ikj} = \frac{\partial g_{ik}}{\partial x^j} + \left(\frac{\partial g_{ji}}{\partial x^k} - \frac{\partial g_{kj}}{\partial x^i} \right)$$

Each observer "sees" that the rate of change in the shape of space-time depends partly on the change toward (in the direction of) j and partly on the difference between the change toward k vs. the change toward i. Clearly the Γ's are built not only from the curvatures at P_1 and P_2, but also from the difference from P_1 to P_2, which is reflected in the added index. Moreover we *let each vector play each role*, which appears as a rotation of the indices; each term allows one of the three vectors to delineate the P_1-to-P_2 shift. This notion brings us to why the index added during differentiation of vectors and tensors is a subscript and not a superscript: The P_1-to-P_2 shift itself behaves more like a gradient, which transforms in co-variant manner.

Since gradients in space-time are linked to gravitational (accelerated) motion, there should be a link between at least some aspect of the metrical connections and such motion. In fact if we restrict ourselves to weak gravitational fields, *one component of the metrical connections, namely the 44-component, is directly proportional to gravitational acceleration*. This is expressed as an equation of motion (in which c^2 is a constant of proportionality but in which no particular direction of motion is specified):

$$c^2\Gamma_{44} = -\frac{d^2x}{dt^2} = -a.$$

The two right-side terms represent accelerated motion, and we will use these in other equations of motion. For now though we note that the rate of change of the metrical tensor, as given in one component of the metrical connections, is proportional to the acceleration in Newton's equations. We already declared that the Γ's circumvent the restriction of the special relativity factor to uniform motion, and now we see a clear link between the Γ's and accelerated motion; *the metrical connections tell us to what extent falling or orbiting objects are deflected from uniform motion.* Conversely uniform velocity is associated with zero Γ's, and indeed Γ's $= 0$ in all rectilinear (Euclidean/Cartesian) coordinates.

Let us recall that the 44-component of the metrical tensor is linked to the gravitational potential, which reflects the strength of a point of a gravitational field. We saw this earlier in

$$g_{44} = 1 + \frac{2\Phi}{c^2}.$$

Now we see that the analogous component of the metrical connections is linked to acceleration, which reflects the strength of Newton's force of gravity. This is clearer if we compare the above equation with $c^2\Gamma_{44} = -a$ or better yet (after some rearrangement) with

$$\Gamma_{44} = -\frac{a}{c^2}.$$

Since these equations connect Γ_{44} to acceleration, why aren't they our preferred equations of motion? Because they do not fully encompass the relativistic nature of gravitation. In terms of our analogy for the metrical connections, it is always possible to find a vantage from which curvature is obscured. For example if an observer could "look" straight down at the surface of space-time, its curvature might not be evident. We will reconsider the mathematical aspects of this issue later, but it shows that observers may disagree on when the Γ's indicate flatness of space-time, indicating that the Γ's are not tensors. That is to say, the Γ's vanish in some coordinates but not in others, and they violate the rules of transformation of tensors. (If we know the components of a Γ in one system, the transformation equations will not necessarily give us its components in another system.) Therefore eventually we must bypass the metrical connections.

Nevertheless the significance of the Γ's is underscored by considering how general relativity treats another of Newton's laws, namely

$$F = ma.$$

First we reframe $F = ma$ into a differential form. Since acceleration is the derivative of velocity (v),

$$F = \frac{d(mv)}{dt}.$$

However this equation neglects relative motion and it does not reflect four dimensions. To answer the first point we recall that the special relativity factor can adjust $F = ma$ to read

$$F = \frac{ma}{\sqrt{1-v^2/c^2}}.$$

To answer the second point let us call F^i the force in four dimensions; it becomes a four-vector, with i = 1 through 4. For brevity let us call t_{srf} "time adjusted by the special relativity factor." Now $F = ma$ reads

$$F^i = \frac{d(mv)}{dt_{srf}}.$$

Still this equation lacks general covariance because it does not accommodate space-time curvature and it is not a tensor equation. Let us therefore invoke metrical connections as an "addition" to a Newtonian law, so that "*F = ma*" adapted to general relativity and to general covariance appears as

$$F^i = \frac{d(mv)}{dt_{srf}} + \Gamma^i_{jk} \frac{dx^j}{dt_{srf}} V^k.$$

The term containing the Γ's admits curvature—*it recognizes curved systems of coordinates.* (If this equation is applied in flat space-time, the Γ-term drops out.) We note the effect of the Γ's: By itself, neither the first term nor the second term of the right side of this equation is a pure tensor, but the combination exhibits general covariance.

We see that Γ's play several diverse roles: In terms of gravitational physics, they signify gravitational acceleration, suggesting that they should appear in relativistic equations of motion. In terms of the gravitational field, the Γ's represent the gradient across a gravitational field—which makes sense since they describe the rate of change of g_{ik}. In terms of the Lorentz contraction, the Γ's reflect the change in angle R in rotational transformations. In terms of coordinate axes, they allow for curvature. In terms of tensor calculus, they allow covariant differentiation of tensors—and this list does not exhaust their significance. Clearly the leap from $\frac{\partial V^u}{\partial x^w} = V^u{}_w$ to

$\frac{\partial V^u}{\partial x^w} + \Gamma^u_{wx} V^x = V^u{}_w$ was a milestone for the mathematics of general relativity. We can sense Einstein's satisfaction when he wrote, "By this means we are able...to formulate generally covariant differential equations." (See "The Principle of Relativity," p. 133.) To anticipate why, let us assume that we have found *a valid law of physics which holds in a frame of reference with only flat space-time, which means that this law manifests special relativity.* Let us further express this law using vectors and tensors. (Plausible candidates for such a law are $R^u{}_{ikl} = 0$ and $R_{hikl} = 0$, but we will probe further.) Provided that we link points in space-time not by ordinary differentiation but by *covariant* differentiation, *this law will hold in any frame of reference, including in frames essential in general relativity.* Incidentally, the concept that a law can be written to hold in all frames is sometimes called the "strong equivalence principle" since it applies to all kinds of forces in a gravitational field. According to this terminology the principle of equivalence we described in Chapter 6 is called "weak" since it is more narrow, referring only to gravitation. (For example see Schutz, pp. 122 and 184, or Kenyon, pp. 10-11.)

In any case the power of the metrical connections suggests another provisional field equation,

$$\Gamma_{ik} = T_{ik}??$$

but it is one which can be promptly rejected, since the Γ's are not tensors. Shortly another objection will emerge, but it is interesting that Einstein once contemplated a similar formulation, though he used different reasoning and other symbols:

$$\Gamma_{uv} = K\Theta_{uv}??$$

(Θ indirectly represents matter, and K is a constant.) However this equation violates the conservation laws; more on this later.

✦

So much for our side-trip into differentiation of tensors and related topics. We cannot forget our goal, the curvature tensor R_{hikl}. To this end we are employing two mathematical tools, the metrical connections and parallel transport. The metrical connections will turn out to be an intermediary step with a "pseudo-rank" of three between the rank-two metrical tensor we started with and the rank-four curvature tensor we seek. We also reiterate that *the metrical connections describe the rate at which the components of the metrical tensor change during parallel transport between points in space-time.* The metrical connections therefore represent the *first derivative* of the metrical tensor; they are functions of—and depend only—on g_{ik} and its partial derivatives.

However since the metrical connections are themselves not tensors, they do not vanish in all frames of reference. For example, as we mentioned, observers disagree when the Γ's = 0. In particular, since observers using the Γ's disagree when space-time between P_1 and P_2 is flat, we have another reason Γ's = T_{ik}?? is inadequate.

In terms of parallel transport in our city-block analogy, proceeding only from P_1 to P_2 "along one street" fails to provide enough information. For instance P_1 and P_2 might be points on a hill, while the surface of the hill is "paved" with a "mosaic" of points so that each point faces in a direction depending on the local terrain. Two adjacent points (one on each side of a crest) could show the curvature of the gap between them by facing in different directions, while two other points (on the same side of a crest) might show no curvature by facing the same way. Therefore observing parallel transport between just P_1 and P_2 is not decisive, and the question remains how intrinsic curvature among points can be conclusively revealed.

The solution is in consummating our plan to transport a vector around a closed loop. Again we may imagine driving around a block on a hilly surface which represents a region of four-dimensional space-time. Here "region" has a precise meaning: It is the space-time covered by our loop. Thus we distinguish a local "point" from a region, and for that matter from a wider "global" area. However we reiterate that there is nothing imperative about a four-sided loop. The mathematics is simplified, but other shapes are acceptable, including round (see Lawden, pp. 102-105; and Einstein in The Meaning of Relativity, pp. 146-147), and many different derivations are possible. We adapt Kenyon's compact method (pp.73-74).

Let us recall (page 289) the pattern for the variation in the components of a vector, δV, during parallel transport between two points:

$$\delta V^u = -\Gamma^u_{ap} V^p dx^a$$

However we can take advantage of the notion of the covariant derivative; after all this equation has that feature. A covariant derivative can be expressed in equations with a parameter, so that

$$\frac{dV^u}{ds} = -\Gamma^u_{ap} V^p \frac{dx^a}{ds},$$

which grants us the more useful differential equation,

$$dV^u = -\Gamma^u_{ap} V^p dx^a.$$

Now a contra-variant vector V is transported in a loop "around the block." We can break the loop down into four parts (four sides), and we note the new components in each part. We will change some indices to more convenient ones, and we will show the terms more explicitly. Diagram T (page 288) again helps, but please ignore the R's in this diagram for now. We are concerned with parallel transport P_1 to P_2 to P_3 to P_4 to P_1.

The coordinate value at P_1 is x, our starting point. We transport vector V for 1 block along "street P_1-to-P_2" to the next corner, P_2, where the coordinate value is $x + dx$ so that $dx = d(P_1 \text{ to } P_2)$. This causes *a change in* V*'s components* which conforms to the above equation but which here equals

$$- \Gamma^u{}_{ip} (x) \quad V^p (x) \quad d(P_1 \text{ to } P_2)^i.$$

Let us dissect this quantity: The term $\Gamma^u{}_{ip} (x)$ provides the rate of change of the surface (i.e. the derivative of the g's) in the u-direction at location x. The term $V^p (x)$ represents the vector transported from x; its components are given by the p. The direction of the "street" (i.e. the direction of the space-time geodesic) to which the vector stays parallel is i. The minus sign means that the vector is contra-variant, which it is because it represents a velocity.

The situation along the second "street P_2-to-P_3" is comparable but the vector is no longer at x; it's at $x + d(P_1 \text{ to } P_2)$. Transport from P_2 to P_3 causes a change in the components equal to

$$- \Gamma^u_{ip} (x + d(P_1 \text{ to } P_2)) \ V^p (x + d(P_1 \text{ to } P_2)) \ d(P_2 \text{ to } P_3)^i.$$

The third "street," P_3-to-P_4, is similar to P_1-to-P_2 but in the reverse direction, as shown by the $+$ sign. Transport from P_3 to P_4 causes a change in the components equal to

$$+ \Gamma^u_{ip} (x + d(P_2 \text{ to } P_3)) \ V^p (x + d(P_2 \text{ to } P_3)) \ d(P_1 \text{ to } P_2)^i.$$

The fourth "street," P_4-to-P_1, is similar to P_2-to-P_3 but in the reverse direction, which is why the $+$ sign. Transport from P_4 back to P_1 causes a change in the components equal to

$$+ \Gamma^u_{ip} (x) \ V^p (x) \ d(P_2 \text{ to } P_3)^i.$$

Incidentally, the similarity of opposite sides of our "block" is related to the aforementioned partial symmetry of the Γ's. It means roughly that we can "start at any corner" but that proceeding one way, for example P_1 to P_2 to P_3 to P_4 to P_1 is not identical to proceeding the other way, e.g. P_1 to P_4 to P_3 to P_2 to P_1. We will say more about symmetry shortly.

If the surface on which our four "streets" are located is "hilly," we expect the components of V^p to have undergone a net change upon circumscribing this "block" or loop; here we call this net change a total derivative, dV^u, though earlier we called it a variation, δV. We treat this "block" as a region consisting of two pairs of similar streets, one pair running "east-west" and one pair running "north-south," so that we still need not specify P_3-to-P_4 nor P_4-to-P_1, but we need indices, k and l, to show the two sides of our "block." Algebraically these directions have opposite signs. The gap from P_1 to P_2 becomes an algebraic difference $(P_2 - P_1)$, and likewise the gap from P_2 to P_3 becomes $(P_3 - P_2)$. Altogether

$$dV^u = \frac{\partial (\Gamma^u_{ip} V^p)}{\partial x^k} \ d(P_3 - P_2)^k \ d(P_2 - P_1)^i - \frac{\partial (\Gamma^u_{ip} V^p)}{\partial x^l} \ d(P_2 - P_1)^l \ d(P_3 - P_2)^i.$$

Two observations about this complicated equation: First, it gives us a total derivative, dV^u, by the summation of the partial derivatives of the Γ's with respect to the x's. Second, the right side contains

$$(P_3 - P_2) \quad and \quad (P_2 - P_1),$$

which multiplied together is the area of our block. This means that if our block has a larger area, the variation in V^p is larger, so that the size of block we select can distort our results. We circumvent this by considering only an infinitesimally small block of points, $d(P_3 - P_2)$ by $d(P_2 - P_1)$.

Since the terms

$$d(P_3 - P_2)^k \, d(P_2 - P_1)^i \quad and \quad d(P_2 - P_1)^l \, d(P_3 - P_2)^i$$

are equivalent, they can be fused. Moreover, since

$$\frac{\partial(\Gamma^u_{ip} V^p)}{\partial x^k} \quad and \quad \frac{\partial(\Gamma^u_{ip} V^p)}{\partial x^l}$$

indicate differentiation, each of these terms can be replaced by *a new term, R, with one added rank.* This R refers to the Riemann tensor, not to angle R. The entire equation, arranged in more lucid order, becomes

$$dV^u = R^u_{ikl} \, V^i \, d(P_3 - P_2)^k \, d(P_2 - P_1)^l.$$

In this equation we note the presence of R^u_{ikl}, which (in mixed form) is the Riemann-Christoffel *curvature tensor.* The equation shows that *the net (total) change in the components given in V^u depends on the components of the curvature tensor given in R^u_{ikl}.*

We can easily justify each part of this equation. First,

$$dV^u$$

describes the overall change in the vector upon its loop-like parallel transport in the region of space-time at $P_1 - P_2 - P_3 - P_4$. The terms

$$d(P_2 - P_1)^l \, d(P_3 - P_2)^k$$

describe the site of transport. The term

$$V^i$$

describes the vector as it is moved. Finally the curvature tensor itself,

$$R^u{}_{ikl},$$

describes how the "surface" of this region affects a parallel-transported vector. We call the components of this tensor "the R's" (again not to be confused with angle R). We can think of the R's as the adjectives which tell us how a vector changes upon parallel transport around a loop, but, as will be more useful to us, these adjectives also tell us about *the shape of space-time* in this region.

We note that the right side of the above equation for dV^u,

$$R^u{}_{ikl} \quad V^i \quad d(P_3 - P_2)^k \quad d(P_2 - P_1)^l,$$

conforms to the pattern in

$$\Gamma^u_{ap} \quad V^p \quad dx^a,$$

though now we find the curvature tensor $R^u{}_{ikl}$ instead of the metrical connections Γ^u_{ap}. In both cases we parallel-transported a vector, but in order to generate the curvature tensor, we moved the vector around a figure with two sides (two d's with indices k and l) rather than across one gap (dx^a).

Though there is no unanimity on the sequence of the indices of the curvature tensor (because of the many ways to derive it) we should also keep in mind that each index represents a vector which has an arrow-like direction. We can illustrate this using diagram S: $R^u{}_{ikl}$ tells us how much the vector which pointed from D toward A (in the i-direction) drifts towards E (in the u-direction) during parallel-transport around a space-time "loop" whose sides run in the BC (k) and CD (l) directions (i.e. around a "block" in the BC-by-CD [k-l] plane of "streets"). We summarize that "i," when parallel-transported around "k-l," drifts toward "u" if space-time at "k-l" is curved. Each time we probe another direction and another dimension (there are 256 combinations of i, k, l, and u) $R^u{}_{ikl}$ gives us a number, and 20 such numbers suffice to describe the regional shape of space-time. If the region is flat, all 20 numbers are zero, because under these conditions a parallel-transported vector does not "drift" in diagram S; E coincides with A, and in equation form, $R^u{}_{ikl} = 0$.

If we abbreviate the partial differentials of the Γ's by use of commas, the R's are such that

$$R^u_{ikl} = \Gamma^u_{il,k} - \Gamma^u_{ik,l} + \Gamma^u_{jk}\Gamma^j_{il} - \Gamma^h_{jl}\Gamma^j_{ik}.$$

In this format we see that *the curvature tensor is defined by differentiation of the metrical connections and that the process is covariant.* Specifically, the first two terms in the right side are first derivatives of the metrical connections. (The reason for the minus sign is that we proceed forward along two sides and back along the other two sides of our loop.) The last two terms contain only metrical connections (not derivatives) whose role is to ensure that differentiation is covariant.

In effect the first two terms meet our goal of obtaining a derivative of g_{ik}, and of course the g's embody the shape of local space-time. The last two terms guarantee that the entire equation holds in any frame of reference. We note that parallel transport along "one street," as reflected in the Γ's, did not ensure general covariance, but parallel transport in the region "around the block," as reflected in $R^u{}_{ikl}$, does so. We can thus aver that covariant differentiation, with help from parallel transport, allows $R^u{}_{ikl}$ to be a tensor. Of course this tensor passes the transformation test.

In fact, in a freely falling frame of reference with locally flat space-time, the metrical connections vanish, leaving only their derivatives (note the commas):

$$R^u_{ikl} = \Gamma^u_{il,k} - \Gamma^u_{ik,l}.$$

The link between the curvature tensor and the metrical connections is clear: The former is based on the first derivatives of the latter. Furthermore, on a regionally flat surface the components of a vector parallel-transported in a loop do not change ($dV^u = 0$). The metrical connections as well as their derivatives vanish and hence all components of $R^u{}_{ikl}$ become zero. We therefore know that space-time is flat if and only if $R^u{}_{ikl} = 0$; more on this later, but we note the efficiency of the curvature tensor in holding all these data.

Keeping in mind that the metrical connections represent first derivatives, a further observation is that during parallel transport around the loop, the metrical connections themselves, the Γ's, *undergo a differential change*. This means that the summation of the changes after one complete loop is a change of a change. It is a "second-order" change, or a "derivative of a derivative," or a *second derivative*. In other words the curvature of the surface in the region around point P_1 induces a second-order change in a vector upon completing its loop on that surface. The derivations also show that the curvature tensor is based upon the g's and their first and second *partial* derivatives. Since we applied covariant differentiation, we may sum up that in the curvature tensor we have *covariant second partial derivatives of the metrical tensor*.

When we lower the index u by use of

$$R_{hikl} = g_{hu}R^u{}_{ikl},$$

we obtain an explicit equation in co-variant form (still for free-fall),

$$R_{hikl} = \frac{1}{2}\left(\frac{\partial^2 g_{hl}}{\partial x^i \partial x^k} - \frac{\partial^2 g_{il}}{\partial x^h \partial x^k} + \frac{\partial^2 g_{ik}}{\partial x^h \partial x^l} - \frac{\partial^2 g_{hk}}{\partial x^i \partial x^l} \right),$$

where each term in the right side is a second partial derivative.

For convenience we will treat R^h_{ikl} and R_{hikl} as interchangeable, though we can tell from the above equations that one of the h's should be different.

Thus, as we had proposed,

$$R_{hikl} = a \ second \ derivative \ of \ g_{ik}.$$

Using a semi-colon to indicate covariant differentiation, the overall strategy can be recapitulated by a "summary-equation" for the curvature tensor,

$$2R_{hikl} = g_{hl;ik} - g_{il;hk} + g_{ik;hl} - g_{hk;il}.$$

We stipulated a freely falling frame of reference. The more general form of this equation, showing explicitly that R_{hikl} is the *covariant* curvature tensor, includes the terms with $\Gamma \Gamma$'s we saw earlier, but we need not be that meticulous. Instead let us compare the above "summary-equation" with the equation

$$2\Gamma_{kij} = g_{ij,k} + g_{kj,i} - g_{ik,j}.$$

We find that the R's are defined by the second covariant derivative of the metrical tensor, while the Γ's are defined by the first ordinary derivative of the metrical tensor. We see that the curvature tensor is a function of the metrical connections and that the metrical connections are functions of the metrical tensor.

Moreover the left side of the equation for R_{hikl} is the curvature tensor while every term of the right side contains the metrical tensor. *The metrical connections have been canceled out of the "summary equation,"* so that their non-tensor nature does not matter; though Γ's are not tensors, their derivatives are. We can think of the metrical connections as essential but finally inconspicuous tools in the construction of curvature tensor.

We also note that the particular vector (V) selected for parallel transport was eliminated from the equation. We can say that parallel transport is another tool for the construction of the curvature tensor which does not appear in the final product. Analogically, our pilots on page 205 can move a gyroscope around the Earth's surface to determine its curvature, but the gyroscopic data do not appear in the final result. But then why did we invoke parallel transport in the process that led up to that final product? Because parallel transport keeps us on geodesics, and geodesics are essential features of motion in space-time. In other words if the curvature tensor is to quantify the curvature of space-time for general relativity, the structure of the curvature tensor must admit geodesics, which means its structure must admit parallel transport.

This notion underscores two additional key features of the curvature tensor applicable to relativity: First, R_{hikl} refers to, and depends only on, the intrinsic property of space-time. In other words the curvature tensor is built *only* from the metrical tensor and its first and second derivatives. For instance in globally flat space-time, the *g*'s have no derivatives because they do not vary, so that all components of the curvature tensor must vanish. Second, since the *R*'s hinge only on the *g*'s and since $g_{ik} = g_{ki}$ (the metrical tensor is symmetrical), the curvature tensor is the *only* curvature tensor that can be built from the metrical tensor and its first and second derivatives. Space-time owns only one curvature tensor.

Incidentally, in a freely falling frame of reference, the components of the metrical tensor are constant and its covariant derivative vanishes. This is called Ricci's theorem:

$$g_{ik;l} = 0.$$

(We tacitly used Ricci's theorem on page 294. During free fall the Γ's also vanish, but again this is not apparent.) In his quest for the field equations, Einstein was temporarily perplexed by Ricci's theorem because he feared that it was impossible to reconcile Ricci's theorem *and* Poisson's equation *and* a link of Φ with the *g*'s, *and* general covariance all at once. In brief, how can $g_{ik;l}$ at some point stand for $\nabla^2\Phi$ while $g_{ik;l}$ is zero? (See Pais, pp. 221-222 and Møller, pp. 277-278, 282.) The resolution depended on using R_{hikl} and a related tensor we shall cover later, R_{ik}.

◆

To appreciate the curvature tensor we recall that the *g*'s depend on two factors: They depend of course on the shape of the surface in question—basically on compliance with pythagoreanism at one point—but they also depend on the system of coordinates we select. Therefore we cannot determine by looking at the 10 *g*'s for any one point whether a surface is flat or curved. Or, in terms of *ds*, we cannot just look at one interval (one *ds*) and certify whether or not the underlying surface has intrinsic curvature. Yes, the Γ's do discern a variation in the *g*'s, but we know that not all observers agree when the Γ's vanish; the Γ's bar general covariance. In effect, at one point the *g*'s are inconclusive, and at two points the Γ's are inconclusive. Therefore we must look "around" at a loop of points and call upon the 20 *R*'s of the curvature tensor.

As we showed already, how the *g*'s *vary* does not depend on the system of coordinates. Now we see that the curvature tensor is based on the *variation* (in the form of a second derivative) of the *g*'s in a region of points, so that the components of the curvature tensor are not altered by the selection of—or transformation between—systems of coordinates. In other words *only a change in intrinsic curvature can change the R's, whereas a change in coordinate systems alone can change the g's.* No system of coordinates can be found which "deceives" the components of the curvature tensor. The *R*'s are conclusive.

For instance it is possible to select a system of coordinates on a curved surfaces so that $g_{11} = g_{22} = g_{33} = g_{44} = 1$ at one point. However it is impossible for the surface to be curved *and* for these *g*'s to equal 1 at *several* points set in a loop. Likewise it is possible to find a system that "deceives" the Γ's. In stricter terminology, if intrinsic curvature exists in this region, the second derivatives of its metrical tensor cannot be made to vanish by any choice of coordinates—they cannot be transformed away—but this is not true of the metrical tensor or its first derivatives. Conversely, vanishing of the curvature tensor in any system of coordinates invariably indicates flatness. This attribute of the curvature tensor (or any tensor) is important enough to repeat and reword: If one system of coordinates shows that $R_{hikl} = 0$, *any other system of coordinates shows the same.* In short, only R_{hikl} is sufficiently non-local and covariant. (Likewise for $R^u_{\ ikl}$.)

These concepts are more meaningful if we select a location in the universe and pose the question, "is there gravity here?" To reply, Newton would have simply looked for gravitational force or weight, but in effect we look for intrinsic curvature of space-time. We can determine the components of the metrical tensor at the point in question, but the problem is that at least one system of coordinates exists in which these g's suggest *no* curvature, even if space-time is indented. In effect the g_{ik} is too local to detect the variations which characterize a gravitational field. A logical alternative is to determine the Γ's, which are less local, but we can still find at least one system in which curvature is undetectable.

Our recourse is to find the curvature tensor—the set of R's for a loop of adjacent points. This technique achieves our goal: If there is intrinsic curvature—if regional space-time is Riemannian rather than Euclidean—our R_{hikl} cannot vanish, indicating assuredly that gravity is present. In terms of our experiment on page 288, once we know enough about the variation of the g's in the region—once we know curvature reliably—we have all the information to ascertain the natural gravitational field. Of course we must still link the curvature of space-time to the amount of nearby matter, and this step will bring in the tensor T_{ik}, which we will detail shortly. Since flat space-time is associated with $T_{ik} = 0$ and with $R_{hikl} = 0$, T_{ik} and R_{hikl} may behave alike in the absence of gravitation. This is one reason we sought R_{hikl} in the first place, though we will identify the exact relationship between T_{ik} and R_{hikl} later.

Let us summarize how the metrical tensor, the metrical connections, and the curvature tensor behave, depending on the answer to our question, "is there gravity here?" (Kenyon presents a similar schema as a flow-chart; p. 70.)

In a frame of reference with no nearby matter, such as in "outer space," space-time is locally and distantly flat, the g's indicate flatness, the g's are constants everywhere, the metrical connections vanish, the curvature tensor vanishes. This is evident in the equation (page 305) which shows how the curvature tensor is derived: The Γ's are zero because the g's are constants, so that every right-side term is zero and $R_{hikl} = 0$. (Here T_{ik} also is zero.) There is no gravity here.

If we are falling freely in a gravitational field, such as in an elevator falling to Earth, space-time is locally flat, the g's indicate flatness, they are locally constant, and the metrical connections may or may not vanish. The curvature tensor cannot vanish because it discerns regional space-time curvature; that is to say, $R_{hikl} \neq 0$ because it embraces a loop of points. Gravity is not detectable locally, but local flatness disappears regionally. There is gravity nearby.

If we widen our frame of reference to encompass a gravitational field, the g's vary point to point, meaning that the metrical connections may not vanish. The curvature tensor again cannot vanish since it embraces any deviation of space-time from flatness, and we posit that $R_{hikl} = T_{ik}$?? Of course objects display accelerated motion along space and time coordinates. There is gravity globally.

Another possibility needs consideration, one which is surprisingly significant: If we are in an *artificially* but steadily accelerating frame, such as our powered elevator, we experience gravity, yet $R_{hikl} = 0$ because there are no second-order irregularities in this inertial field. This means that even if $T_{ik} = 0$ (no nearby matter), the curvature tensor may not vanish. In effect R_{hikl} is so sensitive that it detects space-time variations beyond those of gravity. We must address this issue in detail later.

Let us stand back and consider the words (six paragraphs back) "...all the information to ascertain the natural gravitational field," and let us see how the pieces of the puzzle are falling into place. The primal piece is how the scalar gravitational potential (Φ) is determined by the density (ρ) of nearby matter in Poisson's equation

$$\nabla^2 \Phi = 4\pi K \rho.$$

This Lorentz-invariant equation is doubly important because of the link between gravitational acceleration and the gravitational potential (specifying the j-direction),

$$\frac{d^2 x^j}{dt^2} = \frac{\partial \Phi}{\partial x^j},$$

and because the gravitational potential depends on one component, the 44-component, of the metrical tensor:

$$g_{44} = 1 + \frac{2\Phi}{c^2}.$$

However the derivatives of the metrical tensor form the metrical connections:

$$\Gamma_{ikj} = \frac{1}{2}\left(\frac{\partial g_{ij}}{\partial x^k} + \frac{\partial g_{kj}}{\partial x^i} - \frac{\partial g_{ik}}{\partial x^j} \right).$$

Regarding only the 44-component (and the *j*-direction) reduces this equation to

$$\Gamma_{44j} = -\frac{1}{2}\frac{\partial g_{44}}{\partial x^j}.$$

(The first and second terms in the parenthesis drop out.) Γ_{44j} is clearly a first derivative of g_{44}.

We know that the Γ's are proportional to the acceleration exhibited by falling objects. Thus, recalling $c^2\Gamma_{44} = -a = -\dfrac{d^2x}{dt^2}$ and including the constant of proportionality c^2, we can write the more exact equation (with the *j*-direction specified),

$$\frac{d^2x^j}{dt^2} = -c^2\,\Gamma_{44j}.$$

Nonetheless we know that the Γ's are inadequate and must be replaced, which can be done by further differentiation (with respect to *h*). Still regarding only the 44-components, this gives

$$R_{h4j4} = -\frac{1}{2}\frac{\partial^2 g_{44}}{\partial x^h\,\partial x^j}.$$

R_{h4j4} is clearly a second derivative of g_{44}, which means that certain components of the curvature tensor are proportional to the second derivatives of the gravitational potential, just as Poisson's equation suggests. Stated algebraically,

$$R_{h4j4} = -\frac{1}{c^2}\left(\frac{\partial^2\Phi}{\partial x^h\,\partial x^j} \right).$$

The two preceding equations imply that the full curvature tensor—not limited to only a few components—does allow us to "...ascertain the natural gravitational field." In other words the curvature tensor provides the necessary data to express gravitational behavior in terms of the gravitational potential along a geodesic. Then why do we not rely just on equations with the (scalar) gravitational potential? Because we prefer to work with tensor equations and with a second-order differential of the metrical tensor, namely our curvature tensor, one which we shall soon be able to link to another vital tensor, T_{ik}.

Meanwhile, since our derivation of the curvature tensor relied on parallel transport, we can question the role of transport of a vector "corner to corner around a city block." Does the vector "follow the streets" in the most direct manner? Why can it not weave, dawdle, take a detour, spin around, etc? More to the point, can we trust the metrical connections to connect our corners in the straightest possible most direct manner?

We already have a simple answer: Parallel transport keeps us on geodesics, and geodesics are the straightest and most direct lines. But we also have a more detailed and compelling answer, which we considered on page 233: The laws of Newton and Euclid suggest that the path between each pair of points (between each pair of "corners") ought to be straight, and, mathematically speaking, Newton's first law is a consequence of the principle of least action. We recall that the principle of least action requires maximum efficiency, in that the path of the vector is such that if completing the loop results in a change in the vector, it is the *least possible change*. Of course the least possible change, namely none at all, occurs on a flat (Euclidean) surface, and this zero-change means that the metrical connections and the curvature tensor vanish. This is the same as saying that the metrical connections vanish if a particular geodesic is an ordinary straight line, and it is the same as saying that there is a correlation between parallel transport satisfying the principle of least action and a geodesic staying straightest. Then if the surface is other than flat (if it's non-Euclidean), the curvature tensor yields the optimal measure of non-flatness.

◆

Differential calculus tells us that the derivative of a constant is zero (the change of a non-changing value is nil) and that the derivative of an evenly changing value is a constant. Likewise the derivative of a vector or tensor whose components are constant point-to-point is zero, and the derivative of a vector or tensor whose components change from point to point in an even manner is a constant. On this basis we can summarize that in physical nature, matter is such that...

(1) Nearby four-dimensional space-time does not remain flat. It has intrinsic curvature; otherwise its first derivative would be zero;

(2) Nearby space-time becomes progressively curved; otherwise its first derivative would be a constant and the derivative of the derivative, i.e. the second derivative, would be zero;

(3) Nearby space-time becomes progressively curved but that curvature is uneven, so that its first derivative is not constant but its second derivative is constant; and...

(4) The curvature of nearby space-time is not so uneven that there is a third derivative; i.e. a third derivative is zero. In other words we do not need third-order equations. (The possibility has been explored that a gravitational field may be characterized by the third derivative of the g's, in which case the mathematics of relativity would be even more complex. However it is not necessary to go

beyond the second because the contour of space-time is a second-order curve, just as suggested by the second-order nature of Poisson's equation.)

We must interject that the complexity of space-time curvature addressed by second-order differentiation can be separated from the need for two radii at one point (page 162). By analogy, a football can have two points each of which has two radii, but these radii can be the same for both points with no "second-order" deformation of the surface between them. In other words curved space-time is such that one point has two unequal radii *and* that adjacent points have varying sets of radii.

As long as we understand that we are referring to such a complex change, the 20 independent components of the curvature tensor will eventually answer our question quantitatively: *How much change in the shape of a region of space-time can be attributed to a given distribution of nearby matter?* In case the reader is struck by how complicated the answer has become, we point out that our equations are only descriptions. They represent our efforts to understand and analyze physical reality. For example just as nature is such that falling objects move faster and faster, nature is such that angle R is steeper and steeper. The result is very intricate mathematics, but the basic cause for the complexity is the way the universe behaves.

◆

We must appreciate that the components of the curvature tensor are *not* entirely analogous to the components of the metrical tensor. Yes, both g_{ik} and R_{hikl} are tensors and as such are basically comprised of vectors; both the g's and the R's are tensor components; both transform according to the transformation laws; and both are used to describe space-time. However, even though the g's are relatively easy to link to projections of radii on coordinate axes, we cannot project all the R's in any similar tangible manner. The R's include the *differential variations* of the g's.

The distinction is subtle but deserves emphasis. A local point in space-time does not have 20 independent components. It has 10 independent components, but we can only describe how these 10 vary between adjacent points by means of R_{hikl}, which happens to be a tensor with 20 independent components. In this sense the curvature tensor is somewhat misnamed. It is the curvature-and-its-variation tensor. This feature will be crucial in the next chapter.

On page 195 we criticized the statement that space-time has 10 curvatures; it only has two, but these project as 10 independent components on four coordinate axes. Because of the existence of the curvature tensor we might be tempted to say that space-time can have 20 curvatures, but this is even less accurate. The curvature tensor describes how two curvatures vary, and to do so requires 20 independent components in four dimensions. In other words 10 numbers tell us how space-time curves locally, and 20 numbers tell us how the curvature of space-time varies regionally.

◆

On the other hand we may ask why the curvature tensor has *only* 20 independent components, since a tensor of rank four in four dimensions could have 4^4 or 256 components. This question is not trivial. First of all if the curvature tensor required all 256 components, it would entail 256 equations. Moreover any number other than 20 would undermine the premise of the curvature tensor, especially for general relativity. Finally, explaining how 236 components of the curvature tensor can be deleted exemplifies the intricacies of tensor calculus which Einstein had to master.

To limit the size of the curvature tensor we take advantage of the properties of symmetry and antisymmetry (a.k.a. "skew-symmetry") of tensors. Let us recall our analogy between components and adjectives as it applies to the metrical tensor: Just as the color "a,b" is equivalent to "b,a," so g_{12} is equivalent to g_{21}; the order of the curvatures of space-time does not matter. This is the symmetry of g_{ik}, which can be demonstrated by use of the rule that a rank-two tensor is a sum of a symmetric and an antisymmetric tensor. (See Lanczos, pp. 92-97; Lawden, pp. 111-112; Kenyon, p. 55 and 189, or any text on tensor calculus.) If g_{ik} were not symmetric, it would consist of a symmetric part,

$$\tfrac{1}{2}(g_{ki} + g_{ik}),$$

and an asymmetric part,

$$\tfrac{1}{2}(g_{ki} - g_{ik}).$$

Solving $ds^2 = g_{ik}\, dx^{\,i} dx^{\,k}$ with the latter yields zero, which means that the metrical tensor can only be symmetric:

$$g_{ki} = g_{ik}$$

Similarly the ordinary second derivative of g_{ik} is such that

$$g_{ik,lm} = g_{ik,ml}.$$

The choice of indices is not important but the symmetry is: If the metrical tensor were not symmetric, it would have 3 rather than 10 independent components, and we would call it antisymmetric. When a tensor (say, U) is antisymmetric, we write

$$U_{ab} = -U_{ba},$$

314

and the same rule (negative sign) applies to contra-variant and mixed tensors. Furthermore, if a tensor is symmetric (or antisymmetric or mixed) in one system of coordinates, it remains symmetric (or antisymmetric or mixed) in all systems. (There is an exception; interested readers see Lawden, p. 93.)

We take advantage of these symmetries and our summary equation

$$2R_{hikl} = g_{hl;ik} - g_{il;hk} + g_{ik;hl} - g_{hk;il}$$

to generate certain rules of symmetry and antisymmetry for the curvature tensor. It is easier to appreciate these rules and their consequences in a diagram and with specific numbers for the indices. Our current goal is to use these rules to reduce the number of components of R_{hikl} to the least essential by eliminating components which (1) are redundant, or (2) can be determined from others, or (3) are impossible.

Please re-examine diagram T on page 288, this time noting the R's. Again this diagram is a graphic simplification of our square-shaped "block" or loop. In a diagrammatical manner, R_1... represents the first index in R_{hikl}; $R_{\cdot 2}$... represents the second index in R_{hikl}; etc. Based on this scheme we can say that R_1 stands for the shape of the surface between P_1 and P_2, R_2 stands for the shape between P_2 and P_3, R_3 stands for the shape between P_3 and P_4, and R_4 stands for the shape between P_4 and P_1.

To make full use of diagram T, we must heed four additional rules. These are designed only to make the diagram an effective demonstration, though they have a basis in the conservation of energy. First, we can start at any corner and proceed "around the block" in either direction, but once we select a starting corner and a direction, we cannot change our mind while circumnavigating a block. (U-turns are forbidden.) Second, we cannot cut or skip corners; we cannot leap from, say, P_1 to P_3. Third, we cannot flip the diagram over during the trip. Fourth, once we start, we cannot cover one corner more than once.

Analogically, the block is hilly, and we want detailed driving instructions, such as when to step on the gas pedal, when to coast, which way to turn, etc. We can designate any corner as P_1, and the end result is the same whether we proceed from P_1 to P_2 to P_3 to P_4 and back to P_1 or from P_1 to P_4 to P_3 to P_2 and then to P_1. However the intervening steps, represented by R_1, R_2, etc., are *different* depending on which way we select, simply because the instructions must be different depending on whether we drive down-hill first, up-hill first, to the right first, or to the left first. In other words, going around one way is *not fully symmetric* with going around the other way; our surface is "partially symmetric." [*]

Let us show partial symmetry mathematically by considering the first pair of indices, numbered 1 and 2:

$$R_{1234} = - R_{2134}$$

This means that R_{1234} renders R_{2134} impossible, and that the latter must be zero. Diagrammatically, if we traverse R_2 and then traverse R_1, we would either have to leap across the block or flip the block over to traverse R_3, neither of which is allowed. The same applies for the second pair of indices, 3 and 4:

$$R_{1234} = - R_{1243},$$

so that R_{1243} is eliminated. Verbally, the curvature tensor is partially symmetric; R_{hikl} is antisymmetric with respect to the first pair of indices and with respect to the second pair of indices. In short, "R_{hikl} is antisymmetric in h and i, and it is antisymmetric in k and l." (See Einstein in The Meaning of Relativity, Appendix II.)

However there is another consequence to these antisymmetries: We note that antisymmetry is indicated by a minus sign. Since the only number which can equal its negative is zero, any component such as R_{1134} or R_{2234} is eliminated because of the repeated indices. (As a general rule for antisymmetrical tensors, $U_{ii} = - U_{ii} = 0$.) Likewise considering the second pair of indices, R_{1233} and R_{1244} are eliminated, as well as R_{1111}, R_{2222}, etc. Diagrammatically, we cannot "hop up and down at one corner." Symbolically, R_{iikl} cannot equal $- R_{iikl}$ unless both are zero.

A similar rule applies to the first pair together versus the second pair together:

$$R_{1234} = R_{3412},$$

---

[*] Antisymmetry of the curvature tensor is linked to what is called a lack of commutativity. When the sequence of mathematical terms does not matter (e.g. when $AB = C$ and $BA = C$) we say they commutate. When the sequence of terms matters (e.g. when AB is not the same as BA) we say they do not commutate. Lack of commutativity is an important feature of the curvature tensor; without it, the curvature tensor could not quantify the complicated shape of space-time.

so that R_{1234} renders R_{3412} superfluous. In words, the curvature tensor is symmetric with respect to the first pair together versus the second pair together. Diagrammatically, the end result is the same no matter which corner we start at. R_{1234} means we started at P_1, R_{3412} means we started at P_3, and the same applies for starting at P_2 versus P_4.

The above three rules can be abbreviated and summarized as follows:

$$R_{hikl} = -R_{ihkl} = -R_{hilk} = R_{klhi}.$$

Let us tally the components so far. Although the process is tedious, if we write out R_{hikl} with each index 1, 2, 3 and 4, we find the expected 256 combinations, but upon applying the above rules 235 of these receive a zero. The surviving 21 components can be seen in matrix form:

$R_{hikl} =$

1212	1213	1214	1223	1224	1234*
	1313	1314	1323	1324*	1334
		1414	1423*	1424	1434
		*	2323	2324	2334
	*			2424	2434
*					3434

One more rule is available, which hinges upon a symmetry along the diagonal line "******" in the matrix. The three remaining components along this line, R_{1234}, R_{1324}, and R_{1423}, have in common the feature that in each one, no indices are repeated, unlike in the other 18. (Actually 24 combinations with all different indices are possible, but all except these three have already been eliminated). When any three indices are permutated so that all four indices are different, we find the rule that their "cyclical sum identity" is zero:

$$R_{1234} + R_{1324} + R_{1423} = 0$$

This makes it possible to determine one component with four different indices from the others which have that feature. In other words if we know two of these, the remaining one is not independent, which is written as

$$R_{1234} + R_{1324} = - R_{1423}.$$

Diagraming this "identity" is difficult, but it can be written more generally,

$$R_{hikl} + R_{hkil} + R_{hlik} = 0,$$

and, by virtue of this rule, one more component can be eliminated. (An "identity" is an equation satisfied by every possible value of a variable. E.g. the trigonometric equation $\dfrac{\sin x}{\cos x} = \tan x$ is an identity because it holds for any x.) The symmetry/antisymmetry equations and identities possess general covariance; the rules expressed by them are the same in any system of coordinates.

We see that the mathematical properties of the curvature tensor allow us to disregard most of its components, leaving the 20 independent non-zero components. The concept is not entirely abstract. For example the loop in question could be the orbit of the planet Mercury, and some day we may wish to intercept this planet at several separate points along its orbit. Our flight-plan will be based on the ("hilly") shape of space-time as described by the curvature tensor. Because of the excess precession of the orbit of Mercury, its exact path is not the same between each pair of points, which means that our instructions must be different depending on the sequence of the points to be visited. In other words the shape of space-time encountered going one way is not fully symmetric with going the other way.

◆

We can apply the above methods to demonstrate an interesting point. On page 183 we noted a relationship between K (Gaussian curvature) and the components of the metrical tensor. Indeed Gauss was able to prove that K can be derived solely from the metrical tensor, and he called this finding the "*theorema egregium*," which translates loosely as the "excellent theorem." Given the link we found between g_{ik} and R_{hikl}, we should also find a relationship between K and components of the curvature tensor.

We assume that the Earth is a perfect sphere. Its intrinsic curvature has only two dimensions, restricting us to indices 1 or 2. Let us see how these indices fare in R_{hikl}. Applying our rules,

$$R_{hikl} = - R_{ihkl} = - R_{hilk} = R_{klhi},$$

we write

$$R_{1212} = -R_{2112} = -R_{1221} \quad \textit{and by symmetry} = R_{2121}.$$

and

$$R_{1111} = R_{1122} = R_{2211} = R_{2222} = R_{1112} = R_{1222} =$$

$$R_{1211} = R_{2111} = R_{2221} = R_{2122} = R_{1121} = R_{2212} = 0.$$

Only R_{1212} remains independent and non-zero, occupying one corner of the matrix above. Therefore in describing a surface of a sphere by way of the curvature tensor, one component, R_{1212}, represents Gauss' K, which by definition is $\dfrac{1}{radius^2}$ of that sphere.

To be more precise we recall (page 278) the determinant of the metrical tensor, $|g_{ik}|$. Currently i and k can only be 1 or 2, so that

$$|g_{ik}| = g_{11}g_{22} - g_{12}g_{21}.$$

The numerical value of $|g_{ik}|$ depends on these four g's, so that $|g_{ik}|$ acts as a constant which ties R_{1212} to K according to

$$R_{1212} = -|g_{12}|(K).$$

The fact that a component of R_{hikl} can be equated with a radius means that in theory any surface can be analyzed with the curvature tensor. For example, even if it is "overkill," we could perform an experiment on our flat table top to find the curvature tensor rather than just the metrical tensor. (To find the curvature tensor we would examine its surface while "driving" around in loops.) More worthwhile might be determining the R's for a planet as it is being explored by astronauts. They could easily calculate the radius of the planet from R_{1212} (Kenyon, p. 192).

As R_{1212} portrays the curvature of a sphere, it demonstrates that a sphere is a Riemannian "space" of constant curvature. In contrast, naturally bent space-time is a Riemannian space of variable curvature. However since K and R_{hikl} both yield intrinsic measurements, the information contained in Riemann's R's can be expressed in Gauss' K's. This point recalls our question about why we couldn't just rely on K's and not bother with the R's. The answer is that despite its complexity, the curvature tensor is a more compact and practical tool. Using our analogy between components and adjectives, what Riemann said in 20 words, Gauss could not express in several pages. Indeed one of Riemann's accomplishments is to replace K with a rank-four tensor, namely the curvature tensor, which is well suited for differential tensor equations in four dimensions.

We have arrived at another provisional field equation:

$$R_{hikl} = T_{ik}??$$

In the process we solved some problems but provoked others. To learn about these difficulties, we need to appreciate that T_{ik} is a rank-two tensor with 10 independent components. Just knowing that the curvature tensor R_{hikl} has a of rank four with 20 components warns us that the above provisional equation is flawed.

This issue is tied to an important property of the curvature tensor, one we already touched upon (page 310). While R_{hikl} describes the regional variation of space-time, it encompasses all aspects of that variation, *even those related to tidal change-of-shape effects*. We already know enough about the curvature tensor to understand this feature. Since components of the metrical tensor correspond to Poisson's gravitational potential, the first derivative of the metrical tensor corresponds to the strength of Newton's force of gravity. But once we seek the discrepancy between Newtonian and relativistic predictions, i.e. once we determine the *second* derivative of the metrical tensor, we make our new tensor *so sensitive that it picks up tidal effects whether we want them or not*.

Let us focus on some details about tidal change-of-shape effects. Since tides are a basic feature of natural gravitation, we raised the possibility that by looking for tidal effects, an observer in an elevator may be able to distinguish the gravitation on Earth from acceleration generated by engines. We visualized tidal effects by the behavior of two strings hung near each other. In an inertial field two hanging strings remain parallel, whereas in a natural gravitational field the strings are not strictly parallel. This observation is also possible during free-fall, where an observer is exempt from the feeling of weight but not from the consequences of falling towards a center of mass; for example hanging strings will not drop to the floor if they are released, but they still converge. We were able to neglect this detail till now, but in fact all gravitational effects are eliminated in free-fall *except tidal effects*, which suggests that space-time in free-fall may not be totally flat. This means that inertia and gravity may not be fully equivalent after all, and that by sufficiently careful measurements *we might evade the principle of equivalence*.

First, we need a way of studying tidal effects. We can use diagram U by assuming that the line labeled string A (passing through points P_1 and P_2) is one of our hanging strings. String B, (passing through P_3 and P_4) is another nearby string, and both hang "down" toward the center of the Earth. Since the strings are in a natural gravitational field—they might be hanging from a ceiling—*string A and string B converge slightly* (exaggerated in the diagram). This depicts a tidal effect.

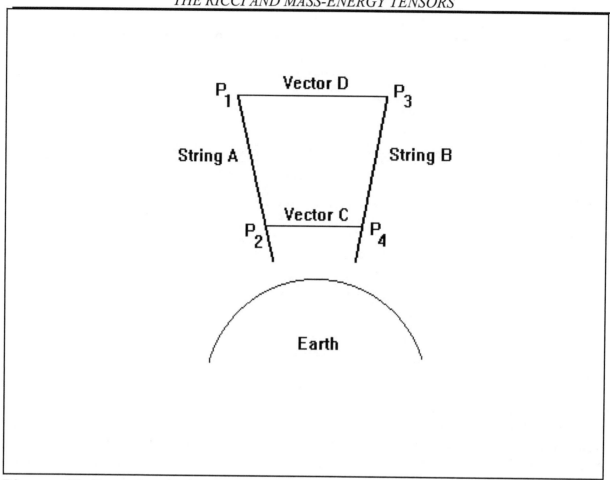

Diagram U. Two strings in the Earth's gravitational field.

The same phenomenon appears when two side-by-side objects fall freely toward the ground; they draw closer to each other as each drops toward the Earth's center. We prefer the hanging-strings scenario to the falling-objects scenario because the former is simpler. In either case converge exists *because of the shape of space-time*, not because the two objects interact with each other gravitationally, and we are reminded that the path each object follows is a *geodesic*, which means that the principles of parallel transport apply.

In diagram U we envision that a vector exists along string A and another along string B. We can also devise a third vector labeled C ("C" for convergence) and a fourth vector labeled D. Both C and D point across the diagram by bridging the gap between the strings. This arrangement lets us quantify our tidal effect, because if the strings converge, vector C shows a shortening compared with vector

D. In other words the relationship between two of the four vectors, C and D, is sensitive to our tidal effect. However since four vectors are operative, we can expect to involve a rank-four tensor.

As the strings converge, the amount of convergence can be expressed by an equation of "geodesic deviation," and in this context deviation means departure from parallelism of strings A and B. The key feature of this equation is that *it contains the curvature tensor R^a_{ikl}.*

By labeling the rate of convergence of the strings as C^a, where "a" is the direction, we can outline our equation of geodesic deviation,

$$C^a = (R^a_{ikl})(\textit{Rate of change of the other vectors}),$$

showing that the rate of convergence is determined by components of the curvature tensor for this region of space-time. This tensor appears in mixed form because the vector in C^a transforms in contra-variant manner, since it is a velocity vector (in the *a*-direction).

We do not need more details about the equation of geodesic deviation (see Synge and Schild, pp. 90-93, or Kenyon, pp. 75-76) but the concept is pertinent: We insert the curvature tensor into the equation by comparing the effect of parallel transport along two paths, P_1 to P_3 to P_4 vs. P_1 to P_2 to P_4 (in diagram U). *The difference between proceeding first "across" along vector D and then "down" along string B, vs. proceeding first "down" along string A and then "across" along vector C is quantified by R^a_{ikl}.* Here is why:

We can picture two observers, each of whom follows one of these paths. Since the strings converge, the first path (across, then down) is longer than the second (down, then across). Moreover each observer can note how the g's change along his path, and each observer can summon the metrical connections. Using various indices to show various directions, one observer quantifies his experience by $\Gamma^a_{il,k}$ and the other by $\Gamma^a_{ik,l}$. Since the scenario is in free-fall, each can apply the equation

$$R^a_{ikl} = \Gamma^a_{il,k} - \Gamma^a_{ik,l}.$$

Let us reinterpret this equation: As long as the shape of space-time appears different along two different paths (as long as $\Gamma^a_{il,k} - \Gamma^a_{ik,l}$ is not zero), space-time in this region has a non-zero curvature tensor—it has a complicated second-order intrinsic curvature—and it generates a tidal effect. This means that *the difference between going across first vs. going down first during parallel transport from P_1 to P_4 quantifies this tidal effect by way of R^a_{ikl}.* This also means that the second derivatives of the metrical tensor vanish in a field with no tidal effects, such as our powered elevator, whereas in a "natural" gravitational field the second derivatives of the metrical tensor do not vanish. In this sense there is a connection between tidal effects, which are second-order phenomena, and the fact that the curvature tensor is based on second-order differential equations.

The difference between the two paths is another case of partial symmetry. In fact if the curvature tensor could not accommodate such symmetry, it would not predict tidal effects, and when we find no such partial symmetry—when it makes no difference which way we go first—we know that there are no tidal effects. Furthermore we note that in this scenario the two vectors sensitive to our tidal effect are in first and last position; *across*, then down, vs. down, then *across*.

The above exercise with strings does not tell us about another physical significance of the curvature tensor. For this purpose we imagine that points P_1-P_3-P_4-P_2 in diagram U outline a solid object, say a box, which is falling freely towards the Earth. As the box falls through the progressively stronger gravitational field, it also shrinks a bit—it loses some volume (its molecules may be drawn together). This *change-of-volume* effect on an object near the Earth is extremely slight (it wouldn't be slight if the Earth were very massive), but it has one pertinent property: *The change in volume depends upon the strength of the gravitational field, and this is not a tidal effect.* In other words we can describe the gravitational field solely by the amount of shrinking of the box, and this shrinking is related to how objects fall, why they feel heavy, why a suspended string points "down," etc. In the case of a star, this change-of-volume effect explains gravitational collapse. The change-of-volume effect can also be quantified by means of the R^a_{ikl}. Incidentally, this change-of-volume effect has an important natural counterpart, gravitational compression and collapse of stars.

But what about tidal change-of-shape of this box? Just as the Earth shows tides because of changes in the gravitational field engendered by the moon, the box becomes deformed—besides shrinking in volume, the box also *changes shape*. We can say that this change in shape is a tidal "convergence," as in the case of our hanging strings. Again, this tidal effect on a falling object like a box is very subtle in the Earth's weak gravitational field (though tidal effects must contribute to the fate of an object caught in the powerful field of a black hole). The point is that we want our provisional field equation

$$R_{hikl} = T_{ik}??$$

to disclose gravitation, such as change-of-volume effects, but it is confounded by tidal change-of-shape effects.

To illustrate why tidal effects are extraneous, let us cite practical examples. First let us try to catch our freely falling box. Even though the box changes volume as well as shape on its way to the Earth's center, the competence of our equations to predict the falling of this object does not rely upon what influences its shape. To know where to stand and when to look up to catch the box, we need only the information about what influences its volume. Yet when we call upon the curvature tensor for the metrical data about space-time, we get more information than we can use. We can say that R_{hikl} (or $R^a{}_{ikl}$) is too inclusive.

The same notion applies where gravitation is not evident. Astronauts experiencing weightlessness cannot tell whether they are in "outer space"—in globally flat space-time—until they check to determine whether nearby strings converge or nearby objects change in shape. In principle, they must use the curvature tensor. But what if other observers wish to correlate the spaceship's orbit with the density of the Earth? I.e., what if they wish to find a universal law of gravitation, one that is generally covariant? *The curvature tensor on the one hand frees all observers from the choice of system of coordinates but on the other hand it includes irrelevant information, namely changes in shape such as tidal convergence.*

We recall that we can link the metrical tensor with the two radii at any one point on a complicated curved surface. In four dimensions this begets two ranks and 10 independent components, and this tensor is symmetrical. Now we have linked the curvature tensor to tidal effects. We find that where a region of a surface is curved and is subject to tidal effects, its tensor has four ranks and 20 independent components, and this tensor cannot be fully symmetrical. Our various scenarios suggest that two of the four ranks of the curvature tensor represent the regular non-tidal shape of space-time, and that another two ranks allow the curvature tensor to encompass the additional tidal deformation of natural space-time. Clearly the former are (indirectly) associated with the change-of-volume, and the latter with the change-of-shape effects. The observation that the former effects usually are dominant does not alter that fact that the 20 components of the curvature tensor embrace both sets of effects. At the same time we assume that the change-of-volume behavior ultimately derives from the 10 components of the tensor T_{ik}, which is symmetrical.

Before proceeding we emphasize that the complexity of this situation is not fictitious or artificial. It is a consequence of the fact that (on a small scale) objects in our universe are not uniformly or regularly distributed but are concentrated at certain places, which we call the centers of gravity. Without these there would be no gravitation as we know it, tidal or otherwise; all of space-time would be one simple frame of reference with no curvature of any sort, and the pertinent mathematics would be much simpler. But nature is not like that, and tidal effects are an inherent part of how the universe behaves.

In a sense we brought the mathematical complexities of tidal effects upon ourselves because of how we look at falling objects. If we study gravitation in frames of reference which are small enough, we find no pesky tidal effects. But the moment we begin to study how the g's vary around a loop, we trespass beyond local-ness, and our mathematical instruments, notably the curvature tensor, detect the natural irregularities. In other words we pay a price for trying to discern the variations in the shape of space-time that explain falling: We encounter tides.

All this means that if we wish to write locally applicable laws of gravitation, as for example about how one string hangs or how one box falls, *we must eliminate half the components of the curvature tensor and we must reduce its rank by two, thereby obtaining a new less inclusive tensor, one which is symmetrical.* We can guess how to manage the indices: The first and last indices in R^{a}_{ikl} should be eliminated. Indeed Einstein found a way by use of a tensor-mathematical procedure called "contraction." However before describing the particulars, we ought to consider in a general way why the subsequent steps absorbed Einstein's attention for several years.

✦

As we mentioned earlier, Einstein's derivations had to heed other laws of physics, including the laws of conservation of mass, energy, and momentum. Since Newton's law of gravitation does not mention conservation, we may ask why Einstein should have bothered. The difference is that Newtonian science admits separate laws of conservation of momentum, of mass, and of energy. Newton's momentum is a three-dimensional quantity; *mv* tells us about an object moving through space, and this space is described as a three-dimensional frame of reference. Mass is described in a separate frame of reference.

However as relativity unifies space and time, the momentum and mass of every object are brought into one four-dimensional frame of reference. (Since kinetic energy [$\frac{1}{2}mv^2$] is reflected in momentum, and since energies can change their form, the conservation of energy in general is drawn into the same scheme.) Therefore if all physical events occur in four-dimensional space-time, all such events, including those with gravitation, must obey a unified law of conservation of momentum, mass, and energy. Later we will examine the mathematics of how the laws of conservation of mass, energy, and momentum are melded into general relativity.

Nevertheless Einstein could not ignore Newton's law of gravitation since it is quite successful in predicting the behavior of falling or orbiting objects in weak gravity and at low speeds. Therefore, even if Einstein formulated a law of gravitation based on relativity, his equations had to produce results very similar to Newton's under ordinary circumstances, and we will verify this later.

Another principle of physics Einstein wished to respect is causality—the idea that effects cannot precede their causes. Adhering to this principle in the framework of general relativity gave Einstein considerable pause. The quandary appeared in following context: Einstein found that many solutions of the field equations of general relativity seemed possible, depending on the choice of coordinate systems, and this insinuated that the entire concept was too arbitrary and uncertain to have scientific validity.

The key argument in this quandary proceeds as follows: Special relativity permits relatively moving observers to disagree on the sequence of events, but it does not allow effects to precede their own causes in any uniformly moving frames of reference (page 106). Conversely if two distant events are truly simultaneous in one frame of reference, they cannot be causally linked since no signal can cross space in zero time. Therefore, Einstein reasoned, the principle of causality applies for all observers in *uniformly* moving (inertial) frames of reference. But then what about *non*-uniform motion? Since a limitless number of frames of reference exists, most of which are not inertial, does this not signify that, in certain frames, observers could find an effect *before* it is caused? Likewise can a frame of reference be found in which distant simultaneous non-uniform events are causally linked? In short, is there a gravitational field in which causality is violated? For some time Einstein believed he had thus proven that general covariance and causality are incompatible, and this predicament seemed so intractable that he temporarily abandoned the principle of general covariance, which of course seriously disrupted his thesis.

However, Einstein, as well as other mathematicians, then realized that *the various solutions are all equivalent* (See Pauli, p. 160, and Clark, p. 198). In fact once this concept was mastered it became an asset rather than a obstacle: As required by the principle of covariance, the choice of the system of coordinates does not matter, and a cause appearing before its effect is impossible even without the restriction to inertial frames. Clark summarizes the resolution of the dilemma this way: "...what appeared to be an infinitely large number of solutions to one problem was really a single solution applicable to each of an infinitely large number of different frames of reference."

Pauli explains this in another way: "The contradiction with the causality principle is only apparent, since the many possible solutions of the field equations are only formally different [are different only in their form]. Physically they are completely equivalent [while physics continues to bar reversed causality]." It thus appears that Einstein had temporarily trapped himself in his own logic. He designed his field equations to hold in a diversity of frames of reference, but he was perplexed that diverse solutions were possible—until it became clear that this feature is essential to general relativity without violating causality. If all frames of reference are equally valid and some forbid reversed causality, then all frames forbid reversed causality.

What about causality under extraordinary conditions? As we shall discuss later, a black hole is an area of space-time that is extremely indented, yet it could house linked events such as A-causing-B. Since a black hole is a system with very curved coordinate axes, is it possible for an observer outside a black hole to detect B before A? No, because signals, including photons, cannot escape a black hole; *neither* event inside the black hole is detectable outside of it. In effect events inside a black hole are bridged to outside events by space-like intervals, which means that the former have no effects elsewhere. Just as a black hole traps light, it "traps events." As said earlier, since a gravitational field bends light in a predictable manner, and since gravitation influences all causal links between events, it also ensures orderly and rational causality, even under the most extreme conditions foreseen by general relativity. (The principle of causality is also challenged by some of the premises of quantum mechanics, a point we will treat in chapter 20.)

These issues were related to various mathematical problems addressed by Einstein. One was that the components of the metrical tensor reflect the system of coordinates as well as the shape of the surfaces. This was resolved by seeking derivatives. But the metrical connections, the first derivatives of g_{ik}, are not tensors. This was resolved by proceeding to the second derivative, R_{hikl}, which as we have seen indeed is a tensor by virtue of covariant differentiation. Another problem was a violation of conservation laws, and as we will find later, this was resolved by yet additional derivation.

In retrospect it seems surprising that Einstein needed several years to settle these issues, but we should regard his mathematical and intellectual ventures as explorations of unchartered territory through which he was guided by instinct as much as by the rules of calculus; more on this issue later. In any case with concentrated effort he eventually unearthed the derivations which satisfied both covariance and causality as well as all the other pertinent laws of physics, and which are supported by experimental results.

◆

Let us return to "contraction" of tensors and to the problem of tidal effects in the proposal that

$$R_{hikl} = T_{ik}??$$

Contraction is a mathematical procedure which reduces the rank of a tensor to form a new less inclusive tensor, one that has fewer indices. To be valid in any system of coordinates—to yield a tensor—contraction is permissible only if one contracted index is co-variant and the other contra-variant. We recall that the distinction between co-variant and contra-variant components vanishes in Cartesian coordinate systems, so by invoking contraction we are essentially conceding that two component-vectors are the same, in which case their derivatives drop out. The result is loss of two ranks (Lieber, pp. 180-182, gives details). The curvature tensor emerges from its derivation in mixed form, e.g. $R^a{}_{ikl}$, making it contractible, and our hanging-strings scenario suggests how to contract this tensor: Eliminate the first and last index, namely a and l. This should yield a rank-two tensor.

Einstein indeed discovered that the 20 components of the rank-four curvature tensor can be contracted into two new rank-two tensors of 10 components each. One of these, R_{ik}, is named the "Ricci tensor" for its originator. In flat space-time

$$R_{ik} = 0,$$

and in curved space-time *tidal change-of-shape effects are not encompassed* in this tensor.

In a detailed equation, contraction of the curvature tensor ($R^a{}_{ikl}$, which is in mixed form) appears as

$$R^a{}_{ika} = \frac{\partial \Gamma^a_{ia}}{\partial x^k} - \frac{\partial \Gamma^a_{ik}}{\partial x^a} + \Gamma^e_{ia} \Gamma^a_{ek} - \Gamma^e_{ik} \Gamma^a_{ea} = R_{ik},$$

wherein all *l*'s (lower case *L*'s) are replaced by *a*'s. In summary,

$$R_{ik} = R^a{}_{ika}.$$

Obviously there are several possibilities as to which pair of indices to equate, just like there are several possibilities which corner of diagram U to start at. This implies that there are several different Ricci tensors. However because of our symmetry rule

$$R_{hikl} = -R_{ihkl} = -R_{hilk} = R_{klhi},$$

only one Ricci tensor is allowed. I. e. if we select two indices with respect to which R_{hikl} is antisymmetrical, h and i or k and l, the result is zero (for clarity ignoring that contraction should be done on only the mixed form such as $R^a{}_{ikl}$). Others choices may give a negative R_{ik} but that does not matter; the point is that this tensor is defined by summing the first and third Riemann indices.

The rationale for contracting the curvature tensor is also visible in terms of the gravitational potential. We recall the equation

$$R_{h4k4} = -\frac{1}{c^2}\left(\frac{\partial^2\Phi}{\partial x^h\,\partial x^k}\right),$$

which shows that the second derivative of the gravitational potential depends on only half of the components of the curvature tensor.

Incidentally, a further contraction of R_{ik},

$$R^i{}_i = R,$$

yields a scalar R of rank zero, which will interest us later.

The Ricci tensor R_{ik} is also symmetrical on the basis of the partial symmetry of the curvature tensor. (Again for clarity we write the co-variant form even though we are considering contraction.) We see from the above that

$$R_{hikl} = R_{klhi},$$

and if we contract by equating, say, l with h, so that

$$R_{hikh} = R_{khhi},$$

we find

$$R_{ik} = R_{ki}.$$

329

pp 169-170.) In the present context, *a Bianchi identity discloses a certain inherent relationship between the components of the curvature tensor, while that relationship, as we will see, implies zero divergence and hence compliance with the laws of conservation.* This conclusion was a mathematical break-through for Einstein.

Our goal, we recall, is to ensure that $G^{ik}_{;i}$, which is the divergence of the Einstein tensor, equals zero. We expect our final equation to hold the term $-\frac{1}{2}R$, as in $K = -\frac{1}{2}R$. More importantly we will have resolved the incompatibility between the Ricci tensor R_{ik} and the conservation laws. There are various ways to proceed, depending upon which rules of tensor calculus are applied in which sequence, and the steps are quite intricate. We cannot go into every detail (references are cited), but we will use an approach that emphasizes explicitness. It will be easier to follow if we assign numbers to most indices.

We start with *the curvature tensor written in co-variant form,* R_{hikl}, portrayed by R_{1234}. (Please be careful to distinguish the letter l from the number 1.) We can state that at a locally flat coordinate point, when the remaining three subscripts are rotated,

$$R_{1234} + R_{1342} + R_{1423} = 0.$$

This is the cyclical sum identity of the curvature tensor, based on the symmetry-nature of the curvature tensor (page 317).

We have available another rule pertaining to the antisymmetry of this tensor: When the first two indices are reversed, a minus-sign appears, so that

$$R_{1234} = -R_{2134}.$$

We also will need the rule that indices can be raised and lowered—i.e. components can be switched from co-variant to contra-variant—by use of the metrical tensor (page 277):

$$A^a = g^{ab}A_b \quad and \quad B_a = g_{ab}A^b$$

However for clarity we will separate this rule into two parts, the raising of an index, and the introduction of the metrical tensor. Other rules of tensor calculus will be cited as we apply them.

In our system the Bianchi identity in co-variant form appears as

$$R_{1234;u} + R_{124u;3} + R_{12u3;4} = 0,$$

where u = 1, 2, 3, and 4. This equation signifies that *the curvature tensor conforms with the Bianchi identities.*

366

We emphasize that the curvature tensor has more components and less symmetry than the Ricci tensor *because* the curvature tensor quantifies more features of curved space-time than does the Ricci tensor. In other words space-time which includes tidal effects takes one partially asymmetric large tensor, while space-time without tidal effects takes one symmetrical smaller tensor.[*] This concept is quite significant. It is an example of the remarkable harmony between the mathematics of relativity and physical nature, for the situation would be far more ambiguous and frustrating if there were several Ricci tensors or if they were asymmetrical. Although Einstein made the statement in another context, we can quote him here (Clark, p. 473, and Pais' title): "Subtle is the Lord, but malicious He is not."

Bergmann (The Riddle of Gravitation, pp. 105-106) summarizes this concept: Only R_{ik} is essential for describing space-time in a frame of reference containing matter and energy, i.e. where T_{ik} is not zero. Recalling that R_{hikl} can be derived using parallel transport, this is equivalent to stating that only the 10 components of R_{ik} contain the relevant effects of parallel transport of a vector around a loop in curved space-time. Conversely in flat space-time all 20 components of the curvature tensor vanish, but for our purposes only the 10 components of the Ricci tensor need to vanish.

The Ricci tensor is also our mathematical reply to the objection that a natural gravitational field and an artificial accelerating field are not entirely equivalent because we can tell them apart (pages 321 and 65). *By relying on R_{ik} rather than R_{hikl}, we treat space-time as a field that is free of tidal irregularities, where the principle of equivalence is inviolate.*

The conclusion is obvious. Is the field equation we seek

$$R_{ik} = T_{ik}??$$

This certainly seems plausible. The left side is based on a second derivative of the metrical tensor, it excludes tidal-change-of-shape effects, and presumably it shares the number of components, rank, and symmetry with T_{ik}. At this juncture we will examine T_{ik} thoroughly.

---

[*]The other 10 components contracted from the curvature tensor form the "Weyl tensor." These components are not affected locally by nearby matter, but they deal indirectly with change-of-shape effects. To be exact, the 20 independent components of the curvature tensor can be split into 10 independent components of the Weyl tensor, 9 independent components of a variant of the Ricci tensor (the trace-free Ricci tensor) and the Ricci curvature scalar, R. Obtaining the Weyl tensor rather than the Ricci tensor involves contraction on a different pair of indices. The Weyl tensor is also known as the conformal curvature tensor. Conformality is a feature of maps on which shapes are faithfully depicted. The Weyl tensor can be used to create such maps. See Wald, pp. 445-447.

✦ ✦ ✦

First of all, why bother with the T_{ik} in $R_{ik} = T_{ik}$? Why not borrow from Newton or Poisson and propose simply that $R_{ik} = m$ (mass) or $R_{ik} = \rho$ (mass-density)? Though in fact we do draw upon of Poisson's approach, the objection to these proposals is that we require equations with versatile invariant quantities, namely tensors, and these tensors should form a four-dimensional field. We must also consider the equivalence of matter and energy as well as the laws of conservation. We will see how T_{ik} meets these requirements.

Depending on the context, T_{ik} goes by various names: the material-energy tensor, the stress-energy tensor, or even the world tensor. It is also commonly called the energy-momentum tensor for reasons we will mention later. Our term for T_{ik} will be the *mass-energy tensor*. It will characterize *the distribution of matter and energy which determines the gravitational field by way of a deformation of four-dimensional space-time.* We anticipate that T_{ik} is a rank-two symmetrical tensor.

At first glance the issue seems simple enough: Since we know about mass-density and about the energy-content of mass, we merely need to identify which attributes of matter and energy affect nearby space-time. However when we try to work out the details, the picture becomes fuzzy. According to relativity, all forms of energy should contribute to the deformation of nearby space-time, but these can be temperature (heat), internal stress (like that in a coiled spring), chemical energy (like that in dynamite), etc. How do we know when these diverse forms of energy have been figured in, how do we know whether all of them relevant, and how do we express them as components of a tensor? We may also question certain assumptions we will make while searching for relativistically suitable models for describing matter.

Consequently we will find less exactitude and precision in the derivation of the tensor T_{ik} compared with R_{ik}, and the former derivation is mainly an empirical one: It is considered valid because it seems to work. Indeed Einstein conceded that the strongest reason to accept his notion of T_{ik} is that it demonstrates compliance with the laws of conservation. On the other hand the derivation of T_{ik} entails far fewer mathematical steps.

✦

We begin our mathematical study of T_{ik} by reiterating that ordinary matter will have to be treated as *a tensor in a field.* It may seem odd to describe matter as a tensor, since this implies that material objects have a complicated vectorial character. However when we accept that moving objects possess energy and that objects move in four-dimensional space-time, a tensorial description of matter becomes more palatable. It also helps to recall the relativistic concept of matter as an area of

331

space-time wherein energy is highly condensed or concentrated—which of course means that mass and energy are equivalent—and we note that Poisson's equation allows the density of matter to be incorporated into a field equation. Hence we can consider the relativistic counterpart of matter to be a tensor field, one which we call a "material field." In fact we will find that as the nature of T_{ik} becomes clearer, the concept of matter in the ordinary and familiar sense, i.e. as tangible weigh-able objects, fades into the background, and an abstract picture made of tensor components emerges.

In order to convert "matter" into a tensor, our approach will be to "build" T_{ik} by adapting the pre-relativity concept of a "fluid." In the ordinary physical sense a fluid is made of moveable particles, such as atoms, molecules, or larger constituents. The meaning of T_{ik} for relativity will be amplified and refined as we progress from one feature of a fluid to another, using three kinds of fluid-models. The first and simplest of these is a "cloud of dust." In this setting a "cloud of dust" is a special type of "general" fluid, which is the second kind of fluid we will consider. A third kind pertinent here is a "perfect" fluid, which will be of interest to us as we near our goal of defining T_{ik}.

A "cloud of dust," for our purpose, is a collection of inert or "incoherent" particles, which means that no particle interacts (coheres) with another. In actuality a cloud of widely dispersed fine dust particles behaves in this manner. "Incoherence" has three features, each of which simplifies our task. First, such a cloud of dust has no pressure—the individual particles do not press, push, or pull each other. Second, this cloud of dust has no viscosity—the individual particles slide or glide by each other freely, with no anti-slipping forces. (Viscosity is the resistance to flow. It is the internal friction between particles.) Lastly, this cloud of dust conducts no heat. (Heat stems from random motion of closely spaced atoms or molecules.) However we allow our cloud of dust to move, and we assume that observers are capable of detecting the relativistic effects of such motion.

The obvious question is how dust is relevant to general relativity. The rationale lies in the observation that a moving fluid-like cloud of dust is a reasonable model for a "matter field." Accordingly let us envision a moving cloud of dust, a sample of which has volume V containing n number of particles, each of which has mass m. (We assume all particles are the same.) The *density* of the mass in this volume is simply $\frac{nm}{V}$. We called this quantity ρ in Poisson's equation, and it is a scalar measure of the distribution of matter in this fluid.

Of course ρ is not a rank-two tensor, but it is our starting point. Therefore we will find it useful to specify that the density as measured by an observer moving along with the cloud of dust is ρ_o. The subscript signifies zero relative motion between this observer and the particles of dust; in effect ρ_o is "density at rest." We know that to another observer, one who sees the particles as moving uniformly, their mass appears increased by the special relativity factor while their dimension in the direction of motion appears decreased by the special relativity factor. To accommodate these effects we label the density of the cloud as measured by a relatively moving observer as ρ, which signifies "density during uniform motion." By convention, this ρ carries no subscript. The relationship between ρ and ρ_o is

$$\rho = \rho_o \left(\sqrt{1 - v^2/c^2} \right)^2 \quad (u)$$

where u is the velocity of the cloud and v is the relative velocity experienced by the observers. The reason for the separate "u" is that a moving cloud of dust appears denser (more particles move by), but by itself this is not a relativistic effect. We note that in this equation the special relativity factor is squared—it would appear twice if written out explicitly, once for mass and again for space—simply because $\left(\sqrt{1 - v^2/c^2} \right)^2 = \left(\sqrt{1 - v^2/c^2} \right)\left(\sqrt{1 - v^2/c^2} \right)$. An observer in relative motion thus reports that the dust particles gain mass but that their volume shrinks; both effects increase density. This also means that velocity v appears twice in the equation above. Since each velocity has its own vector, we envision that describing the density of matter requires two vectors—*it requires a tensor of rank two*, which we label T_{ik}. Moreover, since these two vectors should be interchangeable, we envision that this tensor is symmetrical. At this stage, before we attribute any of its components to energy, we call T_{ik} the "matter" tensor.

Let us see how our matter tensor "tensorizes" Poisson's scalar density of matter, ρ. The physical meaning of a matter tensor is easy to interpret, since its units are simply mass per volume (e.g. $\frac{grams}{cm^3}$). However we expect this tensor to have components which should express density. Here the rest of Poisson's equation helps us. In

$$\nabla^2 \Phi = 4\pi K \rho$$

the scalar gravitational potential Φ is equated with ρ. Under conditions of weak gravitation and low v (see pages 164 and 169), the 44-component of the metrical tensor g_{ik} predominates, and

$$g_{44} = 1 + 2\Phi/c^2.$$

As long as we disregard all components except g_{44}, we can disregard all components of T_{ik} except T_{44}, and we can apply Poisson's equation to these components:

$$\nabla^2 g_{44} = 4\pi K T_{44}.$$

Therefore the relativistic counterpart of the gravitational potential is one component of g_{ik}, and *the relativistic counterpart of the density of matter is one component of T_{ik}*, namely the time component

T_{44}. Indeed in the frame of reference with no relative motion (such that $\rho = \rho_o$), T_{44} *is the sole indicator of density.* This means that we should find an equation which links T_{44} and ρ. It will not be as simple as $T_{44} = \rho$, but we will return to the details after presenting other principles.

First of all the pivotal role of Poisson's concept is obvious: Though both Φ and ρ are scalars, his equation points the way to expressing both space-time and matter as tensor fields. In particular, Newton's force of gravity, G, is succeeded by Poisson's $\nabla^2\Phi$, which is replaced by the second derivative tensor for the shape of space-time based upon g_{ik}, (provisionally R_{ik}); and Newton's

$K\dfrac{m_1 m_2}{D^2}$ is succeeded by Poisson's $4\pi K\rho$, which can be replaced by at least one component of the

tensor T_{ik}. This notion has two important implications: Once we allow *all* components of g_{ik} to represent total (relativistic) gravitation, in which case they represent the shape of space-time, we must allow all components of T_{ik} to represent total (relativistic) matter and energy. Furthermore *in the absence of matter this tensor vanishes*; $T_{ik} = 0.$

As we planned, we can build upon the matter tensor to obtain a comprehensive mass-energy tensor designated as T_{ik}, T^{ik}, or $T^i{}_k$—in co-variant, contra-variant, or mixed forms—though this kind of tensor is customarily derived as T^{ik} because this tensor transforms naturally in a contra-variant manner (and because this method avoids dealing with certain reciprocals). At this juncture, let us switch to the contra-variant form and lay out our tensor, written as T^{ik}, in a matrix. In four dimensions, if $i = 1$ through 4 and $k = 1$ through 4, the 16 possible components of this tensor are

T^{11}	T^{12}	T^{13}	T^{14}
T^{21}	T^{22}	T^{23}	T^{24}
T^{31}	T^{32}	T^{33}	T^{34}
T^{41}	T^{42}	T^{43}	T^{44}

So far we endowed our T^{ik} with only one component, T^{44}. If we are now asked how much gravitation our cloud of dust can induce, we can say tentatively that the mass-density of the cloud expressed as T^{44} curves nearby space-time by an amount expressed as R^{44}. (We assume that we can accept $R^{ik} = T^{ik}$.)

What about other components of T^{ik}? By restricting our model to a fluid of incoherent dust described only by its density, we neglect other sources of a gravitational field. It is therefore necessary to lift the restrictions and consider not merely a fluid made of incoherent dust particles but a "general" fluid. In a general fluid, its constituent particles affect one another—they *interact* with neighboring particles (according to "Eulerian equations" for fluids). One consequence of this interaction is that a general fluid has internal "stress" which gives rise to several properties, including internal *pressure* and *viscosity*. Another consequence is the capacity of a general fluid to conduct heat. *These interactions add energy to a general fluid.*

We anticipate that our T^{ik} will conform to the pattern we have seen before and therefore has 10 independent components. We can now propose a matrix for our tensor for a general fluid. The reason for the positions of the symbols will also be explained soon:

P			
V	P		
V	V	P	
H	H	H	ρ

In the T^{44}-corner we find our density, represented by ρ. Each of the other diagonal boxes has a P which stands for pressure. V stands for viscosity, and H stands for heat transfer. In other words T^{ik} with its components of one ρ, three P's, three V's, and three H's laid out as above tells us about the mass-density and energy content of a general fluid. (We can use "mass-density" and "density of mass" interchangeably.) Since some forms of energy are represented, we can now call T^{ik} a "*mass-energy tensor.*"

Naturally we must justify such a scheme and clearly define each component. However to do this we pause for certain additional observations. The first relates to why we rely on density. One rationale for using "density of mass" resides in Poisson's equation, but we have more practical grounds. When we calculate how our sun curves space-time at a specific nearby location (as we will do later), density of mass, which is *mass per volume*, is a dependable description of the sun. If instead we were to rely only on the sun's mass while ignoring its volume, the location in question might be inside or it might be outside the sun, which would make a substantial difference. Therefore *distribution* of matter, reflected in density, is a better predictor of the nearby gravitational field. Mass by itself, though obviously important, is insufficient for this purpose.

This point raises further issues: Since we will be considering mass and energy in the four dimensions of space-time, we will need to consider quantities such as "mass per time" and "density of energy." In fact we must accept that density need not be only "mass per volume" where volume is the traditional "length times height times width." In relativity, density can be "mass per space times time" or "mass per four-dimensional space-time." In other words *time can be a dimension of a volume that has density.* Furthermore, since time always flows, we must accept that density can mean "flow in the direction of time" or "flow in the x^4- direction." (A formal term for such flow is "flux.") In this context we can think of all matter and energy as "flowing" through space-time.

As for "density of energy," this notion follows from the equivalence of matter and energy, which means there is such a thing as "energy per volume" even if we mean "volume of space times time." In this context energy is a very wide concept that includes anything that can curve space-time. We also recall the dictum of special relativity that all forms of energy have inertia (page 42), which implies that *anything with inertia has energy and hence anything with inertia affects space-time.* Since momentum is a measure of inertia, *momentum should be a key consideration in the structure of our mass-energy tensor T^{ik}*, and we should be able to decipher the components of this tensor in terms of momentum. As strange as it sounds, we can even interpret energy in relativity as "momentum in the direction of time." Moreover, given that we can quantify energy by its density, we anticipate that we will encounter "density of momentum" in four dimensions. In short, because of the link between mass, energy, inertia, and momentum, and because we use density to express distribution, we can describe matter for the purposes of relativity by means of density of energy and even by density of momentum—all in four dimensions. This, incidentally, is why the term "energy momentum tensor" appears in many dissertations on T^{ik}, though we prefer the broader term "mass-energy tensor."

This also means that we could refine our wording of the principle that the "distribution" of matter determines the shape of space-time; we might say that the "distribution *and motion*" is what counts. The implication is that "matter" in general relativity is a dynamic entity, and the concept of "distribution" encompasses *changes* in distribution that constitute motion and that engender momentum. This concept is important enough to illustrate. Imagine two dance partners twirling around in a waltz. To a Newtonian physicist, a gravitational force attracts the partners if they have enough mass, while a centrifugal force pulls them apart if they twirl fast enough. To a relativist, the distribution of the dancers' "matter," reflected in density, affects nearby space-time, the shape of which determines their physical conduct. Moreover *their twirling motion introduces momentum, which alters the event beyond just adding centrifugal force.* One way of seeing this is as an increase in mass through relative motion, as in

$$\frac{m_{motion}}{m_{rest}} = \frac{1}{\sqrt{1-v^2/c^2}}$$

(though this equation considers only uniform motion). Another way is to imagine that if the dancers twirl more energetically, they add to their density of momentum. Of course Newton knew that twirling entails energy, but he did not know exactly why and by how much, so that his equations do not accommodate the modern relativistic interpretation.

We can now fulfill our intention to spell out the mathematical link between the mass-energy tensor and density, which, as we said, is not as simple as $T_{44} = \rho$. We already defined density in a fluid as $\rho = \dfrac{nm}{V}$ (n number of particles of mass m per volume V), and we can adjust ρ for the relativistic effect of uniform motion by way of $\rho = \rho_o\left(\sqrt{1-v^2/c^2}\,\right)^2$. Since the mass in ρ_o (density with no relative motion) is equivalent to energy via $E = m_o c^2$, $\dfrac{energy}{mass}$ equals c^2. Altogether, written out explicitly,

$$T^{44} = \frac{\left(\dfrac{nm}{V}\right)c^2}{\sqrt{1-v^2/c^2}},$$

but more compactly, $T^{44} = \dfrac{\rho c^2}{\sqrt{1-v^2/c^2}}$, and in an even more abbreviated form,

$$T^{44} = \rho_o c^2.$$

Hence $\rho_o c^2$ is *density of energy*, which can be expressed in the relativistic units of energy per volume such as $\dfrac{E}{cm^3}$ or $\dfrac{mc^2}{cm^3}$. In other words the 44-component of our tensor takes on a certain value, $\rho_o c^2$, based on the presence of mass. The mass can move, so that use of the contra-variant form is more appropriate because velocity transforms naturally in that form. The presence of c^2 means that T^{44} can be a very large number, which correctly implies that *under most conditions density of energy is the main ingredient of the entire tensor*—not surprising since under most conditions g^{44} is the main ingredient of the metrical tensor. Later we will compare this component to the other components of T^{ik} using the above derivation.

We note that the above equations give us a measure of the density of energy represented by our cloud of dust. However we already pointed out that momentum in four dimensions must be considered. By linking momentum (mv, which is p) with the special relativity factor and with $E = mc^2$, we also already derived an equation wherein momentum is explicitly related to rest mass and energy:

337

$$E^2 - c^2p^2 = m_o^2c^4.$$

(The subscript o replaces the subscript r.) We can solve this equation for E, clearly showing that the relativistic energy of an object is the sum of its kinetic energy, expressed explicitly in its momentum mv, and its rest mass, expressed by m_o:

$$E = c \sqrt{(mv)^2 + m_o^2c^2}$$

This means that the density of energy in space-time contains the energy of rest mass as well as other kinds of energy, notably kinetic.

If we keep in mind that (in our system) time is dimension number 4, we can further interpret T^{44}. Since density can be flow in the time direction, and since energy can be momentum in the time direction, T^{44} spells out *the amount of energy per volume flowing in the time direction*. We can say that T^{44} tells us how energy, reflected in momentum, flows from the past into the future. In the mathematical language of relativity, this actually tells us about the density of mass.

To examine the other T's in T^{ik} we must interpret the indices in the mass-energy tensor as it appears during uniform relative motion in all four dimensions. In effect we must consider how a cloud of dust appears in a moving frame of reference. Of course motion of the dust particles at a certain velocity endows them with a certain amount of momentum, but since momentum can exist in one direction while the observer of the cloud moves in another, we must also allow for two separate velocities. This point relates loosely to the requirement that T^{ik} be a rank-two tensor.

In general, T^{ik} means how much momentum-in-the-k-direction is flowing in the i-direction. In other words T^{ik} describes the flow of the k-component of momentum along the i-coordinate axis. This idea lends itself roughly to an analogy. We imagine that our cloud of dust is now a group of soldiers marching in the i-direction; if $i = 1$, they are marching in the X-direction in Cartesian coordinates. At the same time their eyes are turned in the k-direction; if $k = 2$, they are looking in the Y-direction. In that case the equation $T^{ik} = T^{12} = T^{12}$ describes their action. Similarly, marching while $T^{ik} = T^{33}$ means marching and looking in the same direction (the Z-direction). Thus if $T^{ik} = T^{44}$, the soldiers (their density of energy to be more exact) are "flowing" in the time direction from their past into their future as they look to that future. In place of the indices i and k, we will use h and j in the next few pages whenever we focus on motion in the three dimensions of space.

338

In mathematical terms, we revert to our explicit equation $T^{44} = \dfrac{\left(\dfrac{nm}{V}\right)c^2}{\sqrt{1-v^2/c^2}}$ and adapt it to a velocity

which in our analogy is the velocity with which our soldiers are marching through three-dimensional space. That velocity can be labeled as u, and its direction in space can be h; this h can be 1, 2, or 3. We again note the role of the special relativity factor, which accommodates an observer's relative motion. At the same time we introduce a simplification in which we assume that V is a unit volume (i.e. that $V = 1$, such as one cubic centimeter or one cubic meter).[*] These maneuvers yield an equation for three additional components of T^{ik}, namely T^{14}, T^{24}, and T^{34}. This equation,

$$T^{h4} = \frac{nmu^h c^2}{\sqrt{1-v^2/c^2}},$$

contains the velocity u^h. Rewriting the right side as $\dfrac{(mu^h)\,nc^2}{\sqrt{1-v^2/c^2}}$ clearly shows that T^{h4} deals with

momentum (mu) in a certain direction (h). We will fit this equation into the overall scheme of T^{ik} shortly.

The next step follows logically. Two separate velocities can be involved, so that

$$T^{hj} = \frac{(mu^h u^j)\,nc^2}{\sqrt{1-v^2/c^2}},$$

where h and j run from 1 to 3. Thus we have accounted for the remaining components of T^{ik}.

We can now redefine and compare the components of T^{ik} in physical terms, and simultaneously we further explain the previous two matrices (on pages 334 and 335).

T^{44} is *density of energy*, which is usually by far the most significant component. In the marching soldier analogy, even when at rest (even when the soldiers only have rest mass), their density of energy is great (their E is $m_o c^2$). Their density of energy flows in the direction of the time coordinate axis, which means from the past into the future. This component is ρ in the

---

[*] Some systems simplify further by assuming that c = 1.

second matrix, though to be accurate its equation is $T^{44} = \dfrac{\left(\dfrac{nm}{V}\right)c^2}{\sqrt{1-v^2/c^2}}$, or in abbreviated

form, $T^{44} = \rho_0 c^2$.

T^{hh} is *pressure*. Both indices are the same and both range from 1 to 3. In this analogy, the soldiers exert pressure forward because they are marching and looking forward. If we place a fence across their path, then T^{hh} describes how their momentum presses against the fence. These three components, T^{11}, T^{22}, T^{33}, which are the P's in the second matrix, are also significant to us but are usually smaller than T^{44}. Their equation is $T^{hh} = \dfrac{(mu^hu^h)\,nc^2}{\sqrt{1-v^2/c^2}}$, which can also be abbreviated similarly to the above.

T^{hj} is *viscosity*. The two indices, h and j, are not the same but each ranges from 1 to 3 (T^{12}, T^{21}, T^{13}, T^{31}, T^{23}, T^{32}). In this analogy, we imagine that marching one way and looking in another causes the soldiers to rub against each other and to hinder the flow of their momentum. We will find that $T^{hj} = T^{jh}$ (by virtue of symmetry), so that T^{12}, T^{13}, *and* T^{23} drop out, leaving the independent V's in the second matrix. The equation is $T^{hj} = \dfrac{(mu^hu^j)\,nc^2}{\sqrt{1-v^2/c^2}}$, which can also be abbreviated.

T^{4j} is *heat flow*. The first index is time and the others are 1 to 3 (T^{41}, T^{42}, T^{43}). In the marching soldier analogy, "marching through time" while looking in any spatial direction allows heat energy to flow from soldier to soldier. This is H in the second matrix, and the equation is $T^{4j} = \dfrac{(mu^j)\,nc^2}{\sqrt{1-v^2/c^2}}$ or its abbreviation.

T^{h4} is *density of momentum*. The second index is time and the others are 1 to 3 (T^{14}, T^{24}, T^{34}). This does not show well in the marching soldier analogy, but in effect marching in a spatial direction while "looking through time" describes only the density of momentum of the soldiers. Though difficult to imagine, $T^{4h} = T^{h4}$, which renders redundant both heat flow and density of momentum. Hence these components are zero in the second matrix; their equation also appears above.

Incidentally many authors let i and k each equal 0 through 3 (rather than 1 through 4) so that what appears above as T^{44} is labeled as T^{00}. In this convention the same numbering system applies to the components of other tensors (g's, R's, G's). Though not used in this book, the 0-1-2-3 convention does allow explicit identification of time components vs. space components. To make this distinction even clearer, authors using this convention may select a different letter to stand for components 1, 2, and/or 3. For example T^{uv} might mean the space components of T^{ik}, T^{00} might mean the time components of T^{ik}, and T^{u0} or $T^0_{\ v}$ might mean the space-time or time-space components of T^{ik}. It is also acceptable to assign Roman letters to the space components and Greek letters to space-time components. We must be aware of this when referring to various sources.

It is important that T^{ik} is symmetrical; $T^{ik} = T^{ki}$. This is obvious when the two indices are the same (in the first two cases above), but it applies to the entire matrix. The components for pressure and viscosity represent stress, and non-relativistic physics recognizes a stress tensor, written as S_{hj} where h and j each run from 1 to 3. Its components are the 9 upper-left boxes of our matrix. The S_{hj} is symmetric; the reason is that general fluids are treated as collections (fields) of points, and the stress at each point must be symmetric in every direction or else each point could accelerate and/or spin spontaneously, which is forbidden by the laws of conservation. Therefore the symmetry of S_{hj} is an indirect consequence of the conservation of angular momentum. (See Schutz pp. 103-104, and Feynman et al, II-31-11.) This means that the T^{i4} components are symmetric, in part because mass is equivalent to energy. Thus T^{ik} as a whole is symmetric, and the 6 blanks in its matrix are not independent, which, as we anticipated, leaves 10 independent components. *These components tell us all about the mass and energy at each point in space-time. Since the information is in tensor form, any observer of an event at any such point can accurately quantify mass and energy.* (To be precise, we should specify that T^{ik} tells us about *non-gravitational* mass-energy; more on this detail shortly.)

We now return to our fluid-model. By advancing from incoherent dust to a general fluid we have gone too far. The mass-energy tensor for a general fluid has components, namely heat conduction (H's) and viscosity (V's), which are irrelevant for us. Only density of energy (ρ) and pressures (P's) contribute significantly to the curvature of space-time governing gravitational motion. Hence we shall remove the H's and V's from our matrix.

The obvious question is why the P's are acceptable for our T^{ik} while the H's and V's are not. It is easy to accept that internal pressure (like the stress from a coiled spring) adds to the mass-energy of an object, so we anticipate that this pressure enhances the effect of this object on nearby space-time. But why not add in viscosity and transfer of heat? After all, overcoming viscosity requires energy, and of course heat is a form of energy. This question underscores the main problem with describing matter in relativistic terms, i.e. with selecting which attributes belong in the field equations of general relativity. We do not know enough about matter to give an unequivocal reply. However we assume the following, using an astronomic scenario:

We know that the orbit of each planet in our solar system is determined mainly by the sun. If the sun suddenly moved to another galaxy, the planets would be subjected to gravitational "pressures." Presuming such a bizarre event, some planets might accompany the sun to its new location while others might not, but in any case, the event would entail the transfer of some energy, and this energy must be considered in any accounting of all the energy in the system. At the same time the planets would not appreciably resist each others' flow as occurs in a viscous fluid, and no significant amount of new heat would be generated. It therefore is reasonable that in common relativistic scenarios, pressure counts while viscosity and heat conduction are negligible.

Astrophysical evidence supports the notion that pressure (P) ought to be added to density (ρ) in the structure of T^{ik}. In the Newtonian view internal pressure should resist the gravitational collapse of stars, whereas in the relativistic view, pressure adds to the energy-content of such a process and thereby promotes collapse. Observations of collapsing stars support the latter. On the other hand our understanding of stellar evolution is incomplete, and the picture is complicated by the fact that some new heat is generated and that quantum-mechanical processes are implicated; see Kenyon, p. 79 and 121, and Schutz, pp. 111 and 264-266. The situation is also unclear in the case of very small objects such as subatomic particles wherein quantum-mechanical processes dominate the picture; more on this later.

Since the mathematical interpretation of T^{ik} involves momentum, it is important to distinguish between regular ("linear") momentum and *angular* momentum. Clearly an object moving in a straight line—one with linear momentum—has kinetic energy. But what if that object is, say, a small spinning top? It may be spinning in one place so that it lacks the familiar "v" for mv, but it has angular momentum—it must contain some kinetic energy. It is also important to recall that spinning is a form of accelerated motion. Since all motion, uniform or accelerated, adds to kinetic energy, we should consider angular momentum in our assessment of sources of energy. However ordinary objects do not have enough rotating motion to produce relativistic effects; their angular momentum is negligible. On the other hand subatomic particles spin very rapidly, which gives them significant angular momentum, and later we will see that angular momentum is a major feature of rotating black holes.

The question is thus raised whether accelerated relative motion can curve space-time—the answer is yes—and whether accelerated relative motion is equivalent to the presence of nearby matter—again yes—which of course is the principle of equivalence in that either can be responsible for the effects of gravity. Clearly then all forms of relative motion contribute to the deformation of space-time, and of course this necessitates the incorporation of momentum into T^{ik} so that the field equations can inform us fully as to how energy alters the shape of space-time. As we said, T^{ik} implicitly embraces *changes* in the distribution of matter, and hence of energy. Nevertheless under ordinary conditions angular momentum does not merit explicit representation, but this issue complicates the interpretation of T^{ik}.

It may then seem odd that the structure of T^{ik} has no provision for any gravitational potential energy. We could argue that just raising an object off the ground so it can fall further—even without adding angular momentum—may add energy which should merit at least one component. The reply to this point is that general relativity allows curved space-time to account for all gravitational behavior. In this view, falling objects need no gravitational potential energy, since they follow space-time geodesics and heed parallel transport. (We can also question whether curving of space-time itself adds energy, but we will come back to this detail.)

Despite the uncertainties, we believe that we should remedy the "overkill" introduced by a model based on a general fluid. This step calls upon a "perfect" fluid, one in which viscosity and heat conduction are zero. The absence of heat conduction appears mathematically as

$$T^{i4} = T^{4k} = 0,$$

In our previous matrix, the H's represented the T^{4i} components, so that now the H's = 0. Absence of viscosity is more complicated:

$$T^{(i=1,2,3 \ and \ k=1,2,3)} = 0 \ \text{unless} \ i = k,$$

which nullifies the V's. This means that the 10 components of our mass-energy tensor for a perfect fluid are

P			
0	P		
0	0	P	
0	0	0	ρ.

We find one component for density of energy or mass—depending on our interpretation—and three components for pressure—which mathematically means for momentum. The rest are eliminated because they are physically negligible or they are symmetrical with components that are negligible.

Still at this point in our derivation, T^{ik} for a perfect fluid is not comprehensively defined, particularly with respect to relative motion that is not uniform. To remedy this, our first step is to reconsider the above non-zero components as they appear in the frame of reference of an observer who is not moving with respect to the fluid. The subscript o explicitly indicates this case. The T's for pressure are

$$T_o^{11} = T_o^{22} = T_o^{33} = P_o.$$

Each of these can individually defined by the equation $T^{hh} = \dfrac{(mu^h u^h)\, nc^2}{\sqrt{1-v^2/c^2}}$, but this is not adequate here because the special relativity factor deals only with uniform motion. The T for density of energy still is $T^{44} = \dfrac{\left(\dfrac{nm}{V}\right)c^2}{\sqrt{1-v^2/c^2}}$, which again applies only to relative uniform motion. These equations are only Lorentz-invariant, so let us take additional steps to ensure that equations with T^{ik} enjoy general covariance.

If an observer in non-uniform motion with respect to the fluid studies this case, each of these four components must be subjected to the transformation equation for a rank-two tensor,

$$T^{ik} = \frac{\partial x^i}{\partial x_o^{\,u}} \frac{\partial x^k}{\partial x_o^{\,v}} T_o^{uv},$$

where the o-subscripts identify which values belong to the non-moving observer. This step yields a long partial-differential equation (the P's and ρ are relocated for clarity) which shows how the fluid's pressure and density of energy appear to any moving observer:

$$T^{ik} = P_o \frac{\partial x^i}{\partial x_o^{\,1}} \frac{\partial x^k}{\partial x_o^{\,1}} + P_o \frac{\partial x^i}{\partial x_o^{\,2}} \frac{\partial x^k}{\partial x_o^{\,2}} + P_o \frac{\partial x^i}{\partial x_o^{\,3}} \frac{\partial x^k}{\partial x_o^{\,3}} + \rho_o c^2 \frac{\partial x^i}{\partial x_o^{\,4}} \frac{\partial x^k}{\partial x_o^{\,4}}$$

This transformation also affects the metrical tensor, according to

$$g^{ik} = \frac{\partial x^i}{\partial x_o^{\,u}} \frac{\partial x^k}{\partial x_o^{\,v}} g_o^{uv},$$

where g_o^{uv} is the metrical tensor in the coordinates for the non-moving observer and g^{ik} is that tensor for the moving observer. We also assume that in a fluid, velocity in four dimensions can replace the velocity expressed as partial differentials (and again this velocity need not be only uniform):

344

$$\frac{\partial x^i}{\partial x_o^4} = \frac{dx^i}{ds}.$$

The 44-component is now defined (parametrized) by

$$T^{44} = \rho_o c^2 \, \frac{dx^4}{ds} \, \frac{dx^4}{ds}.$$

In like manner we can define the pressure-components of our perfect fluid, T^{11}, T^{22}, T^{33}. For instance

$$T^{11} = P_o \, \frac{dx^1}{ds} \, \frac{dx^1}{ds}.$$

If we gather all four non-zero components and apply some algebra (detailed by Bergmann, p. 129) we find

$$T^{ik} = \left(\rho_0 + \frac{P_0}{c^2} \right) \frac{dx^i}{ds} \, \frac{dx^k}{ds} + g^{ik} P_0.$$

This is the definitive form of *the mass-energy tensor for a perfect fluid*, and this equation is generally covariant—it is independent of the choice of coordinates and the kind of motion. The term $(\rho_0 + P_0 / c^2)$ represents density of energy plus pressure as measured by a non-moving observer. (Dividing P_0 by c^2 evens out the units.) The purpose of the term $g^{ik} P_0$ is more subtle. It brings in the metrical tensor so that this equation can apply in systems of coordinates for curved space-time.

Thus $g^{ik} P_o$ helps assure general covariance, analogous to the way the Γ's bestow covariance upon other equations. In this way the mass-energy tensor reliably quantifies the factors that we believe curve space-time: mass and energy, which embrace momentum and pressure. In other words *this description of a perfect fluid provides general relativity with a model for "matter" in tensor form*. The mass-energy tensor in its co-variant form constitutes the right half of our provisional equation

$$R_{ik} = T_{ik} ??$$

where it stands for *the dynamic distribution of matter and energy even in curved four-dimensional space-time*. Indeed we can say that (given the concept of flux as flow in a direction through space and time) what curves space-time is *mass and momentum flux*, as embodied in T^{ik}.

We reiterate that most of the value for the tensor T^{ik} usually resides in the time-time component T^{44}, while the other non-zero components are far less significant. For instance in our moving-sun scenario, the change in density of energy would be far more consequential than the inter-planetary interaction. Furthermore the familiar equivalent of density of energy is mass, which means that *under conditions of weak gravity almost all of the deformation of space-time is attributed to mass; pressure adds little.* This is predictable since Newton's law of gravitation entails mass but not pressure, and it should be no surprise that of all the components of the mass-energy tensor, T^{44} (or T_{44}) will be of particular use to us when we analyze ordinary planetary motion.

◆

We deciphered the structure of T^{ik} or T_{ik} using one model—a fluid—suggesting that other approaches are plausible. Indeed Einstein reasoned that since matter is made up of electrically charged interacting particles, these tensors can also be constructed on the basis of *electromagnetism*. Though not all details are necessary, we should outline the logic, which also has historical interest.

In the eighteenth century Coulomb (1736-1806) found that the force between two electrically charged objects is proportional to the strength of these charges, divided by the square of the distance between them; this is "Coulomb's law." Meanwhile the existence of a magnetic field had been known to science for centuries, but by the early nineteenth century a link between electric and magnetic phenomena was recognized. The pivotal advance is credited to Oersted (1778-1851) who demonstrated that a moving electric charge can create or "induce" magnetism. At about the same time Ampere (1775-1836) suggested that magnetism is a form of electricity, and Faraday (1791-1867) discovered the converse to Oersted's effect: A moving magnet, which means acceleration of a magnetic field, can induce an electric current. However since accelerated motion of an electrified wire also evokes a magnetic field, there must be an electric field. Accordingly, Faraday concluded that a change in an electric field results in a magnetic field and vice versa. Faraday thereby secured a key role for the electromagnetic field in modern physics.[*]

It fell to Maxwell, a skilled researcher and mathematician, to put all this together into a set of equations for the physical behavior of the electromagnetic field; hence we call these the *electromagnetic field equations*. They can be written in various forms, but the suitable one here is the vector-field form for empty space (in theory with no wires or magnets nearby):

---

[*]The tombs of Newton, Faraday, and Maxwell are located a few yards from each other in Westminster Abbey, London.

$$\nabla \times B - \frac{1}{c}\frac{\partial E}{\partial t} = 4\frac{\pi}{c}j$$

$$\nabla \cdot E = 4\pi\rho$$

$$\nabla \times E + \frac{1}{c}\frac{\partial B}{\partial t} = 0$$

$$\nabla \cdot B = 0$$

In these equations the electric field intensity is E and the magnetic field induction is B. The electric current density is j, and the electric charge density, ρ. We see the nabla, ∇, since these are partial-differential equations. (We should be aware that these equations can be framed with different terms, such as D for electric displacement and H for the magnetic field, and we have taken some liberties for the sake of clarity; see Menzel, pp. 261 and 383-385.)

What do the electromagnetic field equations mean for physics in general? They quantify how moving electricity creates magnetism and how moving magnetism creates electricity. Indeed they justify the term "electromagnetism." In the first equation $\partial E/\partial t$ stands for the rate of change in the electric field as a function of the magnetic field and the electric current at that moment. In the third equation $\partial B/\partial t$ stands for the rate of change of the magnetic field as a function of the electric field at that moment. The second equation relates the strength of the electric field to electric charge density. The fourth equation means that there is no such thing as a magnetic charge; i.e. that an unpaired magnetic pole should not exist.

We note that the first and third equations are not reciprocal. Calculating $\partial E/\partial t$ from B involves j because a steady electric current alone produces some (opposing) magnetic field. Calculating $\partial B/\partial t$ from E is simpler because a steady magnetic field does not induce electricity, but the effect of B on E is opposite. In other words the creation of magnetism out of electricity and creation of electricity out of magnetism are not symmetrical. This detail will be significant shortly.

Why is electromagnetism explained as a field? We have cited difficulties in explaining gravitation with the Newtonian system, and the same problems arise for electromagnetism. An analogy helps here, namely several round magnets rolling freely on our trampoline. Based upon the apparent success of Newton's laws whereby all events are thought to occur because objects or particles are pushed or pulled by forces, the Newtonian interpretation of magnetism is this: The trampoline is flat, and the magnets force each other to move by instantaneous "action at a distance." Of course this explanation falters where magnets are not closely coupled to each other. A force can operate across the gaps *only if there is an intervening field*.

What do Maxwell's equations mean for relativity? As we shall see, they provided firm justification for the field-paradigm of general relativity (introduced on page 66). In particular, Einstein exploited the observation that inertial and gravitational mass appear identical, in which case "something" outside the objects themselves had to be responsible for their equal acceleration. *Encouraged by the successful notion of an electromagnetic field, Einstein proposed that this "something" is the gravitational field, for if accelerated motion of an electric charge generates a magnetic field, and if gravity is a consequence of accelerated motion, then acceleration may evoke a gravitational field.* Thus if a magnet deforms the surrounding electromagnetic field, a weight can deform the surrounding gravitational field. In this view the magnets move not because they are pushed or pulled but because they run down gradients, and their behavior, representing gravitation, is described by a deformed field.

In this context a crucially important term in Maxwell's equations is

$$\frac{1}{c}\frac{\partial B}{\partial t}$$

because it contains a partial derivative with respect to time, and it contains the speed of light, c. These ingredients point to two insights. First, unlike Newton's law of gravitation (and in fact unlike Coulomb's law) Maxwell's equations assume that physical effects do not spread instantaneously—hence $c\partial t$ appears above but not in Newton's law. Rather, electric and magnetic fields influence each other at speed c. Secondly, again unlike anything conceived in Newton's equations, Maxwell's equations indicate that c is constant.

Let us digress briefly on the constancy of c: When Maxwell's equations are applied to how electromagnetism behaves inside matter, two constants are included. One is "ϵ" which is the dielectric constant—a gauge of the electricity a substance can store. The other is "μ" which is the permeability of a substance to magnetism. These constants can be measured, and it is possible to verify that the value of

$$\sqrt{\epsilon\mu}$$

is *equal to c, the speed of electromagnetic radiation.* In words, a constant for electricity along with a constant for magnetism yields a constant for light. No provision is needed *for the speed of the source or for the speed of the observer.*

As we already noted, a constant c is not only unexpected on the basis of, but is contradictory to, Newtonian physics. For example we used an analogic runner who, according to Maxwell and Einstein, can never catch up with a beam of light, no matter how fast he moves. According to Newton, if the runner moves fast enough, the beam will appear stationary to him or he may even overtake it. Both scenarios cannot be correct. Experiments (e.g. by Hertz in the 1880's) appeared to confirm the constancy of c, and a few years later results by Michelson and Morley demonstrated, however indirectly, that c is constant and the rate of time is not. Therefore, while the fathers of relativity did not complete their thesis until decades later, they had ample clues at the end of the nineteenth century that a breakthrough in physics was imminent and that Newton would emerge the loser.

Here Lorentz enters into the picture. We mentioned Coulomb's equation for the force between two electrically charged objects,

$$F = C \frac{q_1 \, q_2}{D^2},$$

which has the same form as Newton's law of gravitation—except that like charges repel rather than attract, whereas gravity, as far as we know, only entails attraction. However, *Coulomb's law, like Newton's law of gravitation, is not Lorentz-invariant.*

In 1904 Lorentz applied the special relativity factor to Maxwell's equations, which led him to his transformation and which proved that Maxwell's equations are invariant under this transformation rather than under the Galilean transformation. (Lorentz' key paper appears on pp. 9-34 of The Principle of Relativity.) This point is critical, for it means that only Maxwell's equations are compatible with a constant c and with relative time, and it means that *Maxwell's equations are Lorentz-invariant.* We can say that if "Maxwell's electromagnetic field corrects Coulomb," the path was open for "Einstein's gravitational field to correct Newton." This is why Maxwell's equations are so prominent in the development of relativity and why Maxwell and Lorentz are heavily cited in Einstein's papers; Maxwell's name appears in the opening sentence of the initial paper Einstein published in 1905 on what was to become special relativity. In short, the first laws for fields made invariant during uniform motion were Maxwell's, the first to show this was Lorentz, but the first to grasp the full meaning was Einstein.

(We have simplified history. Poincaré, Minkowski, and others were grappling with the same issues, and those who recognized the deficiencies in Newton's law did not immediately see the field as the answer. In fact theories were devised to explain gravitation on the basis of only special relativity. For example Nordstrøm used a quite credible approach, one which initially even appealed to Einstein, that c is constant in a gravitational field but that mass depends on the local gravitational potential [Pais, p. 232]. The defect in Nordstrøm's theory is that light really is bent in a gravitational field. Another alternative was Mie's theory, but it did not satisfy the principle of equivalence [Pais, p. 235, and Weyl, p. 206].)

Maxwell's equations between the electric field intensity E and the magnetic field induction B neatly parallel the equations of special relativity. We recall that for relatively moving observers A and B, who are measuring the space and time for some gravitational event,

$$(SPACE_A)^2 - c^2(TIME_A)^2 = (SPACE_B)^2 - c^2(TIME_B)^2.$$

(We recall that $(cdT)^2 = c^2 dT^2$.) Using Maxwell's equations to accommodate relative motion during an electromagnetic event, we find a remarkably similar pattern,

$$(E_A)^2 - c^2(B_A)^2 = (E_B)^2 - c^2(B_B)^2.$$

(See Shadowitz, p. 123, for details.) This parallel means three things:

First, just as relatively moving observers fail to agree on *SPACE* alone or on *TIME* alone but agree on combined *SPACE* and *TIME*, so they fail to agree on the strength of an electric field (*E*) or on the strength of a magnetic field (*B*) but agree on the strength of a combined electromagnetic field. In Einstein's words, "...the electric and magnetic fields lose their separate existences through the relativity of motion." (The Meaning of Relativity, p. 41.)

Second, just as

$$(SPACE)^2 - c^2(TIME)^2$$

is an invariant (it is the space-time interval), so

$$(E)^2 - c^2(B)^2$$

is an invariant; both survive transformations of coordinates.

Third, in both cases (*SPACE-TIME* or *E-B*) this line of reasoning requires accepting the non-existence of absolute motion, it requires dealing in four-dimensional space-time, and it implicates the special relativity factor.

Yet so far we have dealt only with uniform motion. This brings us to another key point: *Maxwell's equations can be restated in four-dimensional tensor form*, and as such they possess not only Lorentz-invariance but general covariance—they retain their form upon transformation to a frame with *accelerated* motion. Here Maxwell's four vector equations can be replaced by two tensor equations, each containing a rank-two tensor. Doing so allows Maxwell's equations to describe the "shape" of the electromagnetic field similarly to how Einstein's field equations describe the gravitational field.

The intervening steps for writing Maxwell's equations as tensors are of insufficient interest here (see Sachs, pp. 103-106, and Einstein in The Meaning of Relativity, pp. 21-23 and 40-41), but we should outline the procedure for the first pair. We use the definition of the ∇ to write out these equations as a series of partial-differential equations. We abbreviate each partial-differential term by a "*B*" or "*E*" according to whether it deals with magnetism or electricity. For example any term structured as

$$\frac{\partial E}{\partial x}$$

is simply E. Maxwell's first equation

$$\nabla \times B - \frac{1}{c}\frac{\partial E}{\partial t} = 4\frac{\pi}{c}j$$

can be expanded into three such equations. If we add zeros and +'s to fill out the equations, we can present these as the first three lines of this pattern:

$$+0 + B - B - E = 4\pi/cj_1$$

$$-B + 0 + B - E = 4\pi/cj_2$$

$$+B - B + 0 - E = 4\pi/cj_3$$

Maxwell's second equation

$$\nabla \cdot E = 4\pi\rho$$

can likewise be expanded, but there is a relationship between j and ρ (current density depends on charge density and velocity) which allows the fourth line to read

$$+E + E + E + 0 = 4\pi/cj_4.$$

The 16 B's, E's and 0's take on the form of a matrix for a four-dimensional tensor in which the B's and E's emerge as non-zero components. A similar procedure melds the third and fourth of Maxwell's vector-field equations into a four-dimensional tensor field equation.

In this context we introduce the rank-two tensor F_{ik}, the Faraday tensor, which symbolizes the electromagnetic field. Judged from the above pattern, F_{ik} is not a symmetric tensor, and we will come back to this point. More importantly F_{ik} allows us to describe in one term the changes that spread continuously through this field, and in four-dimensional space-time F_{ik} can combine and replace the separate spatial vectors for E and B. We also introduce the (rank one) tensor J for the charge-current density. The tensor J descends in like fashion from a means of expressing electric currents and charges as a single "four-current," whereby essentially electricity is described in four dimensions.

Maxwell's four vector equations thus reduce to just two tensor equations (reminding us how tersely the conservation laws appear in their tensor version). The relationship between F_{ik} and J_i in the first of the tensor equations is revealed in its expanded form:

$$\frac{\partial F_{i1}}{\partial x^1} + \frac{\partial F_{i2}}{\partial x^2} + \frac{\partial F_{i3}}{\partial x^3} + \frac{\partial F_{i4}}{\partial x^4} = 4\pi J_i$$

This equation replaces the first pair of Maxwell's vector equations. The second pair is replaced by

$$\frac{\partial F_{ik}}{\partial x^l} + \frac{\partial F_{li}}{\partial x^k} + \frac{\partial F_{kl}}{\partial x^i} = 0.$$

With these equations, observers in any state of motion, including accelerated, can agree on an electromagnetic event. I.e. these equations are independent of all coordinate systems, and hence they posses general covariance. Moreover they are field equations.

Let us contrast the first of Maxwell's vector equations with the first of the tensor forms: In the latter, the left part of the equation contains the tensor F_{ik}, whereas the left part of the former holds the vector-quantities B and E. Similarly in the right parts, vector j is replaced by tensor J. Using the summation convention this can be compacted into

$$\partial^k F_{ik} = 4\pi J_i.$$

If we solve for $J_{(1\ through\ 4)}$ and substitute each of the four terms of the left side by a function of B or E, we obtain 16 components for F_{ik}, of which 4 are zero. Each component represents the intensity of the magnetic or electric field in another direction. Still in abbreviated matrix form, the electrical components E and the magnetic components B of F_{ik} may be presented as follows:

0	B	- B	- E
- B	0	B	- E
B	- B	0	- E
E	E	E	0

Again, F_{ik} is an antisymmetric tensor;

$$F_{ik} = -F_{ki},$$

which means that

$$F_{ii} = -F_{ii} = 0,$$

so that all components on the diagonal vanish. Furthermore F_{ik} cannot be rotated about a diagonal axis without changing its meaning.

Next we shall see how F_{ik} can be used to construct T_{ik} for the field equations of general relativity. The main advantage of defining T_{ik} via the electromagnetic approach is its superior mathematical rigor. The main disadvantage is that it hinges on the assumed similarity between gravitation and electromagnetism. We can now present this similarity in more precise terms: Just as local gravitation can be described by a gravitational potential, the electromagnetic field has an electromagnetic potential. (In the latter context we give Φ subscripts to indicate directions.) Based upon this concept, we can devise tensor equations for the energy in an electromagnetic field in terms of the Φ's,

$$F_{ik} = \frac{\partial \Phi_i}{\partial x^k} - \frac{\partial \Phi_k}{\partial x^i},$$

and for the density of energy of this field,

$$T_{ik} = \frac{1}{4}\Phi_{ab}^2 \delta_{ik} + \Phi_{ia}\Phi_{ka}.$$

(δ, called the Kronecker delta, is a function which connects certain vector/tensor components; see Schutz, p. 11, or Kenyon, p. 47.)

The key element of this line of reasoning is similar to that applied to any other set of particles: The motion (momentum) of energetic particles forms an electromagnetic field which can be viewed in four dimensions. This means that a tensor exists which describes the density of electromagnetic energy (the amount of such energy per unit volume). But since energy and mass can be equated, they can be described in one tensor representing a sum of the "material" energy-density of a collection of particles *and* the effects of their electromagnetic charges. We can call this the electromagnetic model, in which the complicated (contra-variant) mass-energy tensor appears as follows:

$$T^{ik} = \frac{1}{c^2}\left(F^i_{\ a}F^{ak} + \frac{1}{4}g^{ik}F^{ab}F_{ab}\right) + \rho_o u^1 u^k$$

The longer part of the right side of this equation obviously relates T^{ik} to F^{ik}. It represents the electromagnetic aspect of the field, and it is the rationale for accepting electromagnetism as a model for the mass-energy tensor. (We will shortly explain the role of the metrical tensor in the above equation.) The shorter part appeared in the equation for the energy tensor for moving matter, $T^{ik} = \rho_o\, u^i\, u^k$, which encompasses momentum.

The tensor T^{ik} can be re-rendered in terms of "EM" for the electromagnetic field components and in terms of ρ for the density of energy. The contribution of the electromagnetic field can be abbreviated by the EM's in the following equation:

$$T^{ik} = \frac{1}{c^2}(TERM\ CONTAINING\ EM's)\ +\ \rho_o u^1 u^k$$

(See Adler et al, p. 342.) In matrix form T^{ik} now appears as

EM			
	EM		
		EM	
			ρ

We note that the electromagnetic field components, the EM's, occupy the same loci as did the pressure components, the P's, in the matrix associated with a perfect fluid (page 343). Usually ρ is much less than EM, just as usually ρ is much less than P, so that the endowment to this tensor from the electromagnetic field is relatively small. Nevertheless the concept is similar: The EM's act to ensure the proper representation of energy, the ρ inserts mass as its density, and the ratio between them (E/m) is c^2. By selecting appropriate units, angular momentum is also represented (via velocity and density of mass). In fact to facilitate certain solutions of Einstein's field equations which we will mention later, we can categorize the ingredients of T^{ik} built upon the electromagnetic model into three measurable quantities or parameters, namely total mass, electromagnetic charge, and momentum.

A legitimate question is whether the two approaches to the mass-energy tensor, the fluid model and the electromagnetic model, can be unified; i.e. whether the two sets of components can be incorporated into one super-tensor. The answer is that there would be some incompatibility among the components from the two sources, and the practical advantage would be negligible.[*] It is more meaningful to see the relationship between these two approaches: The fluid model starts with the idea that matter can be described as particles of mass (dust) which have energy. The electromagnetic model starts with the idea that matter has energy (charge, magnetism) which then acts upon particles. The former approach is easier to explain, yet the latter is historically more prominent since the pioneers of relativity were primarily studying electromagnetism. The electromagnetic approach is also of greater current interest because many natural phenomena, such as stellar evolution and black holes, involve electric charge; we will return to this point.

An instructive way for us to conclude this topic is via this question (as suggested by Jones on p. 47): Since we have the field equations of electromagnetism, and since there are parallels between electromagnetic and gravitational fields, why do we need the field equations of general relativity? After all, we can envision an observer, one who is not electrically charged or magnetized, studying the behavior of a test object moving through an electromagnetic field. He will find that the deviation of that object from uniform motion in this field is predicted by F_{ik}. The correlation with relativity is self-evident: The deviation of a test object from uniform motion in a *gravitational* field should be predicted by T_{ik}. Then why not consider

$$R_{ik} = F_{ik}??$$

We can cite several objections: For one thing we noted that unlike gravitation, the electromagnetic field can be repulsive or attractive (in the scientific sense). For another, gravitation is much weaker than electromagnetism; the gravitational effect of a proton on an electron is negligible. For a third, unlike the electromagnetic field equations, Einstein's field equations turn out to be non-linear.

Finally, as mentioned earlier, F_{ik} is not symmetric, while gravitation has no equivalent asymmetry. This reflects another basic physical difference between electromagnetism and gravitation: Acceleration induces gravity like gravity induces acceleration, but magnetism does not induce electricity like electricity induces magnetism. We can explain this as follows:

Any measuring device—any clock, ruler or scale—must participate in any nearby gravitational event, simply because they all inhabit the same space-time. Indeed general relativity echoes Galileo's idea that everything that "falls" together accelerates together. However in a magnetic field, electrically neutral or non-magnetized objects or particles need not accelerate together, and this is built into Maxwell's equations. For example inertia can easily be distinguished from electromagnetic attraction, even in an accelerating elevator. This is the same as saying that *the principle of equivalence does not apply to electrically charged objects or particles*, and it provides another reason that

$$R_{ik} = F_{ik}??$$

---

[*]Except perhaps for quantum electrodynamic; see Tolman, pp. 93-95.

cannot replace

$$R_{ik} = T_{ik}??$$

We will discuss other incompatibilities between gravitation and electromagnetism in our chapter on quantum mechanics. Meanwhile we still have the question why the equation relating F_{ik} with T_{ik} includes the metrical tensor. Because (in contra-variant form again)

$$T^{ik} = \frac{1}{c^2} (F^i{}_a F^{ak} + \frac{1}{4} g^{ik} F^{ab} F_{ab}) + \rho_o u^i u^k$$

must hold true in any system of coordinates, which means it must include g_{ik} or g^{ik}. In effect the components of F_{ik} (or F^{ik}) cannot correlate with gravitational behavior unless they are adjusted for curvature of space-time. Once they are, then T_{ik} (or T^{ik}) defined by this equation is the *relativistically valid symmetric mass-energy tensor built upon the electromagnetic model.*

◆

The result of either approach, the fluid model or the electromagnetic model, is a tensor T_{ik} *whose components of are sufficient to represent "matter" in the postulate that nearby matter curves space-time.* We can say that T_{ik} embodies the total energy for gravitation. However of the four vectors of space-time curvature (as reflected in the four ranks of R_{hikl}), only two need to be linked with T_{ik}, and these two comprise R_{ik}; . Indeed one of the gratifying features of T_{ik} is that even though its derivations are so different from that of R_{ik}, it still turns out to be a symmetrical tensor with rank two and 10 independent components. T_{ik} also gives equal significance to the various factors which appear in it, similar to the way the dimensions of space enjoy equal footing with time in the metrical and curvature tensors. Such mathematical symmetry is taken as another piece of evidence for the underlying harmony in nature as well as the over-all legitimacy of Einstein's derivations. However the point that T_{ik} and R_{ik} share certain mathematical features is not merely a matter of aesthetics.

We know that if an equation between similar tensors is valid in one frame of reference, it will remain valid upon transformation to any other frame of reference. In other words compatible tensors build generally covariant tensor equations, and this is a key attribute of T_{ik}.

Incidentally, I mentioned (pages 134, 156) that Einstein and his math tutor Grossmann had temporarily overlooked the point that a tensor does not change even if its components change under a change in the system of coordinates (as indeed they should). The tensor in question at the time was T_{ik} , and the resolution of the hangup was the realization that this tensor does comply with the principle of general covariance.

As we draw close to the core of general relativity, we must pay attention to the laws of conservation. These laws demand that in any physical event, all matter and all energy be accounted for, and we intimated that the relationship between these laws and the mass-energy tensor T_{ik} or T^{ik} is important.

This relationship involves *divergence*. In physics, divergence deals with the flow of energy during events. If the rate of flow of energy per volume from one point in a vector or tensor field is zero in all directions, we say that the *divergence is zero* or that the *divergence vanishes everywhere*. If a point is a source of energy, divergence is positive; if it is a "sink" for energy—more flows in than out—divergence is negative.

An intuitive way to envision divergence is with a small hot object. Flow of heat is a rank-one tensor (a vector; an amount flowing in a direction), and if divergence is zero, no net amount of heat escapes from or enters into this object. Thus "vanishing divergence " means that heat is conserved, and the temperature of the object is constant. But we may ask why we need as complicated a notion as divergence for something so obvious. The reasons are that general relativity treats the cause of gravitational events as field conditions at points in four-dimensional space-time, and that the pertinent equations should be in tensor form. As we shall find, divergence, including that of the mass-energy tensor T^{ik}, meets these requisites.

Meanwhile we demand that at least in theory T^{ik} embody all features of matter and energy which account for the deformation of local space-time. We also require that vanishing divergence be the same upon transformation to any other frame of reference. It should therefore not surprise us that gravitational events heed the conservation laws. Since such events involve matter, energy, and motion across space in finite time, why should they be exempt? Accordingly we envision a certain concentration of matter and energy as allotting a certain amount of gravitation at one point in space-time. A falling object must use up this allotment, no more and no less, and if we eliminate all energy, the object must stop falling. Thus *if a gravitational event at a point in space-time obeys the conservation laws, the divergence of its mass-energy tensor must vanish.* We then say that the gravitational field is "conservative" and that T^{ik} is "divergenceless." In short, zero-divergence of T^{ik} satisfies conservation.

Vanishing divergence or zero-divergence of the mass-energy tensor is written compactly as

$$T^{ik}_{;k} = 0.$$

We should know in some detail how this equation comes about. We recall (page 29) that the basic conservation laws applicable to Newtonian physics,

$$\Delta \Sigma m = 0, \quad \Delta \Sigma E = 0, \quad \text{and} \quad \Delta \Sigma mv = 0,$$

can be reduced to two Lorentz-invariant laws applicable to special relativity:

$$\Delta\Sigma mc^2\sqrt{1-v^2/c^2} = 0$$

and

$$\Delta\Sigma\frac{mv}{\sqrt{1-v^2/c^2}} = 0.$$

Let us focus on the conservation of energy (mc^2 in the previous equation) by imagining an energy-containing cube, like a small box filled with hot air, but now we allow energy to be quantified in the components of the symmetrical tensor T^{ik}. It is a contra-variant tensor because a flow of energy is a velocity (which by nature transforms as a contra-variant vector). We note that since divergence is a question of rate of flow, it is a kind of derivative. To calculate divergence we will find the rate of inflow and the rate of outflow of energy from our cube in space-time, and we will take the partial derivatives for each component. (Earlier we envisioned an energy-containing sphere, but a cube is better for our current purpose.)

It helps to segregate T^{ik} into time components and space components. The former can be written as T^{44} or symbolically as T^{TIME} and the other components as T^{SPACE}. As energy enters or escapes from this system, the rate of change is expressed as partial derivatives in the term

$$Vol\ \frac{\partial T^{TIME}}{\partial t}$$

where "Vol" is the volume of the cube and t is time. Since T^{44} (here called T^{TIME}) is indicative of the density of energy, the above term actually describes the rate of change of the density of energy inside the cube. (Kenyon, pp. 80-81, and Schutz, pp. 104-105, show this derivation in detail.) If energy is flowing into the cube it will cross the surface of the cube, whose area is labeled as "Area," at a rate equal to

$$-\ Area\ \frac{\partial T^{SPACE}}{\partial t}.$$

This quantity is negative because we consider outflow to be positive. *To satisfy conservation the net rate of inflow or outflow across all faces of the cube must equal the rate at which the content of the cube increases or decreases.* Hence we stipulate that

$$Vol\ \frac{\partial T^{TIME}}{\partial t} = -\ Area\ \frac{\partial T^{SPACE}}{\partial t},$$

but we can consider the smallest possible size of the cube—a point—so that *Vol* and *Area* are as close to zero as possible, in which case the numerical value of *Vol* and *Area* can be considered to be 1 (one). Furthermore, still dealing in partial contributions of space and time, we can rearrange the terms so that

$$(1)\frac{\partial T^{TIME}}{\partial t} + (1)\frac{\partial T^{SPACE}}{\partial t} = 0.$$

The zero in this scheme implies zero divergence, but we will come back to this notion. Now we omit the 1's and recombine the space and time components. By applying the same procedure for any index (and recalling that some components of T^{ik} describe momentum), we broaden this concept to include conservation of momentum as well as of energy. Considering flow per volume, the divergence of this system with respect to the index k is given by

$$\frac{\partial T^{ik}}{\partial x^k} = divergence_k \ of \ T^{ik},$$

which can be abbreviated as

$$T^{ik}_{,k} = divergence_k \ of \ T^{ik}.$$

We can be more specific: Recalling the meaning of the various components of T^{ik}, we can say that $T^{4k}_{;k}$ represents divergence of energy, and that $T^{1, \ 2, \ or \ 3 \ k}_{;k}$ represents divergence of momentum; divergence of matter follows automatically.

By virtue of the symmetry of T^{ik}, the like applies to the divergence of this system with respect to i, so that

$$T^{ik}_{,i} = divergence_i \ of \ T^{ik}.$$

Of course we require that this strategy hold under any transformation of coordinates, which means that it must possess general covariance; Lorentz-invariance is not enough. Accordingly we invoke the metrical connections (one set of Γ's per rank, as on page 283), which gives us the equation with respect to k:

$$\frac{\partial T^{ik}}{\partial x^k} + \Gamma^k_{uk}T^{iu} + \Gamma^i_{ui}T^{uk} = divergence_k \ of \ T^{ik}.$$

When we let the semicolon symbolize covariant differentiation, this can also be condensed into

359

$$T^{ik}_{\ ;k} = divergence_k \ of \ T^{ik} \text{ in any coordinates,}$$

and the same holds for covariant divergence with respect to *i*:

$$T^{ik}_{\ ;i} = divergence_i \ of \ T^{ik} \text{ in any coordinates.}$$

Lastly we invoke our intent that *the divergence be zero*—we say that we seek *vanishing divergence*. We recall

$$\frac{\partial T^{TIME}}{\partial t} + \frac{\partial T^{SPACE}}{\partial t} = 0.$$

For the case of *k*, but without demanding covariant differentiation, this is scheme equivalent to

$$\frac{\partial T^{ik}}{\partial x^k} = 0,$$

and

$$T^{ik}_{\ ,k} = 0.$$

To satisfy covariant differentiation we finally write

$$\frac{\partial T^{ik}}{\partial x^k} + \Gamma^k_{uk}T^{iu} + \Gamma^i_{ui}T^{uk} = 0,$$

and

$$T^{ik}_{\ ;k} = 0.$$

This series of equations means that when the tensor T^{ik} is differentiated with respect to one of the space-time component (*i* or *k*) which appear in the tensor, the outcome is zero when the conservation laws are honored. In this manner, *the laws of conservation of energy and momentum emerge as a compact unit in the form of the vanishing divergence of the mass-energy tensor.* We can call this critical concept "Einstein's Law of Conservation"— succinctly stated, "T_{ik} is divergenceless"—and we note that it is expressed mathematically as a tensor equation, one which is applicable in any gravitational field.

As mentioned earlier, the use of the superscripts rather than subscripts hinges on the derivation. Technically (Lawden, p. 112) the divergence of a tensor is found by contraction with respect to the index of differentiation and any (contra-variant) superscript.[*] For expediency we disregard the position of the indices wherever it is not critical. We should add that the bond between divergence and the laws of conservation exists only in four-dimensional space-time; e.g. three-dimensional divergence does not ensure compliance with those laws.

◆

Once again we have solved one problem only to encounter a new one: The Ricci tensor R_{ik} does *not* exhibit vanishing divergences as does the tensor T_{ik}; R_{ik} is not divergenceless! For example (in contra-variant form)

$$R^{ik}_{;k} \; does \; not \; = \; 0.$$

That is to say, even though we considered conservation of energy during geodesical motion (via the principle of least action), the left and right parts of the provisional field equation

$$R_{ik} \; = \; T_{ik}??$$

fail to behave alike with respect to the conservation laws. A simple way to envision this is that space-time has a certain stiffness which must be considered when we quantify the capacity of matter to deform space-time. We can think of R_{ik} as requiring more energy for curving local space-time than T_{ik} can provide. Again we are left with the task of finding an alternative to the left side of $R_{ik} \; = \; T_{ik}??$

Einstein (with help from Ricci's work) was able to derive such an alternative, and it goes by the symbol G_{ik} and by the name "Einstein tensor." The non-zero divergence of R_{ik} can be remedied by the term "- $\frac{1}{2}g_{ik}R$" so that

$$G_{ik} \; = \; R_{ik} \; - \; \frac{1}{2}g_{ik}R,$$

and, as we shall detail, *the divergence of*

---

[*] Divergence is related to gradient and to another feature of a field, the "curl," which has to do with flow in a swirl-like pattern. Though it need not concern us, the relationship between curl, divergence, and gradient is of interest in mathematical analysis.

$$R_{ik} - \tfrac{1}{2}g_{ik}R$$

vanishes as does that of T_{ik}. Our next provisional field equation therefore will become

$$G_{ik} = T_{ik}??$$

Moreover, like R_{ik}, G_{ik} is symmetrical—it inherits its properties from R_{ik}—and both tensors are rank two with 10 independent components. Hence G_{ik} and T_{ik} are mathematically compatible, and we can place an equal sign between them in a generally covariant equation.

However defining G_{ik} is quite complex and, as mentioned earlier, it presented Einstein with another difficult mathematical obstacle. As we did in proposing that $R_{ik} = T_{ik}??$, so had Einstein passed through a similar stage in 1915 (see Weyl, p. 239) at which he had not yet come upon the term -$\tfrac{1}{2}g_{ik}R$. However in the following year, once he realized that without this term the conservation laws are not satisfied, he published a correction. (For a translation see The Principle of Relativity, pp. 145-152 under the title "The Foundation of the General Theory of Relativity.") In fact, as Einstein discovered, it is practically impossible to develop valid equations of general relativity without including the vanishing divergence of T_{ik} and G_{ik}, and as these equations are derived, vanishing divergence follows automatically. In other words physical nature is such that if we unearth the correct field equations, we automatically satisfy conservation of matter and energy!

In the equation that defines the Einstein tensor, $G_{ik} = R_{ik} - \tfrac{1}{2}g_{ik}R$, we recognize g_{ik}, the metrical tensor. The other terms are R_{ik}, which we know stems from contraction of the curvature tensor. R is the "Ricci scalar" or the "scalar curvature" or the "curvature invariant;" it is contracted from R_{ik}. It is a scalar quantity—its value is unaffected by transformations of coordinates—representing the average curvature of space-time. If space-time were perfectly spherical and only two-dimensional (like the case we considered on page 318, where i and k can only be 1 or 2), contraction of R_{11} and R_{22} would yield $R = \dfrac{2}{r^2}$, showing that R is equivalent to the radius. Where space-time is flat, $R = 0$.

As we shall recount shortly, R is related to the Gaussian curvature. (Again we are warned that "R" means the angle of axis rotation, while "R" can mean the contraction of the Ricci tensor, or "R" can be a component of the curvature or the Ricci tensor.)

Einstein used the word "fundamental" for the metrical tensor, and indeed we note that each of the elements of the equation for G_{ik} springs from g_{ik}: R is derived by contraction from R_{ik}, R_{ik} is derived by contraction of R_{hikl} (we show the co-variant form for clarity) and R_{hikl} emanates from g_{ik} by covariant differentiation via the Γ's. We note that just as the g's vary across a field of metrical tensors, so the G's vary across a field of Einstein tensors. In other words G_{ik} forms a tensor field in space-time. Incidentally, if we add that the g's arise from the coefficients of the generalized differential pythagorean equation, we have outlined much of the mathematics of general relativity in one paragraph.

The overall structure of the equation $G_{ik} = R_{ik} - \frac{1}{2}g_{ik}R$ can be predicted from the following approach: R_{ik} can be broken down into an algebraic pattern for a linear function

$$(A)R_{ik} + (B)Rg_{ik} + (C)g_{ik}$$

where coefficients A, B, and C are constants. This pattern is implied by Poisson's equation in that the relationship between G_{ik} and T_{ik} should be linear (page 260). Therefore we should reach an equation of the following form:

$$(A)R_{ik} + (B)Rg_{ik} + (C)g_{ik} \text{ is proportional to } T_{ik}.$$

We can show that R is a reflection or function of the Gaussian curvature K. We recall that

$$R_{hikl} = |g_{ik}|(-K) \quad or \quad K = -\frac{R_{hikl}}{|g_{ik}|}.$$

By use of contraction it is possible to define the Ricci tensor in terms of $|g_{ik}|$, so that

$$K = -\tfrac{1}{2}R,$$

since K is the *average* curvature. I.e., as R_{hikl} is contracted twice to become R, the value of $|g_{ik}|$ becomes $\frac{1}{2}$. (Pauli, pp. 159-161.) We might have guessed that $R = -K$, but we see that $R = -2K$; the Ricci scalar is twice the Gaussian curvature, agreeing with $R = \frac{2}{r^2}$. This finding suggests that in the pattern

$$(A)R_{ik} + (B)Rg_{ik} + (C)g_{ik} \text{ is proportional to } T_{ik}$$

the relation between A and B should be

$$A = 1 \quad and \quad B = -\tfrac{1}{2}.$$

As for C, it is negligible because, as a principle in cosmology, the amount of matter in the universe is relatively very small compared to its overall size (the very long radius of Gaussian curvature). Thus the term $(C)g_{ik}$ can be omitted, leaving

$$(1)R_{ik} + (-\tfrac{1}{2})Rg_{ik}$$

and

$$R_{ik} - \tfrac{1}{2}g_{ik}R \ \ is \ proportional \ to \ \ T_{ik}.$$

In the meantime we recall that in a gravity-containing frame of reference, G_{ik} should likewise be proportional to T_{ik}, again suggesting the validity of the equation

$$G_{ik} = R_{ik} - \tfrac{1}{2}g_{ik}R.$$

However these steps have not addressed the issue of *divergence*. In particular, both sides of

$$G_{ik} = T_{ik}??$$

should be divergenceless, so that

$$T^{ik}{}_{;i} = 0$$

as well as

$$G^{ik}{}_{;i} = 0.$$

This means that to satisfy the conservation laws, a covariant derivative of the G's must vanish. Therefore we must substantiate that $G^{ik}{}_{;i} = 0$, or more explicitly that

$$(R^{ik} - \tfrac{1}{2}g^{ik}R)_{;i} = 0.$$

Before considering the verification of this equation, we should mention a complication about $T^{ik}{}_{;i} = 0$ and $T^{ik}{}_{;k} = 0$, one which we glossed over on page 360. We ignored the effect of the gravitational field upon itself. That is to say, general relativity envisions matter indenting the gravitational field (as reflected in the shape of space-time) but in theory the gravitational field gains some energy in the process, like indenting trampoline gives it some potential energy. If t^{ik} represents the additional energy-components that arise from this process, the total mass-energy tensor—the true total energy of the system subject to the laws of conservation—should equal $T^{ik} + t^{ik}$. The term t^{ik} is not a tensor, since it depends on the choice of coordinates, but we can call it the energy of the gravitational field. This quantity is difficult to ascertain mathematically and to interpret physically, but it implies that we could amend the equation $T^{ik}_{;i} = 0$ to read $(T^{ik} + t^{ik})_{;i} = 0$, which then casts doubt on $(R^{ik} - \frac{1}{2}g^{ik}R)_{;i} = 0$. The issue is resolved, or at least evaded, by the fact that $R_{ik} - \frac{1}{2}g_{ik}R$ appears to include the energy of the gravitational field. Hence we ignore this point. (Ohanian, pp 103-104 and 265, and Kilmister, pp 158-160 and 247-248, show more details.)

We now take up the proof of $(R^{ik} - \frac{1}{2}g^{ik}R)_{;i} = 0$, which calls upon several maneuvers of tensor calculus, including use of the "Bianchi identities." The Bianchi identities constitute another important mathematical "tool" in the development of general relativity, but as Pais points out, the original derivations are not really Bianchi's, and Einstein did not learn about this tool until he had consumed considerable time and effort working around the problem. We can grasp the essence of the Bianchi identities by again inspecting a cube; a small six-sided box is a helpful prop. Let us imagine that we start at any corner and that we *parallel-transport a vector* along four edges so as to outline one of the faces of the cube, like we looped around our city block on page 301. We continue around each of the other 5 faces of the cube but only once per face, and we follow the most expeditious path, progressing only back and forth along the edges. Thus we trace out 6 four-sided loops, traversing each edge an even number of times so as not to travel along any edge without a reverse trip. The opposing faces of the cube need not be identical (so it's not exactly a cube), and *we use the curvature tensor to quantify how opposing faces differ*: Top vs. bottom, front vs. back, left side vs. right side.

As the cube is three-dimensional, we label the dimensions x, y, and z; and we recall that the v-component of the curvature tensor can belong to the vector being parallel-transported, while u can be the direction of the drift of that vector. See page 305, though we used other letters. If the left and right faces of our cube lie in the y-z plane, and if the x-edge connects these two faces, then the difference between them—the change in components upon parallel transport along x from one face to the opposing face—is $R^{u}_{vyz;x}$. (The restriction to only dual-direction trips along each edge excludes irrelevant differences. For instance any effect on the components from transport down the y-edge is nullified when we go up the y-edge.) We assign a term for the difference between the faces of each pair—we have three pairs—and *all the changes in the components should cancel out, leaving a net of zero*. For example the difference between top and bottom of the cube might be 2 units, and the difference between front and back might be 1, in which case the difference between left and right faces should be exactly -3. Though the analogy is not perfect, we can say that if the net amount were not zero, our box could grow or shrink just from being examined, and since the net amount is zero, the box is "conserved." The following equation—in mixed form—describes the entire exercise:

$$R^{u}_{vyz;x} + R^{u}_{vzx;y} + R^{u}_{vxy;z} = 0$$

This equation is one of the "Bianchi identities." In the words of tensor calculus, the Bianchi identities are rules about the derivatives of the curvature tensor when its indices are rearranged in a cyclical manner: The sum of the covariant derivatives of this tensor vanishes. (Algebraically, our 6 loops form 6 Γ's and call for second derivatives. The underlying equation is $\Gamma^{u}_{vz,yx} - \Gamma^{u}_{vy,zx} + \Gamma^{u}_{vx,zy} - \Gamma^{u}_{vx,yz} + \Gamma^{u}_{vy,xz} - \Gamma^{u}_{vz,xy} = 0$. See Bergmann, <u>Introduction...Relativity</u>,

Let us use the antisymmetry rule $R_{1234} = -R_{2134}$ on the third term of the co-variant form of the Bianchi identity, so that the first two indices are reversed and so that this term becomes negative:

$$R_{1234;u} + R_{124du;3} - R_{21u3;4} = 0.$$

Since we aim to explore the behavior of the rank-two Ricci tensor rather than the rank-four curvature tensor, we summon contraction. However contraction should be performed on mixed indices (i.e. only between supra- and sub-scripts), meaning that we must raise some indices. As just mentioned, doing so introduces the metrical tensor, but we will set that part aside temporarily and simply write

$$R^1_{234;u} + R^1_{24u;3} - R^1_{2u3;4} = 0.$$

For this contraction we let component number 1 be the same as 3 in the terms that contain the number 3;

$$R^1_{214;u} + R^1_{24u;1} - R^1_{2u1;4} = 0.$$

Hence

$$R_{24;u} + R^1_{24u;1} - R_{2u;4} = 0.$$

This equation may be called the "contracted Bianchi identities" (as described by Schutz, p. 174, equation 6.93). We have eliminated indices 1 and 3 in the first and third term, and the 3 in the second term is now 1. We note that in effect the second term $R^1_{24u;1}$ has been subjected to finding its divergence.

We use the antisymmetry rule again so that the second term reverses its first two indices and becomes negative, and we raise another index in preparation for further contraction. These steps give

$$R^2_{4;u} - R^{21}_{4u;1} - R^2_{u;4} = 0.$$

Upon contraction for components 2 and 4, we are left with

$$R_{;u} - R^1_{u;1} - R^2_{u;2} = 0,$$

which again implicates finding divergences (in the second and third term).

Since $R^a_{u;a}$ and $R^b_{u;b}$ are equivalent (one of these is redundant because they contract to the same value), the terms

$$- R^1_{u;1} - R^2_{u;2}$$

can be replaced by

$$- 2R^1_{u;1}.$$

Altogether we have

$$R_{;u} - 2R^1_{u;1} = 0$$

or

$$\tfrac{1}{2}R_{;u} = R^1_{u;1}$$

or

$$R^1_{u;1} - \tfrac{1}{2}R_{;u} = 0.$$

This is the result of two contractions on the Bianchi identities, showing us *how the Ricci tensor conforms with the Bianchi identities*; the accessory term turns out to be - $\tfrac{1}{2}R_{;u}$, which *contains our sought-after* - $\tfrac{1}{2}R$.

We still have a "loose end": Having raised indices twice, we must heed $A^a = g^{ab}A_b$ by multiplying through by the metrical tensor twice. We can go back to the Bianchi identity we started with and write

$$g^{13}(R_{1234;u} + R_{124u;3} + R_{12u3;4}) = 0.$$

This is repeated on the "contracted Bianchi identities." Two conclusions flow from these steps. One is that

$$R^1_{u;1} = \tfrac{1}{2}R_{;u}$$

which shows that the divergence of the Ricci tensor (the left side of this equation) indeed is not zero.

The second and more important conclusion is that once we re-adopt our familiar set of indices, our earlier equation,

368

$$R^1_{u;1} - \tfrac{1}{2}R_{;u} = 0,$$

will become equivalent to

$$G^{ik}_{\;;k} = 0.$$

The mechanism for this equivalence is as follows: The equation $R^1_{u;1} - \tfrac{1}{2}R_{;u} = 0$ is multiplied through by g^{iu}. By applying the rule

$$g^{iu}R_u = R^i$$

(which raises an index) the first term $R^1_{u;1}$ becomes

$$g^{iu}R^k_{\;u;k}$$

which reduces to

$$R^{ik}_{\;;k}$$

wherein again we see the divergence of the Ricci tensor. Simultaneously the second term - $\tfrac{1}{2}R_{;u}$ becomes

$$-\tfrac{1}{2}g^{iu}R_{;u}$$

which is equivalent to

$$-\tfrac{1}{2}g^{ik}R_{;k}.$$

The two new terms can be melded into

$$(R^{ik} - \tfrac{1}{2}g^{ik}R)_{;k}.$$

(Multiplying through by g_{iu} leads to the co-variant version. The g's appear in intermediate steps but we can call upon Ricci's theorem, $g_{ik;l} = 0$. This step is complicated by the fact that we are still dealing with covariant derivatives, and there are rules for covariant differentiation of products; see Møller, p. 282; Lawden, p. 101; and Kenyon, p. 61.)

Anyway, the whole equation

$$(R^{ik} - \tfrac{1}{2}g^{ik}R)_{;k} = 0$$

may be written as

$$G^{ik}{}_{;k} = 0,$$

or, by virtue of symmetry, with respect to i

$$G^{ik}{}_{;i} = 0.$$

At last, with help from the Bianchi identities, we have confirmed that *the Einstein tensor has zero divergence*. As the above derivation enjoys general covariance, the statement that "the covariant divergence of the Einstein tensor is zero" appears in the literature. This means that the inclusion of $R_{ik} - \frac{1}{2}g_{ik}R$ rather than R_{ik} alone *allows the field equations to obey the laws of conservation.*

It is interesting how such diverse the steps have led us to this conclusion: Earlier we applied the laws of conservation and the mass-energy tensor to a cube, and we found zero divergence. We then applied parallel transport and the curvature tensor to a cube, and we found the Bianchi identities. Finally we adapted the Bianchi identities to the Ricci tensor, and we found zero divergence in the Einstein tensor.

In the last step we passed through a stage with the Ricci tensor (in mixed form): Since this tensor does not have vanishing divergence, $R^1{}_{u;1}$ does not equal zero. Instead we found that

$$R^1{}_{u;1} = \frac{1}{2}R_{;u}.$$

The physical implication of the non-zero value of $R^{ik}{}_{;i}$ is that "something is being created out of nothing," which nature abhors.

In contrast both $G^{ik}{}_{;i}$ and $T^{ik}{}_{;i}$ equal zero, which of course immediately suggests that indeed

$$G_{ik} = T_{ik}??$$

or at least that

$$G_{ik} \text{ is proportional to } T_{ik}$$

and that in flat space-time where T_{ik} vanishes,

$$G_{ik} - 0.$$

We should add that we have two reasons for paying so much attention to the derivation of

$$G_{ik} = R_{ik} - \tfrac{1}{2}g_{ik}R.$$

First it shows how general relativity respects the laws of conservation. Second, it was Einstein's last major hurdle before announcing the complete field equations of general relativity in 1916. It represents a decisive step in a complex sequence of developments which we summarize next. (See also Pauli, pp. 142-145.)

We recall that Einstein envisioned the gravitational field in general relativity as a "global" mosaic of "local" pieces of space-time. He realized that free-fall represents such a piece, within which gravitation is obscured. He let g_{ik} describe each piece, while each piece—but not a larger region of the mosaic—abides by the equations of special relativity. However looking at each piece alone failed to reveal the regional curvature, like looking at one tile at a time fails to reveal the overall shape of a mosaic. Einstein then found that the relation between adjacent pieces of the mosaic is given by Γ^a_{ik}, though this relation still did not fully reveal regional curvature, like looking at two tiles is not enough to delineate the overall shape of a mosaic. Fortunately parallel transport in a loop allowed the geometric nature of the mosaic to be given in tensor form by R_{hikl}, which fully revealed regional curvature for all observers, like looking at a group of four tiles reveals curvature reliably and consistently. Now "global" curvature could be delineated, allowing the description of frames of reference in which a gravitational field can exist. However to exclude tidal complications R_{hikl} had to be contracted to R_{ik}, and to heed the laws of conservation R_{ik} had to be "Bianchified" into G_{ik}. For further verification, Einstein demonstrated that the field equations of general relativity, which describe the overall mosaic, reduce to equations of special relativity, which describe small locally flat "tiles." All local frames in free-fall abide by $G_{ik} = 0$, and indeed $G_{ik} = 0$ is pivotal, for reasons we will detail later. In summary Einstein found a way to study small parts of the gravitational field while anticipating that the laws for the whole apply to its parts. He deduced valid laws for the whole from the interrelationship of these parts. We can only imagine Einstein's exhilaration as his ideas and derivations finally fell into place![*]

◆

Returning to our goal of equating G_{ik} with T_{ik}, one more element is missing: Space and time appear as G_{ik}. T_{ik} stands for mass and energy. To have a true equation a *constant of proportionality* is needed, which we arbitrarily call "X," so that the units of the two sides are compatible. This X must be a scalar—it must be alike in any system of coordinates. Moreover when gravity is weak and speeds are slow, this result must agree with Newton's. Hence the numerical value of X must allow this agreement. Therefore more complete than

$$G_{ik} \text{ is proportional to } T_{ik}$$

---

[*] He did write about his feelings; see Pais, pp. 177 and 251.

is the equation that includes the constant X:

$$G_{ik} = -XT_{ik}.$$

In fact X contains another constant, the same that appears in Newton's law of universal gravitation by the letter K. (This K is unrelated to the Gaussian curvature, and the X is unrelated to the coordinate axis X.) Specifically, our $X = 8\pi/c^4 K$, and it is negative because of how we define our variables; e.g. $\Phi = -KM/r$. Some authors prefer to spell out X as $-8\pi/c^4 K$ when writing the field equations. Meanwhile we also found that

$$G_{ik} = R_{ik} - \tfrac{1}{2}g_{ik}R,$$

so we can interchange G_{ik} with $R_{ik} - \tfrac{1}{2}g_{ik}R$.

At last, as Einstein had achieved in 1916, we have solved all objections, we have met all criteria, and we have all the ingredients needed for the field equations of general relativity:

$$\boxed{R_{ik} - \tfrac{1}{2}g_{ik}\,R = -XT_{ik}}$$

Here Einstein states that *in all frames of reference the induced curvature of space-time at the site of any physical event*—the left side of the equation—*agrees with the density of mass and energy which is responsible for inducing that curvature*—the right side of the equation. This is nothing more than a mathematical restatement of the second postulate of general relativity: Matter curves space-time.

This deceptively brief equation is actually a very compact master formula. Because of its tensor character it conveys a large amount of information, and it represents an entire family or a "system" of equations. With four possibilities for each i and k, there can be 16 equations in this family, but only 10 are independent. Their more complete name is "covariant partially differential field equations" because they comply with general covariance, because partial differentiation is involved, and because gravitation is treated as a field phenomenon. We note that the relationship between G_{ik} and T_{ik} is linear, as predicted in Poisson's equation, but unlike Poisson's equation and unlike Newton's law of gravitation, the full field equations, $R_{ik} - \tfrac{1}{2}g_{ik}\,R = -XT_{ik}$, are not linear and are second-order (hyperbolic).

Although the field equations are identified with Einstein, other contemporaries were knocking at the same door. In fact a mathematician, Hilbert, derived the same equations and published them earlier (in 1915) than Einstein, but Hilbert had used some of Einstein's previous ideas, and only Einstein carried the concepts they embodied to their full fruition.

To use the equation $R_{ik} - \frac{1}{2}g_{ik} R = -XT_{ik}$, or to be more exact, to use these 10 equations-in-one, they can be solved for g_{ik}, which is the metrical tensor that describes the shape of space-time at the site of the physical event. Knowing g_{ik} allows us to use the equations of motion to predict how nearby objects behave. This behavior, as we have said, is motion along a geodesic which we call the effects of gravity such as falling and orbiting. In effect, nearby matter and energy bend otherwise straight lines into geodesics in space-time *to the extent predicted by the field equations*.

Although we say that the field equations are to be "solved for g_{ik}," the operation—as we will see in the next chapter—is complicated and indirect. However we should not be confused by the appearance of R_{ik} and R in these equations. Both R_{ik} and R are related to g_{ik}; they are not independent unknowns. (I.e., at any location R_{ik} and R are functions of g_{ik}.) In other words g_{ik} and its "offspring" are needed to describe space-time in such a way that we can link the changes in the shape of space-time to the distribution of matter and energy expressed by T_{ik}. In this context the field equations merely specify how g_{ik} and T_{ik} are linked.

✦

Although we have conventions (some invented by Einstein) to tell us what the letters and subscripts in the field equations stand for, these equations can be written in various forms, so that cross-comparisons among different sources can be confusing. For example instead of the R's in the above equations, other letters may be used, and instead of the subscripts *ik*, two other subscripts such as *mu* or *ij* may appear, depending on the author's preferences.

The field equations may also be contracted and rearranged, as for example

$$R_{ik} = -X(T_{ik} - 1/2g_{ik}T).$$

This form elucidates the situation in which matter is absent: If T_{ik} becomes zero, T (which is a contraction of T_{ik}) also becomes zero, leaving

$$R_{ik} = 0 \quad \textit{and hence} \quad G_{ik} = 0.$$

It is also legitimate to write the field equations with mixed or contra-variant tensors, such as for example in contra-variant form

$$R^{ik} - 1/2g^{ik} R = -XT^{ik}.$$

Of course we can write the field equations with the Einstein tensor,

373

$$G_{ik} = -XT_{ik}.$$

✦

General relativity does not furnish us with a succinct "law of gravitation" as did Newton's system. "$G_{ik} = -XT_{ik}$" is as close as Einstein comes to a practical "Einstein's law of gravitation." However in the context of physics as a whole

$$\text{"}G_{ik} = 0\text{"}$$

is a better candidate for this distinction, as it reveals several essential concepts regarding the gravitational field. Here is how: For one thing, $G_{ik} = 0$ is the relativistic equivalent of Laplace's equation $\nabla^2 \Phi = 0$ (page 251) which is requisite for Poisson's equation, so that if $G_{ik} = 0$ would not hold, neither could $G_{ik} = -XT_{ik}$. Moreover G_{ik} depends entirely on the shape of space-time (because R_{hikl} depends entirely on the shape of space-time), while from the vantage of a freely falling observer, $G_{ik} = 0$ always holds and always reflects that space-time is flat. However, by the principle of equivalence, a law which holds where gravitation is obscured must hold in any gravitational field. This means that a gravitation-free frame of reference can be treated as one particular frame among all frames, each of which has equal legitimacy, be it a frame with no matter nearby, or one that is sufficiently small, or one in free-fall. Since G_{ik} vanishes where the gravitational field happens to be "transformed away," $G_{ik} = 0$ is one possible solution for the law of gravitation, and since all falling and orbiting objects are in free-fall, it is the most basic solution. $G_{ik} = 0$ is also a tensor equation, and if a law can be expressed by setting all the components of an appropriate tensor equal to zero—if the tensor vanishes in one frame—then this tensor vanishes in any frame. Thus if the components of tensor G_{ik} vanish (they "zero out") for one observer, all observers in the universe agree that this space-time, assessed by the G's, appears flat (it "zeroes out" for all). We also recall the role of covariant differentiation in the derivation of R_{hikl} and hence of G_{ik}. Since "$G_{ik} = 0$" survives all transformations, *it is a generally covariant law*.

That means that any valid solution—any solution meeting all the requirements that Einstein struggled for years to meet—must comply with $G_{ik} = 0$. Indeed if a solution violates this equation, it means something is wrong. (e.g. a law of conservation is transgressed), in which case a freely falling observer may feel gravity, and in which case Einstein's Principle of Equivalence was a dreadful mistake rather than the cause (in 1907) for great exhilaration and glee.

Of course upon transforming to a wider matter-containing system of coordinates, space-time appears curved non-locally. G_{ik} no longer vanishes but equals $-XT_{ik}$. That is to say, G_{ik} is proportional to T_{ik} because all frames are acceptable, not only ones in which T_{ik} is zero. This means that upon transformation to a wider matter-containing frame or to a frame in accelerated relative motion with respect to free-fall, all observers except strictly local ones agree that G_{ik} does not vanish. Still, $G_{ik} = -XT_{ik}$ is not valid unless $G_{ik} = 0$ is valid. (Now pages 265-266 have more meaning.)

There is nothing novel about a basic law of nature set equal to zero. For example Aristotle believed that resting is a natural condition, and that undisturbed objects tend to stop moving. A law for this idea it might state that

"by nature, motion = 0."

This pattern reappears in Newton's law of inertia: An undisturbed object maintains its state of rest (if at rest) or of uniform motion (if in motion) until something disturbs it. The underlying law states that

"by nature, the change in rest-or-motion = 0."

Einstein taught that undisturbed space-time maintains its flatness until matter is close enough to disturb it. The underlying law states that

"by nature, $G_{ik} = 0$"

and valid laws of physics cannot contradict this equation.

Nevertheless it is fair to ask why not call "$R_{ik} = 0$" a law of gravitation, but the answer is clear: This equation does not heed conservation, in that R_{ik} is not divergenceless. Then (given the importance of the vanishing of R_{hikl}) what about "$R_{hikl} = 0$" as a law of gravitation? The reply is that for R_{hikl} to vanish, all 20 components, those associated with change-of-volume and those associated with tidal changes-of-shape, must be zero. This requirement is too restrictive, since we are only concerned with the former. Indeed the 10 components of G_{ik} can be zero ("no gravity") while R_{hikl} is not zero ("at least some tides"), but not vice versa. $R_{hikl} = 0$ includes $G_{ik} = 0$; "no-change-of-volume-nor-of-shape" includes "no-change-of-volume." As we said, the curvature tensor is so inclusive that in theory it can detect natural space-time curvature even in the absence of gravitational falling or orbiting, and now we add that "no gravity" must mean "no G_{ik}" (as during free-fall) but need not mean "no R_{hikl}." We can express this notion in terms of T_{ik}: Space-time could never be curved if $R_{hikl} = 0$ were a law because T_{ik} is incapable of curving space-time which had been such that $R_{hikl} = 0$. However T_{ik} curves space-time which had been such that $G_{ik} = 0$.

375

The possibility that "$G_{ik} = 0$" while "R_{hikl} is not 0" is informative: We asserted that the proximity of matter gives space-time a certain curvature, but *not* the curvature of a sphere. For example in our trampoline analogy the cross-section of a depression is not simply round. Let us see how R_{hikl} and G_{ik} would behave if gravitationally-curved space-time had a perfectly spherical shape rather than the more complicated shape found in nature. Since a sphere has intrinsic curvature in only two dimensions (*i* and *k* can only be 1 or 2) and has no tidal irregularities, we found that its curvature tensor has one non-zero component, R_{1212}, which by symmetry equals R_{2121}. In this case R^1_{212} and R^2_{121} (in mixed form) contract to non-zero components of the Ricci tensor, R_{11} and R_{22}. Further contraction gives us R, the Ricci scalar, which we already noted is $\dfrac{2}{r^2}$. Carrying out the algebra (Kenyon, pp. 192-193) to solve the equations

$$G_{11} = R_{11} - \tfrac{1}{2}g_{11}R$$

and

$$G_{22} = R_{22} - \tfrac{1}{2}g_{22}R,$$

we find that G_{11} and G_{22} end up as *zero*, which means that under spherical conditions, $G_{ik} = 0$. Once we reason this through, it is not surprising: A sphere has intrinsic curvature, which is why its metrical tensor, its curvature tensor, and its Ricci tensor do not vanish, but the contour of a sphere is *not* representative of gravitationally-curved space-time, so that on a sphere G_{ik} does indeed vanish.

In this context our customary statement that zero G_{ik} indicates the absence of space-time curvature is not precise. R_{hikl} indicates the absence of a specific kind of curvature, one where a non-zero R_{hikl} can coexist with a zero G_{ik}. The prominent (local) instance is a frame of reference falling freely toward the Earth. The point is that $G_{ik} = 0$ always means the same thing to us, namely the absence of gravity, which of course also means the absence of any nearby matter and energy.

✦ ✦ ✦

Looking back at the derivation of

$$R_{ik} - \tfrac{1}{2}g_{ik}R = -XT_{ik},$$

it is clear that far more obstacles and complications arose in obtaining $R_{ik} - \frac{1}{2}g_{ik}R$ than in deriving T_{ik}. Though T_{ik} is by no means simple, finding this portion of the field equations was easier for Einstein because its basis is in special relativity (e.g. the equivalence of mass and energy, the implication of momentum, and Maxwell's preparatory work) which had been discovered several years earlier.

Nevertheless we pointed out that the scientific integrity of T_{ik} is not as sound as that of the left side of the field equations because our understanding of the nature of matter is not complete. This was already apparent in Einstein's day, and Einstein himself likened the left part of the field equations to "fine marble" but the right side to "low grade wood." (Out of My Later Years, p. 83.) He pointed out that T_{ik} is only a phenomenological representation of matter based on our experience and observation rather than an understanding of its essential nature. As we shall see in the chapter covering quantum mechanics, subsequent gains in our knowledge have not always mitigated this situation, since not everything we have learned about matter supports relativity.

In any case here is an overview of the rationale for the various ingredients in

$$R_{ik} - \frac{1}{2}g_{ik}R = -XT_{ik},$$

presented as a series of questions and answers. Let us go back to the metrical tensor g_{ik}:

Why do we need g_{ik}?

Because it adapts the pythagorean equation to non-Euclidean geometry.

Where is geometry non-Euclidean?

Where space-time is curved.

Why not adapt the pythagorean equation simply by means of coefficients such as A, B, C, and D?

Because for clear identification each coefficient needs two indices, each of which runs 1 to 4, to accommodate four dimensions and two curvatures.

What curves space-time?

Nearby matter and energy.

Why is this important?

Because curved space-time accounts for gravitation, and because gravity or its effects are found everywhere in the known universe. Curved—i.e. non-Euclidean—space-time thus explains why objects fall, why planets orbit, why starts evolve as they do, why stars glow, and indeed why the universe is the way it is.

Why reject gravitational force?

Because critical observations support curved space-time but not force.

How do we use g_{ik}?

To quantify how non-Euclidean space-time is locally, which means to express how curved it is.

How do we know that g_{ik} is a tensor?

By how it transforms; that is to say, by how it obeys the transformation laws of tensors.

Why is g_{ik} a rank-two tensor?

Because the shape of space-time is too complicated to be described by a vector.

What makes space-time at one point so complicated?

At each point curved space-time has not one but two intrinsic curvatures, and their radii are not constant.

Why do we involve vectors?

Because space-time intervals and radii of space-time curvature can be considered vectors.

What is the advantage of vectors?

Vectors are invariant, and if vectors are invariant, so are tensors.

Why are vectors invariant?

Because they do not change under a transformation of coordinates. When a set of axes is rotated, we obtain new components for the same vector. Therefore a vector does not vary despite a change of the frame of reference. Only its components may vary.

Why do we need invariance?

Because we seek general covariance. I.e., if vectors survive transformation between frames of reference, tensors survive transformation, and if tensor equations survive any transformation, tensor equations possess covariance. Laws of physics written as tensor equations can be generally covariant.

Why must the vectors be four-dimensional?

Because the three dimensions of space alone are relative and time alone is relative, but space-time is not. I.e., observers may disagree on measurements of space and time but not on measurements of space-time. We need four-dimensionality to have general covariance in general relativity.

Why are space alone and time alone relative?

Because the speed of light is the same for all inertial observers.

How do we know that the speed of light is so constant?

Because we can calculate and measure a constant c.

How do we know that space-time is curved in a gravitational field?

Because light passing near a concentration of mass is bent by the amount of calculated curvatures of space-time in the region.

Why is g_{ik} alone not enough to describe these curvatures?

Because we need the point-to-point variation in the shape of space-time.

Why involve such variation?

Because in Newtonian science, gravitation is based on the point-to-point variation in the gravitational potential, Φ, across a gravitational field.

Why invoke fields?

Because fields account for some properties of gravitation better than action at a distance (even though fields are much more complicated mathematically).

Why exploit Poisson's equation $\nabla^2\Phi = 4\pi K\rho$?

Because it outlines the relationship between Φ *(expressed as* $\nabla^2\Phi$*) and the density of matter* ρ*, and because to some extent it is a field equation.*

Is Φ a tensor?

No, Φ *is the scalar gravitational potential, but it can be equated with one component of the tensor* g_{ik}*.*

Why is Φ expressed as $\nabla^2\Phi$?

In this form Φ *is in a gradient, and gravitational fields have gradients.*

Why not keep $\nabla^2\Phi = 4\pi K\rho$ as the final field equation?

Because $\nabla^2\Phi$ *is a vector quantity and neither* $\nabla^2\Phi$ *nor* ρ *is a tensor quantity. This equation is invariant only in Euclidean systems. However* ρ *can be equated with one component of the tensor* T_{ik}*.*

Why T_{ik}?

Because it allows us to quantify the density of mass as a concentration of mass and energy.

Why must T_{ik} be a tensor?

Because we wish to express mass and energy in a region of space-time in such a way that the entire field equations can be tensor equations and can thus be generallycovariant.

Why both mass and energy?

Because $E = mc^2$*.*

Meanwhile, how do we deal with the fact that g_{ik} only evaluates space-time locally?

We use R_{hikl} *to express how space-time varies regionally.*

Why does R_{hikl} reveal regional variation?

Because R_{hikl} *is based on how* g_{ik} *varies in a loop of space-time during parallel transport.*

Why implicate parallel transport?

Three reasons: Because parallel transport generates geodesics, because it has physical counterparts, and because it can be substantiated mathematically.

Why are geodesics involved?

Because gravitational behavior in essence is motion along geodesics in curved space-time.

How can we validate geodesics?

By means of the principle of least action.

If we have R_{hikl}, why derive R_{ik}?

Because R_{hikl} contains superfluous data. In contrast, R_{ik} tells us how curved space-time is regionally while encompassing only the pertinent curvatures, those tied to T_{ik}, and not tidal effects.

Why do we accept G_{ik} over R_{ik}?

Because R_{ik} alone disregards the laws of conservation; R_{ik} does not have vanishing divergence. Only G_{ik} tells us how curved space-time is regionally while considering the pertinent curvatures as well as the conservation laws.

How do we adjust R_{ik} for the conservation laws?

By subtracting $\frac{1}{2}g_{ik}R$, thereby creating the term $R_{ik} - \frac{1}{2}g_{ik}R$, which we call G_{ik}.

How do we connect T_{ik} with $R_{ik} - \frac{1}{2}g_{ik}R$?

We place an "=" sign between them, along with the conversion factor X.

Why is this conversion factor X in the equation?

Because we are connecting space-time with mass-and-energy.

Why the minus sign (why -X)?

Because of the way we define certain constituents of underlying equations, such as $\Phi = -KM/r$.

How can we abbreviate $R_{ik} - \frac{1}{2}g_{ik}R = -XT_{ik}$?

By $G_{ik} = -XT_{ik}$.

What does this equation mean?

That XT_{ik} *distribution of mass and energy causes* G_{ik} *curvature of space-time.*

What fundamental prerequisite must $G_{ik} = -XT_{ik}$ meet?

That in the absence of a cause for gravity, such as where space-time is flat, $G_{ik} = 0$.

◆

It is also worthwhile to review the progress among our provisional field equations in tabular form, summarizing why each equation is objectionable and how the objections are answered:

PROPOSAL	OBJECTION	SOLUTION
$\nabla^2\Phi = 4\pi K\rho$	Not a tensor equation. Φ represents one component. ρ represents one component.	g_{ik} replaces Φ. T_{ik} replaces ρ.
$g_{ik} = T_{ik}$	Derivative needed. Not valid in non-linear transformations.	Covariant differentiation
$\Gamma\text{'s} = T_{ik}$	Γ's not tensors. Not relativistic.	Second derivative
$R_{hikl} = T_{ik}$	Tidal effects and 20 components included.	Contraction
$R_{ik} = T_{ik}$	Non-zero divergence; i.e. non-compliance with conservation laws.	Contraction Bianchi identity
$G_{ik} = T_{ik}$	Conversion factor needed.	Factor "X"

This table also recounts why the mathematics behind the field equations bestows general relativity with the power to explain how the universe works. The first three rows indicate that the equations should be made of tensors in order to be valid in all frames of reference and for all observers everywhere. The fourth row means that only the pertinent mathematical components of space, time, mass, and energy are included, furnishing the equations with some of their physical validity. The fifth row adds to that validity by ensuring that these equations comply with the laws of conservation of matter and energy. The last row means that the tensor for space and time is compatible with the tensor for mass and energy in one equation.

◆

Part of the reason several years elapsed between the publication of special relativity and general relativity was Einstein's struggle with tensor mathematics, which, he complained, was harder than relativity itself. He was not a professional mathematician, and at first he was unaware of the help available to him in the writings of Gauss, Riemann, and Ricci. The other reason for the delay is that Einstein was also working on other projects, one of which earned him the Nobel Prize. Therefore Einstein's eventual triumph over the mathematics of relativity is doubly astounding. At many junctures his phenomenal instincts and intuition led him to correct conclusions which would have otherwise been practically beyond human capacity.

On the other hand not all biographers are so complimentary. Critics point out that Einstein was more talented in physics than mathematics, so that at many pivotal points he relied on his vision of his goal in order to select mathematics that support that goal, implying that his relativity suffers from circular logic. The most notable example is Einstein's adaptation of covariant tensors to his thesis that laws are covariant; no wonder that his tensor equations sustain the idea of covariance! Others emphasize that several scientists had come across the same derivations at the same time or even earlier. Ironically some of Einstein's foremost colleagues died of various illnesses at young ages (Poincaré at 58, Minkowski at 45, Schwarzschild at 42), and we ponder how relativity might have evolved with their continued contributions.

Even if we admire relativity, we may still yearn for the simplicity of Newton's law of gravitation. Does relativity merely convert something easy and direct into something complex and convoluted? We agree that in its time Newton's formulation was elegant and beautifully uncomplicated, a brilliant insight, a milestone in the history of science, and a monument to the power of human reason to decipher nature. But alas, it was incorrect.

If general relativity is the correction, why is it so elusive? After all, once special relativity is accepted in principle, the mathematical challenges it poses are comparatively limited. For example the familiar pythagorean theorem is sufficient to obtain the special relativity factor, and $E = mc^2$ can be derived with ordinary algebra. Even the leap to general relativity appears rather paltry at first glance. Lorentz had shown how to adjust the laws of physics so that they are invariant despite uniform motion, and Newton had invented the calculus with which handle accelerated motion. True, Einstein imposed stringent demands such as the laws of physics must always be invariant, but we can easily apply Newton's differential calculus and basic vectors to show how space-time intervals are invariant. Why then was the process, as suggested in Weyl's diagram (p. 228), not simply one of

Pythagoras + Newton = Einstein?

Because when we work out the details, it turns out to be more like

Pythagoras + Euclid + Galileo + Gauss + Newton + Lorentz

+ Poisson + Minkowski + Maxwell + Ricci + Riemann = Einstein.

And, as we shall find in the next chapter, even after all these contributions (plus many others) are implemented, we cannot avoid complex metrical equations, equations of motion, and various assumptions just to describe the orbit of one object.

Given our environment, we favor intuitively obvious mathematical tools designed for flat surfaces and for two or three dimensions, yet these tools are ill adapted for general relativity. Perhaps if we lived where very high speeds and huge masses are commonplace, we might naturally prefer geometry in which four curved dimensions are the norm and in which flat and two-or-three dimensional surfaces are the exceptions. Then would we feel at home with laws of science expressed in complex tensor equations, and elementary schools might teach covariant differentiation of the metrical tensor before teaching the area of a rectangle.

In any case we can now be quite specific about the statement (page 198) that the notion of fields causes the modern laws of physics to appear "different" than they do in Newtonian science. Said briefly, Newton wrote laws to be obeyed by scalar masses and forces. Einstein wrote laws to be obeyed by tensor fields. Newton's law of gravitation is one "force equation." Einstein's is a set of ten "field equations."

Let us elaborate: In describing the gravitational behavior of two objects, Newton's law of gravitation invokes properties of the objects themselves, and given the experimental evidence available in Newton's day, simple mathematics sufficed. If a field is invoked for Newton's laws, it is only a descriptive mathematical tool, with no real physical interplay between the objects and that field, and certainly with no provision for objects to alter space or time. To describe gravitational behavior, Einstein's field equations invoke the four-dimensional property of the field, and, supported by modern experimental evidence, these equations give the field a physical reality. The field exists with or without objects, but it is affected by objects, and objects act as they do because of the shape of the field. Unfortunately, as we have seen, describing gravitational fields calls upon far more intricate mathematics.

We further note that Einstein's equations tell us what is happening at the very point at which a gravitational physical event is occurring. Unlike Newton's law, distance does not enter into Einstein's equations, so that the relative nature of space is not pertinent, and the question of how masses can exert force over distance is circumvented. Mass alone does not enter into Einstein's equations, only the density of matter and energy in tensor form, so that the relative nature of mass is no obstacle. The mass of the falling or orbiting object does not appear in his equations, which agrees with the notion that objects of any mass behave similarly. And in theory the selection of a frame of reference or system of coordinates does not enter into the picture, so the effects of the motion of observers do not matter.

We can appreciate additional insights here, which we may categorize into three levels. At the most elemental level, the field equations are just another law of gravity. They tell us how mass determines gravitation, which in principle is what Newton's law and Poisson's equation also tell us. In this context the field equations—though precise and reliable under various conditions—represent nothing new.

On another level however the relationship between T_{ik} and G_{ik} tells us how matter and energy change nearby space-time, and herein lies a major departure from the past. Matter and energy are equated, and space and time are fused so as to define a gravitational field. Still we can argue that nothing new has been equated or fused—the synthesis is innovative, but not the ingredients.

It is on a third level that general relativity rises far above its precursors. By way of the field equations Einstein found how the distribution of matter and energy governs the shape of space-time, and in so doing he discovered that *the physical content of the universe determines its geometry*. This conclusion is Einstein's revolutionary and totally unanticipated stroke of genius.

Let us also reassess why Einstein's conclusion is so important. *To understand the physical nature of our universe, we must know which physical laws apply universally. Such laws must address and predict the interplay between matter, energy, and space-time. The mathematical expression of these laws summons the equations of general relativity which are based on tensors that describe matter, energy, and the geometry of four-dimensional space-time.*

✦

We should add that not everyone is necessarily enamored with relativity. We already mentioned objections to Einstein's mathematical methods, and in chapter 20 we will cite dilemmas in reconciling relativity with another branch of physics. Moreover various scientists have leveled criticisms on more basic levels. For example Fowler posits that the velocity of light is not really constant but that it appeared that way since Einstein failed to consider the Doppler Principle. (Fowler's book is devoted to this point, but see in particular pp. 21-29 and 119-126. In contrast see Reichenbach, pp. 230-232; Shadowitz, p. 55; and Møller, pp. 8-15.)

For a more modern view, see Tom Bethell in The American Spectator (Vol 32, number 4, April 1999) who cites other criticisms of relativity, particularly that gravity appears to propagate faster than light. In interpreting this article the reader should know that relativity does not insist that gravity propagate only at speed c, nor that gravitons (see chapter 18) and photons must be very similar.

De Bothezat raises a similar protest to Einstein's conclusion that the velocity of light is constant regardless of the frame of reference. He points out (p. 9) that this velocity "is not a velocity at all, because it lacks the main characteristic of velocity, namely the relation to its frame of reference."

This is a valid point if we insist upon an absolute frame of reference, which of course Einstein did not. Furthermore Einstein did not stipulate that no frames are involved. He allowed all frames to be legitimate.

A prominent criticism of relativity ensues from the "Sagnac effect." This effect was first observed around 1911 and has been confirmed since. (It enters into the design of the GPS.). Using diagram T, let us imagine that the Earth is in the middle of the square, and that P_1, P_2, P_3, and P_4 are four satellites in orbit around the Earth's equator. The satellites and Earth rotate west to east. Satellite P_1 can give off a bright flash of light, and it is also equipped with a sensitive light detector. The other satellites have reflective surfaces. We now let P_1 emit a flash. This flash, which is a batch of photons, encircles the Earth by moving from P_1 to P_2 to P_3 to P_4 and back to P_1 where the detector records the photons' arrival. Since these photons move in the same direction (west to east) as the rotation of the system, we call them "co-rotating." However another batch of photons encircles the Earth by moving from P_1 to P_4 to P_3 to P_2 and back to P_1, where the detector also records its arrival. Let us call this east-to-west batch "counter-rotating." The critical observation, named for Sagnac, is that the counter-rotating photos arrive more promptly than the co-rotating ones. The former batch covers the same distance in less time; *its photons are faster*. This scenario has invited several sharp questions about relativity; let us outline three of these, with a brief reply to each. (A thorough discussion by G. B. Malykin in the Russian literature [Physics-Uspekhi 43 (12) 1229-1252 (2000)] has been translated on line.)

(1) The Newtonian idea of additive velocities appears to explain the Sagnac effect, in that the speed of contra-rotation can simply be added to the speed of light. Does this mean that relativity does not even exist? Actually the mathematical analysis of the Sagnac effect entails approximations (second-order quantities are ignored) so that the results fall within the Newtonian limit. If experiments were sufficiently sensitive, which is technically difficult, they would reveal a relativistic *excess* in the Sagnac effect beyond Newtonian predictions, like the precession of Mercury shows an excess which is explained by relativity. (2) Since the contra-rotating photons are faster than the co-rotating ones, does this mean that *c is not constant*, and therefore does the Sagnac effect invalidate special relativity? Here we recall that c is invariant in *inertial* frames of reference, i.e. during uniform motion. In the Sagnac effect the key event, an encirclement, is *not inertial*. In technical terms the photons are in accelerated motion—their overall path is bent—*during which c does not appear constant*. In other words the Sagnac effect does not disprove special relativity because it is not limited to the conditions for special relativity. This event belongs to general relativity, in that two different rotations induce—by the principle of equivalence—two different gravitational fields which bend light differently. Therefore some photons complete a rotation sooner than others. (3) However since the Sagnac effect detects rotation without reference to external coordinates, does it allow detection of *absolute* accelerated motion? Here we recall that relativity does not insist that the satellites rotate in a relatively stationary universe. In theory the Sagnac effect exists *even if the universe rotates around the satellites*. The motion is still only relative.

The word "solving" implies a straight-forward procedure: Given a particular distribution of matter, the field equations of general relativity should provide the *g*'s of the metrical tensor. These *g*'s can be used in a metrical equation to yield *ds*'s, which in turn describe the gravitational field. The field determines how objects fall or orbit, how heavy they are, and even how they are gravitationally compressed. In short, the field equations should tell us what matter does to space-time, so that we can tell what space-time does to objects. As we shall see, implementing this scheme is far from "straight-forward." Indeed in some ways the mathematics here is more complex than Einstein's derivation of the field equation, like a difficult piece of music that is harder to perform than it was to compose.

The first successes in solving Einstein's field equations by someone other than Einstein were registered in 1916 by Schwarzschild (1873-1916). Schwarzschild's first solution involved the assumption that all the mass of a gravity-producing object is concentrated at one point in space-time, called a point-mass, not unlike a center of gravity conceived in Newtonian physics. Schwarzschild's second solution considered the space-time of a solid spherical mass of finite external dimensions. These solutions are termed "exterior" or "interior" depending upon whether applied to the gravitational field outside an object, such as the site of a planet orbiting a star, or to the field inside an object, such as the interior of a star. We will concentrate on the more instructive external solution, but the internal solution is currently of great interest because it adds to our understanding of the cosmologic evolution of stars, notably their gravitational collapse. (See Adler et al, pp. 461-482 for details.)

Schwarzschild's contribution has some noteworthy historical aspects: For one, even though Schwarzschild was primarily an astronomer, he was so adept at mathematics that he solved Einstein's equations within a few weeks of the appearance of Einstein's publication of the field equations. Einstein was so impressed that he personally presented Schwarzschild's solutions to the scientific community. For another, Schwarzschild succumbed several weeks later to a disease which today is rarely fatal. For a third, his work suggested the possibility of black holes, but this was not fully appreciated until the 1930's when physicists, among them Oppenheimer and Chandrasekhar, worked out the ramifications. However Schwarzschild was not the only mathematician who solved Einstein's field equations. Others have done so since, and Einstein of course had also succeeded. However Einstein's original (1915) method does not give an exact result because he used certain approximations. Schwarzschild's approach was superior in that it bore a "rigorous" (Pauli, p. 164.) solution; we will examine this point later.

We recall the field equations in abbreviated form,

$$G_{ik} = -XT_{ik},$$

which tell us how a distribution of matter, represented by T_{ik}, affects the shape of surrounding nearby space-time, represented by G_{ik}. Schwarzschild's exterior solution in effect answers a specific question: *If that distribution of matter is our sun, what is the shape of space-time at the location of a planet in orbit around the sun?* The reply had two main parts. The more difficult part was determining how to express G_{ik} for the space-time at the site of the orbiting planet; as we shall see, Schwarzschild found four components of the metrical tensor g_{ik} which indicate the shape of space-time outside the sun, where $T_{ik} = 0$. The easier part was figuring out how to express T_{ik} for the sun; as we will see, Schwarzschild found a form of M which quantifies the sun's mass. In other words he translated the assertion in Einstein's field equations that "this is how T_{ik} affects G_{ik}" into "this is how the sun's M affects the planet's g_{ik}." Schwarzschild's solution therefore yields equations in the pattern of "some form of M is responsible for the components of g_{ik}," or in short,

$$\text{"components of } g_{ik} = \text{some form of } M.\text{"}$$

As a preview, one key equation with this pattern will be $g_{44} = 1 - \dfrac{2KM}{rc^2}$, and this equation "satisfies" $G_{ik} = -XT_{ik}$; we will explain the use of the word "satisfies" later. The components of g_{ik} then guide us—though along a rather indirect route—to a description of the motion of the planet as it orbits the sun's M.

Why was the first part of the task—converting G_{ik} into the proper g_{ik}—so difficult? The answer is that even if we work from the more overt form of the field equations,

$$R_{ik} - \tfrac{1}{2}g_{ik}R = -XT_{ik},$$

we cannot simply insert T_{ik} and solve for g_{ik}. R_{ik} is difficult to calculate, and at the same time we must designate the form of the metrical equation

$$ds^2 = g_{ik}\, dx^{\,i} dx^{\,k}$$

we will use, while each of these two equations actually represents 10 separate equations. We can think of this as trying to predict the path of a billiard ball in four dimensions as it rolls across a table of unknown size by determining the shape of the surface of the table, each point of which has multiple curvatures. In technical terms, a complete solution calls for solving 10 simultaneous differential equations in four dimensions, which means determining 10 unknown functions (the components of g_{ik}) of four variables (X, Y, Z, and T). As we shall see, an elaborate chain of mathematical maneuvers is required in order to reduce the problem to the more manageable case of one unknown function of one variable.

At this juncture we can raise an earlier point (page 161): In four dimensions, an intrinsically curved surface can have 6, not 10, independent Gaussian curvatures. (The number of K's = (n - 1)n/2. Here n = 4.) Though we are mixing Gaussian and vector/tensor terminology, the implication is that a complete solution of Einstein's field equations entails only six components of the metrical tensor. We know that four components depend upon the choice of the system of coordinates, regardless of the intrinsic shape of the space-time being analyzed. By analogy, we must say something about our billiard table before we can predict how the game will proceed. On the other hand once we do so, we are still required to pursue six curvatures. Therefore for clarity let us envision that we are pursuing two types of *g*'s: the "*g*'s belonging to the system of coordinates" and the "*g*'s belonging to intrinsic curvature." All the *g*'s together describe the space-time around an object, but the latter type determines gravitational behavior. We will also find that just two *g*'s of each type suffice, and not surprisingly these four lie on the diagonal of a matrix for g_{ik}.

◆

We shall now cite several important assumptions which Schwarzschild made to simplify solving the field equations. It will be easier to explain how these assumptions are helpful as we go along. One assumption is that the ds^2 of the metrical equation be "time-independent" or "static." At first glance this seems to mean that the system under consideration, such as the sun and an orbiting planet, does not move, but that is not accurate. "Staticism" requires symmetry with respect to the past and future, with *no favored direction in time*. Mathematically, it means that ds^2 remains invariant whether we "move" forward or backward along the time coordinate. Since we let x^4 represent time so that dx^4 is equivalent to a time interval, staticism provides that ds^2 is the same whether dx^4 is positive or negative. By analogy, a motion picture of a "static" gravitational event appears identical whether it is run forward or backward. We believe that time actually has only one direction, but this does not lessen the advantage of staticism as a simplification for Schwarzschild's mathematics.

Another assumption is "spherical symmetry." At first glance this appears to mean that the system under consideration is perfectly spherical, but spherical symmetry actually means that there should be *no favored direction in space*, such as a specific axis of rotation. This notion agrees with the concept that spherically symmetric distributions of matter are surrounded by spherically symmetric space-time. Spherical symmetry means that even in four dimensions, rotation around a mass point does not disturb the invariance of the metrical equation. In more practical terms, spherical symmetry provides that a metrical equation be valid and accurate at all points at distance r from the center of the *M* responsible for the gravitational field. (It does not require that r or ds^2 to be the same for all such points, only that the equation hold at all such points.) In effect, because of spherical symmetry, Schwarzschild's scenario has no "up" or "down," and this also simplifies things.

Staticism and spherical symmetry correlate with the astronomical observation that the universe is "isotropic" and "homogeneous." From our vantage on Earth the universe appears similar no matter which way we look. Hence we say that on a large scale "isotropy" allows no preferred direction or directions in our universe; all directions are equally legitimate. The universe also would appear to have the same properties and laws if observed at any other location. Hence we say that on a large scale "homogeneity" allows no preferred location; all locations are equally legitimate.

Of these two concepts, homogeneity is harder to prove. On a small scale of course no gravitational field is homogeneous, and in fact its variation is vital to its nature. On a large scale homogeneity cannot be verified because widely separated observers cannot compare their observations simultaneously. For example by the time we learn about an experiment on falling objects performed on another galaxy, the information is out of date and could be invalid. Nevertheless we have no indication that the laws of physics differ anywhere in the universe, and we interpret this observation as evidence of a fundamental homogeneity in nature.

Isotropy is more easily manifest: Individual stars, constellations, and galaxies look different to us, but on the whole all directions reveal the same features, and the universe has no "up" or "down." Incidentally, this is not as obvious as it sounds. For centuries science held that all objects in the universe can only fall one way, "down" to our Earth, so that if the Earth were spherical, we seriously feared that objects on the far side would fall off.

On a basic level isotropy justifies the rotational transformation: We are permitted to rotate a system of coordinates in any direction *because* all directions are equally legitimate. Similarly homogeneity justifies the "sliding" translation: We are permitted to "slide" a system of coordinates to any location *because* all locations are equally legitimate. In effect, by virtue of isotropy and homogeneity together, physical laws should not be changed by—should be invariant under—transformations in general. Of course this applies to four dimensions. (Even the use of a perfect fluid as a model in deriving T_{ik} can be justified on the basis that homogeneity is one property of a perfect fluid.)

The point is that staticism and spherical symmetry are mathematical reflections of isotropy and homogeneity which helped Schwarzschild. For example if the universe were not isotropic, r, θ and φ (defined shortly) could be negative or positive, which would mean that ds^2 is not invariant under transformations. Or, if the universe were not homogeneous, the equations for ds^2 would not hold at any location nor in any frame of reference, and no law of physics, even if based in general relativity, could be relied upon to be valid everywhere in the universe. Homogeneity is also reflected in the constancy of the Ricci scalar, R, in the field equations.

We can add that if the mathematics of general relativity could not legitimize all frames of reference—particularly if $G_{ik} = -XT_{ik}$ and $G_{ik} = 0$ did not enjoy general covariance—modern cosmology would be futile. After all, $G_{ik} = -XT_{ik}$ and $G_{ik} = 0$ *are laws of gravitation, and gravity is a crucial determinant of how our universe works*. Of course Newton also issued a law of gravity, but we now realize that its mathematics admits only certain frames of reference, namely those without curvature of space-time, whereas the most interesting sites in the universe have space-time that is very much curved. In fact as we mentioned, this is the fundamental reason we even care to study general relativity in a book such as this: Our ability to understand our cosmic home, its origin, and its fate, relies on the mathematical integrity of general relativity in a homogenous and isotropic universe.

As mathematicians since Schwarzschild have shown, it turns out that it is not even necessary to overtly state both assumptions, that of staticism and that of spherical symmetry. The former assumption follows mathematically from the latter. The details are immaterial here (see Carmelli, p. 156), but the result underscores that these assumptions reflect essential features of relativity as well as basic properties of nature.

Let us continue looking at assumptions helpful in Schwarzschild's method. He envisioned that the spherically symmetrical mass that is responsible for a gravitational field exists in a vacuum. In other words his solution is set in the empty space which surrounds a spherical body with its center at the center of spherical symmetry, and it disregards the effects of that body's interior. This assumption may seem superfluous but we state it explicitly because certain mathematical steps we will cover are be based upon the notion that outside the mass $T_{ik} = 0$. Furthermore we should mention the assumption that the total matter in an astronomic object can be represented simply by its mass. This may be disappointing after we waded through the derivation of the various components of T_{ik}, but we know that under most conditions mass-density is the predominant component. This is why Schwarzschild's solution yields an equation which carries "some form of M" as a substitute for T_{ik}.

Finally it will be important to us that two additional assumptions limit Schwarzschild's solution: The internal scenario Schwarzschild selected for his solution does not provide for rotation of the central mass, nor does it provide for the possibility that an M may carry an electromagnetic charge.

✦

A key aspect of Schwarzschild's strategy is his selection of a system of coordinates as a framework. According to the principle of general covariance, Schwarzschild was free to select any system, but as we shall soon find, a *spherical-polar system of coordinates* (introduced on page 189) is the logical choice. An important preliminary question is what compelled him to restrict himself to *any one particular system.*

The answer is that a solution of the field equations must generate an assessment of space-time by means of the g's. We encountered this notion in other contexts, and here we see its application: A set of g's for a point depends not only on the shape at that point but also on the system of coordinates selected for the circumstances. Therefore a system must be designated before the g's can to provide an unambiguous description of space-time. Based on our remarks at the beginning of this chapter, this designation alone ought to decide some of the g's that appear in Schwarzschild's final set.

We emphasize that the selection of a system of coordinates is a matter of practicality. For example we could describe small two-dimensional distances (ds's) on the Earth's surface using Cartesian X and Y (x^1 and x^2) coordinates with coefficients A, B, and D. Then we can use the equation

$$ds^2 = (A)(dx^1)^2 + 2(B)(dx^1)(dx^2) + (D)(dx^2)^2.$$

However this is an unwieldy method for a sphere. We prefer to express distances on the Earth's surface between points of known latitude and longitude by means of spherical-polar coordinates,

$$ds^2 = r^2 d\theta^2 + r^2 \sin^2\theta \, d\varphi^2.$$

(For interested readers, the trigonometric derivation of spherical-polar coordinates is covered in textbooks of geometry. It is possible to derive spherical polar coordinates from Cartesian coordinates, and the essential equations linking the two are $X = r\sin\theta\cos\varphi$ and $Y = r\sin\theta\sin\varphi$. Lawden outlines the process on p. 86, but our Cartesian-type diagram K can be adapted, where our $CosR = X_A / Y$ becomes $X - r(cos\theta)$, $Y = r(sin\theta)$, etc.) In this setting r is the radius of the Earth, θ is the latitude, and φ is the longitude. Here r, θ, and φ take the role of *coordinates*. For example our point on the equator near Brazil has $\theta = 0$ degrees and $\varphi = 30$ degrees. We *solve the above spherical-polar metrical equation for ds* to figure the distance to another geographic point marked by another set of values for θ and φ. We note that we need three items of information for finding ds: We need the change in θ, which we write as $d\theta$ and which is a north-south distance. We need the change in φ, written as $d\varphi$, which is an east-west distance. And we need r, which is the size of the Earth; looking at a globe is helpful to picture this scenario.

The spherical-polar equation for two dimensions may be written with g's to show that r^2 and $r^2 \sin^2\theta$ take the role of *coefficients* of $d\theta^2$ and $d\varphi^2$:

$$ds^2 = (g_{11})(d\theta^2) + (g_{22})(d\varphi^2)$$

where g_{11} and g_{22} are these coefficients; the parentheses are included for clarity. Now

$$g_{11} = r^2,$$

$$g_{22} = r^2 \sin^2\theta,$$

and by implication

$$g_{12} \quad and \quad g_{21} = 0.$$

The above spherical-polar equation only allows for shifts in two dimensions, but we can add dr^2, so as to include a coordinate for variation in a third spatial dimension:

$$ds^2 = dr^2 + r^2 d\theta^2 + r^2 \sin^2\theta \, d\varphi^2.$$

The implied coefficient for dr^2 is 1, but it can have other values. We can think of dr as differences in altitude on the Earth. Since these spherical-polar coordinates make it easy to describe motion of an object along an imaginary curved three-dimensional surface, *spherical-polar coordinates were adapted by Schwarzschild to describe the motion of a planet along an orbit.*

Of course to heed relativity, the time-dimension must be incorporated. In spherical-polar coordinates the term $c^2 dt^2$ is added, in which t is multiplied by c in order to let time behave like a distance. As before, this term is made to be negative, and in this case we do not multiply it by i (which is $\sqrt{-1}$) because we wish t to remain an actual measure. (It is not essential here that all the terms have the same sign.) Thus the implied coefficient for $c^2 dt^2$ is -1. Our full spherical-polar metrical equation becomes

$$ds^2 = dr^2 + r^2 d\theta^2 + r^2 \sin^2\theta d\varphi^2 - c^2 dt^2,$$

in which we recognize coordinates adapted for four-dimensional spherical shapes: r, θ, φ, and t. These coordinates are difficult to envision; r is a radial coordinate (a radius), but t is a time coordinate (a time dimension), while θ and φ are angles. These correspond to Cartesian coordinates x^1, x^2, x^3, and x^4.

In the differential form we recognize

$$dr, \ d\theta, \ d\varphi, \ \text{and} \ cdt,$$

corresponding to

$$dx^1, \ dx^2, \ dx^3 \ \text{and} \ dx^4.$$

The associated coefficients are

$$1, \ r^2, \ r^2\sin^2\theta, \ \text{and} \ -1,$$

which are represented by the components

$$g_{11}, \ g_{22}, \ g_{33} \ \text{and} \ g_{44}$$

of the metrical tensor g_{ik}. In terms of these g's,

$$ds^2 = g_{11}dr^2 + g_{22}d\theta^2 + g_{33}d\varphi^2 - g_{44}c^2dt^2.$$

The other g's equal zero; we already mentioned that

$$g_{12} \ \text{and} \ g_{21} = 0,$$

but every g drops out except when $i = k$, since the g's off the diagonal of a matrix are all zero.

Let us re-write the equation with the four remaining non-zero terms, showing parentheses for clarity:

$$ds^2 = (1)(dr^2) + (r^2)(d\theta^2) + (r^2\sin^2\theta)(d\varphi^2) + (-1)(c^2dt^2)$$

This is analogous to our underlying metrical equation,

$$ds^2 = (g_{11})(dx^1)^2 + (g_{22})(dx^2)^2 + (g_{33})(dx^3)^2 + (g_{44})(dx^4)^2,$$

which in turn is summarized by the metrical equation

$$ds^2 = g_{ik}\, dx^i dx^k.$$

In other words the equation for spherical-polar coordinates is a specialized form of the metrical equation, one which suited Schwarzschild's purposes. In fact part of the appeal of a spherical-polar system is that it is clearly compatible with the restriction to spherical symmetry. That is to say, a spherical-polar coordinate system is such that the identity of ds^2, namely

$$g_{11}dr^2 + g_{22}d\theta^2 + g_{33}d\varphi^2 - g_{44}c^2dt^2,$$

is that same at any given distance from its center of symmetry. At that center $r = 0$.

♦

Before proceeding, readers consulting other sources, including Einstein (e.g. <u>Out</u> <u>of</u> <u>My</u> <u>Later</u> <u>Years</u>, p. 94) and Schwarzschild, must be warned that this equation

$$ds^2 = dr^2 + r^2d\theta^2 + r^2\sin^2\theta d\varphi^2 - c^2dt^2$$

is often written with the space coordinates negative and the time coordinate positive, and the order of the terms can be varied:

$$ds^2 = -dr^2 - r^2d\theta^2 - r^2\sin^2\theta d\varphi^2 + c^2dt^2$$

or

$$ds^2 = c^2dt^2 - dr^2 - r^2d\theta^2 - r^2\sin^2\theta d\varphi^2$$

or, with the r^2 collected,

$$ds^2 = c^2dt^2 - dr^2 - r^2(d\theta^2 - \sin^2\theta d\varphi^2).$$

These versions share the meaning of our equation, and they conform to the way other related terms and equations can be written. They also avoid having to solve for ds^2 with an imaginary term. (If the term containing time is positive, ds^2 is both time-like and positive [because the other terms are less than c^2dt^2] and then ds is real; see page 109.) Nevertheless for consistency we favor the positive-space and negative-time version except when otherwise imperative.

Readers may also encounter alternative ways in which these coordinates and their coefficients are designated. For example the time term may contain the term dt and the coefficient -c, or c is simply taken to equal 1; the principle is the same but the algebra is slightly different. Thus dx^1, dx^2, dx^3, *and* dx^4 may appear as

$$dr, d\theta, d\varphi, \text{ and dt,}$$

and the associated coefficients g_{11}, g_{22}, g_{33}, *and* g_{44} can be

$$1, r^2, r^2\sin^2\theta, \text{ } and \text{ } -c.$$

Moreover many authors prefer numbering the g's as g_{00}, g_{11}, g_{22}, and g_{33}, and in this convention our g_{44} (the time coefficient) corresponds to g_{00}. An advantage of this convention is that the g's associated with space can be easily distinguished from ones associated with time. As we shall apply later, the explicit labels g_{rr}, $g_{\theta\theta}$, $g_{\varphi\varphi}$, and g_{tt} are also useful.

✦

In any format, what can

$$ds^2 = g_{11}dr^2 + g_{22}d\theta^2 + g_{33}d\varphi^2 - g_{44}c^2dt^2$$

tell us? It is a metrical equation for free-fall of an object some distance from a center of mass-energy. As ds represent a space-time interval, *this equation gives the geodesical property of local space-time at the site of the object, and hence the equation describes the local gravitational field.* Our example with a falling apple showed us that to delineate ds at all locations along the dx's—to compute the tangent vector—the g's of the metrical tensor g_{ik} are essential. Schwarzschild selected a much grander scenario, our solar system, but we still must find g_{ik} to find ds. The relationship between ds and the components of g_{ik} resides in the metrical equation above. Finding Schwarzschild's g's will occupy us now.

In this scenario we treat the sun as one central non-moving symmetrical point, even though in reality it is not, but this assumption facilitates the analysis, and we treat an orbiting planet as an object in free-fall, which it is. We envision the path of the planet as a world-line, the space-time geodesic of which is equivalent to motion on the surface of a large imaginary sphere. (Since astronomic orbits are ellipses rather than circles, the analogy calls for a spheroid, but we can defer this detail.) At any given moment the planet is at distance r from the sun's center, and it is at location θ-by-φ on that sphere. Therefore *the orbiting of the planet can be described as small changes in the values of r, θ, and φ, while this transpires in time dt.*

In particular, the planet's progress through space is given by

$$g_{11}dr^2, \quad g_{22}d\theta^2, \quad and \quad g_{33}d\varphi^2$$

while its progress through time is given by

$$- g_{44}c^2dt^2.$$

Obviously ascertaining g_{11}, g_{22}, g_{33}, *and* g_{44} and inserting these into the metrical equation lets us find ds at the location of an orbiting planet.

Let us now study a matrix showing these components:

1			
	r^2		
		$r^2\sin^2\theta$	
			-1 or -1c^2

We can compare this matrix with the one on page 190. Again, all off-diagonal g's are zero. It is understood that the upper-left box contains g_{ik} when $i = 1$ and $k = 1$; the second box in the second row contains g_{ik} when $i = 2$ and $k = 2$, etc. As we shall do shortly, it is preferable to attach the conversion constant c^2 to the -1 in g_{44}, so that g_{44} appears as $-1c^2$ or simply $-c^2$. (The 1 isn't essential, but we show it for emphasis. The minus sign stems from how the metrical equation is set up, i.e. with a negative time component. The c stems from how time is converted to units compatible with space within the metrical equation, and the square sign stems from the use of c within the pythagorean theorem.)

As the term $g_{44}c^2dt^2$ implies, the time coordinate is associated with coefficient or component g_{44}. Furthermore, under these conditions g_{11} and g_{44} are reciprocal, based on how the relativity of length and time are reciprocal. Therefore, neglecting $-c^2$ as a constant,

$$g_{11} = \frac{1}{g_{44}}.$$

In the current setting, r is a "relativistic" distance in space-time, which is not precisely the same as the distance between the sun and a planet. If we assume that this distance is X in three-dimensional Cartesian coordinates, then the stronger the gravitational field, the wider the discrepancy between r and X. But if we imagine that a planet is exceedingly far from the sun and r is very long, the planet finds itself practically unaffected by gravitation. In that case a planet's location along an ordinary X axis indeed equals r, and the ratio between r and X is 1. That is to say, if r is very long, then $g_{11} = 1$ and $g_{44} = -1c^2$, which is another way of asserting that without nearby mass, relativistic effects vanish.

As is more relevant to our current task, the components in this matrix, notably 1 and $-1c^2$, only pertain to a frame of reference with *flat* (Euclidean) space-time and *no* manifest gravitational field. Since our planet maintains a solar orbit, r is short enough to make the gravitational field manifest at the site of this planet, so that g_{11} and g_{44} cannot be 1 and $-1c^2$. We must therefore perform a *transformation to a frame of reference which has a gravitational field and in which space-time is curved.*

As a result of this process we anticipate a "new" g_{ik}. For the sake of completeness let us go back and consider all possible 16 components of g_{ik},

g_{11}	g_{12}	g_{13}	g_{14}
g_{21}	g_{22}	g_{23}	g_{24}
g_{31}	g_{32}	g_{33}	g_{34}
g_{41}	g_{42}	g_{43}	g_{44}

even though we expect that only g_{11}, g_{22}, g_{33}, *and* g_{44} will be non-zero.

The problem is eased by drawing upon Schwarzschild's assumptions, staticism and spherical symmetry. By respecting staticism, we can eliminate terms which are affected by whether *time* is positive or negative; hence we call it "staticism." In our metrical equation the time-term $(dx^4)^2$ appears as c^2dt^2 (in $g_{44}c^2dt^2$). If ds is to be invariant whether dx^4 is positive or negative, elements in the above matrix that contain dt must be zero. For example g_{42} contains dtθ, so this component vanishes, but then, by virtue of the symmetry of the metrical tensor, g_{24} is also zero, etc. As all components containing $dt\theta$, $dt\varphi$ and $dtdr$ are prohibited,

$$g_{41} = g_{42} = g_{43} = g_{14} = g_{24} = g_{34} = 0.$$

Similarly according to spherical symmetry, if ds is to be invariant whether dθ is positive or negative and whether dφ is positive or negative, then a term such as dθdφ is prohibited. Hence

$$g_{12} = g_{13} = g_{23} = g_{21} = g_{31} = g_{32} = 0.$$

As we anticipated, *the vanishing of these components curtails the non-zero g's to just four*, which lie entirely on a diagonal. Each of these has repeated indices, namely

$$g_{11}, g_{22}, g_{33}, \text{ and } g_{44}.$$

We can summarize that the essential components are such that

$$g_{ik} = 0 \text{ when } i \neq k.$$

But we can apply spherical symmetry to simplify further: If we imagine that gravity somehow increased because the sun gained mass while the solar system remains spherically symmetric, the planet does not shift north-south or east-west. (It may be shifted "down" along *r,* but that is not pertinent at the moment.) By analogy, north-south motion on a globe of the Earth ($d\theta$) is basically no different from east-west motion ($d\varphi$), and this kind of symmetry is not affected by mass.

On this basis we gather that g_{22} and g_{33} are *not* affected by the transformation to a gravitational scenario. In other words these two components only have to do with "sideways" motion along the imaginary surface of our spherical-polar system, so that they are invariant under this transformation. If "old" refers to the absence of gravitation and "new" refers to its presence, we can say that

$$g_{22 \; old} = g_{22 \; new} \quad \text{and} \quad g_{33 \; old} = g_{33 \; new}.$$

We thus solve the identity of g_{22} and g_{33} (written out below), *leaving only two unknown non-zero g's,*

$$g_{11} \text{ and } g_{44}.$$

Here we can apply the scheme that we are pursuing two types of *g*'s. Our g_{22} and g_{33} are "*g*'s belonging to the system of coordinates," while g_{11} and g_{44} are "*g*'s belonging to intrinsic curvature." Our g_{22} and g_{33} appear in the terms $g_{22}d\theta^2$ and $g_{33}d\varphi^2$, and the equations which identify these are invariant:

$$g_{22} = r^2 \quad and \quad g_{33} = r^2 \sin^2\theta$$

Since we have spelled out g_{22} and g_{33}, we can condense our terms as we did on page 394, where we saw

$$r^2(d\theta^2 + \sin^2\theta d\varphi^2),$$

allowing us to streamline our metrical equation into

$$ds^2 = g_{11}dr^2 - g_{44}c^2dt^2 + r^2(d\theta^2 + \sin^2\theta d\varphi^2).$$

The two unknown *g*'s which "belong to intrinsic curvature" stand out.

We noted that if *r* is long enough to dilute away all gravitation, $g_{11} = 1$ and $g_{44} = -1c^2$. Conversely if *r* is short enough so that a planet finds itself in a gravitational field—if we transform to a gravitational scenario—g_{11} is *no longer equal to 1*. By the same token g_{44} is *no longer* $-1c^2$, so that *the presence of gravity alters space as well as time as represented, respectively, by* g_{11} *and* g_{44}. (Of course in relativity altered space-time *is* gravity.) Furthermore the shorter the *r*, the further the value of g_{11} is from 1 and the further the value of g_{44} is from $-1c^2$. Hence g_{11} and g_{44} must be functions of—must depend upon and vary with—*r*. In addition, staticism and spherical symmetry signify that these two *g*'s depend only on *r*; they cannot depend on *t*, θ or φ. These points will help us identify g_{11} and g_{44}.

Readers who consult other sources should be aware of two further issues: First, the signs of the *g*'s may be reversed, depending on how the background equations are set up. For example g_{33} can be negative and g_{44} can be positive. Second, as we mentioned, some authors display the meaning of indices by using

$$g_{rr} \ in \ place \ of \ g_{11}$$

to show that the radius is the critical spatial dimension in Schwarzschild's scenario; by using

$$g_{\theta\theta} \ in \ place \ of \ g_{22} \quad and \quad g_{\varphi\varphi} \ in \ place \ of \ g_{33}$$

to show that these components relate to "latitude" and "longitude" in this scenario; and by using

$$g_{tt} \text{ in place of } g_{44}$$

to show that this component relates to time.

In this method, staticism bars g_{tr}, $g_{t\theta}$, $g_{t\phi}$ and their symmetric partners—no preferred direction in time—while spherical symmetry bars $g_{r\theta}$, $g_{1\phi}$, $g_{\theta\phi}$ and their symmetric partners—no preferred direction in space. The distribution of matter of the sun does not affect $g_{\theta\theta}$ or $g_{\phi\phi}$ because spherical symmetry bars gravitation from changing θ and ϕ. (Again θ and ϕ have to do with "sideways" changes in location which are not affected by the central mass.) The sun does affect g_{rr} and g_{tt} because gravity is felt along r and "along" t. I.e., it affects the radius and time-course of the planet's orbit.

No matter which symbols we use, compared with the 16 g's we started with, the problem is less formidable now. In fact given the arithmetic relationship between g_{11} and g_{44} ($g_{44} = \dfrac{1}{g_{11}}$ neglecting $-c^2$ as a constant), *if we determine how one of these components is linked to the sun's M, we can easily calculate how the other is linked to M.* Nonetheless we will find that pinning down even one of these g's is not simple.

✦

Let us see where we are in our strategy: As we outlined at the beginning of this chapter, because we want to know the shape of space-time near the sun, we are seeking the details of an equation which says that

"components of g_{ik} = some form of M."

However at the moment we have available in a fully derived form only of the equation $G_{ik} = -XT_{ik}$ (wherein $G_{ik} = R_{ik} - \frac{1}{2}g_{ik}R$) which says that

"space-time in the form of G_{ik} = matter in the form of T_{ik}."

On the other hand we already know quite a bit about the g_{ik} in question: This g_{ik} has four non-zero components, g_{11}, g_{22}, g_{33}, *and* g_{44}, each of which has repeated indices. We have our metrical equation set up so that these four g's fit spherical-polar coordinates and so that gravitationally curved space-time is accommodated. We also have clinched g_{22} and g_{33}, so we need only to identify g_{11} and g_{44}. Moreover we know that the latter two g's are interrelated, so that if we figure one out, we know both.

We even have enough information to *guess* at the equation that connects g_{44} with *M*. If we review how

$$g_{44} = 1 + \frac{2\Phi}{c^2}$$

is derived (page 245), we recall that

$$\Phi = -\frac{KM}{r}.$$

By replacing Φ by - *KM/r*, we surmise that

$$g_{44} = 1 - \frac{2KM}{rc^2},$$

which would mean that

$$g_{11} = \frac{1}{\left(1 - \dfrac{2KM}{rc^2}\right)}.$$

Clearly we seem to be close to knowing just how "components of g_{ik} = some form of *M*" in Schwarzschild's scenario; we are almost ready to decipher exactly how the *M* of the sun affects the g_{ik} of space-time at a distance of *r* from *M*.

As we shall see, our guess about the identity of g_{44} is correct, but we still have not taken into account Einstein's field equations. This is to say, since "space-time in the form of G_{ik} = matter in the form of T_{ik}," *we must prove that* $g_{44} = 1 - \dfrac{2KM}{rc^2}$ *is truly based upon—that it "satisfies"—the equations* $G_{ik} = -XT_{ik}$. Here the term "satisfies" means that our equation for g_{44} is mathematically compatible with $G_{ik} = -XT_{ik}$, or said in reverse, given $G_{ik} = -XT_{ik}$, we must confirm $g_{44} = 1 - \dfrac{2KM}{rc^2}$.[*] This

---

[*] A highly simplified case of mathematical "satisfaction" is the following: The proposal that $x = 2$ satisfies the equation $3x = 6$ (whereas the proposal that $x = 3$ does not). In our case the proposal that $g_{44} = 1 - \dfrac{2KM}{rc^2}$ must satisfy $G_{ik} = -XT_{ik}$.

is difficult to do, but we will show sufficient details to demonstrate the feat it represents, since the process was the key to Schwarzschild's solution. Our tactic is as follows: In effect the field equations tell us that

each G = a corresponding $-XT$.

(There are 16 possible G's because G_{ik} is a rank-two tensor in four dimensions.) Let us rewrite the pattern for the field equations in Schwarzschild's scenario to state that

each G = some form of the g's = a corresponding $-XT$.

This pattern embraces the fact that we derived G_{ik} from g_{ik} by way of the Γ's, R_{hikl}, and R_{ik}; see the left column of the table on page 382. For instance (since g_{11} and g_{44} are interrelated) we can assume that

G_{11} = some form of g_{11} and/or g_{44} = $-XT_{11}$.

There are two equal-signs in this pattern, so that we really have two patterns in one. If we can find the details for an equation between the left parts of this pattern such that

each G = some form of g_{11} and/or g_{44},

then we have the details for an equation between the right parts such that

some form of g_{11} and/or g_{44} = a corresponding $-XT$.

From that point we can find our sought-after equations which tell us how

component g_{11} = some form of M,

and how

component g_{44} = some form of M.

(For easier computation, we will designate surrogates for g_{11} and g_{44}, "A" and "B" respectively.)

Clearly this process hinges on equating the g's with the G's. There are many methods available, as for example in Menzel (pp. 393-397), Lawden (pp. 142-147), and more explicitly in Lieber (pp. 233-255). In essence we shall retread the derivation of G_{ik} from g_{ik}, which in Lieber's words is a "colossal undertaking" because of the large number of possible permutations of g's, Γ's, R's (of

Ricci's tensor) and G's. The key to this problem will be to take advantage of what we know about our four g's in order to reduce the number of possible Γ's that apply in Schwarzschild's scenario. We will then use this information, namely the list of remaining Γ's, to reduce the number of possible R's. The remaining R's will be used in the equation,

$$G_{ik} = R_{ik} - \tfrac{1}{2}g_{ik}R.$$

Solving for G_{ik} will give us a set of equations that link the g's with the G's for Schwarzschild's scenario.

A preliminary note on mathematical technique is needed. The approach found in some literature seeks the missing g's by setting up the spherical-polar version of the metrical equation such as

$$ds^2 = e^{\,I}dr^2 - e^{\,J}c^2dt^2 + r^2(d\theta^2 + \sin^2\theta d\varphi^2).$$

This is an "exponential" version in which I and J represent g_{11} and g_{44} indirectly. Once I and J are found as exponents of e, they can be reconverted to g_{11} and g_{44}. Many variants of this approach are possible, but their merit is that certain equations are simpler in an exponential format.

We shall not pursue the exponential approach, though it is presented in detail by Lieber. We prefer a more direct approach, similar to Møller's on pp. 323-325, with the metrical equation written as

$$ds^2 = g_{11}dr^2 + g_{44}c^2dt^2 + r^2(d\theta^2 + \sin^2\theta d\varphi^2).$$

It is easier to work with g_{44} positive and to abbreviate the unknown g's as follows:

$$g_{11} = A$$

$$g_{44} = -B.$$

Here A and -B are surrogates for g_{11} and g_{44} that make computation easier. We are now seeking the identity of A and B, and we expect these to depend on r. When A or B is to be differentiated with respect to r we write $A_{,r}$ or $B_{,r}$. For example

$$\frac{\partial A}{\partial r} = A_{,r}.$$

We will now begin the process by which we deduce which Γ's apply in Schwarzschild's scenario—which means we recount Schwarzschild's approach to making the problem manageable.

In four dimensions we could see 64 (4^3) different Γ's, but meanwhile we have limited our g's to only those with repeated indices: g_{11}, g_{22}, g_{33}, *and* g_{44}. This means we can exclude the 24 Γ's whose three indices are all different, such as $\Gamma^1{}_{23}$; on this basis none of these can fit into Schwarzschild's scenario. Out of the original 64, only 40 Γ's remain in which two or three indices are the same.

These 40 can be distinguished by using the Γ-equation (page 296) in which we again see the origin of the Γ's from the components of the metrical tensor. However, since we can only accept those Γ's that are based on g's with repeated (*ii* and *kk*) indices, the operative version of this equation is:

$$\Gamma^i{}_{kl} = \frac{1}{2}\frac{1}{g_{ii}}\left(\frac{\partial g_{ii}}{\partial x^l} + \frac{\partial g_{ii}}{\partial x^k} - \frac{\partial g_{kk}}{\partial x^i} \right).$$

Each of these Γ's has three indices, which means that from the mathematical point of view we are subjecting g_{ii} and g_{kk} to tensor differentiation. In this context we recall that the Γ's accommodate for the variation of motion from straightness and uniformity, while the variability of the g's in a region of space-time propagates the gravitational field. Therefore *the g's enter into this scenario by way of the Γ's*.

There are three categories of Γ's in which an index appears at least twice, which is easier to appreciate using numbers for the indices:

Category 1: Γ's with all three indices the same, e.g. $\Gamma^1{}_{11}$.

Category 2: Γ's where the indices are, e.g., $\Gamma^2{}_{12}$ *and* $\Gamma^2{}_{21}$.

Category 3: Γ's where the indices are, e.g., $\Gamma^1{}_{22}$.

Category 2 shows the partial symmetry of the Γ's, by which pairs such as $\Gamma^3{}_{32}$ and $\Gamma^3{}_{23}$ are identical, and one member of such a pair can be rejected as not independent, which reduces the possibilities further.

When the Γ-equation is written out and solved with any of the indices repeated, we find further simplifications and we are able to further reduce the number of allowed Γ's. This is because some of the terms in the parenthesis vanish. (This is not a case of contraction.) Solving for the 40 Γ's entails lengthy calculus and trigonometry, which in some ways would have been eased had we elected the exponential method. We will sample three simpler cases applying our preferred non-exponential method, and we will show a sample of other details.

In category 1, as all three indices $= 1$, the last two terms in the parenthesis of the Γ-equation cancel each other ($\dfrac{\partial g_{11}}{\partial x^1} - \dfrac{\partial g_{11}}{\partial x^1} = 0$), leaving

$$\Gamma^1_{11} = \frac{1}{2}\frac{1}{g_{11}}\frac{\partial g_{11}}{\partial x^1}.$$

We can solve this equation by letting A stand for g_{11} and $A_{,r}$ for g_{11} differentiated with respect to r, while recalling that A depends only on r. This r is represented by x^1. These substitutions yield

$$\Gamma^1_{11} = \frac{1}{2}\frac{1}{A}\,A_{,r}$$

or simply

$$\Gamma^1_{11} = \frac{A_{,r}}{2A}.$$

The remaining members of this category vanish, as in

$$\Gamma^2_{22} = \Gamma^3_{33} = \Gamma^4_{44} = 0,$$

because, for example, Γ^2_{22} asks how "θ changes per change in θ" while we only admit functions of r. Thus category 1 only has one member that is of use to us.

In category 2, let us examine only the solution for Γ^2_{12} ($i = 2, k = 1, l = 2$; we must not confuse 1 [one] with l [lower case L]). Outside the parenthesis in the Γ-equation, $\dfrac{1}{2}\dfrac{1}{g_{22}} = \dfrac{1}{2r^2}$, because, as we found earlier, $g_{22} = r^2$. In the parenthesis the first and third terms drop out because we allow the g's to be a function of only r, again represented by x^1. The remaining (middle) term now is $\dfrac{\partial g_{22}}{\partial x^1}$, which equals $\dfrac{\partial r^2}{\partial r}$, which in turn (according to a basic rule in differential calculus) equals $2r$. Hence $\Gamma^2_{12} = (\dfrac{1}{2r^2})(2r) = \dfrac{1}{r}$.

In category 3, let us solve for Γ^1_{22} ($i = 1, k = 2, l = 2$): Here $\dfrac{1}{2}\dfrac{1}{g_{11}} = \dfrac{1}{2A}$, because we let A stand for g_{11}. The first two terms in the parenthesis drop out because we allow the g's to be a function of only r, which is represented by x^1.   The remaining term now is $-\dfrac{\partial g_{22}}{\partial x^1}$, which equals $-\dfrac{\partial r^2}{\partial r}$, which

in turn (again according to basic differential calculus) equals $-2r$. Hence $\Gamma^1_{22} = (\dfrac{1}{2A})(-2r) = -\dfrac{r}{A}$.

We repeat this kind of procedure for the remaining possibilities in each category. In every case we find one of two results: Either a particular Γ is zero (all terms vanish), or it has a certain fairly simple non-zero value. It turns out that only 13 Γ's remain non-zero, and several of these are equivalent or symmetric, so that even fewer truly independent Γ's survive.

Specifically, category 1 only has

$$\Gamma^1_{11} = \frac{A_{,r}}{2A} \; (as \; above).$$

Category 2 only has

$$\Gamma^2_{12} = \Gamma^2_{21} = \Gamma^3_{13} = \Gamma^3_{31} = \frac{1}{r} \; (as \; above)$$

$$\Gamma^3_{32} = \Gamma^3_{23} = \cot\theta$$

$$\Gamma^4_{14} = \Gamma^4_{41} = \frac{B_{,r}}{2B}.$$

Category 3 only has

$$\Gamma^1_{22} = -\frac{r}{A} \; (as \; above)$$

$$\Gamma^1_{33} = -\frac{r \; \sin^2\theta}{A}$$

$$\Gamma^1_{44} = \frac{B_{,r}}{2A}$$

$$\Gamma^2_{33} = -\sin\theta \; \cos\theta.$$

We should keep in mind that we are studying the shape of space-time at some point along an orbit, that the Γ's measure the variation in the g's, and that we are using spherical-polar coordinates. We may also imagine that each Γ with its particular combination of indices represents the vantage of an observer capable of detecting a variation in the g's, so that each Γ represents the variation in the shape of space-time as "seen" from one vantage. (Recall how we interpreted the Γ's on page 298.) *These 13 vantages suffice to fully assess the variations in space-time in this scenario.* By this analogy Schwarzschild found that 13 observers can describe all the changes in space-time in a planetary orbit; 64 observers are available but most are superfluous.

Let us comment on the four Γ's which equal $1/r$. They signify that from four different vantages the variation in the shape of space-time is inversely proportional to r. This is hardly astonishing. Space-time varies less farther from a mass, and if the mass is far enough away, then for some observers space-time ceases to vary altogether. If r is infinitely long, these four Γ's equal zero (since $1/\infty = 0$).

From two other vantages, represented in Γ^1_{22} and Γ^1_{33}, the effect of r is also what we would expect. We can think of these as cases wherein certain observers find that the variation in space-time varies with distance. The value of these Γ's is proportional to r.

The Γ's related to trigonometric functions (sine, cosine, cotangent of θ) seem more complicated until we realize that the space-time variation we are examining occurs along an orbit which, using spherical-polar coordinates, is described in part by angle θ. We further note that the Γ's which *only* hold indices 2 and 3 (not 1 or 4) are *only* related to angle θ (not to A, B, or r). This point reflects our scheme that our g_{22} and g_{33} "belong to the system of coordinates" and deal with "sideways" motion.

As for Γ^1_{11}, it recounts A, which it should, considering that $A = g_{11}$. Lastly Γ^4_{44}, Γ^4_{14}, *and* Γ^4_{41} contain the "time" index 4; hence these Γ's are related to B, which represents g_{44}. This point reflects our scheme that g_{11} and g_{44} "belong to intrinsic curvature."

We note that so far we claimed *no link between the Γ's and the mass itself*, not in terms of its m, its M, or its T_{ik}. Why? Because just as the provisional field equation $g_{ik} = T_{ik}$ is invalid, so is Γ's $= T_{ik}$ not valid. In fact we will not see the link between A and B on the one hand and matter on the other until we reach $G_{ik} = -XT_{ik}$.

Let us move closer to G_{ik}, but we must still do so indirectly. We can skip the contraction of R_{hikl} (our tensors are all rank two already) and we can jump to the Ricci tensor R_{ik}. This step is somewhat simplified because R's with unlike indices, such as R_{12}, R_{13}, R_{23}, R_{24}, R_{34}, etc., are excluded—and their symmetrical partners vanish—on the basis of spherical symmetry. Likewise R_{14} and R_{41} vanish on the basis of staticism, which is what we expect recalling how we reduced the number of non-zero g's on page 398. Thus all R's are eliminated save those with repeated indices, which in fact is also mandatory if the space-time of the universe is isotropic. In other words the combinations of indices which denote non-isotropy have vanished, reassuring us that our mathematical results support our physical perceptions.

In short, the only components of the Ricci tensor of use in Schwarzschild's scenario are such that

$$R_{ik} = 0 \ when \ i \neq k,$$

which means we are left with four non-zero R's: R_{11}, R_{22}, R_{33}, *and* R_{44}. We also have 13 non-zero Γ's, many of which equals some form of A and/or B. (Lest we lose sight of our goal, A and -B stand for the g's we need to identify.) Next we invoke the equation that defines the Ricci tensor but written with commas for derivatives and with parentheses, which makes its application somewhat clearer:

$$R_{ik} = \Gamma^a_{ia,k} - \Gamma^a_{ik,a} + (\Gamma^e_{ia} \Gamma^a_{ek} - \Gamma^e_{ik} \Gamma^a_{ea})$$

This equation allows us to solve for the four remaining components of the Ricci tensor. Upon replacing the Γ's in the right side with our 13 remaining Γ's—for instance Γ^1_{11} becomes $\frac{A_{,r}}{2A}$—and upon additional algebraic rearrangements, we find four equations linking R's to our A and B. We let $,r$ and $,,r$ mean first and second derivative with respect to r. Here are the three key equations out of the list of four:

$$R_{11} = -\frac{B_{,,r}}{2B} + \frac{B_{,r}}{4B}\left(\frac{A_{,r}}{A} + \frac{B_{,r}}{B}\right) + \frac{A_{,r}}{rA}$$

$$R_{44} = \frac{B_{,,r}}{2A} - \frac{B_{,r}}{4A}\left(\frac{A_{,r}}{A} + \frac{B_{,r}}{B}\right) + \frac{B_{,r}}{rA}$$

R_{22} is somewhat simpler:

$$R_{22} = 1 - \frac{r}{2A}\left(\frac{B_{,r}}{B} - \frac{A_{,r}}{A}\right) - \frac{1}{A},$$

but R_{33} depends on R_{22}, so only three non-zero R's are independent and only three equations are needed.

(For readers interested in a more detailed sample of what Schwarzschild accomplished, the rest of this passage shows the solution for R_{11}: The right side of $R_{ik} = \Gamma^a_{ia,k} - \Gamma^a_{ik,a} + (\Gamma^e_{ia} \Gamma^a_{ek} - \Gamma^e_{ik} \Gamma^a_{ea})$ repeats four times as each index is set to 1, 2, 3, or 4, which requires that the terms in the parenthesis be written out four times in each repetition. Thus (what follows is one equation!),

$$R_{11} = \Gamma^1_{11,1} - \Gamma^1_{11,1} + (\Gamma^1_{11}\Gamma^1_{11} - \Gamma^1_{11}\Gamma^1_{11}) + (\Gamma^2_{11}\Gamma^1_{21} - \Gamma^2_{11}\Gamma^1_{21}) + (\Gamma^3_{11}\Gamma^1_{31} - \Gamma^3_{11}\Gamma^1_{31}) + (\Gamma^4_{11}\Gamma^1_{41} - \Gamma^4_{11}\Gamma^1_{41})]$$

$$+ [\Gamma^2_{12,1} - \Gamma^2_{11,2} + (\Gamma^1_{12}\Gamma^2_{11} - \Gamma^1_{11}\Gamma^2_{12}) + (\Gamma^2_{12}\Gamma^2_{21} - \Gamma^2_{11}\Gamma^2_{22}) + (\Gamma^3_{12}\Gamma^2_{31} - \Gamma^3_{11}\Gamma^2_{32}) + (\Gamma^4_{12}\Gamma^2_{41} - \Gamma^4_{11}\Gamma^2_{42})]$$

$$+ [\Gamma^3_{13,1} - \Gamma^3_{11,3} + (\Gamma^1_{13}\Gamma^3_{11} - \Gamma^1_{11}\Gamma^3_{13}) + (\Gamma^2_{13}\Gamma^3_{21} - \Gamma^2_{11}\Gamma^3_{23}) + (\Gamma^3_{13}\Gamma^3_{31} - \Gamma^3_{11}\Gamma^3_{33}) + (\Gamma^4_{13}\Gamma^3_{41} - \Gamma^4_{11}\Gamma^3_{43})]$$

$$+ [\Gamma^4_{14,1} - \Gamma^4_{11,4} + (\Gamma^1_{14}\Gamma^4_{11} - \Gamma^1_{11}\Gamma^4_{14}) + (\Gamma^2_{14}\Gamma^4_{21} - \Gamma^2_{11}\Gamma^4_{24}) + (\Gamma^3_{14}\Gamma^4_{31} - \Gamma^3_{11}\Gamma^4_{34}) + (\Gamma^4_{14}\Gamma^4_{41} - \Gamma^4_{11}\Gamma^4_{44})]$$

Of these Γ-terms, most vanish on the basis of 5 "facts." The first fact is that terms such as $\Gamma^1_{11,1} - \Gamma^1_{11,1}$ and $(\Gamma^1_{11}\Gamma^1_{11} - \Gamma^1_{11}\Gamma^1_{11})$ equal zero, which eliminates the entire first row. We already saw the second fact (in category 1), namely that Γ^1_{11} survives but other Γ's with three repeated indices do not. We also saw the third fact (in category 3), namely that only a few Γ's with two subscripts alike are non-zero, Γ^1_{22}, Γ^1_{33}, Γ^1_{44}, and Γ^2_{33}; all others fitting this pattern vanish. The fourth fact is also familiar, namely that Γ's with all three indices different (e.g. Γ^4_{12}) are out of the running. Finally we use the fact that terms like $\Gamma^2_{11,2}$ drop out because a differential of itself is zero. These eliminations leave the following non-zero terms:

$$\Gamma^2_{12,1} - \Gamma^1_{11}\Gamma^2_{12} + \Gamma^2_{12}\Gamma^2_{21} + \Gamma^3_{13,1} - \Gamma^1_{11}\Gamma^3_{13} + \Gamma^3_{13}\Gamma^3_{31} + \Gamma^4_{14,1} + \Gamma^4_{14}\Gamma^4_{41} - \Gamma^1_{11}\Gamma^4_{14}.$$

These terms are replaceable by terms containing our A, B, or r or their derivatives. We already saw that $\Gamma^2_{12} = \Gamma^3_{13} = \dfrac{1}{r}$, $\Gamma^1_{11} = \dfrac{A,r}{2A}$, and $\Gamma^4_{14} = \Gamma^4_{41} = \dfrac{B,r}{2B}$. The appropriate algebraic substitutions and other manipulations yield the goal written out more explicitly,

$$R_{11} = -\frac{B,_{,r}}{2B} + \frac{A,_r B,_r}{4AB} + \frac{B,_r B,_r}{4B^2} + \frac{A,_r}{rA}.$$

For example the first right-side term arises because $\Gamma^4_{14,1} = \dfrac{B,_{,r}}{2B}$ where the two commas indicate a second derivative. The second term arises because $\Gamma^1_{11}\,\Gamma^4_{14} = \dfrac{A,_r}{2A}\dfrac{B,_r}{2B}$. The third term arises because $\Gamma^4_{14}\,\Gamma^4_{41} = \dfrac{B,_r}{2B}\dfrac{B,_r}{2B}$ by virtue of symmetry of the Γ's. The last term arises because $\Gamma^1_{11}\,\Gamma^2_{12} = \Gamma^1_{11}\,\Gamma^3_{13} = \dfrac{A,_r}{2A}\dfrac{1}{r}$, while the remaining Γ's vanish. [End of passage.] See Kenyon, p. 196, though in his format A and B are reversed. Lieber, pp. 230-248 uses the exponential approach. Adler et al, pp. 190-194 also shows the details.)

Now it is time to remember that R_{ik} as a whole leads to the Einstein tensor via

$$G_{ik} = R_{ik} - \frac{1}{2}g_{ik}R.$$

Therefore we can rewrite the field equations of general relativity more explicitly as

$$G_{ik} = R_{ik} - \frac{1}{2}g_{ik}R = -XT_{ik}.$$

Next we use our non-zero R's $(R_{11}, R_{22}, \text{etc.})$, our non-zero g's $(g_{11}, g_{22}, \text{etc.})$, and the R, to solve

$$G_{11} = R_{11} - \frac{1}{2}g_{11}R = -XT_{11},$$

$$G_{22} = R_{22} - \frac{1}{2}g_{22}R = -XT_{22},$$

etc. For instance we know R_{22} from its equation on page 408, where it is a function of A, B and r; we know that $g_{22} = r^2 \sin^2 \theta$ from page 392; and we know from page 362 that $R = \dfrac{2}{r^2}$. These items of information suffice to solve $G_{22} = R_{22} - \frac{1}{2}g_{22}R = -XT_{22}$ and thus to determine G_{22}.

In similar fashion we end up with a set of non-zero G's, each of which is a function of the A and/or B we are seeking. In gross terms let us say that each non-zero G equals some set of "A and/or B terms," so that we see the pattern (updated here from "each G = some form of the g's = a corresponding $-XT$")

each G = "A and/or B terms" = each corresponding $-XT$.

Again we find that each "A and/or B term" is a function of r, and that G_{33} depends on G_{22}, leaving only three equations that fit this pattern and that link G's with g_{11} and g_{44} (indirectly, via A and B). After more substitutions and rearrangements of terms,

$$G_{11} = -\frac{B_{,r}}{ABr} + \frac{1}{r^2}\left(1 - \frac{1}{A}\right) = -XT_{11}$$

$$G_{22} = -\frac{1}{2A}\left(\left(\frac{B_{,r}}{B}\right)_{,r} - \frac{1}{2}\left(\frac{A_{,r}}{A}\right)\left(\frac{B_{,r}}{B}\right) + \frac{1}{2}\left(\frac{B_{,r}}{B}\right)^2 + \frac{B_{,r}}{Br} - \frac{A_{,r}}{Ar}\right) = -XT_{22}$$

$$G_{44} = -\frac{A_{,r}}{A^2 r} + \frac{1}{r^2}\left(1 - \frac{1}{A}\right) = -XT_{44}.$$

As we might expect from their indices, the equations for G_{11} and G_{44} are of immediate interest, and we note the similarity between them. Meanwhile G_{22} (and G_{33}) have to do with "sideways" location along the imaginary surface of our spherical-polar system. Altogether *we have solved*—at

410

least in terms of A, B, and *r—the three critical field equations that apply in Schwarzschild's scenario.*

At this point we turn our attention to empty space, on the assumption that we are examining the gravitational field in the region outside the matter which is responsible for that field. I.e., we are examining the space "exterior" to the sun itself at the site of a planet. To do so, *we set* T_{ik} *equal to zero*, which means we apply the central equation (one that we called a basic law of gravitation)

$$G_{ik} = 0,$$

because where $T_{ik} = 0$, $G_{ik} = 0$. This step provides the conditions for determining our A and B. Accordingly we set each of the "A and/or B terms" equal to zero, particularly for G_{11} and G_{44}. Thus $G_{11} = 0$ becomes

$$- \frac{B_{,r}}{ABr} + \frac{1}{r^2} \left(1 - \frac{1}{A}\right) = 0,$$

and $G_{44} = 0$ becomes

$$- \frac{A_{,r}}{A^2 r} + \frac{1}{r^2} \left(1 - \frac{1}{A}\right) = 0.$$

Subtracting one equation from the other algebraically leaves

$$\frac{A_{,r} B + AB_{,r}}{A^2 Br} = 0,$$

which means that $AB_{,r} = 0$, which in turn means that AB must be a constant. (The derivative of a constant is zero.) In fact with a very long *r* both A and B approach the value of 1, so that

$$A = \frac{1}{-B} \quad and \quad B = \frac{1}{-A},$$

just as we expect on the basis that $g_{44} = \frac{1}{g_{11}}$ (neglecting $-c^2$ as a conversion factor).

The above equations can be solved for B, so that

$$B = 1 - rB_{,r}$$

or

$$1 - rB_{,r} - B = 0.$$

Now we perform integration, the rationale for which is as follows: The steps for moving from the g's to the G's involved differentiation with respect to r, and now we are in essence working backwards. The reverse of differentiation is integration or the finding of the anti-derivative. The general recipe for this procedure is

$$\int k\ dx = kx + C.$$

(This equation also applies to partial derivatives.) We note the appearance of another constant, C. This kind of constant is called a constant of integration, because when the derivative of a constant is zero, the anti-derivative of zero (which is implied) is a constant. The constant C plays a major role in the problem at hand, that of identifying B. We seek the integral of r, and we recall that the comma stands for differentiation ($B_{,r} = \partial B/\partial r$). Applying the above recipe to $1 - rB_{,r} - B = 0$,

$$rB - r = C$$

or, solving for B,

$$B = 1 + \frac{C}{r},$$

which again makes sense: If $r = \infty$, then B = 1.

Finding C is the last hurdle. This unknown can be seen in two lights. One, in our mathematical pursuit of g_{11} and g_{44} via A and B, this C acts as a constant of integration. Two, in Schwarzschild's scenario *C brings the mass of the object responsible for gravitation* (in this case the sun) *into the picture.* We designate the mass of the sun as its density M, and we fall back upon the approximation that in this kind of scenario the "time-time" component T_{44} of the mass-energy tensor represents the principal source of gravitation.

The reader may be disappointed that despite the elaborate derivations of the mass-energy tensor T_{ik} available to us—derivations which include energy and momentum—*density of mass alone suffices to introduce matter into Schwarzschild's scenario for the exterior solution.* However consideration of all non-zero components of T_{ik} would not only complicate this solution but experiments have shown that the other components would not add appreciably to the value of the results (more on this later). In other words T_{44} is an acceptable characterization of the sun because the orbiting of a planet is a low-energy, weak-gravity event. In fact the maximum error is E/c^2, and since E outside the sun is a much smaller number than c^2, the error is insignificant.

Limiting ourselves to T_{44} allows us finally to identify B. By definition, T_{44} is proportional to mass, which of course determines the local gravitational potential Φ. It helps again to recall that g_{44} is 1 when the amount of nearby matter is zero, and that the addition of matter increases g_{44} from 1 to $1 + 2\Phi/c^2$. We see that the deviation of g_{44} from 1 in Schwarzschild's scenario corresponds to the deviation of g_{44} from 1 in the behavior of the gravitational potential. Since our sought-after B corresponds to g_{44}, we anticipate that B will fit into the pattern of

$$g_{44} = 1 + \frac{2\Phi}{c^2},$$

and indeed the corresponding equation is

$$B = 1 + \frac{C}{r}.$$

If we again note how $g_{44} = 1 + \frac{2\Phi}{c^2}$ is derived, we recall that

$$\Phi = -\frac{KM}{r}.$$

By replacing Φ by $-KM/r$, the constant C obeys the equation

$$C = -\frac{2KM}{c^2},$$

and at long last we see that indeed, as we had guessed,

$$g_{44} = 1 - \frac{2KM}{rc^2}.$$

Moreover since we now know B while $-B = 1/A$,

$$g_{11} = \frac{1}{\left(1 - \frac{2KM}{rc^2}\right)}.$$

The identities of g_{44} and g_{11} fall into place like the last pieces of a jigsaw puzzle. Components of g_{ik} *do* equal some form of *M*, and the key relationship between g_{44} and *M* that we have unearthed *does satisfy* the field equations of general relativity, $G_{ik} = -XT_{ik}$.

We use the word "satisfy" (as explained on page 401) because we can argue that Schwarzschild did not "solve" Einstein's field equations in the way we usually think of an algebraic solution. Instead, as we pointed out on page 388, solving these equations entails 10 unknown functions (the components of G_{ik}), but Schwarzschild reduced the problem to one function (g_{44}) in such a way that the key element of this "solution," $g_{44} = 1 - \dfrac{2KM}{rc^2}$, satisfies $G_{ik} = -XT_{ik}$. Nevertheless, as we will discuss later, his method is superior to Einstein's and it represents a major advance in general relativity. If only in Schwarzschild's honor, we still call his work a "solution."

In any case we note that if r is very large, in which case the gravitational field at the end of r is very weak, g_{11} approaches 1 and g_{44} approaches $-1c^2$, just as suggested earlier. This also means that if r is large enough, the gravitational potential falls to zero; only "nearby" M deforms space-time, and of course with no M at all these g's are 1 and $-1c^2$. If we disregard the constant $-c^2$, these g's = 1, which implies just what we expect: Empty space-time is flat. Conversely the greater the M, the farther g_{11} and g_{44} depart from 1. It is noteworthy (for a point we will make later) that unless M is very large, g_{44} is positive and g_{11} is negative. Not surprisingly, g_{22} and g_{33} are unaffected by M; as we already know, these equal r^2 and $r^2\sin^2\theta$ respectively.

Since we allowed A and -B to represent g_{11} and g_{44}, we can go back to the three field equations we solved in terms of A, B, and r, and we can make the substitutions. For example we can replace A and -B in

$$G_{11} = -\frac{B_{,r}}{ABr} + \frac{1}{r^2}\left(1 - \frac{1}{A}\right)$$

with g_{11} and g_{44}, clearly showing a link between this G's and these g's. Moreover we can then replace g_{44} with $1 - \dfrac{2KM}{rc^2}$ and g_{11} with its reciprocal, revealing the identity of G_{11} even more concretely. However these substitutions, though of theoretical interest, have little practical value. As we shall see later, the identity of the g's—and to an even greater extent the identity of the Γ's—are more useful.

We can now complete the spherical-polar metrical equation *adapted for the curved space-time and the significant gravity found in Schwarzschild's scenario*; i.e. for the conditions of general relativity. In the equation

$$ds^2 = g_{11}dr^2 + g_{22}d\theta^2 + g_{33}d\varphi^2 - g_{44}c^2dt^2$$

our metrical tensor g_{ik} (in matrix form showing the non-zero g's) is such that

$1/(1-2KM/rc^2)$			
	r^2		
		$r^2\sin^2\theta$	
			$1-2KM/rc^2$

By inserting the identity of g_{11} and g_{44}, by rearranging the order of the terms, and by changing the signs, the equation can be written as it usually appears in the literature, and it is called the *Schwarzschild metrical equation*:

$$ds^2 = \left(1 - \frac{2KM}{rc^2}\right)c^2dt^2 - \frac{dr^2}{\left(1 - \frac{2KM}{rc^2}\right)} - r^2(d\theta^2 - \sin^2\theta d\varphi^2).$$

This is the harvest of Schwarzschild's work: It is his "exterior" solution of Einstein's field equations; it is valid outside of a static and spherically symmetric chunk of matter. *This equation tells us about the gravitational field, characterized by* ds^2, *in the empty space at distance r from the center of a distribution of matter of density M.* In this equation M is in units of weight, usually grams or kilograms, per volume; r is in units of distance, usually centimeters; and θ and φ are angles. However the term $\frac{2KM}{rc^2}$ converts the expression of mass into units of length, which simplifies the arithmetic.* The key non-zero g's, g_{11} and g_{44}, are

$$1 - \frac{2KM}{rc^2} \quad \text{and} \quad \frac{1}{\left(1 - \frac{2KM}{rc^2}\right)}.$$

As expected, M only appears in these two terms, as they "belong to intrinsic curvature." In effect

---

* Thus $\frac{2KM}{rc^2}$ for the Earth is 4.44 millimeters and for the sun is 1.47 kilometers.

$$\left(1 - \frac{2KM}{rc^2}\right) c^2 dt^2$$

in the Schwarzschild metrical equation tells us what mass does to the time-course (*t*) of a gravitational event, while

$$\frac{dr^2}{\left(1 - \frac{2KM}{rc^2}\right)}$$

tells us about gravitational behavior in the radius-direction (r). In particular, a planet near the *M* of the sun moves through space and "moves into the future" through time at a rate set by $1 - \frac{2KM}{rc^2}$.

Let us pause to consider a key point: Special relativity told us that time expands and space shrinks by the factor $\sqrt{1 - v^2/c^2}$, while these effects are attributed to uniform relative motion (*v*). General relativity, and in particular Schwarzschild's solution of the field equations, predict that time expands and space shrinks by the factor $1 - \frac{2KM}{rc^2}$, while these effects are attributed to the proximity (*r*) of mass—i.e. to the relative location in a gravitational field. Clearly the value of

$$\frac{2KM}{rc^2}$$

is critical: In response to the goal we laid out at the beginning of this chapter, *this term tells us what matter does to space-time*, calculated for a spherically symmetrical static case.

The other two terms in the Schwarzschild metrical equation, combined into

$$r^2(d\theta^2 - \sin^2\theta d\varphi^2),$$

tell us about the spherical-polar trigonometry of the scenario (*r* for size, θ and φ for location), but as we already noted, these do not contain *M*; they "belong to the system of coordinates."

Incidentally, it is possible to calculate the effect of matter directly on the Gaussian curvature of space-time:

$$K_{Gaussian} = \frac{KM}{r^3c^2}.$$

(This equation is derived using the curvature tensor, but the K's have two unrelated meanings here, the Gaussian curvature and a constant; c^2 is a conversion factor.) While "$K_{Gaussian}$" is not an efficient articulation of the shape of space-time, this equation clearly shows how the intrinsic curvature depends on M at distance r. For instance unless M is enormous or r is very short, $K_{Gaussian}$ is very small. In other words, as we showed earlier, the sun's effect on the space-time near the Earth is comparatively weak, which is why, on a cosmic scale, the orbiting of the Earth is a rather leisurely process.[*]

◆

Besides the exterior solution, Schwarzschild also provided us with an "interior" solution which, as the term implies, predicts gravitation inside a massive body. It is of particular interest in cosmology and astrophysics, since it helps explain the evolution and eventual collapse of stars and pulsars. Most of the logic and mathematics mirror the exterior solution, but other assumptions are made, such as that the mass-density of the body is uniform and that it is a sphere made of a perfect fluid, which is a sound model for many astronomic objects (page 343). The applicable tensor is defined (in contra-variant form) by

$$T^{ik} = (\rho + P)\, \frac{dx^i}{ds}\, \frac{dx^k}{ds} + g^{ik}P.$$

The solution of the field equation in the interior setting involves additional components of T_{ik}, not just T_{44}. Again the equations are solved for G's because in non-empty space G_{ik} is proportional to T_{ik}. The result is a metrical equation similar to the one for the exterior case; see Wald, pp. 125-128.

◆

Considering the importance of the term $\dfrac{2KM}{rc^2}$ and of the equation

$$g_{44} = 1 - \frac{2KM}{rc^2},$$

we should see their verification. These do hold up in astronomical predictions, such as the orbit of Mercury, to which we will return. More compelling experimental confirmation can be found in the einstein red-shift. We know that when a photon moves through a gravitational field, some energy

---

[*] Planetary orbiting is not the only effect of the sun's M. The solar contribution to the tides is another, though on Earth the tidal effect of the M of the moon, which is at a much shorter r, is more prominent. The Ricci tensor in Schwarzschild's solution neglects tides.

is expended (page 244), which we should perceive as a change in how much time each wave consumes. As we pointed out, this observation carries great significance because relativity demands that any cause-to-effect process, including a gravitational event, involve the transfer of some "signal" in the form of matter and/or energy. We can say this another way: This transfer occurs in space-time, and since space-time is less curved further from the mass of the sun, time appears to pass faster—it is not as slowed—further from the sun. Frequency means wave-undulations or wave-crests per time (page 72), so that "faster" time translates into lower frequency.

We underscore that the einstein red-shift owes its existence to curved space-time, and it demonstrates that the principle of equivalence includes optical events. Otherwise clocks would not be affected by their location in a gravitational field, and waves of light could not act as "clocks." Hence the importance of the einstein red-shift in the history of general relativity, as we mentioned on page 130. On the other hand the red-shift is not unique to general relativity. Any mathematical system for explaining gravitation which accepts curved space-time and which relies on the principle of equivalence will anticipate a red-shift (see Will, pp. 49 and 147-159). However no other system has withstood scrutiny as well as general relativity.

In terms of Schwarzschild's model, we investigate the red-shift by asking what happens to the frequency of a wave of sunlight received on Earth if the Earth is considered to be at distance r from the sun's M. Since it is easier to work with the time intervals between wave crests than with frequency, our question becomes whether indeed the measurements of time intervals, labeled as dt's, are accurately predicted by using $\dfrac{2KM}{rc^2}$.

We already derived an equation for the relationship between Φ and two dt's:

$$\frac{dt_o}{dt_r} = 1 + \frac{\Phi}{c^2}.$$

For our current scenario we can re-write this as

$$\frac{dt_{weaker\ field\ (Earth)}}{dt_{stronger\ field\ (sun)}} = 1 + \frac{\Phi}{c_2}.$$

We also have the relationship between Φ and g_{44},

$$g_{44} = 1 + \frac{2\Phi}{c^2},$$

and we can equate the gravitational potential with an expression of mass, M. At distance r

418

$$\Phi = -\frac{KM}{r},$$

so that *a red-shift experiment hinges on confirming that*

$$g_{44} = 1 - \frac{2KM}{rc^2}.$$

Therefore we need to see that indeed

$$\frac{dt^2_{weaker\ field\ (Earth)}}{dt^2_{stronger\ field\ (sun)}} = 1 - \frac{2KM}{rc^2}.$$

In the red-shift experiment we deploy the term $1 - \dfrac{2KM}{rc^2}$ as follows: The time interval between wave crests of sunlight measured on Earth is given by

$$``dt\ on\ Earth^2" = (1 - \frac{2KM}{rc^2})_{Earth}$$

where M represents the mass of the Earth and r is its radius. Likewise the time interval between wave crests of sunlight measured on the sun is

$$``dt\ on\ sun^2" = (1 - \frac{2KM}{rc^2})_{sun}$$

where M represents the mass of the sun and r is its radius. (Lieber, pp. 294-296, provides more details.) We can then set up a ratio for these intervals,

$$\frac{dt\ on\ Earth^2}{dt\ on\ sun^2},$$

which accommodates the contrary effects on the red-shift. The mass of the Earth diminishes the red-shift because its effect on the gravitational field adds to a photon's energy; i.e. the photon is "gliding" into the depression in space-time that surrounds the Earth. Meanwhile the mass of the sun increases the red-shift as the photon is losing energy "working" to climb out of the deeper depression in space-time around the sun. The arithmetic is simplified by the much greater mass of the sun, so that the value of $(1 - \dfrac{2KM}{rc^2})_{Earth}$ is practically equal to 1. Inserting known values for the sun's M and r, we expect that

$$\frac{1}{(1 - \dfrac{2KM}{rc^2})_{sun}} = 1.00000212.$$

Therefore the frequency of a wave of sunlight reaching the Earth should be red-shifted by 0.000212%, and with sufficiently sensitive instruments we find, in fact, that it is. Several experiments (notably by Pound, Rebka and Snider cited in Kenyon, pp. 17-18, and in Will, pp. 52-54) have corroborated this result. Obviously 0.000212% is very little, even though to us the sun's M is enormous, because c^2 is a very large number. The point however is that the validity of $\dfrac{2KM}{rc^2}$ in Schwarzschild's solution is demonstrated, and this conclusion can be generalized to all effects of the gravitational field surrounding M.

We mentioned that relativity has a practical application in the design of the U.S. global positioning system (GPS).[*] This system consists of several satellites in orbit around the Earth, but their orbit is not geosynchronous, so that there is relative motion between any GPS satellite and a GPS device (receiver) on Earth. Each satellite carries several extremely accurate atomic clocks, and each satellite emits periodic radio messages that are transmitted at the speed of light and that contains two items of information, the position of the satellite and the exact time. The receivers on or near the Earth's surface, for example on a boat at sea, detect these two data. Since we know the speed of light, the receiver can calculate the distance to the satellite. With data assembled by computers from several such satellites, the position of the receiver—and therefore of the boat—can be calculated and presented to a navigator with an astounding degree of accuracy. (Receivers can also give altitude, speed, direction of motion, and of course the exact time.)

From what we already know about relativity, two issues pertinent to our topic are obvious: The satellites are at some distance from the Earth (about 20,000 km), so that the strength of their local gravitational field is less than on the Earth's surface, and they are in relative motion with respect to anything on the Earth's surface. This means that the system must incorporate compensations for the effects of both general and special relativity. The details are quite complicated because many factors enter into consideration. However in general the mathematics is similar to that for the red-shift but must also accommodate relative velocity. Since the ratio $\dfrac{dt_{satellite}}{dt_{Earth}}$ indicates the shift in time intervals of the radio waves emanating from the satellites, the form of the equation is

---

[*] Russia has a similar system in place, called Glonass.

$$\frac{(dt_{satellite})^2}{(dt_{Earth})^2} = \frac{\left(1 - \dfrac{2KM}{r_{satellite}\,c^2}\right) - v_{satellite}^2}{\left(1 - \dfrac{2KM}{r_{Earth}\,c_2}\right) - v_{Earth}^2}.$$

The right side of this equation contains what we expect: We see terms with the structure of $1 - \dfrac{2KM}{rc^2}$, which represent the effects of the two different distances, r, from a mass as predicted by general relativity, and we see v's for two different velocities as noted in special relativity. Solving the above equation shows the relativistic effects. Interestingly, these are actually two concurrent but *opposite and very real effects*; the larger is explained by a satellite's clock in a weaker gravitational field running faster than an identical in a receiver on or near the Earth's surface. Meanwhile, there is a lesser slowing of each satellite's clock relative to its counterpart on or close to the Earth's surface by virtue of relative motion. When the first satellite with its clock was launched in 1977, skeptics doubted whether the effects of relativity would be real, but after several days in orbit the clock had gained time by the amount predicted relativistically, and if this had been ignored, the system would be useless. This is why, for example, the clocks in GPS satellites are built to run slower, and of course the amount of slowing represents an excellent example of a physically tangible application of the mathematics of relativity. Also please recall the Sagnac effect, which appears in the GPS.

We have fulfilled a promise: Armed with the appropriate rank-two metrical tensor and metrical equation, we described space-time at the site of a gravitational event; that event is the orbiting of a planet around the sun. The plan from this point on is to use this result to describe this event in ordinary terms—which means to find the path of the planet. This plan calls for the use of *the equations of motion*. The underlying reason we invoke separate equations of motion is that we relied on the gravitational potential to derive the field equations (page 245). As we pointed out, while Φ is a field quantity, it says nothing about motion. Therefore calculating how objects fall or orbit calls for two sets of equations, one for fields, which we just solved, and another for motion, which we will solve later in this chapter. In other words Schwarzschild showed us how mass curves space-time but not *how bent space-time determines motion*.

It makes a mathematical difference whether the object in the gravitational field is a photon or a material object. The first case presupposes a null geodesic, as in the 1919 observations on a bent beam of light whose photons are at speed c. The second case, which presupposes a time-like geodesic, can be a planet in a solar orbit. We are now concerned with the latter case, representing any ordinary freely falling object. We also assume that our planet is not subjected to non-gravitational influences, that it does not alter space-time on its own (i.e. it is a "test object"), and that it is the only nearby object (i.e. the other planets are too far to influence the outcome).

We already met some equations of motion (pages 200, 227, 236, and 293). Their basic version

$$\delta \int_A^B ds = 0$$

is important because its derivation ties world-lines of falling of orbiting objects to geodesics. Given the pivotal significance of this equation and of the space-time interval ds, it is tempting to lay out the following scheme: Since we have already identified the four non-zero g's for the Schwarzschild metrical equation $ds^2 = g_{11}dr^2 + g_{22}d\theta^2 + g_{33}d\varphi^2 - g_{44}c^2dt^2$, why not simply solve for ds and then use this ds to calculate gravitational motion by means of the above equation of motion? Because $\delta \int_A^B ds$ does not provide data for describing this event. Then why not use the four g's themselves, namely $1/(1-2KM/rc^2)$, r^2, $r^2\sin^2\theta$, and $1-2KM/rc^2$, to describe this event? Because g's tell us only about the local curvature of space-time. The best way to reach our goal is to revert to our Γ's and then insert them into a more useable equation of motion,

$$\frac{d^2x^j}{ds^2} = -\Gamma^j{}_{ik}\frac{dx^i}{ds}\frac{dx^k}{ds},$$

which we shall call it our "main" equation of motion and which can lead us to a description of an astronomic orbit. As we shall see in more detail, this approach works because the Γ's tell us how the curvature of space-time changes point-to-point, which is the key to understanding gravitational motion. (In theory we could use the G's or the R's for this purpose, since these also tell us about changes in the shape of space time, but doing so is far more complicated.) We will return to how the Γ's are inserted into our main equation of motion, but first we need to know more about this equation.

We found on pages 227, 227 to 233 that both preceding equations of motion are also geodesic equations compatible with the principle of least action. However when we reached this conclusion we had not yet introduced the metrical connections, so that we could not explain why the Γ's appear in our main equation of motion. Let us now rectify this omission, and in the process we will cover the derivation of this equation from a vectorial approach. Altogether we will have arrived at our main equation of motion by three means, via the principle of least action, via Newton's equation of motion, and (as follows now) via a method that articulates the role of the Γ's.

As we shall see, the left side of our main equation, $\dfrac{d^2x^j}{ds^2}$, represents accelerated motion, which is what makes it so useful. This term stems form the most elementary "equation of motion," $RT = D$,

422

better written as $vt = x;$ v is velocity, t is time, and x is distance. (In this context $RT = D$ is a more basic equation of motion than $F = ma$.) A change in location is dx, so in differential terms v is the derivative (rate of change) of distance through time;

$$v = \frac{dx}{dt}.$$

Of course motion is a vector, i.e. a magnitude in a direction, and we ordinarily consider three component-directions, X, Y, and Z, to match three dimensions. The components of velocity measured along these dimensions are $\frac{dX}{dt}$, $\frac{dY}{dt}$, and $\frac{dZ}{dt}$, but we can collect these into x^j where j runs from 1 to 3 (measurements along x in the three j-directions). Because velocity naturally transforms in contra-variant fashion we assign superscripts. The "velocity vector equation" therefore is

$$v = \frac{dx^j}{dt}.$$

However orbiting and other modes of gravitational motion are forms of *acceleration*; in these cases velocity changes over time. Accordingly we differentiate the components of the velocity vector so as to yield a new vector,

$$a = \frac{d^2x^j}{dt^2}.$$

Now $\frac{d^2x^j}{dt^2}$ is acceleration in three dimensions, which is the derivative of velocity (how fast velocity changes), which in turn is the derivative of motion (how fast location changes).

To accommodate relativity, we must let j run from 1 to 4, and we differentiate through space-time, not just time. The vectorial expression for velocity becomes $\frac{dx^j}{ds}$, and for four-dimensional acceleration it becomes

$$\frac{d^2x^j}{ds^2},$$

which is the left side of our main equation of motion, and which describes relativistic gravitational motion. (For small velocities the difference between dt and ds is negligible, but let us set this point aside for now. We can also use a generic parameter, customarily designated by λ.)

We now turn to the right side of our main equation of motion, using different indices for clarity. The terms

$$\frac{dx^a}{ds} \quad and \quad \frac{dx^b}{ds}$$

reveal in vectorial language what occurs in space-time. Let us recall how: On a flat surface, motion occurs along—tangent to—that surface. On a curved surface, motion is also tangential; see diagram O on page 139, as well as page 145. Since we also consider motion to be a chain of infinitesimally close *ds*'s, the direction of the path is tangent to the surface at each point, and that motion lends itself to description by vectors.

For clarity let us designate the tangent vector—the vector for motion in the *a*-direction along the surface at any one point—by the letter *U*. If the parameter for the surface is *s* and the coordinates are *x*'s, we define *U* by the equation

$$U^a = \frac{dx^a}{ds}.$$

Let us also designate the vector which is parallel-transported by the letter *V*. By the definition of parallel transport at any point on the surface,

$$\frac{dV^a}{ds} = 0,$$

which can be written as $V^a{}_{,s} = 0$ and which implies that if transport were not parallel, no zero would appear here; recall page 208. (Use of the letters *U* and *V* is not mandatory; compare Schutz, p. 166, with Kenyon, pp. 186-188.) If the surface on which parallel transport occurs is only flat, we write the equation for any point,

$$U^a V^b{}_{,a} = 0.$$

The comma in , *a* indicates ordinary differentiation with respect to *a*.

However we expect the same principle to hold for covariant differentiation, which means where the surface can vary and can be described by geodesics. That is to say, we must accommodate other than only inertial frames of reference. We do this by augmenting the equation with a semicolon, so that

$$U^a V^b{}_{;a} = 0.$$

Since we expect the surface to be describable by geodesics, the transported vector V is interchangeable with U, which means that we can simplify the above equation to read

$$U^a U^b_{\ ;a} = 0.$$

This equation is an efficient way to state that in curved space-time, motion is tangent to a geodesic. We note that the representation of the parallel-transported vector (the V) is no longer needed. As we commented on page 307, parallel transport is a useful mathematical tool but it does not appear in the final product.

Next we call upon the notion that covariant differentiation entails the metrical connections (the Γ's). This crucial step allows us to rewrite the equation $U^a U^b_{\ ;a} = 0$ in a more explicit form,

$$U^a U^b_{\ ,a} + \Gamma^b_{\ ca} U^c U^a = 0,$$

in which we revert to the comma but we insert the term with the metrical connections. If we write this out without abbreviating the U's, we obtain

$$\frac{dx^a}{ds} \frac{dx^b}{ds}_{,a} + \Gamma^b_{\ ca} \frac{dx^c}{ds} \frac{dx^a}{ds} = 0,$$

which indicates explicitly that *since the behavior of our vectors is affected by the point-to-point change in the curvature of space-time, we need the Γ's in order to quantify these changes.* Hence we expect the right side of our main equation of motion to take the form

$$\Gamma^b_{\ ca} \frac{dx^c}{ds} \frac{dx^a}{ds}.$$

(The summation convention applies; Σ is implied.) This term reflects our comments on page 298: Our main equation of motion should contain Γ's, which after all are based on the g's according to the Γ-equation,

$$\Gamma^b_{\ ik} = \frac{1}{2} g^{ba} \left(\frac{\partial g_{ik}}{\partial x_a} + \frac{\partial g_{ai}}{\partial x_k} - \frac{\partial g_{ka}}{\partial x_i} \right).$$

The choice and position of indices is unimportant here. The point is that the Γ's allow our main equation of motion to tell us *how a falling or orbiting object moves in a gravitational field.* We note that g_{ik} does not enter into the picture directly but via the Γ's, because gravitational acceleration is not determined by the curvature at one point but by how that curvature varies point-to-point. We also see that the mass of the test object in gravitational motion does not enter into the main equation of motion, just as Galileo had suggested, and we already jettisoned symbols for parallel transport. (We also recall the paradigm of general relativity introduced on page 66: The properties of the field, and not the mass of a falling or orbiting object, determine its gravitational behavior.)

Meanwhile $\dfrac{dx^a}{ds}\dfrac{dx^b}{ds},a$ (as in the equation $\dfrac{dx^a}{ds}\dfrac{dx^b}{ds},a + \Gamma^b_{\ ca}\dfrac{dx^c}{ds}\dfrac{dx^a}{ds} = 0$) represents

acceleration through space-time, which we can condense to $\dfrac{d^2x^b}{ds^2}$. This step lets us assemble the

equation (with previously used indices),

$$\frac{d^2x^j}{ds^2} + \Gamma^j_{\ ik}\frac{dx^i}{ds}\frac{dx^k}{ds} = 0,$$

which is our main equation of motion. It states that *acceleration* $(\dfrac{d^2x^j}{ds^2})$ *is set by how the shape of*

space-time ($\Gamma^j_{\ ik}$) *affects motion tangential to space-time* ($\dfrac{dx^i}{ds}\dfrac{dx^k}{ds}$) *while heeding parallel transport.*

Since parallel transports keeps us on geodesics, this is another way of saying that a geodesic in space-time satisfies this equation of motion. We note that this equation is set for zero (it ends with "= 0").

A more intuitive arrangement—one which more clearly shows that acceleration is set by the shape of space-time—is

$$\frac{d^2x^j}{ds^2} = -\Gamma^j_{\ ik}\frac{dx^i}{ds}\frac{dx^k}{ds}.$$

However setting the equation for zero better reflects its significance. In a frame of reference with no gravity the metrical connections vanish. Motion is only uniform; acceleration is zero:

$$\frac{d^2x^j}{ds^2} = 0.$$

The "addition" of gravitation is then indicated by adding the term

$$\Gamma^j_{\ ik}\frac{dx^i}{ds}\frac{dx^k}{ds}$$

426

to $\dfrac{d^2x^j}{ds^2}$, which reconstitutes our equation of motion in its standard set-for-zero format. Of course "adding" gravitation is equivalent to transforming to a frame with a gravitational field. Thereupon the Γ's do not vanish (they are not zero) but instead reflect the deviation from uniform motion.

This chain of reasoning invites several observations regarding our main equation of motion, which we will enumerate 1 to 5.

(1) In many publications these terms and equations appear in other patterns which differ on the use of co-variant rather than contra-variant forms and/or on the sign (+ or -). E.g. compare equation 46 on p. 143 of The Principle of Relativity (Einstein at al.) with equation 80 on p. 40 of Pauli's Theory of Relativity.

(2) While substituting *ds* for *dt* simply replaces time with space-time, this step is tied to the precept that the equations for the laws of physics can be written so as to be invariant under the Lorentz transformation—and subsequently so as to be invariant under any transformation. When there is no relative motion, replacing *dt* by *ds* is elementary: For example if an observer can accompany a moving object, dx^1, dx^2, *and* dx^3 (dX, dY, and dZ) for this observer are all zero; there is no relative motion between the observer and object in any direction. This leaves only dx^4 (dT), so that

$$dx^j \;=\; dx^4 \;=\; ds \;=\; dt.$$

Likewise when dealing with small velocities, *ds* is practically the same as—and can be interchanged with—*dt*. (To be precise, a constant, c^2, is required in most equations; when *v* is small, $ds^2 = c^2 dt^2$.)

But when *v* is substantial we summon (from page 178)

$$ds \;=\; \frac{dt}{\sqrt{1-v^2/c^2}}.$$

Since $\dfrac{d^2x^j}{dt^2}$ represents Newtonian acceleration, *substituting dt by ds via this equation recasts gravitational acceleration so that it is invariant under the Lorentz transformation.* (Moreover, other parameters besides *s* are feasible.)

(3) We recall that it is not possible to derive valid equations of general relativity without heeding the laws of conservation. In this regard Pauli (p. 217) addresses an interesting point: "The fact that the...[conservation] law for matter...is a consequence of the field equations for gravitation alone leads one to expect that the...[equations] of motion for material particles...must also follow from these field equation without further assumptions." That is to say, just as the laws of conservation are contained

within relativity, so the equations of motion are deducible from the equations of relativity. Thus general relativity provides the proper tools for its practical application in one self-contained unit (even if the manner in which we derive the relativistic field equations forces us to separate out the equations of motion.) In contrast the equations of motion cannot be derived from within Newton's gravitational equation but rather appear as separate entities.

This idea is another aspect of the remarkable aesthetic harmony and mathematical self-sufficiency within general relativity. We can say that when we "buy" Newton's law of gravitation, it does not come with the equations of motion nor with the laws of conservation; we must procure these separately. But when we "buy" the field equations of general relativity, the equations of motion and the laws of conservation follow automatically.

(4) Not only are the equations of motion part and parcel of relativity, but they actually disclose the principle of equivalence. Newton's law of inertia states that with no force impinging on it, an object exhibits only uniform motion; acceleration is zero. I.e., if $F = 0$ in $F = ma$, then $a = 0$. In a space-time terms,

$$a = \frac{d^2x^j}{ds^2} = 0,$$

and this is the condition we find in a freely falling frame of reference. The term $\frac{d^2x^j}{ds^2}$ is the left side of our main equation of motion, and, by virtue of the "a" in $F = ma$, it embodies *inertia*. Meanwhile, locally during free-fall (in the absence of gravitation)

$$\Gamma^j_{ik} = 0.$$

Here inertia, represented in $\frac{d^2x^j}{ds^2}$, is canceled by gravity, represented by the Γ's. By virtue of their relation with the shape of space-time, the Γ's embody *gravity*, which of course vanishes in free-fall. The Γ's too appear in our main equation of motion. Upon the "addition" of gravitation, the Γ's do not vanish. According to the principle of general covariance, the same equation must hold, as is clearer in this format:

$$\frac{d^2x^j}{ds^2} = -\Gamma^j_{ik}\frac{dx^i}{ds}\frac{dx^k}{ds}.$$

Here then is our point: The equal-sign in our main equation of motion denotes that inertia is equal to gravity. That is to say, this equation, half inertia and half gravity, affirms their equivalence!

(5) Eötvös and others have verified experimentally that inertia equals gravity, but we raised the question on pages 60 and 63 as to why such experiments hold in *all* frames of reference. Einstein, in The Meaning of Relativity (p. 82), illuminates this point: By itself $\dfrac{d^2x^j}{ds^2}$ is not a tensor and does not appear the same in all frames of reference. By itself Γ is not a tensor and does not vanish in all frames of reference. But they each contribute to our main equation of motion which, though not a tensor equation, *possesses general covariance*. Since this equation confirms the equivalence of inertia and gravity, these experiments should work in any frame of reference. (As mentioned, non-tensor equations can be generally covariant.)

The point that our main equation possesses general covariance is doubly important when we recall that this equation satisfies a geodesic in space-time. This means that a geodesic is independent of the system of coordinates. In contrast a straight line may not appear straight in another system of coordinates, and this is one reason Newton's law of gravitation lacks covariance.

✦

Let us return to our mission, to use the results of Schwarzschild's solution in an equation of motion. This process is quite lengthy, so we will only highlight it enough to acknowledge the major mathematical hurdle: Schwarzschild provided us with g's, whereas our main equation of motion

$$\frac{d^2x^j}{ds^2} + \Gamma^j{}_{ik}\frac{dx^i}{ds}\frac{dx^k}{ds} = 0,$$

calls for Γ's and $\dfrac{dx^i}{ds}\dfrac{dx^k}{ds}$. Furthermore this equation can be solved for $\dfrac{d^2x^j}{ds^2}$, which is acceleration, whereas we prefer that the end result be a description of motion in more familiar terms.

We keep in mind that we are picturing a planet as it accelerates (orbits) through a certain point on an imaginary sphere of radius is r. The location of that point is given by a "latitude" θ, and a "longitude" φ, as if the sphere came with a map, except that this map is four-dimensional; it gives location along four coordinates, including along r and along t. Accordingly we let j run from 1 to 4, which yields what we will call four "component-equations" of motion, one for each component of motion. For clarity we replace some of the j's with the explicit symbols r, θ, φ, *and* t. These component-equations start with

$$\frac{d^2r}{ds^2} + \Gamma^1_{ik}\cdots, \qquad \frac{d^2\theta}{ds^2} + \Gamma^2_{ik}\cdots, \qquad \frac{d^2\varphi}{ds^2} + \Gamma^3_{ik}\cdots, \quad and \quad \frac{d^2t}{ds^2} + \Gamma^4_{ik}\cdots.$$

(The pattern is similar for each equation, but the indices are different.) In effect we are dissecting motion into four directions corresponding to four dimensions. Obviously to solve these component-equations we must identify Γ^1_{ik}, Γ^2_{ik}, Γ^3_{ik}, *and* Γ^4_{ik}, which is simplified by the fact that only a few Γ's are non-zero. To recall this simplification, see page 406 where we determined which Γ's are appropriate for Schwarzschild's scenario. We identified 13 such Γ's (each has a pair of repeated indices), but the material on those pages is based on the link between the Γ's and the g's which we noted on page 296.

It is easier to begin with the second component-equation, containing Γ^2_{ik} and representing "north-south" motion along a longitude. (A globe of the Earth is again helpful here.) Since $j = 2$, only three Γ's are eligible, namely Γ^2_{12}, Γ^2_{21}, *and* Γ^2_{33}. Without using the summation convention, this component-equation limited to these Γ's becomes

$$\frac{d^2\theta}{ds^2} + \Gamma^2_{12}\frac{dr}{ds}\frac{d\theta}{ds} + \Gamma^2_{21}\frac{d\theta}{ds}\frac{dr}{ds} + \Gamma^2_{33}\frac{d\varphi}{ds}\frac{d\varphi}{ds} = 0.$$

(We note that the second and third terms are equivalent to each other.) Substituting the Γ's by use of

$$\Gamma^2_{12} = \Gamma^2_{21} = \frac{1}{r} \quad and \quad \Gamma^2_{33} = -\sin\theta\cos\theta,$$

we obtain a solvable component-equation of motion, which we write out to show the algebra:

$$\frac{d^2\theta}{ds^2} + \left(\frac{2}{r}\right)\frac{dr}{ds}\frac{d\theta}{ds} - \sin\theta\cos\theta\frac{d\varphi}{ds}\frac{d\varphi}{ds} = 0$$

However this can be simplified further. We can assume that the planet orbits only in one plane, so that we need not consider changes in latitude. The remaining variables are the radius (how far from the center) and the longitude. The latter is angle φ which the planet sweeps out during its orbit. In this way we limit motion to a plane along φ at r. (This restriction is a consequence of Schwarzschild's assumption of spherical symmetry: Objects orbiting in spherically symmetric space-time will not depart from a plane. This is like saying that objects falling "down" freely do not move "sideways.") In geometric terms this means that $\theta = \frac{1}{2}\pi$, so that $\cos\theta = 0$, and any change in θ

drops out. At any one point or instant $\frac{d\theta}{ds} = 0$. Thus every term is zero and we can ignore this component-equation; it vanishes.

Another component-equation, the first in our set, is built upon

$$\frac{d^2r}{ds^2} + \Gamma^1_{11}\frac{dr}{ds}\frac{dr}{ds} + \Gamma^1_{33}\frac{d\varphi}{ds}\frac{d\varphi}{ds} + \Gamma^1_{44}\frac{dt}{ds}\frac{dt}{ds} = 0.$$

Here $j = 1$, which represents a radius, and we know that

$$\Gamma^1_{11} = \frac{A_{,r}}{2A}, \quad \Gamma^1_{33} = -\frac{r\sin^2\theta}{A} \quad and \quad \Gamma^1_{44} = \frac{B_{,r}}{2A}.$$

We need not write out these substitutions, but we note that A and B enter into the picture. Since we allowed A and - B to represent g_{11} and g_{44}, *we have succeeded in incorporating these g's into our equation of motion.* Moreover since

$$A = g_{11} = \frac{1}{\left(1 - \frac{2KM}{rc^2}\right)}$$

and

$$- B = g_{44} = 1 - \frac{2KM}{rc^2},$$

we have brought mass into the picture. Here M quantifies matter as a point that gives rise to a gravitational field, but it will be easier to work with mass expressed as m, where $m = \frac{KM}{c^2}$, so that

$$- B = g_{44} = 1 - \frac{2m}{r}.$$

Thus the first of our component-equations does not vanish, as it deals with mass.

The third component-equation of motion, in which $j = 3$ stands for "east-west" motion, is built upon

$$\frac{d^2\varphi}{ds^2} + \Gamma^3_{13}\frac{dr}{ds}\frac{d\varphi}{ds} + \Gamma^3_{31}\frac{d\varphi}{ds}\frac{dr}{ds} = 0.$$

We know that

$$\Gamma^3_{13} = \Gamma^3_{31} = \frac{1}{r}.$$

Therefore this component-equation does not vanish, as it considers the radius.

The fourth component-equation, with $j = 4$ for "time," is based on

$$\frac{d^2t}{ds^2} + \Gamma^4_{14}\frac{dr}{ds}\frac{dt}{ds} + \Gamma^4_{41}\frac{dt}{ds}\frac{dr}{ds} = 0.$$

Here

$$\Gamma^4_{14} = \Gamma^4_{41} = \frac{B_{,r}}{2B}.$$

Again, writing out the substitutions is unnecessary, but it is important that B is in the picture, which means that this component-equation does not vanish, as it also deals with mass.

We see that two of the three surviving component-equations of motion hinge upon mass. The first component-equation means that the motion of the planet along the radius-dimension is affected by the mass of the sun, which is to be expected; for instance if the sun lost mass, the planet's orbit would widen. The fourth component-equation means that "motion" of the planet through the time-dimension is affected by the mass of the sun. For instance if the sun lost mass, each orbit would take longer and nearby clocks would speed up (which, as we have seen, is relativity's unprecedented inference).

The next step is integration, which we can think of as constructing a complete orbit from its parts. Our third and fourth component-equations are in identical differential form, and on integration they reduce to

$$r^2\frac{d\varphi}{ds} = h \quad and \quad g_{44}\frac{dt}{ds} = k,$$

where *h* and *k* are constants of integration.[*] We implicate mass as *m* rather than as *M* for algebraic convenience. The fourth equation, written as $\dfrac{dt}{ds} = \dfrac{k}{1 - 2m/r}$, is inserted into the first—both contain $\dfrac{dt}{ds}$—so that now we need only two equations to describe the orbit. (See Tolman, p. 208, Menzel, pp. 397-398, or Bergmann in Introduction...Relativity for details.) The major one of these remaining equations holds the key factors, *r, φ* and *m*, but it still has the constant *k*:

$$\frac{dr}{ds}\frac{dr}{ds} + r^2\frac{d\varphi}{ds}\frac{d\varphi}{ds} - \frac{2m}{r} = k^2 - 1 + 2mr\frac{d\varphi}{ds}\frac{d\varphi}{ds}.$$

Since we are studying the motion of the planet as it sweeps through angle φ, the last step is differentiation with respect to φ, which eliminates *k*. The algebra is simplified if we invert the radius. We let

$$u = \frac{1}{r}.$$

The end result is a relativistic equation for planetary orbits in terms of ordinary quantities like mass and radius,

$$\frac{d^2u}{d\varphi^2} + u = \frac{mc^2}{h^2} + 3mu^2,$$

and we bring forward our third component-equation, which here is also an equation for planetary orbits,

$$r^2\frac{d\varphi}{ds} = h.$$

This pair of "orbital equations," as we call them, forms a functional unit. In the first member of this pair, the term

$$\frac{d^2u}{d\varphi^2}$$

[*] When finding an antiderivative by integration, an arbitrary constant can be added to the result. (The derivative of a constant is zero, and the antiderivative of zero, which can be arbitrarily added to a function, is a constant. The "*h*" here is not related to that used for Plank's constant.

is an acceleration, except that here we mean acceleration along u with respect to change in angle φ. We can rewrite the first equation of this pair more explicitly ($u = \dfrac{1}{r}$) and with parts in reverse order, so that

$$\frac{1}{r} + \frac{d^2\left(\dfrac{1}{r}\right)}{d\varphi^2} = \frac{mc^2}{h^2} + 3m\left(\frac{1}{r}\right)^2.$$

From this orbital equation it is clear that *we have a link between r (distance), φ (location along the orbit), and m (the sun's mass)*. The right side has two terms. The first holds m modified by two constants, c^2 and h^2; of course c^2 makes this term a very large number. The second term, $3mu^2$ or $3m\left(\dfrac{1}{r}\right)^2$, which we can think of as an "excess" in the same sense we have before, is a much smaller number, but we will examine its significance later.

In the second member of the pair of orbital equations, the term

$$r^2\frac{d\varphi}{ds}$$

defines the angle φ, and the h in this equation is a constant. We note that if r^2 increases (longer distance from m), $\dfrac{d\varphi}{ds}$ decreases (less change in the angle per ds), and vice versa. Therefore the constant h represents the law of conservation of angular momentum, in that h is angular momentum per mass of the planet. That is to say, if h varied the planet could take non-geodesical paths; see page 240. (Likewise conservation of energy is represented by the constant k we saw on the previous page, reaffirming that Schwarzschild's solution conforms to the conservation laws.)

Given this pair of orbital equations, the orbital path can be seen as the variation in r as the planet progresses through φ. Moreover comparable pieces of information, r and φ, appear in the two-dimensional equation (page 189)

$$ds^2 = dr^2 + r^2d\varphi^2,$$

where φ is the bearing (an angle) and r is the range (a distance). We also recall the four-dimensional equation for spherical-polar coordinates,

$$ds^2 = dr^2 + r^2d\theta^2 + r^2\sin^2\theta d\varphi^2 - c^2dt^2,$$

and we note that the two equations are structured alike except for the last two terms in the latter. We absorbed the fourth term c^2dt^2 by considering m, and we eliminated the third term $r^2\sin^2\theta d\varphi^2$ by assuming that the path does not deviate from the equatorial plane. *Thus we reduced the problem to only two values, equivalent to range and bearing.*

Let us imagine we are astronomers tracking a planet in solar orbit. Knowing the sun's m, we can calculate what motion it is in (the change in bearing, $d\varphi$) at the range r. As we pursue the planet, we find that the range shows an additional change-within-a-change, represented by the second-differential term $\dfrac{d^2(\frac{1}{r})}{d\varphi^2}$; i.e. the course of the planet marks a complicated curve. If we follow the planet long enough, that curve turns out to be almost an ellipse. We will explain what is meant by "almost" later. Suffice it that *we have found the planet's orbital motion using a method that heeds general relativity.*

By reducing the problem to bearing and range, we have defined one kind of gravitational behavior, namely orbiting, in terms of location in the gravitational field. This is part of the concept cited on pages 73 and 372-3: Special relativity maintains that measurement of time depends upon relative motion. General relativity tells us that measurement of time also depends on relative *location*. Thus the rate of time is set by the gravitational potential at the site of a clock in a gravitational field. Even the behavior of light is determined by the location of the measurement, i.e. by the site of photons moving through a gravitational field, as shown in the einstein red-shift and in the bending of a beam of light. Indeed *all gravitational behavior, be it that of a photon or a planet, hinges on location in space-time.*

Let us review the steps we took to apply this key concept to a planet that is orbiting the sun: We wrote the general metrical equation,

$$ds^2 = g_{ik}\, dx^i\, dx^k,$$

in a spherical-polar form,

$$ds^2 = g_{11}dr^2 + g_{22}d\theta^2 + g_{33}d\varphi^2 - g_{44}c^2dt^2.$$

We assumed staticism and spherical symmetry, and we assumed that exterior to the sun, the field equations

$$G_{ik} = -XT_{ik}$$

are such that

$$T_{ik} = 0.$$

.

435

We admitted only certain components of g_{ik}, which limited our possible Γ's by virtue of the Γ-equation

$$\Gamma^b_{ik} = \frac{1}{2} g^{ba} \left(\frac{\partial g_{ik}}{\partial x_a} + \frac{\partial g_{ai}}{\partial x_k} - \frac{\partial g_{ka}}{\partial x_i} \right).$$

This procedure opened the door to determining the identity of the g's, including the crucial

$$g_{11} = \frac{1}{\left(1 - \dfrac{2KM}{rc^2}\right)} \quad \text{and} \quad g_{44} = 1 - \frac{2KM}{rc^2},$$

which describe the shape of space-time at the planet's location in relation to the sun's M. By using the allowed Γ's, we then solved the equation of motion,

$$\frac{d^2 x^j}{ds^2} + \Gamma^j_{ik} \frac{dx^i}{ds} \frac{dx^k}{ds} = 0,$$

but for clarity we fractured this equation into four equations, one for each component of motion. This method left us with equations that describe the orbit in relatively simple terms of bearing and range.

✦ ✦ ✦

We mentioned that Schwarzschild's exterior solution yields results very similar to those based on Newton's methods, and in fact the small difference—the "almost"—can be traced to one certain term in Schwarzschild's solution. This finding falls under the concept that *Einstein's field equations can lead to (or "reduce to") Newton's equation for gravitation.*

An older wording of this idea is that Newtonian gravitation can be found by means of a "first approximation" of gravitation within general relativity. This statement implicates Einstein's "first" and "second" approximations. The terminology is somewhat inconsistent (e.g. The Principle of Relativity, pp. 157-158, vs. The Meaning of Relativity, pp. 80-81) but we already used both approximations on pages 252-226. We made the first approximation when we assigned the "flat" values to f_{ik} in the equation

$$g_{ik} = f_{ik} + h_{ik},$$

thereby neglecting space-time curvature.* We made the second approximation when we allowed some deviation from flatness in the term h_{ik} but neglected non-linear terms. These steps allowed us to equate g_{44} with the gravitational potential, and they prompted us to consider only the 44-components of T_{ik} under conditions of weak gravity and slow speeds. Einstein used these approximations to show that g_{44} can explain Newtonian gravitation (The <u>Principle</u> of <u>Relativity</u>, p. 159, says more, and we shall elaborate shortly). Given the success of Newton's law in our environment, these approximations served as a mathematical verification of general relativity several years before experimental confirmations became feasible.

Let us reinterpret these approximations by using the notion of the "Newtonian limit." An alternative term might be the "Newtonian abbreviations," for when the field equations are abbreviated in certain ways, a "limited" facet of general relativity remains which turns out to be identical to Newton's law of gravitation.** The underlying assumption in the Newtonian limit is that the gravitational field is weak, so that the curvature of space-time is negligible and the resulting speeds are nowhere near those of light. The gravitational field is also regarded to be nearly static, which means that the distribution of matter does not change quickly. We already know what these assumptions mean: g_{44} is the dominant component of the metrical tensor; ds is practically the same as dt; T_{44} is the dominant component of matter; and only linear terms in the field equations need consideration.

Let us recall these field equations in this form:

$$R_{ik} - 1/2 g_{ik}\, R \;=\; -XT_{ik}$$

The term $-1/2 g_{ik} R$ addresses the conservation laws, which we need not consider here, and neither do Newton's laws. Hence we are left with

$$R_{ik} \;=\; -XT_{ik}.$$

We agreed to ignore all R's as well as T's except where i and k are 4, so the remaining "limited" field equation is

$$R_{44} \;=\; -XT_{44}.$$

R_{ik} descends from partial derivatives of the Γ's according to

---

*The "h" in this equation is not related to the constant of integration.

** The term "limit" stems from the fact that Newton's treatment of gravitation implies an infinite speed of light. We then say that if c could reach infinity, Newton's law of gravitation becomes the limit of general relativity.

$$R_{ik} = \frac{\partial \Gamma^j_{ij}}{\partial x^k} - \frac{\partial \Gamma^j_{ik}}{\partial x^j} + \Gamma^e_{ij} \, \Gamma^j_{ek} - \Gamma^e_{ik} \, \Gamma^j_{ej},$$

but by ignoring the last two non-linear terms, and by letting Γ^j_{ij} vanish (it has no 44-component), we find

$$R_{44} = -\frac{\partial \Gamma^j_{44}}{\partial x^j}.$$

With only the 44-component, the Γ's descend from partial derivatives of the g's according to

$$\Gamma^j_{44} = \Gamma_{44j} \cong \frac{1}{2} \frac{\partial g_{44}}{\partial x^j}.$$

(We let Γ^j_{44} equal Γ_{44j} because the distinction between co-variant and contra-variant forms fades when ordinary Cartesian coordinates are feasible, which they are in this setting.) We now need a link between R_{44} and Newtonian acceleration. This is available from the equation of motion,

$$\frac{d^2 x^j}{ds^2} = -\Gamma^j_{ik} \frac{dx^i}{ds} \frac{dx^k}{ds}.$$

In concert with our assumptions, wherein dx^4 predominates and is practically the same as ds, the components of vectors found in this equation of motion are approximately

$$\frac{dx^1}{ds} = \frac{dx^2}{ds} = \frac{dx^3}{ds} = 0 \quad and \quad \frac{dx^4}{ds} = 1.$$

Under these conditions

$$\frac{dx^i}{ds} \frac{dx^k}{ds} = 1,$$

so that our equation of motion reduces to

$$\frac{d^2 x^j}{ds^2} = -\Gamma^j_{ik}.$$

Furthermore under these conditions ds equals dt. The essential equation of motion becomes

$$\frac{d^2x^j}{dt^2} = - \Gamma^j_{44},$$

where j runs only 1 through 3. Using the last equation on the previous page, we make a substitution so that

$$\frac{d^2x^j}{dt^2} = -\frac{1}{2}\frac{\partial g_{44}}{\partial x^j},$$

in which $\frac{d^2x^j}{dt^2}$ is the Newtonian form of acceleration "*a*" as in

$$a = \frac{K\, m_2}{D^2}.$$

Clearly the partial derivatives of the metrical tensor limited to its 44-component determine Newtonian gravitational behavior.

Meanwhile Newton's $F = ma$ can be adopted by replacing F with the gravitational force, G:

$$G = ma \ \ \text{or} \ \ \frac{G}{m} = a$$

By elementary algebra (we substitute for "*a*")

$$\frac{G}{m} = \frac{K\, m_2}{D^2},$$

and finally we attain the equation for Newton's law of gravitation,

$$G = K\,\frac{m_1\, m_2}{D^2}.$$

That is to say, *by "limiting" the field equations of general relativity, we have replicated Newton's law of gravitation.*

We can now link this finding to Schwarzschild's solution by recalling the spherical-polar equation

$$ds^2 = g_{44}c^2dt^2 - g_{11}dr^2 - g_{22}d\theta^2 - g_{33}d\varphi^2.$$

We place the preponderant 44-component-term first for clarity. The terms can be positive or negative depending on how we set up the underlying equation, but we ignore this detail here. More pertinently, we can omit the last three terms, "limiting" the spherical-polar equation to

$$ds^2 = g_{44}c^2dt^2.$$

In these conditions

g_{44} is very nearly ("almost") 1,

supporting our conclusion that under everyday conditions (with c^2 as a conversion constant)

ds^2 is practically the same as c^2dt^2.

However *when measured carefully enough,*

$$g_{44} = 1 - \frac{2KM}{rc^2},$$

which can be inserted into the solved spherical-polar equation,

$$ds^2 = \left(1 - \frac{2KM}{rc^2}\right)c^2dt^2 - \frac{dr^2}{\left(1 - \frac{2KM}{rc^2}\right)} - r^2(d\theta^2 - \sin^2\theta d\varphi^2).$$

The first right-side term is magnified by the c^2. Again we can omit the last three terms, leaving us with

$$ds^2 = \left(1 - \frac{2KM}{rc^2}\right)c^2dt^2.$$

We call this equation the "Newtonian limit of Schwarzschild's solution," and we can say that this one term,

$$\left(1 - \frac{2KM}{rc^2}\right)c^2dt^2,$$

embodies Newtonian gravitation.

Indeed this "limit" leads to the outcome found by way of Newton's law of gravitation, because when we apply the steps we took on pages 432 and 434, we do not end up with our orbital equations

440

$$\frac{d^2u}{d\varphi^2} + u = \frac{mc^2}{h^2} + 3mu^2 \quad \text{and} \quad r^2\frac{d\varphi}{ds} = h.$$

Rather we see

$$\frac{d^2u}{d\varphi^2} + u = \frac{mc^2}{h^2} \quad \text{and} \quad r^2\frac{d\varphi}{dt} = h.$$

We note the dissimilarities: The first equation of the latter pair lacks the "relativistic" term $3mu^2$ (or $3m/r^2$), reflecting the omission of certain terms earlier. The second equation of this pair contains dt rather than *ds*, reflecting that Newtonian equations do not accommodate a distinction between *ds* and *dt*.

This finding means that what we just omitted from Schwarzschild's full equation constitutes the impact of relativity. True,

$$- r^2(d\theta^2 - \sin^2\theta d\varphi^2)$$

belongs to the trigonometry of the system of coordinates, but it makes little numerical difference here and contains no *M* anyway. However

$$\frac{dr^2}{\left(1 - \frac{2KM}{rc^2}\right)}$$

predicts the deviation of actual observations from Newtonian calculations. This one term embodies *the difference relativity makes*. Of course the smaller the *M*, the more subtle this difference, and under familiar conditions $\frac{2KM}{rc^2}$ is too small to have a significant impact.

In this sense Newton's gravitation is but a component (here identified by the index 44) of certain tensors, and the Newtonian limit is that part of general relativity which agrees with Newtonian physics. We can then say that Newton's law occupies one corner of general relativity, but in the weak gravitational field of our location in the universe, that corner happens to be by far the most prominent—so prominent that if we neglect the other corners, the two interpretations of gravitation yield identical outcomes.

The fact that Newtonian and relativistic results on weak fields are the same suggests another question: Why couldn't both interpretations be correct? Could gravity be a force *and* space-time be curved? No, because relativity by itself replaces the Newtonian system. For example a certain gravitational problem solved by Newton's methods gives a numerical answer of 1, while the field equations may give 1.00000212. Though the excess is small, relativity alone provides the entire—and the correct—result.

◆

Two words of caution are needed here. First, we must not confuse the assumptions and approximations designed to ease the solutions of the field equations. In Schwarzschild's solutions G_{ik} is not abbreviated; all four non-zero components are embraced. Einstein's approximations exclude all components save the one identified by the index 44. As a result Einstein's solutions give correct results only under conditions of weak gravitation, whereas Schwarzschild's exterior solution provides more accurate predictions in weak gravitation, and his interior solution gives much truer results in strong gravitation, including in the extreme case of black holes. (This notion suggests further refinements by the inclusion of other components of the mass-energy tensor. We shall study this possibility soon.) In short, by admitting only the 44-part of the metrical tensor, Einstein chose to consider only weak fields. By assuming spherical symmetry, Schwarzschild chose to consider only simply-structured fields, but he was able to admit some 11-, 22-, and 33-components.

The second caution has to do with the idea that the relativistic equations for gravitation do not implicate the "test" object (e.g. a planet), only the object responsible for the gravitational field (e.g. the sun). That is to say, even though both objects contribute to the deformation of space-time, Einstein's field equations appear to admit only the latter and to disregard the input from the former. In contrast, Newton's law of gravitation contains "m_1" and "m_2," which correspond to these two objects, so that this approach seems to be more realistic, especially when both objects are very massive. We might thus conclude that general relativity fails to regard both m's.

The reply is this: We recall (page 60) that the intensity of the gravitational field is independent of "m_1." Therefore relativity assigns a set of field equations to each mass. When more than one object significantly affects the local gravitational field, in theory *each contribution is treated individually*, although the actual calculation becomes extremely complicated. An analogy here is two mutually attracting magnets: The "Newtonian" strategy for calculating their behavior is to multiply the strengths of the two magnets and divide by the square of their distance—which is easy but not fully reliable, since distance (space) is relative. The "Einsteinian" strategy is to calculate the individual effect of each magnet on the surrounding magnetic field and to blend the results—which is mathematically difficult but theoretically sound.

◆

Schwarzschild's exterior solution can serve as the basis for an accurate computation of the orbit of the planet Mercury. We recall that the "relativistic" term $3mu^2$ disappears when we solve "limited" equations, i.e. when we omit the contribution of general relativity. This point is useful in explaining the "excess" in Mercury's precession.

We also recall that planets have "almost" elliptical paths. In particular we must consider the semimajor axis "a" (2a is the widest extent of an ellipse) and we must consider a measure of eccentricity "e" (e depends on the widest and narrowest extent of an ellipse). As Mercury orbits, we think of it as rotating through angle φ at distance r, which were our "bearing" and "range" earlier, while "*a*" and "*e*" describe the ellipticity of that orbit. If an orbit is exactly spherical (not an ellipse) $e = 0$, $a = r$, and there can be no precession. In reality the orbit of Mercury is almost elliptical, which is why we envision orbital motion as occurring on the surface of a spheroid rather than a sphere. However, using any system of coordinates to plot the entire orbit, we find that after each revolution, the expected location has over-shifted: there is *excess* precession.

Though we can omit the trigonometry (See Kenyon, pp. 91-92), solving the orbital equations

$$\frac{d^2u}{d\varphi^2} + u = \frac{mc^2}{h^2} + 3mu^2 \quad \text{and} \quad r^2\frac{d\varphi}{ds} = h$$

for *u* (or *r*) but *without the term* $3mu^2$, we obtain an equation for an elliptical orbit,

$$u = \frac{1}{r} = \frac{1 + e \cos \varphi}{a\,(1 - e^2)}.$$

This result applies to the orbit *without considering relativity*. However if we include the term $3mu^2$, we find a more complicated but astronomically trustworthy equation,

$$u = \frac{1}{r} = \frac{1 + e \cos\,[(1 - 3m/a(1 - e^2)\varphi]}{a\,(1 - e^2)}.$$

The crucial difference between the latter two equations, calculated with more trigonometry, is the "excess" shift (per revolution) in the angle φ of the orbit:

$$\frac{6\pi m}{a\,(1 - e^2)}$$

This is the *excess precession*. Using known values of *m, a,* and *e* for Mercury in this term yields about 43 seconds of arc per century, which is very close to actual observations. This is a tiny amount (just one orbit of this planet is 1,300,000 seconds of arc) but much larger excess-precessions have been found, such as in the orbits of pulsars, which manifest this aspect of general relativity even more convincingly.

We should be aware that most of Mercury's precession, about 530 seconds of arc per century, is predicted by purely Newtonian methods. The manifestation of general relativity resides in only the *excess* 43 seconds. In other words Newton himself could have figured out that Mercury's orbit is almost an ellipse, but his "almost" would not have been quite correct. Schwarzschild's method is better in part because it has fewer omissions—and those omissions encompass the improvement provided by general relativity.

✦

While Mercury has an orbit which reveals relativistic precession, this astronomic object does not actively emit photons—it does not glow. But if an object orbits *and* glows, does it show excess precession as well as a red-shift? The answer is found in orbiting *pulsars* which have three convenient features: Their glow is strong, they pulsate at a regularly frequency, and the value of

$$\left(\frac{2KM}{rc^2} \right)_{pulsar}$$

is quite large. The red-shift of a pulsar's glow can be predicted using a general-relativistic ratio

$$\frac{1}{(1 \, - \, 2KM/rc^2)_{pulsar}} \quad ,$$

but because pulsars move very rapidly, the special relativity factor must be included. The ratio becomes

$$\frac{1}{(1 \, - \, 2KM/rc^2)_{pulsar} \; (\sqrt{1 \, - \, v^2/c^2})_{pulsar}} \quad ,$$

exemplifying the cooperative roles of general and special relativity in explaining natural phenomena. The former provides for the effects of matter (represented in *M*) and the latter for the effects of relative motion (represented in *v*). Moreover the behavior of the pulsar demonstrates the various effects of relativity at a location in the universe far from our solar system.

Schwarzschild's mathematics can be used to calculate the bending of light as it passes close to the sun. If *r* is the closest a ray comes to the sun, the deflection is $\dfrac{4KM}{rc^2}$. This gives about 1.75 seconds of arc, which of course was the critical finding in the solar-eclipse experiments of 1919. Likewise an experimentally bent radar pulse requires more time. This prediction can be made with similar calculations.

✦ ✦ ✦

We pointed out that when *M* is small (gravity is weak), $\dfrac{2KM}{rc^2}$ is a very small number, so that $1 \, - \, \dfrac{2KM}{rc^2}$ is practically equal to 1. But let us consider what happens if *M* is very great. *M* can be large enough so that $\dfrac{2KM}{rc^2}$ reaches 1, which occurs when r exactly equals $\dfrac{2KM}{c^2}$ (as

$$\frac{2KM}{(2KM/c^2)c^2} = \left(\frac{2KM}{c^2}\right)\left(\frac{c^2}{2KM}\right) = 1).$$

In that case we see that in Schwarzschild's metrical equation,

$$ds^2 = \left(1 - \frac{2KM}{rc^2}\right)c^2dt^2 - \frac{dr^2}{\left(1 - \dfrac{2KM}{rc^2}\right)} - r^2(d\theta^2 - \sin^2\theta d\varphi^2),$$

the first term vanishes ($1 - 1 = 0$). More remarkably the second term becomes infinite (dr^2 is divided by zero), giving ds^2 an infinite value. For many years physicists discounted these findings as mathematical anomalies, reminiscent of the initial response to Lorentz' results which suggested that time is relative. Einstein (Pais, p. 289) also initially rejected the possibility that r can equal $2KM/c^2$, because it meant that matter can be infinitely concentrated and that the particles which make up such matter can reach velocity c. It also meant that the red shift can be infinite and that electromagnetic waves can be infinitely long.

Today these possibilities no longer seem irrational, and we accept their astrophysical implication: An r that is equal to $2KM/c^2$ is called the Schwarzschild radius or the Schwarzschild singularity. We label such an r as r_S. A singularity is the condition of extreme density and extreme curvature of space-time. Modern cosmology recognizes Schwarzschild's "coordinate" singularity at $r \le 2KM/c^2$, which depends on the choice of coordinates, and a truer "physical" or "continuum" singularity at $r = 0$, which is an invariant point is space-time. (See Wald, pp. 124 and 152-157, or Schutz, p. 290.) We will say more about the Schwarzschild radius r_S shortly, but for now we need to know that if some object is so concentrated that its $2KM/c^2$ equals or exceeds r_S, its gravity is too strong for the escape of light. In other words the object's mass-density creates an indentation of space-time so steep that photons cannot climb out. Borrowing a term from space-technology, even c is too low to be an "escape velocity." This condition constitutes the infamous *black hole*.

The notion of photons unable to "escape" may seem paradoxical: Does this mean that light can stop moving? In reply, let us imagine an "event," the emission of photons from a glowing object. These photons have a speed (c), a wavelength, and a (reciprocal) frequency, which means that the glowing object acts like a clock that "ticks" with that frequency. To a local observer, these "ticks" appear very frequent, and they account for the color of the glow. Meanwhile, if the object becomes compressed into a black hole, according to an external observer, the wave length of emitted light increases, the frequency decreases, and the time interval between "ticks" becomes longer and longer. *When the object's r becomes $2KM/c^2$, the time between "ticks" is "forever," so that the external* *observer "waits forever" for the next "tick."* The photons still have speed c, but they escape with "zero frequency," they appear infinitely red-shifted, and their wave length is endless. Of course if no photons escape ("zero ticks") the object looks black. In other words no matter how long he waits, the external observer receives "zero signals" from the event; for him, the object has ceased to glow and the "event" has ceased to exist. *The "escape" of signals requires infinite time even at velocity c.*

Meanwhile, to the observer within the Schwarzschild radius, where r is less than $2KM/c^2$, the g_{44} has turned negative and g_{11} positive (by algebra). This means that his space-time path is such that r never increases and can only become smaller. Here, inside the black hole, photons still move, but neither the observer nor the object nor any photon can escape. Indeed with the inevitable passage of time, r becomes zero and they become a "physical" singularity! The speed of light in space-time is not zero, but space-time becomes zero.

Clearly the mathematical key to black holes is the possibility that $r = \dfrac{2KM}{c^2}$, and this equation merits several comments. First we recall that a radius in ordinary Cartesian coordinates differs from r in curved space-time, and of course the r in Schwarzschild's metrical equation applies to such space-time. Nevertheless we express these r's, and even an r_S, in ordinary distance- or length-units such as centimeters or meters, and we express the mass in M in grams or kilograms.

Moreover to interpret r_S we must consider size and density, and we must appreciate that an object undergoing gravitational compression can have a size that exceeds this r, that just equals this r, or that is less than this r. In the first case the object is not very dense and merits no special consideration. In the second case, if an object becomes so dense that $2KM/c^2$ equals r, the entire object just fits inside r, which is what we call the Schwarzschild radius r_S. We assume that the object has become *a black hole*; it is now a "coordinate" singularity. If we further assume that stars are spherical, they become black holes when their density reaches $\dfrac{M}{(4\pi/3)(r_S^3)}$ (the volume of a sphere of radius r is $4/3\pi r^3$), in which case M represents the "weight" of a black hole. In this context the Schwarzschild radius r_S designates the "size" of the black hole, and the sphere of radius r_S forms an "event horizon." Here "weight" and "size" are difficult to interpret, since we are dealing in highly curved four-dimensional space and time, and of course no human observer has yet used a scale or ruler on a black hole. The "event horizon" is the edge or surface of a black hole; anything inside is invisible to outside observers, and any observer (if surviving) inside a black hole is cut off from communication with the rest of the universe. Therefore we can redefine a spherically symmetric black hole as a region of space-time surrounded by an event horizon which has size r_S. (For the sun to become a black hole, it would have to be compressed to fit an r_S of about 3 km or 2 miles; $\dfrac{2KM_{sun}}{c^2} = 3 \ km$. [We noted that $\dfrac{2KM}{c^2}$ is in units of length.])

If the density rises more (it is smaller than r_S) the object is still a black hole, and it is destined for a "physical" singularity in which it fits into an r of zero. As difficult as it is to imagine something compressed enough to fit into a zero distance, we believe that this occurs naturally when especially massive stars undergo complete gravitational collapse.

What about the gravitational field outside r_S? It is still determined by the M at its center, and unless an observer in this field gets very close to that center, he feels nothing unusual. In fact if our sun became a black hole, the planets would probably continue to orbit as they do now. Contrary to popular opinion, black holes do not voraciously suck in all theirs surroundings. However if the observer does get close enough to a black hole, he will be crushed as he "falls into a depression in space-time" which is shaped somewhat like a funnel and which carries him inexorably to a point—the "physical" singularity of vanished space-time.

We can now comment on the intriguing point that in a black hole space-time may vanish and become "zero." This possibility leads to two inferences: First, despite our assurances, "time"—along with space-time—can cease. The second is the even more provocative prospect that all of time, along with the entire universe, has a finite life span. In other words our universe and its time very likely had a clear beginning (in a singularity followed by the "big bang") before which time did not exist, and time may yet have an equally decisive end (in another singularity following a "big crunch") after which it no longer exists. We note that these events, particularly the hypothetical fate of the universe and the associated "end of time," are a direct consequence of the interplay—extreme gravitational collapse to be exact—between matter and space-time as predicted via the field equations of general relativity. Hawking has written extensively on this issue, and of course it has physical, cosmological, philosophical, and even theological facets. However, dodging the obvious question what exists "before and after time," let us focus on a narrower consideration: For our purposes we can redefine "time:" It is simply a dimension of space-time, one that can be relative, can be curved, can have end-points, can have measurable intervals, and can have a fixed length along which objects move.

✦

We do not imply that Schwarzschild was the first to hint at the possibility of black holes; this was done in the eighteenth century with Newtonian equations. Schwarzschild however was the first to apply the mathematics of Einstein's relativity to this phenomenon. As we shall see next, his solution became the "parent" of newer more comprehensive solutions which open the door to new facets of astrophysics and cosmology, including the possibilities of anti-gravity, negative space, "white" holes, and other universes. Many books, articles, and web sites cover this provocative subject, but see Kaufmann and Gribbin for overviews; also Hawking, p. 91 of A Brief History of Time, Kenyon, p. 115, and Wald, p. 308. Let us touch on the mathematical fundamentals:

The Schwarzschild solution itself is not adequate for the study of black holes as we currently envision them, mainly because it simplifies the mass-energy tensor T_{ik}. To appreciate this issue we need to categorize the components of this tensor into three parameters: *Total mass, charge, and angular momentum.* (Recall page 354.) We envision that all effects of matter upon the gravitational field which surrounds a black hole can be ascribed to these three parameters. Of course the first parameter is obvious, even if we treat it as density-of-mass. The second is an electromagnetic property. The third stems from rotation of the mass. Since Schwarzschild could afford to ignore rotation and electromagnetism, his solution deals only with the extent to which mass affects the gravitational field. For example the rotation of the sun is negligibly slow, all members of our solar system appear to be electrically neutral, and all magnetic poles are weak as well as neutral (north and south magnetic poles balance). The implication of Schwarzschild's assumptions for the study of black holes is that we should be able to calculate...

(1) to what extent the *charge* of whatever was compressed into a black hole affects the surrounding gravitational field,

(2) to what extent the *angular momentum* of whatever was compressed into a black hole affects that field, and

(3) to what extent *both* the charge and angular momentum of whatever was compressed into a black hole affect that field.

Let us first consider a black hole based on the compression of a "charged mass," by which we mean a sample of dense matter which is *not electromagnetically neutral*. This possibility may be unrealistic since astronomic objects appear to be electromagnetically neutral, or any net charge may be quickly neutralized in the process of forming a black hole. Nevertheless several features of this possibility are noteworthy:

Its mathematics is not much more difficult than that for Schwarzschild's solution, and indeed a solution was promptly achieved by Reissner and Nordstrøm (and others) in 1916-1918. Their solution still assumes spherical symmetry, but it holds an additional term which we can call "EM" for the electromagnetic parameter, and its T_{ik} is based on Maxwell's equations and the electromagnetic model. Not surprisingly, if EM = 0, Reissner-Nordstrøm's solution reduces to Schwarzschild's. One peculiar point however is that this solution suggests the existence of a magnetic monopole—a north or south magnetic pole with no mate. (We can imagine one end of a long bar-magnet protruding from a black hole; energy from the other end is trapped in the hole.) In fact magnetic monopoles have since been observed experimentally, implying that one of Maxwell's equations is inaccurate. (Kaufmann, <u>Cosmic</u> <u>Frontiers</u>, p. 156, and Carmelli, p. 155.)

Next we take up the case of a black hole arising from a *rotating* mass. Rotation adds the parameter of angular momentum which we can label as "AM." Since astronomic objects routinely appear to be rotating, and since the shrinking of the diameter of a rotating object tends to increase the speed of rotation (like a figure skater increases spin), we believe that angular momentum is a common feature of black holes. Although the significance of rotation had been suspected for several decades, the computation is far more complicated; it took until 1963 before Kerr published a solution. A primary source of mathematical difficulty arose from the fact that Kerr could not assume spherical symmetry, only axial symmetry. I.e., a black hole rotating on one axis loses its spherical symmetry. We can visualize this through a feature of the Earth: Because of rotation, the Earth is not a perfect sphere; it is flattened at the poles and it bulges at the equator. In the case of a rotating black hole, this bulge can be thought of as a place in which space-time is being "dragged around by the rotation." (See Gribbin's Figure 26 on p. 126, or Kaufmann's Figure 11-4 on p. 176. of <u>Cosmic Frontiers</u>...) Kerr's solution provides a mathematical description of space-time "dragged" by motion under conditions as complicated as rotation around a non-spherical black hole. More on "dragging" shortly.

In geometry a flattened and bulging rounded object is said to be oblate, and analysis of an oblate spheroid calls for elliptical coordinates, in contrast to the simpler spherical-polar coordinates in Schwarzschild's scenario. Therefore the metrical equation for Kerr's solution contains more angles in order to accommodate the fact that an oblate structure appears different from different vantages. Kerr's metrical equation of course contains M for mass but it also has AM to accommodate rotation. One version appears in Schutz, p. 297, but of note is that the equation has five main terms on the right side, compared with Schwarzschild's four, and the associated matrix has a component off the diagonal. These complications correspond to a more comprehensive T_{ik}, one which houses not only density of mass but also angular momentum. If AM = 0, the off-diagonal term vanishes so that Kerr's equation reduces to Schwarzschild's.

Finally we touch upon the most complicated case, a black hole formed from matter that is rotating *and* electromagnetically charged. The solution was announced by Kerr and Newman in 1965. In principle we see the features of both preceding solutions: No spherical symmetry is assumed, the metrical equation is based on an elliptical coordinate system, and the overall equation contains terms for EM and AM. Conversely if rotation is ignored, we find ourselves in the Reissner-Nordstrøm solution. If the terms associated with rotation and charge are both eliminated, we again recover Schwarzschild's solution.

✦

We just mentioned "dragged" space-time. This is an important topic in modern relativity. If uniform motion can drag space-time as envisioned in special relativity (page 116), so can *accelerated motion*, and therefore so can *rotation*. This rotational "frame dragging" [*] can be envisioned on our trampoline which has a grid imprinted on its surface and also has a heavy ball firmly glued onto the middle of that surface. The ball indents the trampoline, which of course represents the major effect of mass on space-time, as considered by Schwarzschild. Because of indentation, four nearby points on the grid may appear as they do in diagram U (page 322); note how "P_2" and "P_4" converge. This convergence is another aspect of the geodetic effect (pages 211 and 216) which, we recall, reveals curvature of space-time; the convergence would not appear if space-time were flat. However we can grab the glued-on ball and *twist* it (like we would a doorknob). The twisting, evidenced as a distortion of the surface, represents *the rotational dragging of space-time* by virtue of angular momentum, as considered by Kerr. Again using diagram U, when the ball ("Earth" in this diagram) is twisted clockwise, the lower ends of "String A" and "String B" are dragged to the right, so that "P_2" and "P_4" are displaced to the right. In theory, if we let two strings dangle near the north pole of the Earth, not only would they converge slightly toward the Earth's center, but they would also

---

[*] A.k.a the Lense-Thirring effect after its 1918 investigators, though Einstein proposed it earlier.

drift very slightly toward the east because the Earth "twists" eastwardly, dragging space-time along in that direction. The drift, though less extensive than the convergence, interests us now.

If we imagine that diagram U is superimposed on the surface of our trampoline, the two effects—convergence of points from indentation, and displacement of points from twisting—are at right angles from each other. This detail suggests that an experiment with a gyroscope in a rotating gravitational field can isolate each of the two effects. Moreover we can add an important facet: By analogy, we can hold the ball to keep it stationary, but we can *rotate the entire trampoline.* Lense and Thirring calculated that even under these conditions *a drag of space-time should be discernible.* In short, even dragging should be relative, and this notion supports the term "frame dragging."

Rotational frame dragging of space-time is a subtle but nonetheless critical prediction of general relativity. It clarifies the nature of inertia while supporting Einstein's assertion that accelerated motion can only be relative. We recall the argument on page 63: The surface of the water in a spinning bucket is concave, which is an inertial effect. If we disallow absolute accelerated motion and consider the bucket to be stationary, what accounts for the concavity except *rotation of the universe around the bucket?* That is to say, *in order to explain the inertia demonstrated by the concavity, we should find nearby space-time to be dragged.*

Detecting frame dragging is difficult. It requires extremely sensitive gyroscopes in an experiment able to distinguish several different effects of motion (page 213), and this has been tried with limited success. The reason for the difficulty is that easily observable astronomical objects do not rotate enough to generate measurable frame dragging; their T_{ik} houses far more density of mass than angular momentum; their nearby space-time is mainly indented rather than twisted. Of course rotating black holes may have enough angular momentum but are difficult to study. A new experiment, the Stanford-NASA Gravity Probe B, is under way to explore frame dragging ensuing from the Earth's rotation. Gyroscopes will be placed in Earth orbit, and changes in their orientation relative to a distant star will be measured.

If dragging of space-time is found, another feature of general relativity will have been revealed, namely *its ability to account for inertia without invoking absolute accelerated motion.* Moreover a profound principle of modern cosmology will have been maintained, that the content of the universe is always in relative motion—everything has kinetic energy—see page 40—and that *inertia is the result.* This concept also entreats relativity's principle of equivalence: *Since gravity is the same as inertia, gravity exists because the universe is full of mass, energy, and motion, while these determine the shape of space-time.*

✦   ✦   ✦

Despite the obviously pivotal importance of relativity in cosmology, it is ironic that during the first few years after their enunciation, Einstein's field equations

$$G_{ik} = -XT_{ik}$$

seemed to be at odds with the accepted cosmologic hypothesis that the universe is "static"—neither enlarging nor shrinking. In their original form these equations favored an enlarging universe, and this prompted Einstein to "staticize" them by inserting an adjustment in the form of a "cosmological constant." The revised equations, compatible with a "static" universe, read

$$G_{ik} - \Lambda g_{ik} = -XT_{ik}.$$

The term Λg_{ik} appears in place of $(C)g_{ik}$ in the pattern for $(A)R_{ik} + (B)Rg_{ik} + (C)g_{ik}$; see page 363.

In the 1920's Hubble and others found evidence that the universe is indeed expanding, presumably as a consequence of its creation in the "big bang," the theorized moment our universe came into existence. In response, Einstein happily withdrew the cosmological constant, although recent findings explained by quantum mechanics suggest that there may be a rationale for such a constant after all. (See Kenyon, p. 83, and Wald, p. 99. Incidentally, Λg_{ik} requires that the divergence of g_{ik} [like that of G_{ik}] be zero, which it is.)

Does general relativity have any meaning or potential use beyond the study of gravitational phenomena? To answer this, let us use the hypothetical case of visitors from a distant planet who have a very fast spaceship and with whom we made a date to meet at 4 pm at a certain address in New York. We assume that neither the visitors nor we have considered relativity. During their journey, the visitors' clocks and navigational instruments appear to work as expected aboard their spaceship, and the visitors agree on the "correct" time among themselves. Meanwhile, to us our clocks in New York also appear to behave normally, and we seem to agree on the "correct" time with other nearby (local) observers.

Nonetheless we and the visitors are fated to miss each other. To us, the visitors are late. Our clocks indicate 4 pm, but the visitors believe it is, say, 3:45 pm, and even though we agreed to meet (simultaneously) we no longer concur on when an event (our meeting) happens. In fact once we get to see them, the visitors, their clocks and all time-keeping devices, and all their belongings are "younger" than we think they should be. A sophisticated relativist can easily explain the discrepancy. The visitors underwent motion relative to us, and even if that motion is uniform, measured time slowed during the trip. Knowing the speed of the visitors' motion (v) and the speed of light (c), we can calculate the amount of discrepancy. However, this calculation may predict a slowing by a few minutes, *not* the whole quarter hour.

It turns out that the visitors passed close to the sun, and relativists know that this slowed the visitors' clocks even more. To calculate the additional delay is more difficult because near the sun the visitors' course was affected by a curved deformation of space-time, but with enough geometry and calculus (and knowing the mass of the sun), the entire quarter hour can be accounted for.

Not only are the visitors late, but they miss New York entirely. Again relativists know why. During the part of the trip far from the sun, while the visitors maintained uniform speed, they reckoned position in space by use of a map showing how far their home planet is from New York. The visitors knew their speed and the time span on their clocks, so they naïvely calculated how far they had traveled on the basis that $RT = D$. As they neared the Earth they shut down their engines at 3:55 pm on their clocks. Of course this brought them to rest many miles past their goal, and we proclaim that they went too far.

As they hurl by the Earth in amazement at the error in navigation, the visitors begin to doubt their map, which had been made by a much slower-moving cartographer. After all, conclude the visitors, the distances are shorter than the map indicates, so the map may be wrong, or their clocks may be defective, or some "force" drew them further than expected. Altogether the visitors are doubly humiliated. *They are late in time and lost in space. They missed our date in space-time!* (There is even an error in their estimation of the mass of the Earth needed to ensure a soft landing. Thus there are errors in measurement of time, space, and mass.) But as the visitors read this book, the embarrassment fades. It was all the fault of space-time, which slowed their clocks and bent their path, while the visitors used Newtonian equations which neglected these excess effects on their trajectory. Of course space-time isn't to blame either. *Motion* dragged space-time and *mass* indented it. The cartographer is also innocent, for he measured space while making the trip slowly. When the cartographer reads the visitors' log, the former concludes that the visitors' reckoning of time and distance was wrong.

There is a sad post-script: On the way home, the visitors come too close to the sun. Their engines cannot prevent a crash, and their last message claims they are falling inexorably because of a "force." The relativist knows better: it's curved space-time.

✦

Considering the superiority of general relativity, should Newtonian physics be discarded? Not at all. Newton's equations have quite acceptable accuracy for practical purposes at ordinary velocities. Special relativity is appropriate only for very fast but uniform motion, especially when dealing with small particles which have high speeds but negligible gravity. For rapid accelerated motion, especially near very large masses and in strong gravitational fields, general relativity is needed to make accurate predictions.

This is why ordinary engineering and technology still relies on Newton's laws, why atomic and subatomic physics need special relativity, and why astronomy uses general relativity. Cosmology has become a very sophisticated branch of science in the twentieth century, and general relativity is at its core; general relativity provides the modern explanation of gravity, while gravity is the major factor in the large-scale dynamics of the universe. Relativity is even a basic aspect of biology because the sun is the source of all energy for the processes of life on Earth. Gravity compresses the matter in the sun so that a prolonged thermonuclear reaction is sustained. That reaction releases the required energy according to $E = mc^2$, and in this sense relativity is a prerequisite for our very existence.

The practical applications of relativity are multiplying. Besides the obviously revolutionary importance of nuclear science and weaponry, we mentioned technology wherein relativistic effects are significant, such as accurate navigation and extra-terrestrial radar. The basis of these effects is the concrete and measurable relativity of time.

Despite its merits, general relativity does not tell us—except as a general principle—why matter distorts local space-time. One specific possibility is that accelerated motion of objects causes undulations of the gravitational field which constitute "gravity waves" or "gravitational waves." Another possibility is that objects emit subatomic particles dubbed "gravitons." We will return to the role of gravitons shortly. Gravity waves might convey energy capable of distorting space-time. For example gravity waves generated by the spinning universe could explain why water is concave in Newton's bucket even if we consider the bucket to be stationary. The results of the Stanford-NASA Gravity Probe B experiment (mentioned in the previous chapter) sheds some light on this issue, and other recent results (mentioned below) are also very encouraging.

The concept of gravity waves leads to other inferences, such as that energy should constantly be used up by gravitational events. This notion helps explain the einstein red-shift, and it also predicts that an orbit of a celestial body should gradually shrink (undergo an "orbital decay"). The latter effect must be very subtle and slow, but it has been observed in certain orbiting stars. (E.g., as binary pulsars loose energy by radiating out gravity waves, their rate of rotation wanes by amounts predicted by general relativity.) A further inference is that the gravitational field around our Earth should be constantly fluctuating, simply because the distribution of matter and energy in the rest of the universe is constantly shifting as galaxies and stars move, stars "die," new stars are formed, etc. The process is analogous to how moving electrical charges generate electromagnetic waves or how a flickering lightbulb causes a variation of radiated energy. In fact this is why our analogy (page 243) between a lightbulb emitting light and matter affecting space-time is not as far-fetched as it sounds: Uneven "emissions" of gravitation, as from violent cosmic events like the merging of black holes, ought to appear as unevenness ("ripples") of surrounding space-time. Furthermore if gravity waves transmit changes of space-time, we expect that this effect is not instantaneous but requires a finite amount of time, just as there is a maximum velocity for physical processes in nature. As Reichenbach points out (p. 264) this means that if we build a perfectly circular structure now, it will not be perfectly circular in the future (at least according to an observer some distance away). More importantly, if we devise an apparently cogent law of physics now without considering general relativity, in theory that law will become invalid within a finite period of time.

Another motivation to find gravity waves is that their origin is presumed to help confirm the big bang theory of cosmology. In particular, gravity waves are best explained by the rapid expansion of the universe (its "inflation") shortly after the big bang. As general relativity shows how gravity acts on photons, this process should affect the observed cosmic microwave (photonic) radiation.

Thus, for a variety of reasons, proving the existence of gravity waves is highly significant for general relativity, and we might expect gravity waves to be readily detectable. After all, every bit of the tremendous amount of matter and energy in the universe contributes to the deformation of space-time, and all gravitation is "positive." (In the case of electromagnetism the negativity and positivity cancel each other, so that on a large scale the universe is electromagnetically neutral.) However many factors make gravity waves extremely elusive. First of all on a cosmic scale, gravity is very weak (as a force, it obeys the inverse square rule while cosmic distances are enormous), and masses that bend space-time are thinly dispersed. Moreover electromagnetic waves cause motion that is bipolar, whereas motion from gravity waves is "quadrupolar," which makes the mathematics and physics more complicated. Indeed, after Poincaré and others laid the groundwork, Einstein derived a complicated quadrupole equation for this purpose in 1918 (see Pais, pp. 279-280), but it is difficult to verify; more on this on the next page. Another problem is that spherically symmetric objects, no matter how massive and rapid their spin, do not emit gravity waves; we can say that stellar objects must be "lumpy" to emit gravity waves, while stars tend to assume smooth "unlumpy" spherical shapes. No wonder then that despite the development of very sensitive gravity-wave detectors, clear direct evidence of gravity waves has not been abundant found. It turns out that to perceive gravity waves, we must be able to spot the effects of abrupt asymmetric changes in the distribution of matter such as the aforementioned violent events, which on a cosmic scale are quite faint, or which—fortunately for our safety—are very far away.

We may challenge the previous paragraph in that each orbit of the moon around the Earth is a conspicuous "abrupt change in the distribution of matter," and that the Earth and moon together are an asymmetric unit. Why not simply consider our ocean tides to be evidence for gravity waves? The answer is that tides can be explained without general relativity. Tides would constitute experimental testimony for gravity waves only if we could detect and measure an "excess" in tidal excursions beyond Newtonian predictions, which at present is technically very difficult.

◆

Given the attention that gravity waves enjoy in modern experimental physics, we should be aware that their quadrupolarity is not tied to the four-dimensionality of space-time. Rather, it is related to the complex nature of frame dragging (diagram U). An analogy helps: A small object on a table top is caused to move slightly when the table is bumped. Obviously we can say something about the bump by observing how the object moves; a hard bump causes more motion, etc. The table top can represent an electromagnetic field, the bump can represent an electromagnetic wave in that field, and the object can be a small magnet located at a point in that field. In order to say something precise about an electromagnetic wave, we need *two* small magnets at *two* points. An equation of motion that describes the relative motion of two small objects will suffice. But if our table represents *a gravitational field* and the objects are small weights, the situation is more complicated. (The objects do not just move back and forth relative to each other in a bipolar fashion; they also twist and turn relative to each other.) To say something precise about a gravity wave that causes the objects to move, we need at least *four* objects at *four* points, akin to how we need four points in diagram U to describe how strings move. Of course an equation for such motion is also more complicated, which compounds the problems in studying gravity waves.

Nevertheless recent experiments, such as the Cosmic Extragalactic Polarization 2 study done at the south pole (Scientific American, March 17, 2014) reveal polarization of background light that strongly supports the existence of gravity waves. Not only does this finding represent an elegant and convincing confirmation of general relativity, but it also bolsters of our ideas on early post-big-bang cosmology. Recently evidence for gravity waves has been further confirmed by by world-wide research (such as "Ligo," a huge laser interferometer gravitational-wave search), and thus we may have found the explanation of why space-time is deformed as envisioned by Einstein.

We mentioned that light has the properties of particles. If so, gravity waves should have these features as well, and thus we anticipate the existence of "gravitons" which should be for gravity what photons are for light. This concept is also suggested by today's standard model of particle physics, wherein certain other particles transmit other forces and do so vary rapidly. Thus gravitons might have to exist so that they can transmit the effects of space-time curvature to nearby objects at a very high speed, possibly that of light. In this context gravitons are "good news and bad news." The good news is that their properties, though largely presumed, could account for many otherwise unexplained observations. For example, like photons (and some other subatomic particles) gravitons are thought to have no mass, which is why, again like photons, they could act on matter at great distances—i.e. why they can travel at very high speed over billions of miles and still have some influence. The bad news is that general relativity requires gravitons to determine or at least to alter the shape of space-time, yet so far such gravitons have not been conclusively identified experimentally. We will say more about gravitons later, as they may play a role in quantum mechanics, and we will also look a recent development in experimental physics that may shed light (pardon the pun) on them.

✦

Unanswered questions about gravity waves and gravitons are linked to other issues in modern cosmology: Our current assessment of the total amount of matter and energy in the universe does not explain certain observations about how the universe behaves. In particular, given what we know about general relativity, the universe should have more mass and more energy than we can currently detect. For example when we insert the total amount of mass into the field equations, we calculate less gravitation than we actually find in cosmic gravitational events. Here the validity of the cosmological constant mentioned earlier plays a role, but the gist of the modern explanation of the discrepancy is that the universe has "dark matter" and perhaps even "dark energy" which so far has eluded definitive detection. These considerations arise more directly in quantum mechanis, whose chapter follows afer a brief overview on relativity.

Beyond their scientific and technologic value, the elements of special and general relativity have an intellectually admirable and satisfying unity, a tightness of logic and an elegant internal harmony. Relativity springs from only a few simple postulates and observations, and yet by linking these together with astute and dynamic reasoning, it culminates as an enormously complicated and far-reaching synthesis of knowledge. The drawback to the compact logic of relativity is its fragility, in the sense that if any link is found to be defective, the entire scheme is refuted. For example, we pointed out how important it is for relativity that neither absolute uniform motion nor absolute accelerated motion be detectable. In this spirit, the reader may enjoy a very brief summary of relativity. The main contributors other than Einstein are in parenthesis, and the most revolutionary ideas are in **bold print.**

1. The laws of nature are invariant, and all observers in the universe should obtain the same result from the application of these laws.
2. Uniform motion is not detectable in mechanical experiments.
3. Mechanical laws of physics are not changed by uniform motion. (Galileo, Newton)
4. **Uniform motion is not detectable in optical experiments.** (Michelson, Morley)
5. **The speed of light is finite, constant, and not additive.** (Maxwell)
6. **Neither mechanical nor optical laws of physics are changed by uniform motion.**
7. Relative uniform motion between frames of reference has unexpected results on measurements of time and space. (Lorentz)
8. **Motion is never absolute, only relative.**
9. **Relative uniform motion changes measured time, space, simultaneity, and mass.**
10. Motion of mass requires energy.
11. Any frame of reference can have kinetic energy.
12. 9, 10 and 11 together mean that $E = mc^2$.
13. Gravity causes equal acceleration. (Galileo) Acceleration causes the effects of inertia. (Newton)
14. **Gravity is inertia,** so acceleration can cause gravitation; acceleration accounts for the effects of gravity
15. Accelerated motion causes inertia which bends light, so **gravity bends light.**
16. Gravity is a property of matter (Newton), so **mass bends light.**
17. **Accelerated motion also changes measured time and space.**
18. **The presence of matter changes measured time and space.**
19. **Motion in space through time is an event in four-dimensional space-time.** (Minkowski)
20. In space-time, events are not changed by relative motion.
21. Relative motion changes ordinary measurements of the space and time for events (see 9.).
22. The essence of special relativity is that relative motion changes space-time.
23. Relative motion, uniform or accelerated, deforms space-time.
24. Mass changes measurements of time and space, so **mass distorts the shape of space-time.**
25. The essence of general relativity is that nearby matter changes space-time.
26. **The effects of gravity are a consequence of the altered shape of space-time.**

27. Mathematical analysis of physical events in space-time bypasses the unreliability of ordinary measurements of space, time and mass as obtained by relatively moving observers.

28. The invariant space-time interval is a vector, and tensors are multiple vectors. (Riemann)

29. The tensor for mass/energy is linked to the tensor for the shape of space-time.

30. **The link between mass/energy and the shape of space-time is contained in the field equations of general relativity.**

31. **Tensor calculations of the curvature of space-time predict the effects of gravity for any observer in any state of motion.**

32. Solving the field equations is feasible only with certain assumptions and approximations, but these equations successfully explain the large-scale dynamics of the universe.

33. The main outcome of **special relativity** (1-12 above) is that motion alters time, space, and mass. This means that mass generates energy, which is a basic process in our universe. The main outcome of **general relativity** (13-32 above) is that mass curves time and space. This means that curved space-time generates gravity, which is also a basic process in our universe.

I can even contract relativity into one paragraph, five sentences and 39 words:

Uniform relative motion dilates time and contracts space. Therefore uniform relative motion alters space-time. Since uniform relative motion alters space-time, accelerated relative motion curves space-time. Gravitational motion, such as falling, is accelerated relative motion. Therefore gravity is curved space-time.

Besides relativity, the other great scientific advance of the twentieth century is *quantum mechanics*. In this chapter we consider how the principles of quantum mechanics relate to relativity, although here the distinction between special and general relativity is prominent and critical. We will see that modern physics includes, and in fact demands, close cooperation between *special* relativity and quantum mechanics, whereas, as I explained earlier, the incorporation of *general* relativity into the rest of physics remains somewhat elusive. The cooperation between special relativity and quantum mechanics is particularly evident in modern particle physics, and—ironically—the discord between general relativity and quantum mechanics is also evident in modern particle physics. Though this sequence is somewhat illogical, let us look at the latter problem first because it is less technical.

Here is helps to know that today's particle physics recognizes twelve fundamental particles of matter, as well as four forces that are in play when these particles interact; this scheme is the well-accepted and very successful "standard model." For example, the electromagnetic force between electrons is "mediated" or transmitted by "force particles," the photons—the very same quanta of energy that are crucial in special relativity—and this pattern elegantly explains our experimental observations. But here is the issue: While each of these four forces has a know strength, the force of gravity (the purview of general relativity) is *much weaker*; so weak in fact that it cannot contribute appreciably to any inter-particle interactions and does not contribute to the explanation of experimental observations in particle physics. Of course in the eyes of general relativity gravity is not truly a force, though this fact is immaterial at the moment; the point is that particles of gravity should act like a discernible and significant force, as do other force-transmitters in the standard model.

But there is more to this issue: Given the effectiveness of the standard model in explaining the workings of the universe, we expect the "force" of gravity—even if very weak—to be mediated by a specific force particle, the "graviton," as discussed in the previous chapter. However, said again, *the existence of gravitons has not been proven*. This circumstance in effect leaves a gap in the standard model.

Furthermore, quantum mechanics very effectively describes the submicroscopic world, while general relativity superbly describes the macro world, and yet is seems impossible to render these two branches mathematically compatible within the standard model. For instance, the Heisenberg uncertainty principle, vital in describing particle interactions via quantum mechanics, finds no counterpart in general relativity. Thus quantum mechanics insists that it is impossible to compute the path of an electron the way we predict the path of a planet using general relativity. Rather we can detect only a probability or statistical likelihood of locating an electron, simply because quantum mechanics insists that it is impossible to determine the position *and* the motion (momentum) of a subatomic particle—and theoretically of anything else—at the same time. Since velocity is "distance per time" and position means "location in space," we can say that in glaring contrast to general relativity, quantum mechanics bars the simultaneous determination of "time" and "space" for one event. It thus seems that different forms of physics apply, while by intuition alone we do not expect to need more than one explanation for our universe, and Einstein in particular resisted the possibility that uncertainty should dominate in one realm of nature and yet is absent in the other.

A related problem is that space as envisioned in general relativity is "continuous;" there seems to be no theoretical limit to how small a space (or space-time) interval can be, and if it were visible, the surface of space-time would appear glossy and without a grainy texture. Meanwhile, quantum mechanics gets its very name form the concept of quantization—for example particles, including photons, are quantized and thus have a minimum size—which suggests that space intervals also have a definite minimum size, and which should compel the surface of space-time to be rough and bumpy or grainy. Worse yet, the quantum bumps or grains are constantly and randomly changing in size and location, which has been likened to the surface of a restless stormy sea, with no counterpart in the formulation of general relativity.

True, this scenario is purely theoretical, as the minimum size is extremely small (the grain would be extremely fine), but the concept interferes with devising mathematics valid for a unified theory of "quantum gravity." Think of a small falling object: The math of general relativity treats this event as the object gliding into a smooth dip in space-time, while quantum math describes space-time as rough, allowing only a chaotic and bumpy ride. Moreover, the quantum bumps fluctuate unpredictably, so that if the object were very small, such a molecule or smaller, its dominant form of motion would be a random "jiggling" rather than a predictable and orderly trajectory. This problem would be aggravated if the entire scenario involved only light-weight particles, none of which has mass sufficient to indent space-time appreciably. The motile behavior of such particles would be oblivious to Einstein's general theory, it would reflect only quantum theory, and the prospects for any unification with quantum mechanics would appear to be nil.

But there is hope. Today in 2014 a viable candidate for such unification is "*string theory*," which has been in development for the past four decades. To describe this approach we should elaborate that according to the standard model, all matter is made of various elementary particles called fermions; for instance an electron is a fermion. Furthermore, we envision four basic forces, also known as "interactions," that affect fermions, and we believe that four additional force particles called bosons mediate or transmit each of these forces. (Most bosons have integer spin [1, 2, etc.] and obey Bose-Einstein statistics. Any number of identical bosons can exist in the same quantum state, which is crucial in determining the statistical behavior of large numbers of boson. Thus the Pauli exclusion principle does not apply to photons, but it does to electrons. The spin of electrons will interest us later.) In particular the bosons account for the weak nuclear force (which deals with radioactive decay), for the strong nuclear force (responsible for nuclear energy), for the electromagnetic force (which we already discussed), and perhaps for gravitation as defined by general relativity. Like the boson of electromagnetism is the photon, this scheme requires a quantized boson of gravity, namely the graviton as mentioned earlier. The existence of a new boson, named for Higgs, has recently been tentatively confirmed; also more on this later, as this finding may change the picture.

In any case string theory holds that all constituents of matter—fermions as well as bosons—are in reality extremely small strings. These strings are band-like units which can vibrate, oscillate, and rotate. The difference between the various particles resides in the behavior of the strings; how a string moves determines what kind of particle it is. Obviously, the strings are strange entities that are difficult to imagine and have never been observed experimentally. They apparently have only one dimension but they have energy and, by virtue of just their state of vibration, they compose all things and explain all matter in the universe, including the interactions between fundamental particles. (See Hawking, pp. 159-163, Jones, pp. 313-314, Kenyon pp. 181-182 and 183-185, and Ferris, pp. 328-334.)

The key point for us is that *the mathematics of string theory unites all bosons.* String theory calls upon the same features (zero mass, etc.) of one of the bosons that quantum theory can ascribe to the graviton. In other words the graviton may be merely one kind of vibration of a string, and in this way gravitation can be brought into a unified system shared by all fermions and all other bosons. In this way string theory may dispel the incompatibility between general relativity and quantum mechanics, and it can eliminate the pesky duality of laws that we feared might be needed to explain the physics of the universe. One string theory suffices.

In fact the mathematics of string theory leads to the equations of special as well as general relativity, so that not only is there no conflict with quantum mechanics. Indeed all of relativity is an essential constituent of this approach. For instance $E = mc^2$ remains essential, and string theory insists, in stronger terms than even Einstein used, that matter and energy are interchangeable. Likewise space-time curvature remains a necessary ingredient, while string theory allows tiny quantum undulations of space-time to be spread out in such a way that Einstein's field equations continue to hold. This idea makes validating general relativity doubly important because general relativity actually gives support and credence to string theory.

Not astonishingly, string theory does introduce serious obstacles: Its mathematics invokes not four but at least 10 dimensions. This proposal holds that in addition to the space-time we know and measure as length, width, height, and time, there exists a six-or-more-dimensional part of "space-time" that is inaccessible to our experience. To explain this tenet we speculate that the universe started out with 10 or more dimensions and that upon the big bang, it began to expand. Four of these dimensions continued to enlarge along with the universe, and they are still enlarging today. The remaining dimensions stopped expanding shortly after the big bang and remain "tightly curled up." We can think of every point within our familiar space-time as possessing a system of coordinates with not four but 10 or more axes. Four clearly evident axes are usually straight or smoothly curved, but others are coiled up so tightly that they occupy an extremely compact space, too small for any known measuring device and much too subtle to affect our lives.

String theory even assigns numbers to these ideas: The six-plus extra dimensions stopped expanding 10^{-43} seconds after the big bang, and a curled-up dimension occupies 10^{-33} centimeters of space, so that it behaves like a tiny point. One implication of this notion is that points in space-time may not be infinitely small, which avoids the thorny issue—restless graininess—we mentioned earlier. The solution becomes simple: Points which make up space-time but which have a minimum size can behave in manner that is compatible with general relativity—for instance they can house *ds's*—*and* with quantum mechanics—for instance they can be governed by probability and quantization.

In subatomic physics the aforementioned 10^{-33} centimeters is the "Planck length," the size below which quantum mechanics predominates. Planck length can be derived from three elements, Planck's constant h (linking energy with frequency in quantum mechanics), the speed of light c (the key to special relativity), and Newton's universal constant of gravitation (which appears in the field equations of general relativity). The aforementioned time interval of 10^{-43} seconds is Planck's length divided by c. We note how neatly the constants of quantum mechanics, relativity, and gravitation meet here, implying that string theory deals with very fundamental aspects of nature united into one paradigm. (See Schutz, pp. 52 and 210.)

Famous experiments* suggest that quantum events which are ostensibly separated by a significant distance can behave as if they are adjacent, seemingly allowing instantaneous cause-to-effect links. Such events may indeed be separated in terms of our familiar four dimensions but may be adjacent along one or more of the other six undetectable "curled up" dimensions. In this way for example, a 10-dimensional string theory can allow particles that seem "entangled" to be adjacent to each other along at least one dimension, so that a "signal" can traverse the distance between them apparently in zero time. This possibility would satisfy relativity—cause-to-effect signals need not have infinite speed in four dimensions—and it would satisfy quantum mechanics—events can be adjacent in other dimensions, allowing the immediate cause-to-effect linkages that characterize entanglement.

✦

While string theory remains controversial largely because of a lack of experimental evidence, another recent and more tangible development has received wide publicity: the 2013 Nobel Prize in physics awarded to Francois Englert and Peter Higgs for their work on the Higgs particle. (Please see w w w.nobelprize.org/nobel_prizes/laureates/2013, and Scientific American, "The Higgs Boson at Last?," Oct. 8, 2013. [2014]) For an introduction we need to revisit the standard model in quantum mechanics and to know something about the Higgs boson and its field. As I discussed, in the standard model bosons are elementary particles that transmit basic forces, possibly including gravity. (For instance, a photon is a boson, one which carries the electromagnetic force.) As was proposed several decades ago, the Higgs boson and its associated field may fill a gap in the standard model with regard to why particles have mass, which of course is a key feature of gravitation. Thus (equating mass with weight) we may at last understand why any object has gravitational weight, which is a principal concern of general relativity.

Early in the history of this concept (around 1960) Higgs, Braut and Englert proposed the existence of a certain field, which is another distribution of energy that pervades the universe as envisioned in modern quantum mechanics. After all, science already accepted the notion that space is filled with the gravitational field and electromagnetic field. Moreover, according to this conjecture *if there is a Higgs field, there should be a Higgs boson.* Hence this potential new member of the standard model became a very important hypothetical but still experimentally unsubstantiated quantized entity. In short, the nascent Higgs-et-al. theory suggested that *with mediation by Higgs bosons, the Higgs field gives mass-possessing particles their masses.*

---

* Please consider an internet search for "EPR " and the "EPR paradox," a huge separate topic.

In more detail, according to the "Higgs mechanism" particles ultimately acquire weight (mass) only through contact with the Higgs field. How much weight they acquire presumably depends on the type of particles. Some, like the photon, seem to be unaffected by the field and do not acquire mass. However, electrons do interact with the field via its Higgs bosons, accounting for their mass. Thus if the Higgs field were to somehow vanish, the suddenly weightless electrons would speed away, and all atoms would collapse. There would be no gravity, and there would be no matter! So presumably the Higgs field always exists and persists, even though it is very elusive. (The Higgs field is a scalar field; it has strength but no direction, and it entails zero spin.) On the other hand, when this field is disturbed (as can be other quantum fields), waves should be generated which travel through it, which is a notion obviously suggesting gravity waves. And because of the nature of Higgs field, any fundamental particles in this field cause Higgs bosons to clump around the particles. More clumping should cause particles to acquire more energy and therefore more mass (via $E = mc^2$), and—presto!—therefore more gravitational weight.

To try to visualize this, we can imagine that our universe brims with the energy of the Higgs field that acts like viscous or syrupy "honey," impeding the particles' motion and thus imbuing them with mass. (This is not as odd as it sounds. Impeding motion can mimic inertia, and mass is a measure of inertia.) Even more fancifully, perhaps a particle becomes massive by "ingesting" a Higgs boson, as if a particle can gain weight by "eating." Since not all particles are "hungry," the Higgs boson and its field might explain why some objects such as quarks (a constituent of protons) possess mass, while others, notably photons (a constituent of light), have only energy and can move around the universe at the speed of light, as if unrestrained from hauling a load.

Even without such whimsical elucidations, the Higgs mechanism is a strange notion, claiming that the entire universe must be filled with a newly envisioned field, physically different from anything previously known. Still, the theory adds an extra field to gauge theory (please see page 232 and footnote here*) which is crucial for the symmetry that is now accepted in quantum mechanics, and its scientific appeal cannot be denied. In any case, this idea was presented in a paper in 1964 by Englert, who now [in 2014] is 81 years old, by Higgs, now 85, and by Brout, who died in 2011 and was thus not eligible for a Nobel prize. In 2012, forty-eight years after this paper, upon many attempts at CERN's Large Hadron Collider, firm evidence for the Higgs particle was uncovered, and the scientific community immediately grasped the profound significance of this development.

Incidentally, the purpose of colliders is to accelerate particles (e.g. protons) to as close to light speed as possible—in the case of the LHC, 99.9999991% of c—and then cause the protons to collide, releasing energies that simulate the conditions just after the big bang. This event produces more elementary particles, virtual ones, that flash into and promptly out of existence, but long enough for instruments to detect them. If a Higgs boson exists, it might appear in that short time

---

* I oversimplifying greatly, ignoring grueling intricacies: Spheres look the same from any angle, a change of angle is a "transformation," and this is a "symmetry." General relativity looks the same to any observer, changing the observer is a "transformation," and this is a "symmetry." Particle interactions look the same "under gauge transformations," and this is "gauge symmetry." Ergo, general relativity shows gauge symmetry and *it fits into modern particle physics.*

interval, when the evidence would consist of a spike in energy in the range of 115 to 135 billion electron volts—or gigavolts (GeV)—with the expected value of about 125 GeV. (We note the application of the relativistic concept that a boson can be recognized by measuring energy.)

"We observe in our data clear signs of a new particle, at the level of 5 sigmas, in the mass region of 126 GeV," announced Fabiola Gianotti, head of one of the LHC research teams, on July 4, 2012. The reference to sigmas indicated the level of certainty in the findings, and a score of five is quite high." Ever since, a series of tests on the particle have yielded confirmation, establishing Higgs's and Englert's work as a new constituent of modern particle physics. The existence of the Higgs field and Higgs particle appeared to have been proven by time the Nobel committee made its announcement in 2013, clearly appreciating what Higgs et al. had contributed to modern science: a description of how the world is constructed. The Nobel committee declared that "according to the [newly expanded] Standard Model, everything [in the universe] consists of just a few building blocks: matter particles. These particles are governed by forces mediated by force particles that make sure everything works as it should."

Alas, what the committee omitted in its statement is the concept that the Higgs mechanism can account for gravitation! Space-time is deformed by mass, and Higgs' paradigm seems to account for that mass. Furthermore, the whole notion of gravitons may be rendered pointless, with Higgs bosons assuming their roles. No wonder then that gravitons as such have not been found; we were looking for something that does not exist. Moreover, any gravity waves that can be detected may be fluctuations in the Higgs field. *Thus, via the Higgs mechanism, general relativity can become an unquestionably vital element of particle physics, of the standard model, of gauge field theory, and of modern quantum mechanics.*

Another point pertinent in our context was omitted by the Nobel committee: What principle of nature allows the energy in Higgs field to confer mass on particles? The answer is simply Einstein's concept that *E equals mc squared*, which of course is at the heart of special relativity, and which is dramatically revealed in the Higgs paradigm and in the updated standard model. This thought, gratifying to students of Einstein's accomplishments, brings us to the next section.

◆

While a comprehensive unification of relativity and quantum mechanics may be on the horizon, modern particle physics has already turned out to be a stage on which *special relativity in particular* and quantum mechanics act out vital roles in harmony and mutual interdependence. Two names stand out in this development, Richard Feynman (1918-1988) and Paul Dirac (1902-1984), though as usual in scientific progress many contributors were essential. In this context let us pick up an important thread from the end of chapter 15, where we saw the origin of the Euler-Lagrange equation of motion. With this equation we concluded that the alliance of the principle of least action and the calculus of variations provides a secure foundation for the notion that a geodesic is the correct description of the actual world-line of an object in gravitational motion. (Please see page 233.) This key inference is logical and nearly self-evident: From the vantage of energy physics, *motion entails as little "action" as possible.* From the vantage of mathematics, *the variational principle is respected.* From the vantage of relativistic space-time geometry, *motion is as straight as possible, which means it is geodesical.*

Of course in the realm of special relativity, motion which is by definition straight is also geodesical, even though this term is overkill here (the straightest possible geodesic is a straight line; of course it has zero curvature). More to the point, the principle of least action—based on the Lagrangian—applies in special relativity. We also raised the possibility that an object can move from one point to another on any one of a theoretically infinite number of paths. (Pages 221 - 223.) Feynman latched onto this concept (in the 1930's and 1940's) and molded it into an entire novel approach to quantum mechanics, one which indeed today is the preferred form of quantum mechanics, especially as it is applied to particle physics. This approach or system or version or generation of quantum mechanics deserves the following digression, and we will see how it leads to the unification with special relativity.

By way of historical background, the development and evolution of quantum mechanics can be thought of as occurring in stages. One such stage was dominated by the work of Heisenberg and Schrödinger, and the central equations were the time-independent and time depended Schrödinger's wave equations. The latter is a kind of equation of motion, in that it provides the probability that a particle will be found at some location after the passage of some time.[*] The fact that the equation only yield a probability—not a certainty, as do for example the equations of Newton, Euler, Lagrange and, alas, Einstein—reflects a basic and inescapable feature of quantum mechanics: The motion of particles has an element of randomness which I call "jiggling," and the smaller the particle, the more prominent is its jiggling. Therefore, the entities of greatest interest to quantum research and experimentation—very small objects and particles, notably subatomic particles—are also those whose behavior is most affected by probabilities. This probabilistic "fact of life" in quantum mechanics was one of the key factors that made relativity and quantum mechanics seem incompatible, much to Einstein's dismay. However, the current stage of quantum mechanics, dominated by the work of Feynman and Dirac, contains exciting progress in resolving the incompatibilities, and in fact some of the facets of quantum mechanics demand cooperation, particularly with special relativity. With this in mind, I am ready to summarize Richard Feynman's Nobel-prize-winning approach to quantum mechanics. (See Bibliography, particularly Greene (p. 108-12), Polkinghorne (p. 66-68), and of course Feynman.)

Feynman's formulation rests on the concept of *path integrals*, a.k.a. the sum over histories approach. In this setting it helps to think in terms of events. A fundamental (indivisible) event may be a particle moving from point A at some time t_0 to point B at some later time t_1. Classically the particle always traverses only one and the same path—one trajectory in space and time—and thus this event has only one "history." As envisioned by Feynman (and earlier notably by Hamilton), quantum mechanics allows the particle to traverse *many potential paths during the same time interval*, and therefore in theory this event has multiple simultaneous histories, one for

---

[*] Solving this equation gives a "probability amplitude" Φ, the square of which equals the actual probability. The Φ is a complex number in the form a+bi where $i = \sqrt{-1}$ (imaginary). In Feynman's work this allows probability amplitudes to interfere with each other.

each path. Indeed the event may involve an infinite number of possible paths and histories. Some paths seem absurd (like going from a photon-source on Earth to a nearly screen via Jupiter), but all possible paths are included in the theory and in its mathematics.

Please note that it is not necessary to envision a particle physically traversing all possible paths and hence actually occupying all locations in the universe at the same time. The theory only requires a possibility of many—and even an infinite number of—paths and locations. Each possibility occupies a spot in a mathematical superposition of "quantum states," and each state has an associated probability (a probability amplitude in the equation) for finding a particle in some physical location or path.* To appreciate quantum mechanics, particles need not be imputed with mysterious properties, but their whereabouts and motions are governed by probabilities. For example no one has ever found the same photon everywhere at once, but there is always a likelihood of finding it anywhere. The location of a particle is not universal, but its location-probability is universal. Likewise no one has ever observed one photon pursuing many paths at once, but in theory it could be on any possible path in the universe. (These paths can be called "virtual.") My interpretation is that these are "paths of probability" and that each probability is a wave-like property of a path. Only the *probabilities* for all paths can coexist anywhere.

This brings me to the obvious question of how the actual observed path (the usual trajectory; the classical history; the Newtonian path) is selected naturally from all possible paths. Looking ahead, the general idea is that the likelihood for deviant paths is reduced, whereas the likelihood for commonplace paths is raised. In particular, by way of destructive and constructive wave interference, certain paths acquire low probabilities, whereas one path—the one with stationary "action"—acquires maximum probability. This is a powerful application of the principle of least action; nature favors maximum efficiency, but here nature also favors the most likely.

While the paths are probabilistic, of course the probabilities can be calculated. Each history or path contributes to the probability for their common outcome—the event. In particular, the probability for the event is obtained after adding together the probability of all the histories within that event. Ergo "sum-over-histories." In other words, the likelihood for an event is built from the combined effect of every path; ergo the concept of "sum-over-paths" or the "path integral." In this paradigm, the most probable path dominates the event (e.g. for a sizable object) as we ordinarily witness it in every-day life, and as Newton had analyzed.

Here I call upon the notion that any two wavy paths can be analyzed as waves of probability which can be in phase with each other or which can be out of phase with each other to any extent. The concept of phase differences is crucial in this picture, but the idea of paths being out of phase is the same that explains wave interference in familiar two-slit experiments. The phase increases at a fixed rate along each path as the phase cycles repeatedly; the final phase also depends on the length of the path, since a longer path—which also must be faster to comply with its time restraint—provides more opportunity for phase gain. Since the phase of the contribution of each

---

*This concept coincides with what I cover in more detail in my book, "Quantum Mechanics that Makes Sense:"

path is proportional to its length, one way to visualize the process is that longer and less direct paths readily attain markedly different phases compared with the phases of shorter and undeviating paths. In effect, straying promotes dissonant phasing. Conversely, in space-time the shortest and most direct paths—the straightest possible ones—are resistant to slight variations between the paths; they have nearly identical lengths, so they tend to be consonant waves in phase with each other. This is why, for example, a thin light beam routinely coincides with the shortest distance between two points. It is also why "the shortest distance between two points is a straight line," and it is exactly what the Euler-Lagrange equation demonstrates.

Interference is often stressed in explaining Feynman's approach, so let me refocus on it in the terms usually found in the literature. When paths are "summed up" in the path integral, the paths that are out of phase with each other tend to cancel each other by means of *destructive interference*; imagine a rise in the value of one investment simultaneous with a greater fall in value of another. In this same way the probabilities for most paths are reduced. Conversely in-phase paths are amplified; a rise in both investments is a case of *constructive interference*. Deviant paths, which are out of phase, correspond to non-Newtonian highly unlikely histories, and thus non-classical behavior of large objects is practically impossible. On the other hand, even the remote possibility of non-classical histories is significant because it accounts for the counter-intuitive probabilistic subatomic behavior seen in quantum physics research.

Since I interpret paths as waves of probability, the process can be stated rather simply: When two waves of probability interfere because they are out of phase, their net probability is suppressed. The history with the highest net probability recounts the actual path. For large objects the augmentation is overwhelming, so that by far the most probable path is the straightest possible one, which is what we normally observe in our environment and in large-scale experiments.

I think of Feynman's approach via the following analogy: An infinite number of different flight paths can lead an airplane from New York to London. Of course based on our experience, most of these are extremely unlikely if not absurd and practically impossible; e.g., flying from New York to London via Jupiter. Only one flight path, or at least one set of very similar paths, is ideal. Feynman's triumph was to show mathematically that by far the most probable path is the one that agrees best with our experience, and it is the path obeying the principle of least action. In other words, our reality—notably our every-day observations rather than what is revealed in subatomic experiments—consists of what is overwhelmingly the most likely.

Meanwhile, quantum mechanics had developed sophisticated equations to predict the probability amplitudes for particles to behave in some manner. Around 1933 Dirac was working on the time-dependent version of Schrödinger's equation which did just that, but in which energy is expressed by the Hamiltonian (page 221) rather than the Lagrangian, though the latter remains inherent in Feynman's formalism. In the process Dirac used a particular exponential term (this form reflects the wave-nature of Schrödinger's equations):

$$e^{iS/\hbar}$$

This term has two interesting features. First, Dirac's work was rather obscure at the time, and his exact intentions were not clear. Second and more momentously, it contains *S*, which is "action" as seen in the context of the Euler-Lagrange equation of motion and the variational principle. Here please see page 219, recalling that every point on a path of motion has a Lagrangian L and that the sum of the *L*'s gives *S* as in the integral

$$S = \int_{t_1}^{t_2} L(q, \dot{q}) \, dt \, .$$

Taking the hint, *Feynman grasped a connection between Dirac's concept of the action of each path and his (Feynman's) concept of the sum-over-paths where each path has a probability amplitude.*

As Feynman worked out the math, some of which was rather controversial at the time, he devised the term K(b,a) which represented the total probability amplitude between the a, the start of a path, and b, its end. The final probability for that path (in keeping with the quantum rule that the square of a probability amplitude equals the actual probability) would then be

$$P = K(b,a)^2$$

In that case,

$$K(b,a) = \int_a^b e^{iS(b,a)/\hbar} \, Dx(t) \, .$$

Here we see integration (in effect, summation) from a to b; we have a path integral. Moreover, in the integrand we see the exponent *S*, the two-hundred year old classical measure of action; Euler and Lagrange would have been delighted. The *Dx* term is quite complicated and controversial, in part since the number of paths is theoretically infinite and yet the total probability must be finite, but in the end this equation became the key to Feynman's approach to quantum mechanics.

Beyond the significance of this approach in physics in general, we are particularly interested in the fact that modern particle physics, as introduced at the beginning of this chapter, relies heavily on Feynman's work in quantum mechanics. In fact if only older forms of quantum mechanics were available, this branch of physics would not succeed—it would fail to account for our experimental observations. For instance the venerable equations of Schrödinger are not adequate for today's needs, but the successor, in the form of Dirac's equation, works very well.

Obviously particle physics studies the nature and behavior of elementary particles, but these entities "do things" *which require not only modern quantum-mechanical explanations but relativistic explanations as well.* We can list five prominent facets of special relativity that are thus invoked, and in each case I show a key pertinent equation or concept.

(1) Events in particle experiments occur at high speeds in four dimensions. Hence we need

$$ds^2 = dX^2 + dY^2 + dZ^2 - (cdT)^2$$

wherein the three spatial dimensions are on the same footing as the time dimension (page 104). We recall that this equation enters into making an equation Lorentz-invariant.

(2) In experiments and in theory, particles with certain masses and momenta appear and disappear spontaneously by "borrowing" energy. Hence we need

$$E = mc^2 \quad \text{and} \quad E^2 = c^2p^2 + m_r^2c^4$$

wherein (rest) mass, energy and momentum are equated (pages 37-39). Before proceeding I should detour to show how energy can be "borrowed," and for this goal I must start with the Heisenberg Uncertainty Principle. Paraphrasing Heisenberg, "the more precisely the location is determined, the less precisely the momentum is known, and vice versa." In equation form (published in 1927 as part of quantum mechanics), this principle is now commonly written as

$$(\Delta p)(\Delta x) = \hbar$$

where the uncertainty in momentum is "(Δp)" and the uncertainty in location is "(Δx)." (These terms are unconventional uses of "Δ" which usually stands for "change.") Whenever Δx increases, Δp decreases, and conversely whenever Δx decreases, Δp must increase; we see that these uncertainties are reciprocal (arithmetically because Plank's constant $\hbar$ is not zero, implying that this uncertainty exists because of quantization). Therefore, perfect accuracy of simultaneous (or nearly simultaneous) measurements of location and momentum are impossible, and this is an inviolate fact of quantum mechanics.

It is not even essential that we take measurements in the usual sense. For example we say that when a particle has a definite location, it must have indefinite momentum. Therefore, if we could measure a definite location for a particle—though we need not—we would encounter infinitely uncertain momentum for that particle at that same moment. In theory, uncertainty is inherent in the behavior of particles whether or not they are being measured.

However, besides location and momentum, other pairs of measurements—called complementary or conjugate or incompatible pairs of quantum observables—show the same kind of reciprocal uncertainty! Another such case involves angular location and angular momentum, though this is not pertinent in the present context. Similarly, and very importantly, *we cannot determine exact energy in an exact time frame for a particle.* In equation form, this kind of quantum-mechanical uncertainty principle (uncertainty of energy vs. time) can be written as

$$(\Delta E)(\Delta t) = \hbar.$$

468

In words, if we wish to determine the energy in a particle exactly, we must measure for a long time. Conversely, a brief measurement (small Δt) yields an inexact assessment of energy (large ΔE). Thus we can quickly approximate strong energy, and we can gradually measure weak energy, but we cannot determine weak energy quickly. I stress that like the uncertainty of location vs. momentum, this too is fundamental to quantum mechanics.

The uncertainty principle dealing with energy and time finds paramount application in particle physics. One of the key interactions among elementary particles is the spontaneous creation and annihilation of various subatomic particles, apparently "out of nothing." The principle is that because of energy-time uncertainty, spontaneous creation is possible by temporarily obtaining energy from a quantum energy field with the proviso that the "debt" is promptly repaid. The more energy is "borrowed" (the greater ΔE), the sooner it must be paid back (the smaller Δt).

Clearly, another principle is essential to this picture: The masses of the temporary ("virtual," not visible in cloud chambers) particles can be calculated because of, and is determined by, special relativity's equally inviolate $E = mc^2$. That is to say, if mass were not concentrated energy (page 33), critical events seen in particle physics would not occur. End of this detour; continuing....

(3) In research, particles are accelerated to very high "relativistic" speeds close to c, and given that $E = mc^2$, the mass of these particles (small as they are) becomes a major consideration.

(4) Particles turn out to have antiparticles, and indeed matter turns out to have antimatter. For instance we find electrons and positrons (anti-electrons). This phenomenon is anticipated when $E = mc^2$ is simply squared. Then $E^2 = m^2 c^4$ allows the equations

$$-E = mc^2 \ (\text{since } (-E)^2 = E^2) \text{ and } E = -mc^2$$

wherein we see either negative energy or negative rest mass (as on page 38, including footnote).

(5) The last facet to discuss here is that elementary particles exhibit "spin," which is a kind of magnetic phenomenon that has no counterpart in classical physics. (See Penrose, pp. 264-265, and Gribbin [Q is for Quantum], pp. 371-372.) Quantum mechanics forbids two associated electrons from having the same spin; if one has "$+\frac{1}{2}$ spin" the other can only have "$-\frac{1}{2}$ spin," and no other types of spin exist; their spin is "quantized." In alternate terminology, these are "spin up" and "spin down." The units of quantum spin are Plank's constant $\hbar$. Pauli worked hard on explaining this issue, but later it turned out that when Dirac's equation is derived and solved—as is possible only with the aid of special relativity and $E^2 = c^2 p^2 + m_r^2 c^4$ in particular—then quantized spin arises automatically.

The full mathematical reasoning for Dirac's derivation is very subtle and tricky. (See w w w.eng.fsu.edu, section D. 71 "Emergence of spin from relativity." [2014]) However the essence, substantiated experimentally, is that spin is anticipated by inserting special relativity into modern quantum mechanics. Next let us therefore look at Dirac's equation, as it is obviously very important in modern science. However, I must interject that if I were to relate the complete tale and its mathematical detours, this chapter could quadruple in length. In fact Bernd Thaller wrote a 360-page book on just this topic, "The Dirac Equation," published by Springer in 1992, and the story of Dirac's equation occupies chapters in almost all extensive treatments of quantum mechanics and special relativity.

Here we will concentrate on electrons as representatives for other elementary particles, as this was Dirac's initial focus. Electrons are studied with quantum electrodynamics (Feynman's "QED"), which is also the prototype for other related topics in particle physics.[*] Dirac wished to find an equation to describe electrons and their behavior accurately, which means that a new equation of motion was in the offing. We note that an electron can appear in *four combinations*, the particle and its antiparticle, and each of these with $+\frac{1}{2}$ spin and $-\frac{1}{2}$ spin (spin up and down). This circumstance correctly implies that such an equation must be very complicated, housing a key four-component term. Let's see how Dirac approached this problem.

He was aware that in the early 1920's Schrödinger had tried to make his own wave equation compatible with relativity, but his result did not treat space and time on equal footing—time was a first derivative and space was a second derivative—and Schrödinger did not know about spin. A more promising improvement was the Klein-Gordon equation of 1926 wherein all four dimensions appear as first derivatives, but this equation still did not address $+\frac{1}{2}$ and $-\frac{1}{2}$ spin.[**] It fell to Dirac in 1928 to draw upon the works of Heisenberg and Pauli (and others) who had applied matrix math that better accommodated the spin of electrons, while also applying some clever, novel and difficult math (e.g. Clifford algebra) that effectively accommodated four-dimensional space-time.

The outcome was the famous Dirac equation, which is really four matrix equations in one, by means of which *electrons can be completely described* for use (later in the 1930's) in quantum electrodynamics and in particle physics in general. This triumph of applied math and modern science respected both quantum mechanics (particularly à la Feynman) and special relativity. A breakthrough in Dirac's equation was that when certain matrices act on four column vectors—to accommodate the four combinations of the states of an electron—they are acting on "spinors," the ones I mentioned on page 150 as being more comprehensive than tensors. As the term implies, spinors were needed to encompass the two kinds of electron spin. Dirac's notoriously complicated equation also provided for electrons moving as relativistic speeds, as well as for both conventional negatively charged electrons and positively charged positrons (anti-electrons). The latter particles, hypothesized by Dirac earlier, were found experimentally in 1932, and, sure enough, he was awarded a Nobel prize a year later. (Dirac also wrote his doctorate thesis on quantum mechanics, which had never been done before; see Kumar, pp. 197-200.)

---

[*] This large subject is called quantum field theory (QFT). It includes quantum chromodynamics (QCD) which deals with quarks (labeled by fictitious "colors.")
[**] Not all elementary particles have $+\frac{1}{2}$ and $-\frac{1}{2}$ spin; electrons do but photons do not.

Before I show what Dirac's equation looks like, I should say more about spin, since it clearly is a major feature of quantum mechanics in general and Dirac's work in particular. At first glance spin should be a simple idea, readily grasped by any player of serious tennis or American football or baseball: The motion of a ball has a direction in space—which way it's going—as well as a speed—how fast it's going—and these two data constitute a vector, a velocity in this case, that specifies a direction and a magnitude. (If somehow the ball had more than one velocity, the data would constitute a tensor.) Ordinary space has three dimensions, so that velocity can be depicted on a graph as a thin line pointing in some direction, while the line has a length that indicates the amount—the magnitude—of speed. In this case the ball's motion can be described by three numbers, corresponding to the three dimensions of space, and we say the ball-describing vector has three components (ignoring 4-D space-time for now).

However here I add a complication: *while it is moving, the ball is also spinning.* To keep this simple, I assume that the ball has one certain axis (more obvious for an American football because it's not a sphere but has two polar points) and its spin takes place only around that one axis. Now a graphic depiction is also more complicated but not baffling: Instead of a thin line, the motion of the ball can be represented by a ballpoint pen that has a clip on it. This pen can be spun so that the clip can face in any direction, independent of which way the pen is pointing or how long it is. Now the ball's motion through space can be described by four numbers, with the additional number specifying which way the clip faces, and we say that this ball-describing *spinor* has four components. (In the spinning case, it may be more intuitive to use a spherical rather than a Cartesian system of coordinates, but this option makes no difference to us.)

Next I must add three technical complications, two of which are somewhat taxing.
(1) In our context the speed of ball may be close to that of light, and its motion takes place in the four dimensions used in relativity, i.e. in space-time. This addition obviously makes an accurate description of the spinning ball's motion more complicated, though it is "easy for algebra" to handle several dimensions.

(2) To adapt this description to the math of quantum mechanics and particle physics, we consider motion through space and time to be a "differential" event; a change along x, along y, along z, and along t. Furthermore, each change is considered separately, so that we use partial differentials, a.k.a. partial derivatives; one value is allowed to change while the others are held constant. (Please see page 249.) By virtue of the trigonometry used to study cyclical quantities, some components have imaginary numbers in them (as in complex numbers; please see appendix). This method leads to the condition that Dirac's equation contains *four "Dirac matrices."* Dirac's matrices, one for each space-time dimension, are often called gamma (γ) matrices, identified by superscripts μ which range 0 to 3 (not 1 to 4); thus we will find γ^μ in the literature to represent the four matrices in equations. (Careful: gamma is also used in other math contexts.)

(3) The notion of spin is quite logical and intuitive when the spinning object is a ball or pen. It is tempting to envision the spin of electrons or other particles the same way, and indeed some of the

math crosses over. However spin in the context of quantum mechanics is different, and the differences are neither logical nor intuitive. The main issue is that when a ball or pen spins once around, we expect it to look the same as when it started. In the case of an electron, *two* rotations are required to restore it to the same orientation. (We can say that electron spin is +½ or −½ spin because [contrary to common sense], one rotation brings the electron only half-way back to its starting orientation.) Classical physics cannot explain this phenomenon, but an internet search on particle spin will reveal many examples of various contortions that mimic the need for two rotations to bring something back to its original state. These analogies are sometimes startling or even amusing, but they really never represent particle spin as it appears in experimental demonstrations, and no analogy does full justice to the strange nature of quantum spin.

Another feature of particle spin that I mentioned above is that it is quantized. For example electron spin can only be "half integer," +½ or −½ spin but never 1/4 or 2/3. Obviously a mathematical description of an electron must reflect this restriction, and the quantized nature of spin is also impossible to explain with classical physics.

Dirac matrices meet various subtle mathematical requirements which I will list in this paragraph, as they reflect Dirac's formidable math skills and the effort he exerted. To meet an objection to Schrödinger's wave equation, all derivatives have to be first-order. To be properly interconnected, they had to be unitary matrices, so that $(\gamma^0)^2 = 1$ and the other three $(\gamma^\mu)^2$ each $= -1$. As Heisenberg had worked earlier for his matrix system of quantum mechanics, they had to "anticommute," so that when i does not equal j, $\gamma^i \gamma^j + \gamma^j \gamma^i = 0$. Already noted for his exclusion principle (no two electrons in one atom have the same four quantum numbers), Pauli was on the same trail in describing spin with matrices, but he was trying to explain spin as it appeared in the experiments of his day, but he stopped at three spatial dimensions and three smaller matrices. Dirac took the extra steps, admitting both spin and relativistic space-time. In keeping with special relativity, Dirac's equation is Lorentz invariant (aka "covariant;" please see page 102 and recall the aforementioned $ds^2 = dX^2 + dY^2 + dZ^2 - (cdT)^2$), meaning that Dirac's matrices appear the same despite a change of frames of reference, a detail that pleased Einstein.

I should touch on other sources of inspiration for Dirac. As I mentioned, he knew that Pauli had worked on matrices that apply to spin. Meanwhile Heisenberg developed an entire math system for the quantum mechanics of his day, based on matrix algebra (unlike the more intuitive Schrödinger's wave-algebra approach). Heisenberg capitalized on his uncertainty principle that requires that matrices do not commute, and as I mentioned above, Dirac realized that this concept was essential for his efforts to extend quantum mechanics. Here I reemphasize the central role of Dirac in bringing together several very sophisticated concepts and in generating even more advanced ones. This feat is often under-appreciated, in part because of Dirac's introverted nature compared with the more vivid public personae of, say, Einstein, Heisenberg or Schrödinger.

Finally I need to discuss the role of operators in quantum mechanics. Please imagine that you wish to multiply some mathematical object, even just the number 5, by three. In the context of quantum mechanics you could say that "times three" or "x 3" is an operator, and you obtain 15 by letting this operator "act on" 5. I envision that by "acting," the "x 3 operator" has "extracted" or returned a solution. Thus an operator is a mathematical tool which, in this case, performs an action by multiplying a number by 3. This tool is selected according to the information to be extracted, in this case the product upon multiplication by 3. Similarly the location operator is used to extract (or return) location, the momentum operator is used to extract momentum, the total energy operator is used to extract total energy, etc. Still, the obvious question is why bother with operators? E.g., why "operate" when "multiplying" will do? There are three pertinent reasons. First, one operator may perform several different and/or complicated mathematical procedures in order to "do its acting." For example

$$\frac{\hbar \partial}{i \partial x}$$

is an operator that is to multiply by $\hbar$, divide by i, and obtain a partial differential ∂ with respect to x. *A key point here is that something as complicated as a Dirac matrix can also be an operator.*

Second, operators can have mathematical abilities that simple procedures (like multiplication) lack, such as returning a different result if used in a certain order. As a case in point, when A times B cannot be the same as B times A—when A and B must not commute—*operators can be selected that heed this restriction.* And in our setting, Dirac's matrices do not commute, which was a critical point in Dirac's derivation of his equation wherein his matrices form operators.

Third, *operators can act on functions* rather than just on simple numbers. This is very important in quantum mechanics where systems are described by functions, notably wave functions such as Schrödinger's and such as parts of Dirac's equation. I should add that in quantum mechanics, what is described about particles is their behavior, particularly the likelihoods thereof. For instance a quantum description may reveal a probable location of an electron or the probability for it to interact with a positron, but it does not deal with features like color or texture. Quantum behavior is generally described by means of functions, and appropriate operators are up to these tasks.

So now finally I am ready to present one exploded form of Dirac's equation, in order to show how it is constructed and how it works in principle. The following scheme appears in many forms and sources, but I composed it this way for clarity in what I wish to explain:

$i\frac{\partial}{\partial t} - m$	0	$i\frac{\partial}{\partial z}$	$i\frac{\partial}{\partial x} + \frac{\partial}{\partial y}$	$\psi(x)1$		0
0	$i\frac{\partial}{\partial t} - m$	$i\frac{\partial}{\partial x} - \frac{\partial}{\partial y}$	$-i\frac{\partial}{\partial z}$	$\psi(x)2$	$=$	0
$-i\frac{\partial}{\partial z}$	$-i\frac{\partial}{\partial x} - \frac{\partial}{\partial y}$	$-i\frac{\partial}{\partial t} - m$	0	$\psi(x)3$		0
$-i\frac{\partial}{\partial x} + \frac{\partial}{\partial y}$	$i\frac{\partial}{\partial z}$	0	$-i\frac{\partial}{\partial t} - m$	$\psi(x)4$		0

The 16 cells inside the double-line border together are a 4 by 4 matrix, representing a compilation of the four Dirac gamma matrices. *The entire 16-cell matrix is an operator.* In these cells we see imaginary numbers. More importantly we see partial first derivatives of the four dimensions of space-time and we see *m*'s, which here are the mass of an electron or positron.

We note that each row deals with each of the four dimensions *x, y, z* and *t*, and with the mass *m*. In the first two rows the algebraic sign of *m* is negative, as it should be for an electron, and in the last two rows it is positive, as it should be for a positron. (I could show each gamma matrix separately [texts and papers on Dirac's equation usually do so] and explain more about how each of the above 16 cells is filled in, but that is a very tedious process of little pertinence.)

The fifth (shaded) column of four ψ's contains the four functions which represent the four possible states of the electron, as determined by the two possible charges and two possible quantized spins:

The first ψ is for the negatively charged electron with +½ spin (up).
The second is for the negatively charged electron with −½ spin (down).
The third is for the positively charged antielectron (positron) with +½ spin (up).
The fourth is for the positively charged antielectron (positron) with −½ spin (down).

This column in Dirac's equation is a four-component complex vector and it is a *spinor*. We note that the quantity "four" in this case does not refer to four dimensions, but to the four combinations of charge and spin. (In more detail, each ψ is a wave function that is the product of a spinor term and a plane wave term. Like electrons' peculiar quantum spin, a spinor must be "rotated" 720° [twice] to return to its initial orientation. Each component of ψ is a function of location *x*.)

The reason for the seventh column of zeros is that this is the standard form of wave equations, somewhat like an accountant calculating whether the difference between assets and debits is zero. Arithmetically, every row of four cells of the matrix has at least one 0 in it.

To solve the Dirac equation, *the operator (a Dirac matrix) acts on the four ψ's (spinors),* and the result gives the probability amplitude for finding the electron in this state at some location. This probability amplitude, when squared, gives the corresponding probability. I stress that four solutions become available, and these are in the form of only probabilities. We should also know that the methodology of today's particle physics, worked out laboriously by Feynman and others, draws upon the data we see above, plus what we know about the interactions between elementary particles; parts of Dirac's equation appear in Feynman's very successful particle equations. We thus can predict the probabilities for such interactions, and this is a huge accomplishment in science.

Let us consider just one row. The five cells of the first row describe a negatively charged electron with +½ spin, and every issue has been addressed: the four dimensions of special-relativistic space-time, the mass of the electron or positron, its charge and its quantum spin. *The operator-matrix has acted on the spinor and extracted a description of the quantum and relativistic behavior of the electron or positron.* In this sense, Dirac's equation is really a coupling of four interrelated equations, yielding one solution that has four components. This equation can be abbreviated in several formats, and here is a compact one which summarizes the above exploded version. (This form is "covariant" in the geometric sense, that indicates how it transforms.)

$$(i\gamma^{\mu}\partial_{\mu} - m)\psi = 0$$

Obviously what is in the parentheses summarizes the matrix-part, while the ψ recaps the 4-component spinor. Yet lest we lose sight of the forest among the trees, Dirac still needed special relativity for the ratio of energy to mass, and he needed quantum mechanics for the probability of that ratio to be manifest.

✦

I do not wish to imply that the above developments represent the only instances where quantum mechanics and special relativity, so to speak, live symbiotically. Let me cite more examples which signify that they can coexist in nature, and that both are needed to explain certain phenomena.

When a piece of metal is slowly heated, it glows in certain colors as it is given more and more energy; first it appears red, then orange, then yellow, etc. The amount of energy determines the frequency of emitted light, which in turn determines its color. This is surprising; by common sense, energy should determine amplitude, not frequency. For example when a key on a piano is struck harder, the sound becomes louder, not higher in pitch. Quantum mechanics provides part of the explanation for why energy sets frequency: Energy from a glowing body is emitted in discreet quanta, and the quantum of electromagnetism is the photon. Light is a stream of particles, and the amount of energy in each particle determines the frequency of its wave. Now relativity enters the picture: Since a gravitational field affects ("bends") time, gravity affects the frequency of a photon's vibration. In effect both quantum mechanics and relativity must be invoked to account for the behavior of emitted light. (A cardinal equation in quantum mechanics is $E = fh$, where f is wave frequency and h is Planck's constant. If a photon receives more energy, its frequency rises but not its speed. See Ferris, p. 286.)

The notion that physical entities may exist in only certain discrete amounts is not an invention of quantum mechanics. Science already knew about the "quantization" of matter into subatomic particles. What is distinctive of quantum mechanics is the quantization of *energy*. Relativity makes this concept lucid in the notion of $E = mc^2$. Since m comes in indivisible packets, why shouldn't E? In fact Einstein won the Nobel Prize by revealing that in experiments on the "photo-electric effect" one photon is completely absorbed for each emitted electron. (In this "effect" an object which is exposed to light emits electricity. Such experiments manifest the equation $E = fh$ and demonstrate energy quantization. The amount of energy absorbed or emitted by an atom is nfh, where n is always an integer such as 1, 2, 3, etc., so that energy cannot be absorbed or emitted in fractions of quanta, and the intervals between quanta must be multiples of Planck's constant h.)

Another illustration of an alliance between relativity and quantum mechanics is in the phenomenon of "Hawking radiation" in which black holes create charged subatomic particles and eject them into space. How can that be? Two principles come together here: Special relativity allows energy to be converted into particles of matter, and quantum mechanics allows particles to exist in pairs—e.g. electrons and anti-electrons—one of which is trapped in a black hole while the other is released. (See Gribbin, the second citation, pp. 138-140, and of course the writings of Hawking himself.)

We may also ask, why doesn't the speed of light rely on the uniform speed of its source? We again have an explanation based on special relativity as well as quantum mechanics. First, let us envision a lightbulb glowing at a certain temperature, which means it emits light with a certain amount of energy. Next, let us think of each photon of light as an ordinary particle of matter being "thrown" from the glowing object, and each particle has a certain amount of kinetic energy and momentum. Finally, we know that the amount of energy possessed by such a photon just suffices to propel it at speed c. How much energy? We recall that a photon with no rest mass ($m_o = 0$) still has energy and momentum, as in $E = c\sqrt{p^2} = cp = c\,m_{motion}v$. (Please see page 40.) But what if the source of the light, the lightbulb itself, is in uniform motion at speed v, and what if an observer is measuring a beam emitted by this moving object? Common sense expects some photons in the beam to move at speed $c + v$. However the motion of the glowing object requires some energy, and the supply of energy is not infinite. This event leaves less energy for the emission of each photon. The amount of diverted energy reduces the speed of the photons by v, so that in the end

$$net\ speed = c + v - v,$$

which of course is merely c. Although harder to envision, the result is the same if the source is receding; *net speed* $= c - v + v$.

This principle appears in atomic physics: If all the available mass in a piece of plutonium is converted to energy, an enormous amount of electromagnetic radiation (including photons) is released, but the speed of the emission is still c. If the piece of plutonium is in a moving rocket, some of the total available energy is diverted into its motion, and the net speed of emission remains c. Furthermore it makes no difference whether we consider light to consist of energy or mass. For example our sun provides us with energy by constantly spewing photons in all directions, but as a result it is slowly and inexorably losing mass.

We may ask why other missiles (e.g. darts, bullets, rockets) do not routinely behave like light, even if there were no air resistance. Why is all electromagnetic radiation, including light, so "exclusive"? (Please see page 4.) Quantum mechanics helps with the answer: We can again think of the emission of light as a "throwing" of discreet bundles of energy in the form of photons. All available energy is devoted to accelerating them to speed c, and the fixed speed c reflects the available energy. Each photon behaves as pure energy, and because it has no mass, it can have c. Yet when a photon enters a gravitational field it functions as a mass, and like any ordinary missile it conforms to the shape of space-time.

We may still raise an objection: Doesn't *everything* in nature share the property that it can, at least in theory, function as pure energy and as pure mass? Other missiles should therefore behave like

476

light, so that if (which is a big "if") they could reach c, they too should be unaffected by the speed of their sources. Here special relativity has an affirmative answer. Indeed the very few other "missiles" found in nature that move at c, such as neutrinos from stellar explosions, do reveal this "exclusive" behavior.

We may raise yet another objection, one which is easier to word for the case where the source of light is a moving lightbulb: Since there is no absolute motion, it is not necessary to consider the bulb to be moving. We are free to consider it to be stationary and to treat an observer as approaching the bulb at speed v. In that case, from the vantage of the moving observer, the bulb itself requires *no* extra energy. Why then does the beam not exceed speed c? In other words if v is ascribed to the observer and c to a photon, isn't the net speed $c + v$? Does this scenario refute the idea that the measured speed of light does not rely on the uniform speed of its observer?

Quantum mechanics enters into the explanation: Light has a particle-and-wave nature, so that it can also be treated as a wave (of probability) progressing at speed c, while the frequency of that wave is linked to its energy by $E = fh = pc$, as noted by Einstein (page ?). Then even if we consider the source to be motionless, the approaching observer sees the crests of the wave crowded together, like a faster speedboat bounces over the waves more frequently, so that the color of the emitted light appears to be shifted toward the blue end of the spectrum (a Doppler effect, not of quantum mechanics). This is the same as saying that from the vantage of the approaching observer the wave of this light has a higher frequency or shorter wave length—which means it has more energy *but not more speed*. In yet other words, this frame of reference contains additional energy, specifically the kinetic energy of relative motion between the source of light and the observer. However the extra energy is revealed as higher frequency, still with no alteration in the measured speed of light.

The quantum-mechanical notion that energy is limited also helps explain why a moving clock appears slower. Motion through space-time is an energy-requiring event, which is why, for example, we find orbital decay. However we treat the dimensions of space on equal footing with the dimension of time. Hence we can assign the property that "motion requires energy" to time, in which case "motion through time" alone becomes an energy-requiring event (as applied on page 72). A relatively faster-moving object, including a clock, expends more of its energy for crossing space, leaving less energy for "crossing time." Consequently its time-measuring activity—such as ticking—slows down. We raised a parallel argument in another setting: Light has a fixed and maximum speed because it has a fixed and maximum amount of energy, and light has no infinite speed simply because nothing has infinite energy. In this context the speed of light ceases to be so impressive or mysterious: On a cosmic scale nature provides so little energy per photon that even without diversions, light needs aeons to cross just a small part of the universe, somewhat like our space vehicles that require years to cross a small part of our solar system, given the tiny amount of energy we can muster.

A similar concept can be expanded into a comprehensive system for illustrating relativity—as Epstein does elegantly in his book (pp. 79-81). His theme is that with respect to space-time, everything is always moving at the speed of light (since even an object stationary in space progresses through time). As a clock moves through space-time at a fixed maximum speed, it may cover distance as well as time. Therefore, faster motion through space "diverts" speed, leaving a smaller component to carry the clock through time. Epstein extends the argument into general relativity via the idea that a conical deformation of space-time explains gravity [pp. 147-152].

The concept of limited energy has concrete examples. We know that an electron and a positron can be produced by a photon (as per particle physics; please see page 469), and we know the rest mass of each, about 9×10^{-31} kg. However, the process requires energy (as per special relativity; page 32). How much energy (students please note)? Yes, $E = 2mc^2$, which comes to about 1.6×10^{-13} Jules. We can also calculate the frequency of the original photon by using this result to solve $E = fh$ (as above on page 477) which will show that the photon is in the x-ray range.

✦   ✦   ✦

Let us step back for a wider historical view. Throughout this book we contrasted Newton's physics and relativity. A similar contrast exists between Newton's system and quantum mechanics because the former also rests on causality and determinism. We can therefore say that relativity rejected the Newtonian paradigm in the realm of very massive and rapid objects—whose gravitational behavior complies with fields and curved space-time rather than with force—while quantum mechanics rejected the Newtonian paradigm in the subatomic realm, where phenomena comply with quanta and probability.

For instance wherever he used the instruments and telescopes available to him, Newton found no exceptions to $F = ma$. He taught that only undisturbed motion is uniform, and neither relative speed nor the act of observation enters the picture. Hence Newton assured us that F always equals ma. Einstein later added that only undisturbed time is uniform, and he discovered that motion disrupts the covariance of $F = ma$. Newton's conclusion therefore holds only when velocities are negligible and events are local. Hence Einstein informed us boldly but meticulously that sometimes F does not equal ma. The faster we move, the greater the inexactitude; the farther we look, the more exceptions appear. But then Heisenberg came along and disputed both: The randomness in nature and the act of observation nullify the covariance of $F = ma$. In fact he dared to assert that F never certainly equals ma. The closer we observe and the smaller our subject, the greater the uncertainty. In short, F does not equal ma near the speed of light because m becomes almost infinite, and it does not equal ma inside an atom because a becomes random.

This notion has other far-reaching ramifications. A criticism of relativity is that in theory its laws do not prohibit time running backward, at least in the sense that relativity's equations hold no matter which way time is considered to be flowing. Quantum mechanics does not permit such reversibility; randomness can not be undone. Thus if by chance a planet abandoned its orbit because of random subatomic motion, and then suddenly time ran backwards, the planet would not necessarily jump back into its original path; its behavior would still be random.

Even philosophical and theological arguments arise here. Another criticism of relativity is that its determinism implies that all future events are caused by and are predictable from past events, in which case there is no room for free will or divine intervention. This is not a new idea. For

478

example Laplace believed that given enough information about the position of all objects in the universe and about the nature of all forces acting upon these objects, all past events can be reconstructed and all subsequent events can be anticipated. The task would be exceedingly complex but theoretically possible.

Heisenberg's uncertainty principle provides an appealing and reassuring alternative: Even with all the data at hand, and even with the ability to handle all the complexity, reconstructing the past and calculating the future is inherently impossible. Though Einstein resisted this concept, predictability and causality in nature are not absolute, so that on an elemental level there is no incompatibility between physical laws and non-physical realities. This inference perplexes the relativist/mathematician but it delights the theologian, as it implies that science and God coexist.

In any case Einstein's objections to quantum mechanics damaged his reputation in the scientific community. In this context the schism was not "Newton vs. Einstein." It was "Newton-and-Einstein vs. Heisenberg," and it identified Einstein with a stage of scientific thinking that was rapidly becoming obsolete. As quantum mechanics exhibited more theoretical merits, more experimental confirmation, and more practical applications, it became clear that probability and randomness indeed are at the heart of physics, at least in the realm of very small events. As Pais cites (p. 443), physicists with whom Einstein differed were devastated; they found themselves reaching apparently legitimate conclusions disavowed by the world's greatest scientist.

It thus seemed during Einstein's lifetime (he died in 1955) that there are two families of basic physical laws, one for quantum mechanics and another for relativity, and the relationship between them was ambivalent and at times antagonistic. The question that was beginning to surface is obvious: If relativity replaced the notion of a force of gravity with curvature of space-time but relativity cannot similarly redefine basic subatomic forces, is there another more comprehensive theory which covers everything? If we agree that relativity is a generalization of Newtonian physics, can we agree on another even more general set of laws that might encompass relativity *and* quantum mechanics?

Indeed in his later years Einstein worked to find a unifying theory, essentially by trying to explain subatomic physics on a geometric basis similar to the way he explained gravity. For him an acceptable outcome would have been a comprehensive structure in which quantum mechanics and relativity are but ingredients, and he hoped to prove that the former is a mathematical consequence of the latter. He failed to reach these goals, as nothing seemed to meld relativity into a grand scheme of nature, and this, like his dissension against some of the tenets of quantum mechanics, also diminished his reputation.

Of course we can debate whether there actually is a need to explain very large "relativistic events" and very small "quantum-mechanical events" together. Why not concede that the two systems are each correct in their own domain, even if it means that one universe has two such domains? The answer is this: First, we have the somewhat philosophical position that such a dichotomy should not exist in nature; one universe should not need two sets of laws. Second, we believe that we can cite two situations involving masses that are large enough but distances that are small enough so

479

that both domains contribute. One such situation is the primordial universe around the time of the big bang. The other is at the center of a black hole. Given current technology, we cannot build an accelerator powerful enough to faithfully reproduce such situations experimentally, because we do not know how to compress a huge mass into a very small space. Nonetheless we see that one way or another relativity and quantum mechanics can be interrelated and unified, and today the future seems more promising than ever.

✦ ✦ ✦

If we trace the history of science, the emergence of relativity and quantum mechanics in the twentieth century seems to be destined. Centuries ago we were convinced that the Earth is at the center of the universe, that stars and planets move in perfect circles, that heavier objects fall faster, that all moving objects seek a state of rest, and that the laws of nature need not be universal. We had magnificent philosophy but defective physics. (We should acknowledge Greek and Hellenistic scientists who managed to discover or at least propose "modern" concepts, such as the true shape and size of the Earth, the atomic nature of matter, and the idea that physical laws appear the same everywhere. We also cannot neglect the obvious importance of Euclid and Pythagoras in the mathematics of relativity. Furthermore ancient Oriental scientists correctly recognized many principles which their Western counterparts once resisted.)

In the early Renaissance, European science entered into a crisis when contradictory observations began to surface. For example the more measurements were made, the harder it became to accept a circular astronomic orbit, to explain a geocentric solar system, and to ascribe an absolute quality to resting objects. At the same time the evolution of the scientific method encouraged a renewed confidence in the ability of human intellect to explain nature. Somewhat later, using newly discovered devices (telescopes, microscopes, scales and clocks) Galileo, Copernicus, Kepler and of course Newton rescued the situation for the next few centuries. In the Newtonian view, motion, not rest, is the norm in nature, and the laws of physics are clear, decisive, and inescapable. The universe seemed to be merely one gigantic machine whose workings are revealed in terse and immutable mathematical formulæ, and Newton's system was so successful that further improvement seemed out of the question.

However around the start of the twentieth century, as our observations and mathematics became even more sophisticated, new scientific crises appeared. In the realm of large objects, observations about the ether, about the orbit of Mercury, and about the nature of the speed of light disputed the legitimacy of Newton's equations. In the realm of small objects, accepted physical laws seemed to suggest that atoms should self-destruct and that matter should not exist. Today's reply to the latter is quantum mechanics. Today's reply to the former is special and general relativity. As we have discussed, modern versions of quantum mechanics appear to be compatible and even closely allied with special relativity. Currently the issue is how general relativity can be reconciled and united with quantum mechanics, but as we have also seen, string theory provides an appealing explanation. Moreover, the Higgs boson and Higgs field appears to constitute a major part of the solution, one that lends itself to experimental confirmation.

480

APPENDIX

Here is a compilation of the essentials in mathematics and geometry which appear throughout the text. The topics are in **bold** print.

Multiplication, division, addition, and subtraction:

Multiplication of two numbers represented by A and B is AB or $(A)(B)$, division is $\frac{A}{B}$, addition is $A + B$, and subtraction is $A - B$.

Equalities:

A = B means A and B are equal, A ≈ B means A and B are approximately equal, A ≠ B means A and B are not equal, A > B means A is greater than B, A < B means A is less than B, and A>>B means A is much greater than B.

Numerator and **denominator:**

In division or in **fractions**, the parts are called $\frac{Numerator}{Denominator}$.

Cancellation, rearrangement, and substitution:

We use various algebraic procedures, many of which have names. Cancellation (of c in this case):.

$$\frac{cA}{cB} = \frac{A}{B}.$$

Rearrangement and solving for A:

$$If \ A + B = C, \ then \ A = C - B.$$

$$If \ \frac{A}{B} = C, \ then \ A = BC.$$

Substitution:

$$If \ A = B \ \ and \ \ D = \frac{A}{C}, \ then \ D = \frac{B}{C}.$$

Distributive, commutative, and associative rules:

$$AB + AC = A(B + C).$$

$$AB = BA.$$

$$(AB)C = A(BC).$$

Ratio rule (cross multiplication):

$$\text{If } \frac{A}{B} = \frac{C}{D}, \text{ then } AD = BC.$$

Binomial multiplication:

$$(A + B)(A + B) = (A + B)^2 = A^2 + 2AB + B^2.$$

Squares and square roots:

$$AA \ (A \ multiplied \ by \ A) = A^2.$$

$$\sqrt{A^2} = A.$$

The imaginary number i is such that $i = \sqrt{-1}.$

In exponential form, $\sqrt{D} = D^{1/2}.$

Manipulations with zero (where ∞ means an infinite number and where A is not zero):

$$\frac{A}{0} = \infty.$$

$$\frac{0}{A} = 0.$$

$$If \ A = -A, \ then \ A = 0.$$

Rate (which also is called speed or velocity) multiplied by time equals distance. Written in differential terms where "*v*" means "velocity," "*dx*" means "change in distance," and "*dt*" means change in time,

$$v = \frac{dx}{dt}.$$

Summation, indicated by $\sum$, means that all the values of a quantity are added together.

For example if *x* can be 1, 2, 3, or 4, while *y* can be 1 or 2, then

$$\sum (xy) = (1)(1) + (2)(1) + (3)(1) + (4)(1) + (1)(2) + (2)(2) + (3)(2) + (4)(2) = 30.$$

Triangles and **angles:**

The pythagorean theorem applies to a "right" triangle, which means a triangle in which one angle has 90 degrees and hence in which two sides are perpendicular. If the longest side, called the hypotenuse, is *H* and the other two sides are *A* and *B*, the pythagorean theorem states that

$$A^2 + B^2 = H^2.$$

In such a triangle if C is the angle in question, **sine**, **cosine**, and **tangent** are defined as follows:

$$The \ sine \ (a.k.a. \ sin) \ of \ C = \frac{Opposite \ side}{Hypotenuse},$$

$$The \ cosine \ (a.k.a. \ cos) \ of \ C \ = \ \frac{Adjacent \ side}{Hypotenuse},$$

$$The \ tangent \ (a.k.a. \ tan) \ of \ C \ = \ \frac{Opposite \ side}{Adjacent \ side}.$$

Sin of zero degrees = 0. (Sin0 = 0.) Sin of 90 degrees = 1. (Sin90 = 1.)

Cos of zero degrees = 1. (Cos0 = 1.) Cos of 90 degrees = 0. (Cos90 = 0.)

The pythagorean theorem does not hold if the triangle has no "right" (90 degree) angle. If that angle is R,

$$H^2 \ = \ A^2 \ - \ 2cosR \ (AB) \ + \ B^2.$$

In **differential calculus** the derivative of a constant is zero.

In **integral calculus** the integral of zero is a constant.

A series of *ds*'s integrated between *A* and *B* is such that

$$s \ = \ \int_A^B ds.$$

Graphs and **axes:**

On a graph, the *x* or X axis (a.k.a. the *x* or X coordinate) is horizontal, the *y* or Y axis is vertical, and if there is one, the *z* or Z axis protrudes from the surface that is bounded by *x* and *y*.

In such a graph, $X_2 \ - \ X_1$ or $X_1 \ -to- \ X_2$ means the distance between X_1 and X_2. If that distance is infinitesimal, as it is in differential calculus, we label it as *dx*.

In such a graph, a location that is X units to the right and Y units up can be expressed as X by Y. Therefore X by Y is a specific point on the graph. In three dimensions, X by Y by Z is a point in a space.

A fourth dimension, which is usually the time-dimension in relativity, is labeled as *t* or T. In four dimensions, X by Y by Z by T is a point in space-time.

Circles and **spheres:**

In such a graph, a circle of radius r is such that $r^2 = X^2 + Y^2$.

In such a space, a sphere of radius r is such that $r^2 = X^2 + Y^2 + Z^2$.

The area of a circle is $4\pi r^2$.

The volume of a sphere is $\dfrac{4}{3}\pi r^3$.

Very large numbers such as 10,000,000,000,000 are written as 10^{12} which means 10 followed by 12 zeros.

Very small numbers as $\dfrac{1}{10,000,000,000,000}$ are written as 10^{-12} which is 1 divided by 10 followed by 12 zeros.

Example of **calculus of variations:**

Assume that the path of a line is described by the equation $,x = \dfrac{dx}{dt}$ on a surface whose coordinates are x and t. (The surface could be that of space-time.) Thus the path is a function $x(t)$ and it depends on time t. Consider a to be the start time and b to be the end time. The calculus of variations can answer the following question: Among all functions that start at a and end at b, which path $x(t)$ of the line on this surface is the shortest?

This concludes the appendix.

Adler, R., Bazin, M. and Schiffer, M. Introduction to General Relativity. New York: McGraw-Hill Book Company, 1965.

Al-Khalili, Jim. Quantum. London: Weidenfield & Nicolson, 2003.

Barrow, J. D. and Tipler, F. J. The Anthropic Cosmological Principle. Oxford: Oxford University Press, 1986.

Bergmann, Peter G. Introduction to the Theory of Relativity. New York: Dover Publications, Inc., 1976.

Bergmann, Peter G. The Riddle of Gravitation. New York: Charles Scribner's Sons, 1968.

Bishop, R. L. and Goldberg, S. I. Tensor Analysis on Manifolds. New York: Dover Publications, 1980.

Borisenko, A. I. and Tarapov, I E. Vector and Tensor Analysis. New York: Dover Publications, 1968.

Calder, Nigel. Einstein's Universe. New York: Penguin Book, 1979.

Carmelli, Moshe. Classical Fields: General Relativity and Gauge Theory. New York: John Wiley & Sons, 1982.

Clark, Ronald W. Einstein: The Life and Times. New York: Avon Books, 1984.

Cox, Brian and Forshaw, Jeff. The Quantum Universe. Boston: Da Capo Press, 2011.

de Bothezat, George. Back to Newton: A Challenge to Einstein's Theory of Relativity. New York: Stechert & Co., 1936.

Einstein, Albert. The Meaning of Relativity. Princeton, New Jersey: Princeton University Press, 1922.

Einstein, Albert. Relativity: The Special and the General Theory. New York: Crown Publishers, 1959.

Einstein, Albert. On the Generalized Theory of Gravitation. In Scientific American, pp. 40-45, Vol. 3 Number 1, 1991.

Einstein, Albert. Out of My Later Years. New York: Citadel Press Book, 1956.

Einstein, A. and Infeld, L. The Evolution of Physics. New York: Simon and Schuster, 1966.

Epstein, Lewis Carroll. Relativity Visualized. San Francisco: Insight Press, 1981.Ferris, Timothy. Coming of Age in the Milky Way. New York: Doubleday, 1988.

Feynman, R.P. and Hibbs, A. R. Quantum Mechanics and Path Integrals. Mineola, NY: Dover Publications, Inc., 1965 and 2005.

Feynman, R.P. QED. Princeton University Press, 1985.

Feynman, R.P., Leighton, R.B., and Sands, M. The Feynman Lectures on Physics (Three volumes). Reading, Mass.: Addison-Wesley, 1963.

Fowler, Shannon. The Equations of Light: A Refutation of Relativity Theory. New York: Carlton Press, Inc. 1990.

Gamow, George. The Great Physicists from Galileo to Einstein. New York: Dover Publications, 1961.

Gardner, Martin. The Relativity Explosion. New York: Vantage Books, 1976.

Greene, Brian. The Elegant Universe. New York: Vantage Books, 1999.

Gribbin, John. Q is for Quantum: An Encyclopedia of Particle Physics. New York: The Free Press, 1998.

Gribbin, John. Unveiling the Edge of Time. New York: Harmony Books, 1992.

Hawking, Stephen W. A Brief History of Time. New York: Bantam Books, 1988.

Hoffmann, Banesh. Albert Einstein Creator and Rebel. New York: Plume (Penguin Group), 1972.

Holton, Gerald & Elkaka, Yehuda, ed. Albert Einstein: Historical and Cultural Perspectives. Princeton U. Press, 1982.

Hopf, L. Introduction to the Differential Equations of Physics. New York: Dover Publications, 1948.

Isaacson, Walter. Einstein: His Life and Universe. New York: Simon & Schuster, 2007.

Internet. The reader may search the "world wide web" for information on relativity. Though they vary in quality and factuality, many sites deal with special relativity, general relativity, and specific details. The internet is also a forum for questions about relativity, for criticisms of relativity, and for discussions of alternative systems.

Jagerman, Louis S. Quantum Mechanics that Makes Sense. Available on Amazon.com. ISBN 9 781492 257462, 2012, Revised May 2014.

Jones, Roger S. Physics for the Rest of Us. Chicago: Contemporary Books, 1992.

Kaufmann, William J., III. Black Holes and Warped Spacetime. San Francisco: W. H. Freeman and Company, 1979.

Kaufmann, William J., III. The Cosmic Frontiers of General Relativity. Boston: Little, Brown and Company, 1977.

Kenyon, I. R. General Relativity. Oxford: Oxford University Press, 1990.

Kilmister, C. W. General Theory of Relativity. Oxford: Pergamon Press, 1973.

Kumar, Manjit. Quantum. New York: WW Norton & Co., 2008.

Lanczos, Cornelius. Albert Einstein and the Cosmic World Order. New York: Interscience Publishers, a division of John Wiley & Sons, Inc., 1965.

Lawden, D. F. An Introduction to Tensor Calculus, Relativity and Cosmology. Chichester: John Wiley & Sons, 1982.

Lieber, Lillian R. The Einstein Theory of Relativity. New York: Holt, Rinehart and Winston, 1936.

Lorentz, H.A., Einstein, A., Minkowski, H., and Weyl, H. The Principle of Relativity. New York: Dover Publications, 1952, reprint of 1923 translation.

Lovelock, D., and Rund, H. Tensors, Differential Forms, and Variational Principles. NY: Dover Publications, 1975.

Menzel, D. H. Mathematical Physics. New York: Dover Publications, 1961.

Møller, C. The Theory of Relativity. Oxford: Oxford University Press, 1952.

Mook, D.E. and Vargish, T. Inside Relativity. Princeton: Princeton University Press, 1987.

Newman, James R., ed. The World of Mathematics. (Four volumes) New York: Simon and Schuster, 1956.

Ohanian, Hans C. Gravitation and Spacetime. New York: W W Norton & Co. Inc., 1976.

Pais, Abraham. Subtle is the Lord...The Science and the Life of Albert Einstein. Oxford: Oxford University Press, 1982.

Pauli, W. The Theory of Relativity. New York: Dover Publications, 1958

Penrose, Roger. The Emperor's New Mind. New York: Penguin Book, 1989.

Reichenbach, Hans. The Philosophy of Space and Time. New York: Dover Publications, Inc., 1958.

Rucker, Rudolf v. B. Geometry, Relativity and the Fourth Dimension. New York: Dover Publications, 1977.

Sachs, Mendel. General Relativity and Matter. Dordrecht, Holland: D. Reidel Publishing Co., 1982.

Schutz, Bernard F. A First Course in General Relativity. Cambridge: Cambridge University Press, 1990.

Shadowitz, Albert. Special Relativity. New York: Dover Publications, 1968.

Stewart, Ian. Does God Play Dice? Cambridge: Basil Blackwell Inc., 1989.

Synge, J. L. and Schild, A. Tensor Calculus. New York: Dover Publications, 1949.

Tolman, Richard C. Relativity Thermodynamics and Cosmology. New York: Dover Publications, 1987.

Wald, Robert M. Space, Time, and Gravity. Chicago: University of Chicago Press, 1977.

Wald, Robert M. General Relativity. Chicago: University of Chicago Press, 1984.

Weyl, Hermann. Space-Time-Matter. New York: Dover Publications, 1952.

Will, Clifford M. Was Einstein Right? New York: BasicBooks, 1986.

INDEX

488

489

493

495

496

497

498

500

504

LIST OF DIAGRAMS with main page numbers

CPSIA information can be obtained
at www.ICGtesting.com
Printed in the USA
BVHW02s1344201217
503314BV00004B/651/P